Interactive Quantum Mechanics

S. Brandt • H.D. Dahmen • T. Stroh

Interactive Quantum Mechanics

Quantum Experiments on the Computer

Second Edition

With CD-ROM, 128 Figures, and 344 Exercises

Siegmund Brandt
Physics Department
Siegen University
57068 Siegen
Germany
brandt@physik.uni-siegen.de

Hans Dieter Dahmen
Physics Department
Siegen University
57068 Siegen
Germany
dahmen@physik.uni-siegen.de

Tilo Stroh
Physics Department
Siegen University
57068 Siegen
Germany
stroh@sirs02.physik.uni-siegen.de

Additional material to this book can be downloaded from http://extras.springer.com

ISBN 978-1-4939-3824-7 ISBN 978-1-4419-7424-2 (eBook)

DOI 10.1007/978-1-4419-7424-2
Springer New York Dordrecht Heidelberg London

Printed on acid-free paper

Springer is part of Springer Science+Business Media (www.springer.com)

Preface to the Second Edition

For the present edition the concept of the book and of INTERQUANTA, the accompanying *Interactive Program of Quantum Mechanics*, (**IQ**, for short), was left unchanged. However, the physics scope of the text and the capabilities of the program were widened appreciably.

The most conspicuous addition to **IQ** is the capability to produce and display *movies* of quantum-mechanical phenomena. So far, **IQ** presented time dependence as a series of graphs in one frame. While such plots (which can, of course, still be shown) lead to a good understanding of the phenomenon under study and can be examined quantitatively at leisure, the new movies give a more direct impression of what happens as time passes. For such movies, as for the conventional simulations, the parameters defining the physical phenomenon and the graphical appearance can be changed interactively. Movies can be produced and also stored for many quantum-mechanical problems such as bound states and scattering states in various one-dimensional potentials, wave packets in three dimensions (free or in a harmonic-oscillator potential), and two-particle systems (distinguishable particles, identical fermions, identical bosons).

Concerning the physics scope, these are the main additions: One-dimensional bound and scattering states are now discussed and computed also for *piecewise linear potentials*. These, as opposed to the usual step potentials (which are piecewise constant), allow for much better approximations of arbitrary smooth potentials. Another interesting addition to one-dimensional quantum mechanics is the juxtaposition of quantum-mechanical wave packets with classical phase-space distributions. The treatment of quantum mechanics in three dimensions is extended by the hybridization of bound states and by the simulation of magnetic resonance.

In the present edition the number of data sets (we call them *descriptors*), defining a complete simulation – either presented as conventional plot or as movie – is more than tripled. In this way users have a much richer choice of ready-made examples from which to start their exploits. Moreover, solution descriptors are now provided for the exercises.

Siegen, Germany
May 2010

Siegmund Brandt
Hans Dieter Dahmen
Tilo Stroh

Preface to the First Edition

This book can be regarded as a concise introduction to basic quantum mechanics: free particle, bound states, and scattering in one and in three dimensions, two-particle systems, special functions of mathematical physics. But the book can also be seen as an extensive user's guide for INTERQUANTA, *the Interactive Program of Quantum Mechanics*, which we will abbreviate henceforth as **IQ**. The book also contains a large number of exercises. The program can be used in two ways. By working through (at least a part of) these exercises, the user of **IQ** explores a *computer laboratory in quantum mechanics* by performing computer experiments. A simpler way to use **IQ** is to study one or several of the ready-made *demonstrations*. In each demonstration the user is taken through one chapter of quantum mechanics. Graphics illustrating quantum-mechanical problems that are solved by the program are shown, while short explanatory texts are either also displayed or can be listened to.

INTERQUANTA has a user interface based on tools provided by the *Java* programming language. With this interface using the program is essentially self-explanatory. In addition, extensive help functions are provided not only on technical questions but also on quantum-mechanical concepts. All in all using INTERQUANTA is not more difficult than surfing the Internet.

The modern user interface is the main improvement over older versions of **IQ**.[1] Moreover, new physics topics are added and there are also new graphical features.

The present version of INTERQUANTA is easily installed and run on personal computers (running under Windows or Linux) or Macintosh (running under Mac OS X).

We do hope that by using INTERQUANTA on their own computer many students will gain experience with different quantum phenomena without having to do tedious calculations. From this experience an intuition for this important but abstract field of modern science can be developed.

Siegen, Germany
February 2003

Siegmund Brandt
Hans Dieter Dahmen
Tilo Stroh

[1] S. Brandt and H. D. Dahmen, *Quantum Mechanics on the Personal Computer*, Springer, Berlin 1989, 1992, and 1994; *Quantum Mechanics on the Macintosh*, Springer, New York 1991 and 1995; *Pasocon de manebu ryoushi nikigacu*, Springer, Tokyo 1992; *Quantenmechanik auf dem Personalcomputer*, Springer, Berlin 1993

Contents

1. Introduction

1.1 Interquanta

The language of quantum mechanics is needed to describe nature at the atomic or subatomic scale, for example, the phenomena of atomic, nuclear, or particle physics. But there are many other fields of modern science and engineering in which important phenomena can be explained only by quantum mechanics, for example, chemical bonds or the functioning of semiconductor circuits in computers. It is therefore very important for students of physics, chemistry, and electrical engineering to become familiar with the concepts and methods of quantum mechanics.

It is a fact, however, that most students find quantum mechanics difficult and abstract, much more so than classical point mechanics. One easily detects the reason for this by recalling how students learn classical mechanics. Besides learning from lectures, they draw on experience from everyday life, on experiments they perform in the laboratory, and on problems they solve on paper. The important concept of a mass point is nothing but that of a very small stone. The experience with throwing stones helps to understand mechanics. Additional experiments are very direct and simple and there is a wealth of problems that are easily solved.

All this is different with quantum mechanics. Although – for all we know – elementary particles are point-like, the concept of the trajectory of a mass point breaks down and has to be replaced by a complex probability amplitude. This function cannot be measured directly; its properties have to be inferred indirectly from experiments involving optical spectra or counting rates, and so forth. Finally, nearly all nontrivial problems pose severe computational difficulties and require approximative or numerical methods. Thus, students can do only a few problems.

Many quantum-mechanical problems can, however, be quickly solved numerically by computer. The answer is often very easy to analyze if presented in graphical form. We have written an interactive program taking *alphanumeric input* defining quantum-mechanical problems and yielding *graphical*

S. Brandt et al., *Interactive Quantum Mechanics: Quantum Experiments on the Computer*,
DOI 10.1007/978-1-4419-7424-2_1, © Springer Science+Business Media, LLC 2011

output to produce a large number of illustrations for an introductory textbook on quantum mechanics.[1]

Here we present an improved and generalized version of this program, which we call INTERQUANTA (**IQ** for short) – the interactive program of quantum mechanics. The program has a convenient, essentially self-explanatory user interface written in the *Java* programming language[2] and running on a *Java virtual machine* (VM). Moreover, there is extensive online help. **IQ** can be used in two rather different ways. While studying ready-made *demonstrations*, the user is only an interested onlooker. In the *interactive mode*, on the other hand, the user determines what happens. He or she can solve quantum-mechanical problems of his or her own choice. Users can work through a complete *computer laboratory in quantum mechanics*. This can be done at leisure at home on one's own computer or on that of a friend. Often it is more fun if two or three students join to define their problems, solve them, and discuss the solutions. In the computer laboratory students define and solve a quantum-mechanical problem, analyze the result, change one or several parameters of the original problem, study the next result, and so on. Using the program on a large variety of problems of different types, students gain experience in quantum mechanics because they performed computer experiments and at the same time solved problems by numerical methods.

Of course, **IQ** can also be used in organized courses run by a tutor or for demonstrations in lectures.

1.2 The Structure of This Book

This book consists of a main text and several appendixes. Chapters 2 through 10 of the main text are devoted to the various physics subjects covered by **IQ**. Each chapter begins with a section "Physical Concepts" in which the relevant concepts and formulae are assembled without proofs. Although this book is in no way intended to be a textbook, this section is needed to allow a precise definition of what the program does and also a clear formulation of the exercises. The following sections of each chapter are devoted to specific *physics topics* that can be tackled with **IQ** and give details on the user interface as far as they are of particular interest for the physics topic at hand. Each chapter is concluded by a collection of exercises. Chapter 11 is devoted to the special functions of mathematical physics relevant to quantum mechanics. In Chap. 2 there is in addition a section "A First Session with the Computer", which in an informal way provides a minimum of general knowledge of **IQ**.

---

[1] S. Brandt and H. D. Dahmen, *The Picture Book of Quantum Mechanics*, 3rd edition, Springer, New York 2001

[2] Java is developed by Sun Microsystems, Inc. For details see `http://java.sun.com`.

Chapter 12 contains hints for the solution of some exercises. It begins with a section on the different possibilities for choosing units for physical quantities. Its content is useful in many exercises for determining numerical values of input parameters and for correctly interpreting the numerical results of computations by **IQ**.

Appendix A is a systematic guide to **IQ**. Appendix B contains technical information on the installation of the program.

1.3 The Demonstrations

IQ provides ready-made demonstrations for the following physics topics corresponding to Chaps. 2 through 11 of this book:

- free particle motion in one dimension,
- bound states in one dimension,
- scattering in one dimension,
- two-particle systems,
- free particle motion in three dimensions,
- bound states in three dimensions,
- scattering in three dimensions,
- spin and magnetic resonance,
- hybridization,
- functions of mathematical physics.

There are two further demonstrations which are meant to be tutorials on the use of **IQ**. Their titles are

- Introduction,
- Creating and Using Movies.

Each demonstration contains many example plots and explanatory text (written or spoken). See Appendix A.1.13 for information about how to run a demonstration, and Appendix A.12 for how to prepare your own demonstration files.

1.4 The Computer Laboratory

The course itself consists in working through (some of) the *exercises* given in the different chapters of this book. For most of the exercises an *initial descriptor* is provided with properly chosen graphics and physical parameters. The students are asked to run **IQ** with this descriptor, study the graphical output, and answer questions for which they usually have to change some parameter(s), run **IQ** again, and so on. At any stage they can store away their changed descriptors for later use. They can also take hardcopies of all graphical output to perform measurements on or simply file them, preferably with some comments. In many exercises it is intended to draw the attention of the student to a particular feature of the plot. This is usually attempted by asking a question that can in most cases be answered by qualitative arguments. Most of these are answered in Chap. 12. Of course, students may define and solve problems not contained in the lists of exercises given.

1.5 Literature

Because in the introductory sections "Physical Concepts" we present only a very concise collection of concepts and formulae, the user of **IQ** is urged to study the physics topics in more detail in the textbook literature. Under the heading "Further Reading" at the end of our introductory sections we refer the user to the relevant chapters in the following textbooks:

Abramowitz, M., Stegun, I. A. (1965): *Handbook of Mathematical Functions* (Dover Publications, New York)
Alonso, M., Finn, E. J. (1968): *Fundamental University Physics*, Vols. 1–3 (Addison-Wesley, Reading, MA)
Kittel, C., Knight, W. D., Ruderman, M. A., Purcell, E. M., Crawford, F. S., Wichmann, E. H., Reif, F. (1965): *Berkeley Physics Course*, Vols. I–IV (McGraw-Hill, New York)
Brandt, S., Dahmen, H. D. (2001): *The Picture Book of Quantum Mechanics* (Springer-Verlag, New York)
Feynman, R. P., Leighton, R. B., Sands, M. (1965): *The Feynman Lectures on Physics*, Vols. 1–3 (Addison-Wesley, Reading, MA)
Flügge, S. (1971): *Practical Quantum Mechanics*, Vols. 1,2 (Springer-Verlag, Berlin, Heidelberg)
Gasiorowicz, S. (2003): *Quantum Physics* (John Wiley and Sons, New York)
Hecht, E., Zajac, A. (1974): *Optics* (Addison-Wesley, New York)
Merzbacher, E. (1970): *Quantum Mechanics* (John Wiley and Sons, New York)
Messiah, A. (1970): *Quantum Mechanics*, Vols. 1,2 (North-Holland Publishing Company, Amsterdam)
Schiff, L. I. (1968): *Quantum Mechanics* (McGraw-Hill, New York)

2. Free Particle Motion in One Dimension

Contents: Description of a particle as a harmonic wave of sharp momentum and as a wave packet with a Gaussian spectral function. Approximation of a wave packet as a sum of harmonic waves. Probability-current density. Quantile motion. Classical phase-space description. Analogies in optics: harmonic light waves and light wave packets. Discussion of the uncertainty principle.

2.1 Physical Concepts

2.1.1 Planck's Constant. Schrödinger's Equation for a Free Particle

The fundamental quantity setting the scale of quantum phenomena is *Planck's constant*

$$h = 6.626 \times 10^{-34}\,\mathrm{J\,s} \quad , \quad \hbar = h/2\pi \quad .$$

A *free particle* of mass m and velocity v traveling in the x direction with momentum $p = mv$ and kinetic energy $E = p^2/2m$ has a *de Broglie wavelength* $\lambda = h/p$. The harmonic *wave function*

$$\psi_p(x,t) = \frac{1}{(2\pi\hbar)^{1/2}} \exp\left[-\frac{\mathrm{i}}{\hbar}(Et - px)\right] \tag{2.1}$$

is called a Schrödinger wave. It has the *phase velocity*

$$v_{\mathrm{ph}} = E/p = p/2m \quad .$$

Schrödinger waves are solutions of the *Schrödinger equation for a free particle*

$$\mathrm{i}\hbar\frac{\partial}{\partial t}\psi_p(x,t) = -\frac{\hbar^2}{2m}\frac{\partial^2}{\partial x^2}\psi_p(x,t) = T\psi_p(x,t) \tag{2.2}$$

with

$$T = -\frac{\hbar^2}{2m}\frac{\partial^2}{\partial x^2}$$

being the *operator of the kinetic energy*.

S. Brandt et al., *Interactive Quantum Mechanics: Quantum Experiments on the Computer*,
DOI 10.1007/978-1-4419-7424-2_2, © Springer Science+Business Media, LLC 2011

2.1.2 The Wave Packet. Group Velocity. Normalization

Since the equation is linear in ψ_p, a *superposition*

$$\psi(x,t) = \sum_{n=1}^{N} w_n \psi_{p_n}(x,t) \tag{2.3}$$

of harmonic waves ψ_{p_n} corresponding to different momenta p_n each weighted by a factor w_n also solves the Schrödinger equation. Replacing the sum by an integral we get the wave function of a *wave packet*

$$\psi(x,t) = \int_{-\infty}^{\infty} f(p)\psi_p(x-x_0,t)\,\mathrm{d}p \quad , \tag{2.4}$$

which is determined by the *spectral function* $f(p)$ weighting the different momenta p. In particular, we consider a *Gaussian spectral function*

$$f(p) = \frac{1}{(2\pi)^{1/4}\sqrt{\sigma_p}} \exp\left[-\frac{(p-p_0)^2}{4\sigma_p^2}\right] \tag{2.5}$$

with *mean momentum* p_0 and *momentum width* σ_p.

Introducing (2.5) into (2.4) we get the wave function of the *Gaussian wave packet*

$$\psi(x,t) = M(x,t)\mathrm{e}^{\mathrm{i}\phi(x,t)} \tag{2.6}$$

with the *amplitude function*

$$M(x,t) = \frac{1}{(2\pi)^{1/4}\sqrt{\sigma_x}} \exp\left[-\frac{(x-x_0-v_0t)^2}{4\sigma_x^2}\right] \tag{2.7}$$

and the *phase*

$$\phi(x,t) = \frac{1}{\hbar}\left[p_0 + \frac{\sigma_p^2}{\sigma_x^2}\frac{t}{2m}(x-x_0-v_0t)\right](x-x_0-v_0t) + \frac{p_0}{2\hbar}v_0t - \frac{\alpha}{2} \tag{2.8}$$

with *group velocity*

$$v_0 = p_0/m \quad , \tag{2.9}$$

localization in space given by

$$\sigma_x^2 = \frac{\hbar^2}{4\sigma_p^2}\left(1 + \frac{4\sigma_p^4}{\hbar^2}\frac{t^2}{m^2}\right) \quad ,$$

and

$$\tan\alpha = \frac{2}{\hbar}\frac{\sigma_p^2}{m}t \quad .$$

The initial spatial width at $t = 0$ is thus $\sigma_{x0} = \hbar/(2\sigma_p)$. In terms of σ_{x0} the time-dependent width becomes

$$\sigma_x^2(t) = \sigma_{x0}^2 \left(1 + \frac{\hbar^2}{4\sigma_{x0}^4} \frac{t^2}{m^2}\right) \quad . \tag{2.10}$$

The absolute square of the wave function, which is the product of the wave function ψ with its *complex conjugate* ψ^*,

$$\varrho(x,t) = \psi(x,t)\psi^*(x,t) = |\psi(x,t)|^2 \quad , \tag{2.11}$$

is interpreted in quantum mechanics as the *probability density* for observing the particle at position x and time t. It fulfills the *normalization condition*

$$\int_{-\infty}^{\infty} \varrho(x,t)\,\mathrm{d}x = 1 \quad , \tag{2.12}$$

which states that the probability of observing the particle anywhere is one.

Computing the probability density for the wave function (2.6) we find

$$\varrho(x,t) = \frac{1}{\sqrt{2\pi}\sigma_x(t)} \exp\left[-\frac{(x - \langle x(t)\rangle)^2}{2\sigma_x^2(t)}\right] \quad . \tag{2.13}$$

This is the probability density of a *Gaussian distribution*, Sect. 11.1.11, with time-dependent *mean* or *expectation value*

$$\langle x(t)\rangle = x_0 + \frac{p_0}{m}t \tag{2.14}$$

and the time-dependent *width* (2.10).

The widths σ_x and σ_p of the wave packet are connected by *Heisenberg's uncertainty relation*

$$\sigma_x\sigma_p \geq \hbar/2 \quad , \tag{2.15}$$

the equality holding for a Gaussian wave packet and $t = 0$ only.

2.1.3 Probability-Current Density. Continuity Equation

Although in this chapter we deal with a free particle, i.e., the motion of a particle in the absence of any force, the concepts introduced in this and the following sections can also be applied to the motion of a particle under the influence of a force. We therefore generalize the *Schrödinger equation* to

$$\mathrm{i}\hbar\frac{\partial}{\partial t}\psi(x,t) = -\frac{\hbar^2}{2m}\frac{\partial^2}{\partial x^2}\psi(x,t) + V(x)\psi(x,t) \quad , \tag{2.16}$$

where $V(x)$ is the *potential energy* or simply the *potential* of the force $F(x)$ acting on the particle, $F(x) = -\mathrm{d}V(x)/\mathrm{d}x$.

Because the total probability (2.12) does not change, the probability density $\varrho(x,t)$ has to be connected to a *probability-current density* $j(x,t)$ by a *continuity equation*

$$-\frac{\partial\varrho(x,t)}{\partial t}=\frac{\partial j(x,t)}{\partial x} \quad . \tag{2.17}$$

The probability-current density is

$$j(x,t)=\frac{\hbar}{2mi}\left(\psi^*\frac{\partial\psi}{\partial x}-\psi\frac{\partial\psi^*}{\partial x}\right) \quad . \tag{2.18}$$

[This equation is derived by computing the left-hand side of (2.17) using (2.11) and taking the time derivatives of ψ and ψ^* directly from the Schrödinger equation (2.16) and its complex conjugate.]

If $\varrho(x,t)$ is known, then $j(x,t)$ can be computed directly from (2.17). For a Gaussian wave packet one obtains

$$j(x,t)=\left[\frac{\mathrm{d}\langle x(t)\rangle}{\mathrm{d}t}+\frac{1}{\sigma_x(t)}\frac{\mathrm{d}\sigma_x(t)}{\mathrm{d}t}(x-\langle x(t)\rangle)\right]\varrho(x,t) \quad . \tag{2.19}$$

In relations (2.13) and (2.19) we considered a free Gaussian wave packet. If a force acts on the packet, in general, it loses its Gaussian shape. For two special forces, namely, the constant force, Sect. 4.1.19, and the harmonic force, Sect. 3.1.8, their general form remains unchanged. Only the time dependences of the mean $\langle x(t)\rangle$ and the width $\sigma_x(t)$ change.

2.1.4 Quantile Position. Quantile Trajectory

In classical mechanics the *position* $x(t)$ of a point particle is well-defined and so is its time derivative $v(t)=\mathrm{d}x(t)/\mathrm{d}t$, the *velocity*. In quantum mechanics only the probability density $\varrho(x,t)$ of the position x is known. From this we have already derived the probability-current density. We will now show that mathematical statistics allow us to define a position $x_P(t)$, the *quantile position*, and its time derivative, the *quantile velocity*. [This section is based on the following publication: S. Brandt, H. D. Dahmen, E. Gjonaj, T. Stroh, Physics Letters A249, 265 (1998).]

For any probability density $\varrho(x)$ the *quantile* associated with the probability Q is defined by

$$Q=\int_{-\infty}^{x_Q}\varrho(x)\,\mathrm{d}x \quad . \tag{2.20}$$

For the time-dependent probability density $\varrho(x,t)$ and the time-independent probability P, $0\le P\le 1$, we define the time-dependent *quantile position* by

$$\int_{x_P(t)}^{\infty}\varrho(x,t)\,\mathrm{d}x=P \quad . \tag{2.21}$$

At the time t the probability to observe a particle described by $\varrho(x,t)$ *to the right of* $x_P(t)$, i.e., at $x > x_P(t)$, is P. The probability to observe it *to the left of* $x_P(t)$ is $Q = 1 - P$.

The function

$$x_P = x_P(t) \tag{2.22}$$

defines the *quantile trajectory* of the quantile position $x_P(t)$ in the x, t plane. For a Gaussian wave packet (2.13) it can be easily calculated. Using the complementary error function $\operatorname{erfc}(x)$, Sect. 11.1.11, we have

$$P = \int_{x_P(t)}^{\infty} \varrho(x,t)\,\mathrm{d}x = \frac{1}{2}\operatorname{erfc}\left(\frac{x_P(t) - \langle x(t)\rangle}{\sqrt{2\pi}\,\sigma_x(t)}\right) \quad .$$

This equation can be used to determine $x_P(t_0)$ at a given time t_0 for a given probability P. At time t the quantile position is

$$x_P(t) = \langle x(t)\rangle + \frac{\sigma_x(t)}{\sigma_x(t_0)}(x_P(t_0) - \langle x(t_0)\rangle) \quad . \tag{2.23}$$

We consider the time derivative of (2.21). Taking into account the time dependence of the lower limit and the time independence of the integral we obtain

$$-\frac{\mathrm{d}x_P(t)}{\mathrm{d}t}\varrho(x_P(t),t) + \int_{x_P(t)}^{\infty} \frac{\partial \varrho(x',t)}{\partial t}\,\mathrm{d}x' = \frac{\mathrm{d}P}{\mathrm{d}t} = 0 \quad .$$

The continuity equation (2.17) allows us to replace $\partial\varrho/\partial t$ by $-\partial j/\partial x$. The integration can then be performed yielding

$$\frac{\mathrm{d}x_P(t)}{\mathrm{d}t} = \frac{j(x_P(t),t)}{\varrho(x_P(t),t)} \quad . \tag{2.24}$$

This is the *differential equation for the quantile trajectory*. For a specific solution an initial condition is needed. It is the initial quantile position $x_P(t_0)$ for a given probability P at time t_0.

2.1.5 Relation to Bohm's Equation of Motion

We differentiate (2.24) once more with respect to time and multiply it by the particle mass m,

$$m\frac{\mathrm{d}^2 x_P(t)}{\mathrm{d}t^2} = m\frac{\mathrm{d}}{\mathrm{d}t}\frac{j(x_P(t),t)}{\varrho(x_P(t),t)} = -\left.\frac{\partial U(x,t)}{\partial x}\right|_{x=x_P(t)} \quad . \tag{2.25}$$

By writing the right-hand side as the negative spatial derivative of a potential $U(x,t)$, we have given the equation the form of Newton's equation of motion.

The potential $U(x,t)$ is determined by using the expressions for ϱ, (2.11), and j, (2.18), and making use of the Schrödinger equation (2.16) to eliminate expressions of the type $\partial\psi/\partial t$ and $\partial\psi^*/\partial t$. The result is

$$U(x,t) = V(x) + V_{\mathrm{Q}}(x,t) \quad .$$

Here $V(x)$ is the potential appearing in the Schrödinger equation and

$$V_{\mathrm{Q}}(x,t) = \frac{\hbar^2}{4m\varrho(x,t)}\left(\frac{\partial^2\varrho(x,t)}{\partial x^2} - \frac{1}{2\varrho(x,t)}\left(\frac{\partial\varrho(x,t)}{\partial x}\right)^2\right) \tag{2.26}$$

is the time-dependent *quantum potential* introduced by David Bohm. For a Gaussian wave packet (2.13) it is

$$V_{\mathrm{Q}}(x,t) = -\frac{\hbar^2}{2m}\frac{1}{2\sigma_x^2(t)}\left[\frac{(x-\langle x(t)\rangle)^2}{2\sigma_x^2(t)} - 1\right] \quad . \tag{2.27}$$

Bohm did not use the quantile concept. He wrote (2.25) in the form

$$m\frac{\mathrm{d}^2x(t)}{\mathrm{d}t^2} = -\frac{\partial U(x,t)}{\partial x} \tag{2.28}$$

and thus formally connected quantum mechanics with Newton's equation of classical mechanics. The price to pay for this connection is twofold. (i) The existence of a quantum potential $V_{\mathrm{Q}}(x,t)$ has to be assumed. (ii) For the solution of (2.28) the two initial conditions, the particle's position $x_0 = x_0(t)$ and velocity $v_0 = v(t_0)$ at a time t_0, have to be known. In quantum mechanics this is impossible because of the uncertainty principle.

We would like to stress here that Bohm's particle trajectories are identical to our quantile trajectories, which are defined within the conventional framework of quantum mechanics.

2.1.6 Analogies in Optics

We now briefly consider also a harmonic electromagnetic wave propagating in vacuum in the x direction. The *electric field strength* is (written as a complex quantity – the physical field strength is its real part)

$$E_k(x,t) = E_0 \exp[-\mathrm{i}(\omega t - kx)] \quad , \tag{2.29}$$

where the *angular frequency* ω and the *wave number* k are related to the *velocity of light in vacuum* c and the *wavelength* λ by

$$\omega = c\,|k| \quad , \quad \lambda = 2\pi c/\omega \quad . \tag{2.30}$$

In analogy to (2.3) and (2.4) we can again form superpositions of harmonic waves as a weighted sum

$$E(x,t) = \sum_{n=1}^{N} w_n E_{k_n}(x,t) \qquad (2.31)$$

or as a wave packet

$$E(x,t) = \int_{-\infty}^{\infty} f(k) E_k(x,t)\,\mathrm{d}k \quad . \qquad (2.32)$$

As spectral function for the wave number k we choose a Gaussian function that is slightly different from the form (2.5),

$$f(k) = \frac{1}{\sqrt{2\pi}\sigma_k} \exp\left[-\frac{(k-k_0)^2}{2\sigma_k^2}\right] \quad . \qquad (2.33)$$

Integration of (2.32) yields

$$E = E_0 \exp\left[-\frac{\sigma_k^2}{2}(ct-x)^2\right] \exp[-\mathrm{i}(\omega_0 t - k_0 x)] \qquad (2.34)$$

and

$$|E|^2 = E_0^2 \exp\left[-\frac{(ct-x)^2}{2\sigma_x^2}\right] \quad , \qquad (2.35)$$

with the relation

$$\sigma_x \sigma_k = 1/2 \qquad (2.36)$$

between the spatial width σ_x and the width σ_k in wave number of the electromagnetic wave packet. The importance of (2.35) becomes clear through the relation

$$w = \frac{\varepsilon_0}{2} |E|^2 \quad , \qquad (2.37)$$

where w is approximately the *energy density* of the electromagnetic field (averaged over a short period of time) and ε_0 is the electric field constant.[1]

2.1.7 Analogies in Classical Mechanics: The Phase-Space Probability Density

In conventional classical mechanics the position x and the momentum p of a mass point are assumed to be known exactly. In reality this is, of course, not the case. There are experimental errors $\Delta x = \sigma_x$ and $\Delta p = \sigma_p$. To take these

[1] This statement only holds for short average wavelengths, $\lambda_0 \ll \sigma_x$.

into account we define a classical probability density $\varrho^{\rm cl}(x, p)$ in phase space (spanned by x and p). For the initial time $t = 0$ we write it as an uncorrelated bivariate Gaussian distribution (Sect. 11.14) with expectation values x_0, p_0 and widths σ_{x0}, σ_{p0},

$$\varrho_{\rm i}^{\rm cl}(x_{\rm i}, p_{\rm i}) = \frac{1}{\sqrt{2\pi}\sigma_{x0}} \exp\left\{-\frac{(x_{\rm i}-x_0)^2}{2\sigma_{x0}^2}\right\} \frac{1}{\sqrt{2\pi}\sigma_{p0}} \exp\left\{-\frac{(p_{\rm i}-p_0)^2}{2\sigma_{p0}^2}\right\} \quad . \tag{2.38}$$

For a free particle the momentum stays unchanged, $p = p_{\rm i}$, whereas the position changes linearly with time, $x = x_{\rm i} + v_{\rm i}t = x_{\rm i} + (p_{\rm i}/m)t$. We can also write

$$x_{\rm i} = x - (p/m)t \quad , \qquad p_{\rm i} = p \quad . \tag{2.39}$$

Inserting these relations into (2.38) we obtain the time-dependent classical phase-space probability density

$$\begin{aligned} \varrho^{\rm cl}(x, p, t) &= \varrho_{\rm i}^{\rm cl}(x - pt/m, p) \\ &= \frac{1}{2\pi\sigma_{x0}\sigma_{p0}} \exp\left\{-\frac{1}{2}\left[\frac{(x - x_0 - pt/m)^2}{\sigma_{x0}^2} + \frac{(p-p_0)^2}{\sigma_{p0}^2}\right]\right\} \\ &= \frac{1}{2\pi\sigma_{x0}\sigma_{p0}} \exp L \quad . \end{aligned} \tag{2.40}$$

The exponent can be rewritten as

$$\begin{aligned} L &= -\frac{1}{2}\left\{\frac{(x - [x_0 + p_0t/m] - (p-p_0)t/m)^2}{\sigma_{x0}^2} + \frac{(p-p_0)^2}{\sigma_{p0}^2}\right\} \\ &= -\frac{1}{2}\frac{\sigma_{x0}^2 + \sigma_{p0}^2t^2/m^2}{\sigma_{x0}^2}\left\{\frac{(x - [x_0 + p_0t/m])^2}{\sigma_{x0}^2 + \sigma_{p0}^2t^2/m^2}\right. \\ &\qquad \left. - \frac{2(x - [x_0 + p_0t/m])(p-p_0)}{(\sigma_{x0}^2 + \sigma_{p0}^2t^2/m^2)m/t} + \frac{(p-p_0)^2}{\sigma_{p0}^2}\right\} \quad . \end{aligned}$$

Comparing this expression with the exponent of the general form (11.59) of a bivariate probability density we find that $\varrho^{\rm cl}(x, p, t)$ is a bivariate Gaussian with the expectation values

$$\langle x(t)\rangle = x_0 + p_0t/m \quad , \qquad \langle p(t)\rangle = p_0 \quad , \tag{2.41}$$

the widths

$$\sigma_x(t) = \sqrt{\sigma_{x0}^2 + \sigma_{p0}^2t^2/m^2} \quad , \qquad \sigma_p(t) = \sigma_p = \sigma_{p0} \quad , \tag{2.42}$$

and the correlation coefficient

$$c = \frac{\sigma_{p0} t}{\sigma_x(t) m} = \frac{\sigma_{p0} t/m}{\sqrt{\sigma_{x0}^2 + \sigma_{p0}^2 t^2/m^2}} \quad . \tag{2.43}$$

In particular, this means that the marginal distribution $\varrho_x^{\rm cl}(x,t)$, i.e., the spatial probability density for the classical particle with initial uncertainties σ_{x0} in position and σ_{p0} momentum, is

$$\varrho_x^{\rm cl}(x,t) = \frac{1}{\sqrt{2\pi}\sigma_x(t)} \exp\left\{ -\frac{(x - [x_0 + p_0 t/m])^2}{2\sigma_x^2(t)} \right\} \quad .$$

We now consider the classical probability density $\varrho^{\rm cl}(x,p,t)$ of a particle with initial uncertainties σ_{x0} in position and σ_{p0} in momentum, satisfying the minimal uncertainty requirement of quantum mechanics, $\sigma_{x0}\sigma_{p0} = \hbar/2$. In that case the spatial width of the classical probability distribution is

$$\sigma_x(t) = \frac{\hbar}{2\sigma_{p0}} \sqrt{1 + \frac{4\sigma_{p0}^4}{\hbar^2} \frac{t^2}{m^2}}$$

and therefore identical to the width of the corresponding quantum-mechanical wave packet. Also the expectation values for x and p and the width in p are identical for the classical and the quantum-mechanical case.

Thus it is demonstrated that the force-free motion of a classical particle described by a Gaussian probability distribution in phase space of position and momentum yields the same time evolution of the local probability density as in quantum mechanics if the initial widths σ_{x0}, σ_{p0} in position and momentum fulfill the relation

$$\sigma_{x0}\sigma_{p0} = \frac{\hbar}{2} \quad .$$

In the further development of the quantum-mechanical description of particles we shall see that this finding does not remain true for particles under the action of forces other than constant in space or linear in the coordinates.

We can generalize our considerations from the free motion of a Gaussian phase-space distribution to that of an arbitrary distribution under the influence of a potential energy $V(x)$. A trajectory in phase space is described by

$$x = x(x_{\rm i}, p_{\rm i}, t) \quad , \qquad p = p(x_{\rm i}, p_{\rm i}, t) \quad . \tag{2.44}$$

The point $(x_{\rm i}, p_{\rm i})$ is assumed at the initial time $t = 0$, i.e.,

$$x_{\rm i} = x(x_{\rm i}, p_{\rm i}, 0) \quad , \qquad p_{\rm i} = p(x_{\rm i}, p_{\rm i}, 0) \quad . \tag{2.45}$$

The functions (2.44) satisfy the equations

$$\frac{{\rm d}x}{{\rm d}t} = \frac{1}{m} p(x_{\rm i}, p_{\rm i}, t) \quad , \qquad \frac{{\rm d}p}{{\rm d}t} = -\frac{{\rm d}V}{{\rm d}x} = F(x(x_{\rm i}, p_{\rm i}, t)) \quad . \tag{2.46}$$

We denote by $\varrho_{\rm i}^{\rm cl} = \varrho_{\rm i}^{\rm cl}(x_{\rm i}, p_{\rm i})$ the classical phase-space probability density at the initial time $t = 0$. Solving the equations (2.44) for $x_{\rm i}$ and $p_{\rm i}$ we obtain the initial phase-space points

$$x_{\rm i} = x_{\rm i}(x, p, t) \quad , \qquad p_{\rm i} = p_{\rm i}(x, p, t) \tag{2.47}$$

in terms of x and p. The time-dependent phase-space probability density is then obtained as

$$\varrho^{\rm cl}(x, p, t) = \varrho_{\rm i}^{\rm cl}(x_{\rm i}(x, p, t), p_{\rm i}(x, p, t)) \quad . \tag{2.48}$$

For the force-free Gaussian distribution treated at the beginning of this section the equations (2.47) take the form (2.39). In this case (2.40) is obtained from (2.48) if the initial distribution is an uncorrelated Gaussian.

Since $\varrho_{\rm i}^{\rm cl}(x_{\rm i}, p_{\rm i})$ does not depend explicitly on t, its total time derivative, i.e., its partial time derivative for fixed values $x_{\rm i}$, $p_{\rm i}$, vanishes,

$$\frac{{\rm d}\varrho^{\rm cl}(x, p, t)}{{\rm d}t} = \left(\frac{\partial \varrho^{\rm cl}(x(x_{\rm i}, p_{\rm i}, t), p(x_{\rm i}, p_{\rm i}, t), t)}{\partial t}\right)_{x_{\rm i}, p_{\rm i}} = \frac{\partial \varrho_{\rm i}^{\rm cl}(x_{\rm i}, p_{\rm i})}{\partial t} = 0 \ . \tag{2.49}$$

The explicit calculation of the total time derivative yields

$$\begin{aligned} \frac{{\rm d}\varrho^{\rm cl}(x, p, t)}{{\rm d}t} &= \frac{\partial \varrho^{\rm cl}(x, p, t)}{\partial t} \\ &+ \frac{\partial x(x_{\rm i}, p_{\rm i}, t)}{\partial t}\frac{\partial \varrho^{\rm cl}(x, p, t)}{\partial x} + \frac{\partial p(x_{\rm i}, p_{\rm i}, t)}{\partial t}\frac{\partial \varrho^{\rm cl}(x, p, t)}{\partial p} \ . \end{aligned}$$

Because of (2.49) and (2.46) we obtain

$$-\frac{\partial \varrho^{\rm cl}(x, p, t)}{\partial t} = \frac{p}{m}\frac{\partial \varrho^{\rm cl}(x, p, t)}{\partial x} - \frac{\partial V}{\partial x}\frac{\partial \varrho^{\rm cl}(x, p, t)}{\partial p} \ , \tag{2.50}$$

the *Liouville equation* for the classical phase-space probability density.

This equation remains correct also in quantum mechanics for a potential that is a polynomial in x of up to second degree,

$$V(x) = V_0 - mgx + \frac{m}{2}\omega^2 x^2 \quad .$$

For $\omega = 0$ this corresponds to the free fall of a particle of mass m in a gravitational force field $F(x) = mg$. For $g = 0$ it describes the harmonic motion with circular frequency ω of a particle in the force field $F(x) = -m\omega^2 x$ of a harmonic oscillator.

Further Reading

Alonso, Finn: Vol. 2, Chaps. 18, 19; Vol. 3, Chaps. 1, 2
Berkeley Physics Course: Vol. 3, Chaps. 4, 6; Vol. 4, Chaps. 5, 6, 7
Brandt, Dahmen: Chaps. 2, 3, 7
Feynman, Leighton, Sands: Vol. 3, Chaps. 1, 2
Flügge: Vol. 1, Chap. 2
Gasiorowicz: Chaps. 2, 3
Hecht, Zajac: Chaps. 2, 7
Merzbacher: Chaps. 2, 3
Messiah: Vol. 1, Chaps. 1, 2
Schiff: Chaps. 1, 2

2.2 A First Session with the Computer

In this section we want to give you a first impression of how the program is used. For a systematic guide to **IQ** see Appendix A.

We assume that **IQ** is already installed on your computer. If it is not, follow the Installation Guide in Appendix B. Also, most probably the **IQ** symbol featuring the letter $\hbar$ is displayed on the computer's desktop.

2.2.1 Starting IQ

You simply start **IQ** by clicking (once or twice – depending on your operating system) on the **IQ** symbol on the desktop. For the case that the **IQ** symbol is not present on the desktop, the start-up procedure is described in the file `ReadMe.txt` on the CD-ROM.

The **IQ** *main frame*, Fig. 2.1, appears on your desktop. It carries the *main toolbar*, i.e., a row of buttons. In its title bar it carries the name of the currently open *descriptor file*.

Fig. 2.1. **IQ** main frame with main toolbar

A *descriptor* is a set of data that completely defines a plot produced by **IQ**. Each *descriptor file* contains several descriptors. (If the **IQ** main frame displayed on your desktop does not contain the descriptor file name `1D_Free_Particle.des`, then press the button **Descriptor File**. A *file chooser* opens in which, with the mouse, you can select that file.)

Usually, a small frame is now displayed offering you an *introductory demonstration* of how to use **IQ**. Just press the **Start** button.

2.2.2 An Automatic Demonstration

Press the button **Run Demo**. A file chooser opens. Select the file

`1D_Free_Particle(Sound).demo`

if you have loudspeakers at your computer. Otherwise select

`1D_Free_Particle(NoSound).demo.`

Next, a *dialog box* is displayed giving you the choice between *automatic* or *step by step*. Select **automatic**. Lean back, relax, and watch (and listen to) our demonstration.

2.2.3 A First Dialog

2.2.3.1 Selecting a Descriptor Press the button **Descriptor** in the **IQ** main frame. A *descriptor selection panel* opens containing a list of the titles of all descriptors on the selected descriptor file. Move the mouse cursor to the first line of the list and press the mouse button, i.e., select the first descriptor on the descriptor file. A *graphics frame* opens. It contains the plot corresponding to the descriptor, the descriptor title in the title bar, and six buttons.

2.2.3.2 Changing Parameters Press the button **Parameters** in the graphics frame. A *parameter panel* appears that carries all the information contained in the descriptor. It is composed of several subpanels. The plot corresponds to Fig. 2.2. It shows the motion of a free quantum-mechanical wave packet. That is the topic of Sect. 2.3. Here we just want to give an idea of the dialog. So, press the *radio button* **Real Part of Wave Function** near the bottom of the parameter panel. Then press **Plot** either in the parameter panel or in the graphics frame and observe the change in the plot.

Next, press again **Descriptor** in the **IQ** main frame and again select the first descriptor from the list. You then have on your desktop two graphics frames, one showing the time development of Re ψ, the other showing that of $|\psi|^2$, and you can compare the two plots directly. You may discard the plot with Re ψ by pressing the *close* button in its title bar.

For the remaining graphics frame open the parameter panel and in it the subpanel **Wave Packet**. Near the middle there is a field labeled **p_0** carrying the numerical value 1. Using mouse and keyboard change that to 3, press **Plot**, and observe the change.

Now, for the third time, select the first descriptor on the descriptor file thus opening another graphics frame. In its parameter panel press **Graphics**, then **Accuracy**. In the field **n_y** replace 9 by 19 and plot. The function $|\psi(x,t)|^2$ is now shown for 19 values of the time t. Then, in the parameter panel, press **Physics** followed by **Multiple Plot**. There, set **Number of Rows** and **Number of Columns** to 2 and plot again. The graphics frame contains four plots each with a different set of parameters for the mean momentum p_0 and the momentum width σ_p of the wave packet.

2.2.3.3 Creating and Running a Movie Select the second descriptor on the descriptor file. On the graphics frame, which opens, you will find an additional bottom toolbar with four buttons under the heading **Movie:**. It signals that the graphics frame has *movie capability*. Press **Create** to create a movie which is stored in core. Once creation is completed you can start and stop the movie by using the ⇢ and ⏸ buttons, respectively. You can change parameters in the corresponding parameter panel and recreate the changed movie by pressing **Plot** and then **Create** or simply **Create**.

Now again, select the first descriptor on the descriptor file. It has no bottom toolbar. But its top toolbar contains a button **Prepare for Movie**, indicating *indirect movie capability*. When it is pressed, another graphics frame with direct movie capability is opened. This represents the same physics situation but graphically adapted to be shown as a movie.

2.2.3.4 Help Press the button **Help** in the **IQ** main frame. A window is opened in which a special version of the 'Acrobat® Reader' displays the text of this book. It is opened on the page with the beginning of Appendix A, *A Systematic Guide to* **IQ**. Using the tools of the reader you can access all of the information in the text.

2.2.3.5 Context-Sensitive Help The contents of the parameter panel depend very much on the physics topic chosen. Therefore, special help is provided for it. Select a subpanel of the parameter panel. (Selection is indicated graphically in the *tag field* by a thin frame around the name of the subpanel, e.g., **Physics** or **Comp. Coord.**.) Now, press the `F1` key on the keyboard. (On some keyboards press `Fn` plus `F1`.) A page relevant to the selected subpanel is displayed.

2.2.3.6 Closing IQ Press the *close* button in the title bar of the **IQ** main frame.

2.3 The Free Quantum-Mechanical Gaussian Wave Packet

Aim of this section: Demonstration of the propagation in space and time of a Gaussian wave packet of Schrödinger waves ψ, (2.6). Illustration of the probability-current density $j(x,t)$, (2.19), and Bohm's quantum potential $V_Q(x,t)$, (2.27). Illustration of quantiles $x_P(t)$, (2.21), and the quantile trajectory.

Because this is the first physics topic for which the use of the program is explained, we will present the explanation in a rather detailed way.

The *parameter panel* is displayed (once you press the button **Parameters** in the graphics frame, see Sect. A.1.8) with the subpanel **Physics—Comp. Coord.** visible. This subpanel is present for every physics topic chosen and is always visible when the parameter panel is first displayed. Usually there are more **Physics** subpanels (in the present case there is a total of four). All of them are explained in sections like this one, devoted to the physics topic. The other subpanels, **Graphics**, **Background**, and **Format**, are more or less the same for all physics topics and are explained in Sects. A.6 through A.8.

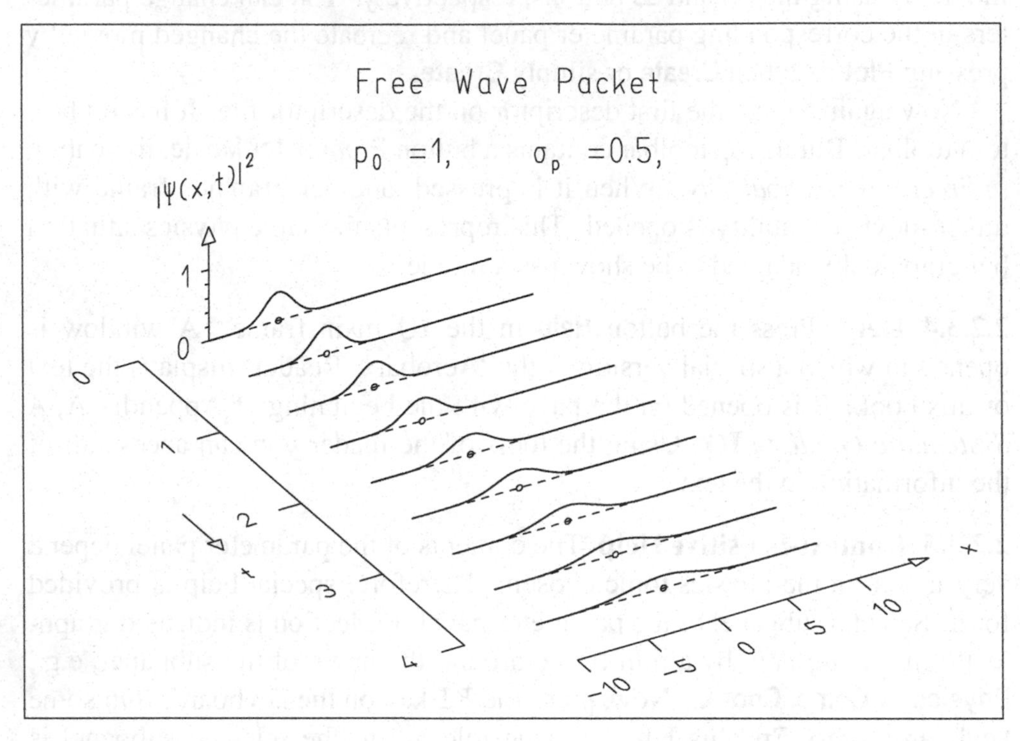

Fig. 2.2. Plot produced with descriptor `Free quantum-mechanical Gaussian wave packet` on file `1D_Free_Particle.des`

2.3.1 The Subpanel Physics—Comp. Coord.

This subpanel contains three items:

- The meaning of the **Computing Coordinates**. There are three lines of text explaining the meaning of the computing coordinates x, y, z in terms of physics. (In our case x is position x, y is time t, and z a function of x and t determined in the third item below.)
- The **Ranges of Computing Coordinates**. This item consists of six numbers. Four of them, **x_beg**, **x_end**, **y_beg**, and **y_end** define the ranges of the computing coordinates x and y, whereas **z_beg** and **z_end** (together with **Z_beg**

and **Z_end** on the subpanel **Graphics—Geometry**) define scale and offset in z, see Sect. A.2.

- Under the heading **Function Shown is** you find a set of *radio buttons* (i.e., buttons of which one and only one is on) allowing you to select one of the following functions:
  - **Absolute Square of Wave Function** $|\psi(x,t)|^2$,
  - **Real Part of Wave Function** $\mathrm{Re}\,\psi(x,t)$,
  - **Imaginary Part of Wave Function** $\mathrm{Im}\,\psi(x,t)$,
  - **Probability-Current Density** $j(x,t)$,
  - **Bohm's Quantum Potential** $V_{\mathrm{Q}}(x,t)$.

  Immediately after selection, the selected function is shown as computing coordinate z in the field **Computing Coordinates**. To produce a plot of the selected function, you have to press the **Plot** button.

2.3.2 The Subpanel Physics—Wave Packet

The Gaussian wave packet is completely determined by three values characterizing it at the initial time $t = 0$, the *mean position* x_0, the *mean momentum* p_0, and the *momentum width* σ_p. Instead of p_0, the energy $E_0 = p_0^2/2m$ may be given as input value. Instead of σ_p, the fraction frac $= \sigma_p/p_0$ may be given as input value if $p_0 \neq 0$.

The subpanel **Physics—Wave Packet** begins with two sets of radio buttons allowing the choice of input quantities for p_0 and σ_p as discussed above. Under the heading **Wave Packet** you find the numerical values of five quantities. The first is the mean position x_0. It is followed by p_0 and σ_p (or, possibly, E_0 and frac, depending on the choices of input variables). The next two, Δp_0 (or ΔE_0) and $\Delta\sigma_p$ (or Δfrac), are used in a *multiple plot* only in which p_0 is incremented horizontally from plot to plot and σ_p is incremented vertically.

Under the heading **Constants** you many change the numerical values of *Planck's constant* $\hbar$ and the *particle mass* m. Usually, they are best left at their default values $\hbar = 1$, $m = 1$. Under the heading **Graphical Items** there are two more numerical values, one for the *dash length* of the zero line drawn for each function graph and one for the *radius of the circle symbolizing the position of the classical particle* on that line. Both quantities are given in world coordinates and are usually left at their default values.

2.3.3 The Subpanel Physics—Quantile

The contents of this subpanel are used only if the function **Absolute Square of Wave Function** was chosen on the subpanel **Physics—Comp. Coord.**. Under the heading **Quantile Motion** you can select either **Not Shown** or **Shown**. The **Quantile** is defined by the probability $P = 1 - Q$. If you choose to show the quantile motion, the area $x > x_P(t)$ under the curve $\varrho(x,t) = |\psi(x,t)|^2$ is hatched and the quantile trajectory $x_P = x_P(t)$ appears in the x, t plane.

2.3.4 The Subpanel Movie

The graphics frame created with some descriptors has *movie capability*. It contains a bottom toolbar with movie buttons. Pressing the button **Create** on the graphics frame results in the production of a movie. You can start and stop its display by using the [start] and [stop] buttons, respectively. The corresponding parameter panel contains a subpanel **Movie** with parameters like the time interval, over which the movie extends, and more technical details (Sect. A.4).

Other descriptors give rise to a graphics frame with *indirect movie capability*. Its top toolbar carries a button **Prepare for Movie**. An example is the descriptor `Free quantum-mechanical Gaussian wave packet` which leads to the plot shown in Fig. 2.2, where the time development is shown as a series of graphs for different values of the time t. Technically, Fig. 2.2 is a plot of the type "surface over Cartesian grid in 3D" (see Sect. A.3.1). For the presentation of a movie, in which time really runs, we use the plot type "2D function graph" (see Sect. A.3.3). The conversion to this plot type is done by pressing the button **Prepare for Movie**. A new graphics frame with full movie capability results, displaying the first frame of that movie. The corresponding descriptor can be changed and/or appended to the descriptor file. Such an adapted descriptor is `Movie: Free quantum-mechanical Gaussian wave packet`.

Example Descriptors on File `1D_Free_Particle.des`

- `Free quantum-mechanical Gaussian wave packet` (see Fig. 2.2)
- `Movie: Free quantum-mechanical Gaussian wave packet`

2.4 The Free Optical Gaussian Wave Packet

Aim of this section: Demonstration in space and time of a Gaussian wave packet of electromagnetic waves E, (2.34).

This section is the analog in optics to Sect. 2.3 and is therefore kept short.

On the subpanel **Physics—Comp. Coord.** you can select to display the **Absolute Square**, the **Real Part**, or the **Imaginary Part** of the complex electric field strength.

On the subpanel **Physics—Wave Packet** you find the parameters x_0 (initial mean position), k_0 (mean wave number), and σ_k (width in wave number), as well as the increments Δk_0 and $\Delta\sigma_k$ (used in a *multiple plot* only).

Example Descriptors on File `1D_Free_Particle.des`

- `Free optical Gaussian wave packet` (see Fig. 2.3)
- `Movie: Free optical Gaussian wave packet`

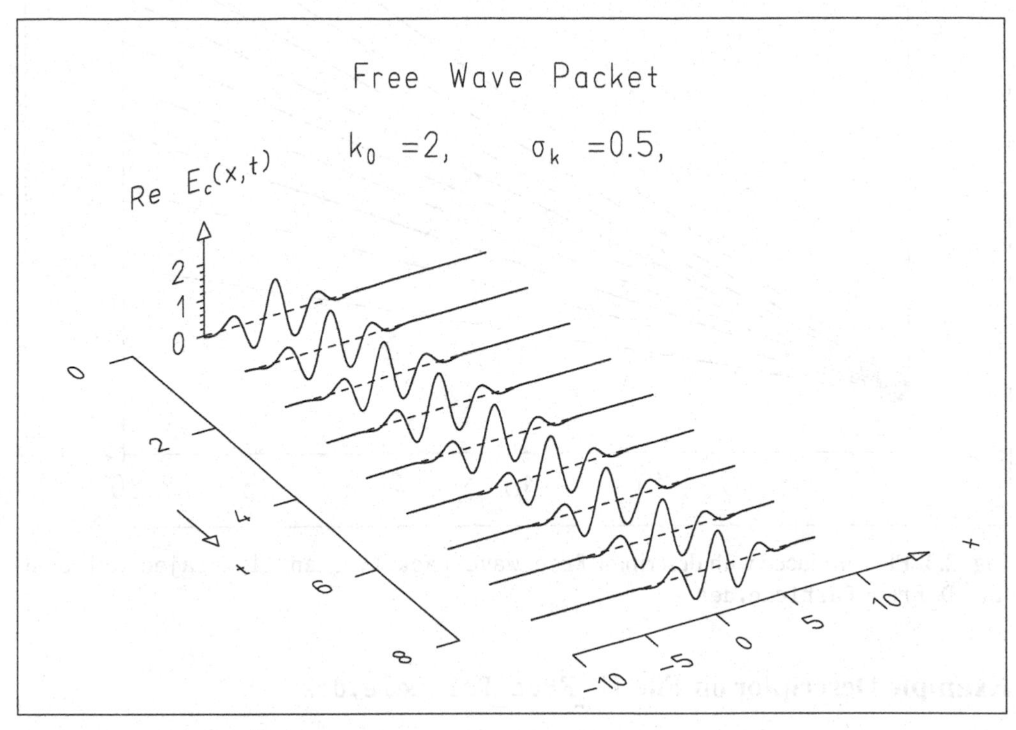

Fig. 2.3. Plot produced with descriptor `Free optical Gaussian wave packet` on file `1D_Free_Particle.des`

2.5 Quantile Trajectories

Aim of this section: Illustration of the quantile trajectories $x_P = x_P(t)$, (2.23), for various values of the probability P as a set of 2D function graphs in the x_P, t plane.

On the subpanel **Physics—Quantiles** you find the **Range of P** defined by the two values **P_beg** and **P_end** and the **Number of Trajectories** defined by **n_Traj**. Trajectories are drawn for $P = P_{\text{beg}}$, for $P = P_{\text{end}}$, and for other values of P placed equidistantly between these two limits. The subpanel offers the possibility either to draw all trajectories in the same color or to draw one trajectory, corresponding to a given value of P, in a special color.

The contents of the subpanel **Physics—Wave Packet** are as described in Sect. 2.3.2.

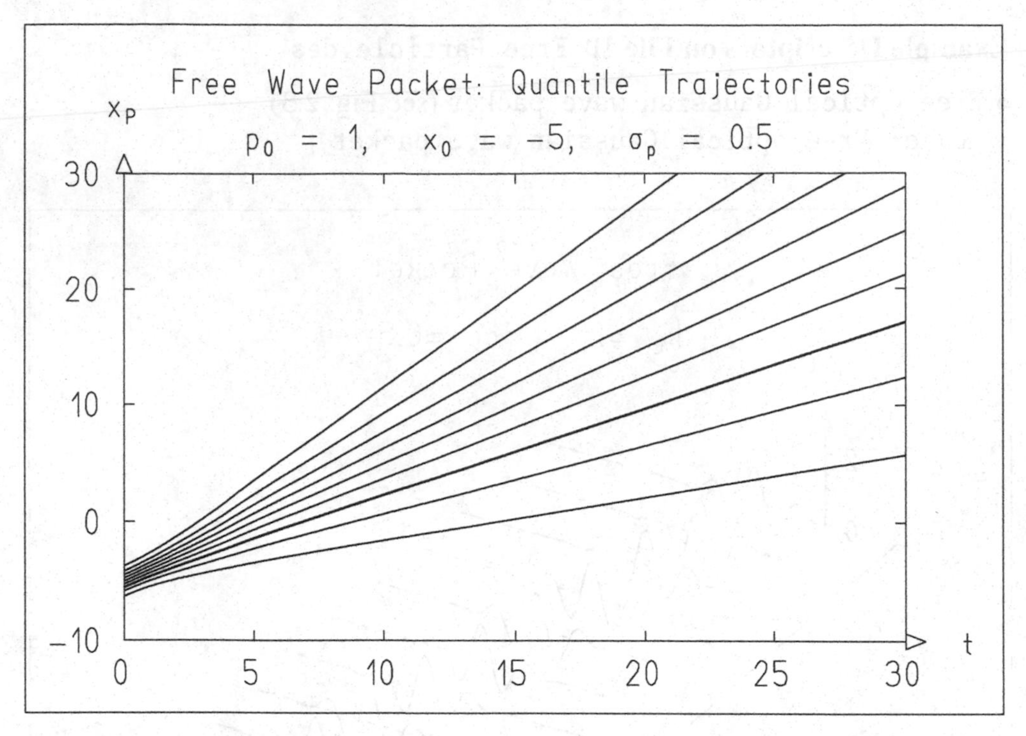

Fig. 2.4. Plot produced with descriptor `Free wave packet, quantile trajectories` on file `1D_Free_Particle.des`

Example Descriptor on File `1D_Free_Particle.des`

- `Free wave packet, quantile trajectories` (see Fig. 2.4)

2.6 The Spectral Function of a Gaussian Wave Packet

Aim of this section: Graphical presentation of the spectral wave function (2.5) and (2.33) for both quantum-mechanical and optical Gaussian wave packets.

On the subpanel **Physics—Comp. Coord.** you can select to show the spectral function $f(p)$ of a **quantum-mechanical** or the spectral function $f(k)$ of an **optical** wave packet. The parameters defining the wave packet are given on the subpanel **Physics—Wave Packet**. They were described in detail in Sect. 2.3.2 (for the optical wave packet the wave number k takes the place of the momentum q).

Example Descriptor on File `1D_Free_Particle.des`

- `Spectral function` (see Fig. 2.5)

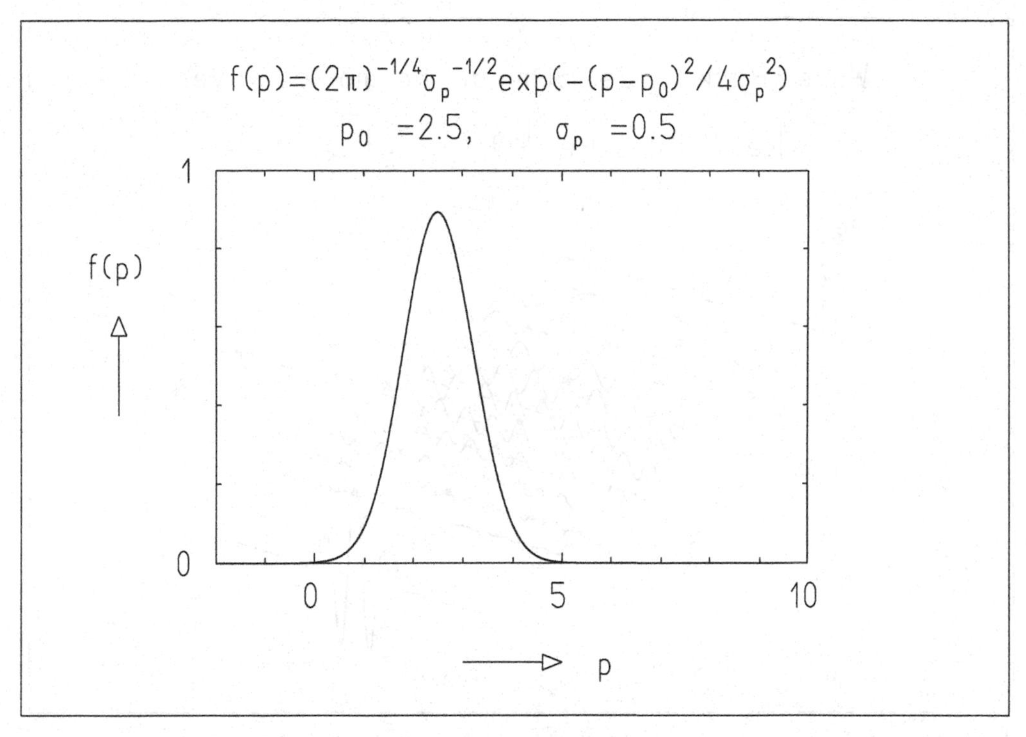

Fig. 2.5. Plot produced with descriptor `Spectral function` on file `1D_Free_Particle-.des`

2.7 The Wave Packet as a Sum of Harmonic Waves

Aim of this section: Construction of both quantum-mechanical and optical wave packets as a superposition of harmonic waves of the form (2.3) and (2.31), respectively.

Figure 2.6 in the background shows a series of harmonic waves ψ_n of different momenta p_n and different weight factors w_n. In the foreground the sum (2.3) of all wave functions is shown. Technically, the figure is a plot of the type *surface over Cartesian grid in 3D*, Sect. A.3.1. The grid is formed only by n_y lines along the x direction. (The number n_y is given as **n_y** on the subpanel **Graphics—Accuracy**.) Two of these lines are missing to clearly separate the graph of the sum $\sum_n \psi_n$ in the foreground from the graphs of the individual terms ψ_n in the background. Accordingly, the values for p_n are

$$p_n = p_0 - f_\sigma \sigma_p + n\Delta p \quad , \quad n = 0, 1, \ldots, N \quad ,$$

with

$$\Delta p = \frac{2 f_\sigma \sigma_p}{N - 1} \quad , \quad N = n_y - 3 \quad .$$

(For an optical wave packet replace p by k.) Here f_σ is a positive number of reasonable size, e.g., $f_\sigma = 3$.

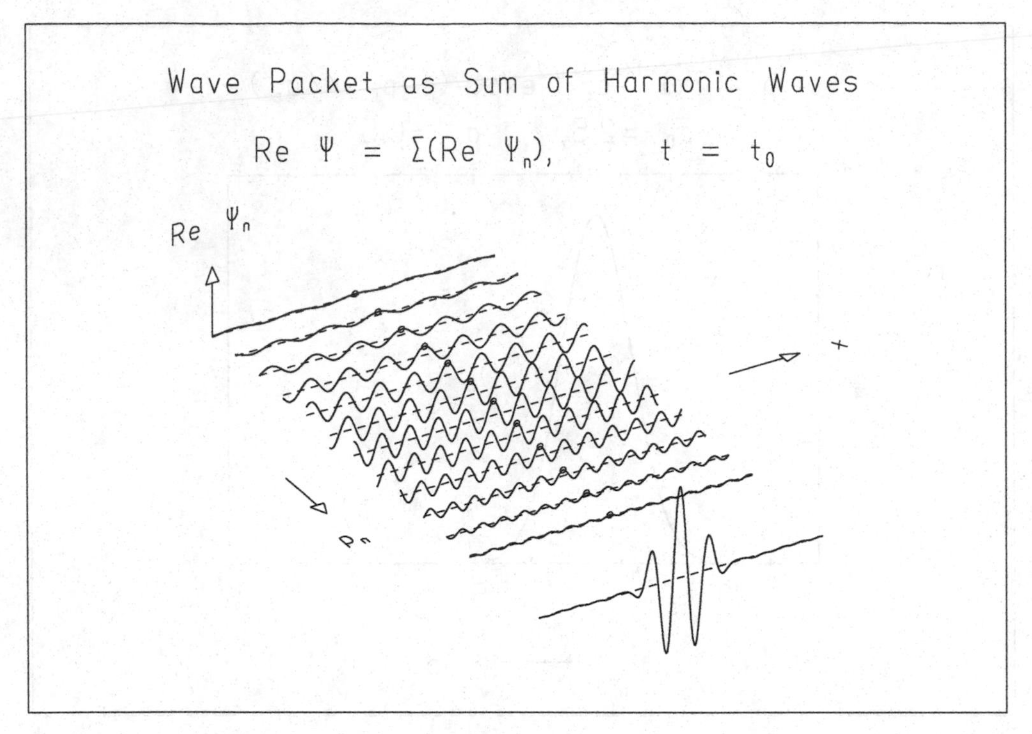

Fig. 2.6. Plot produced with descriptor `Wave packet as sum of harmonic waves` on file `1D_Free_Particle.des`

The weight factor w_n is chosen to be

$$w_n = \exp\left[-\frac{(p_n - p_0)^2}{4\sigma_p^2}\right]$$

for the quantum-mechanical wave packet and

$$w_n = \exp\left[-\frac{(k_n - k_0)^2}{2\sigma_k^2}\right]$$

for the electromagnetic wave packet, where p_0 and σ_p are mean value and width in momentum of a quantum-mechanical wave packet and k_0 and σ_k the mean value and width of the wave number for an optical wave packet.

On the subpanel **Physics—Comp. Coord.** you can choose between a **quantum-mechanical** and an **optical** wave packet and ask to display either the **Real Part** or the **Imaginary Part** of the wave functions.

The subpanel **Physics—Wave Packet** contains the parameters of the packet and is essentially identical to the one explained in Sect. 2.3.2. It is different, however, with respect to a *multiple plot*. Now the time is varied: For the first plot the time is $t = t_0 = 0$. It is increased by $\mathbf{\Delta t}$ for every successive plot. Moreover, the subpanel **Physics—Wave Packet** carries the parameter

f_σ needed for the approximation of the Gaussian wave packet as a sum of harmonic waves.

All waves ψ_n at $t = t_0 = 0$ have the same phase at $x = x_0$. These points are marked by little circles of radius **R**. The waves ψ_n move with time and so do the points of constant phase. Whereas quantum-mechanical waves ψ_n have different velocities $v_n = p_n/m$, the electromagnetic waves all have the velocity c. Therefore, the optical wave packet moves without dispersion, whereas the quantum-mechanical one disperses.

Remarks: 1. The finite sum $\psi = \sum_{n=0}^{N} \psi_n$ ($E = \sum E_n$) is an approximation of a Gaussian wave packet if N is large and f_σ is at least equal to 3. No attempt has been made to normalize this wave packet. For very small N it becomes evident that ψ (or E) is periodic in x. For larger N the period is longer than the x interval plotted. 2. Do not use a scale in y because here it is not meaningful.

Movie Capability: Direct. The time, over which the movie extends, is $T = \Delta t(N_{\text{Plots}} - 1)$; N_{Plots} is the number of individual plots in a multiple plot and Δt is the time interval between two plots. For a single plot $T = \Delta t$.

Example Descriptor on File `1D_Free_Particle.des`

- `Wave packet as sum of harmonic waves` (see Fig. 2.6)

2.8 The Phase-Space Distribution of Classical Mechanics

Aim of this section: Graphical presentation of the phase-space probability density $\varrho^{\text{cl}}(x, p, t)$ of Sect. 2.1.7 which initially (at time $t = 0$) is uncorrelated and fulfills the condition $\sigma_{x0}\sigma_{p0} = \hbar/2$. The marginal distributions $\varrho_x^{\text{cl}}(x, t)$ and $\varrho_p^{\text{cl}}(p, t)$ are also shown.

On the subpanel **Physics—Comp. Coord.** you have to select

- Linear (or constant) potential

and to set g equal to 0 (ensuring that the potential is constant, i.e., that there is no force).

On the subpanel **Physics—Phase-Space Distr.** you enter the initial parameters x_0, p_0, σ_{p0} as well as a time t_0 (normally 0) and a time difference Δt. The distribution $\varrho^{\text{cl}}(x, p, t)$ is shown for $t = t_0$. If you ask for a multiple plot, it is shown for $t = t_0$ in the first plot, $t = t_0 + \Delta t$ in the second, etc. At the bottom of the subpanel there are 3 check boxes. By enabling them you may show

- the *expectation value* as a circle in the x, p plane for the times $t = t_0, t = t_0 + \Delta t, \ldots$ including the time of the particular plot and the trajectory of the expectation value,
- the *covariance ellipse* as a line $\varrho^{\mathrm{cl}}(x, p, t) = \mathrm{const}$,
- a rectangular *frame* enclosing the covariance ellipse (not normally wanted).

Movie Capability: Direct. The time, over which the movie extends, is $T = \Delta t\,(N_{\mathrm{Plots}} - 1)$; N_{Plots} is the number of individual plots in a multiple plot and Δt is the time interval between two plots. For a single plot $T = \Delta t$.

Example Descriptor on File `1D_Free_Particle.des`

- `Classical phase-space distribution, free particle` (see Fig. 2.7)

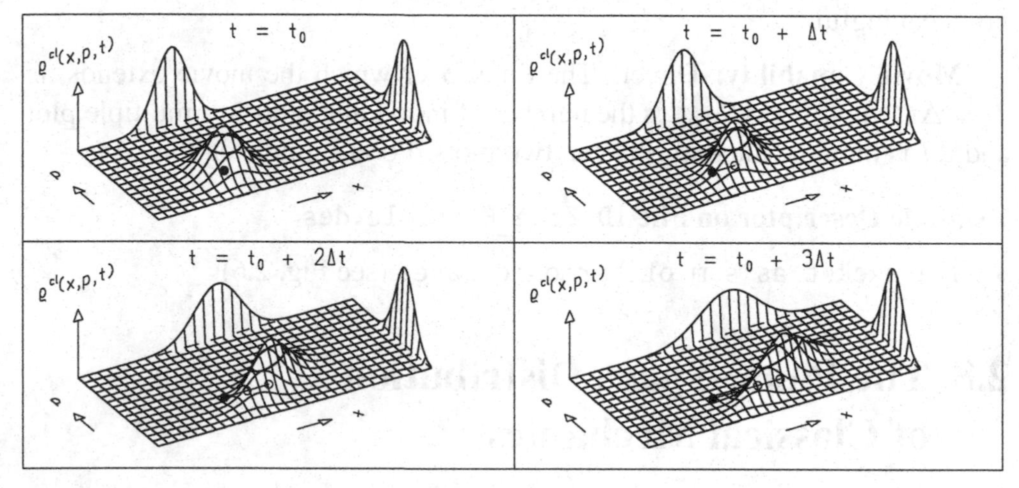

Fig. 2.7. Plot produced with descriptor `Classical phase-space distribution, free particle` on file `1D_Free_Particle.des`

2.9 Classical Phase-Space Distribution: Covariance Ellipse

Aim of this section: Graphical presentation of the covariance ellipse, characterizing the phase-space probability density $\varrho^{\mathrm{cl}}(x, p, t)$ of Sect. 2.1.7, which initially (at time $t = 0$) is uncorrelated and fulfills the condition $\sigma_{x0}\sigma_{p0} = \hbar/2$.

On the subpanel **Physics—Comp. Coord.** you have to select

- Linear (or constant) potential

and to set g equal to 0 (ensuring that the potential is constant, i.e., that there is no force).

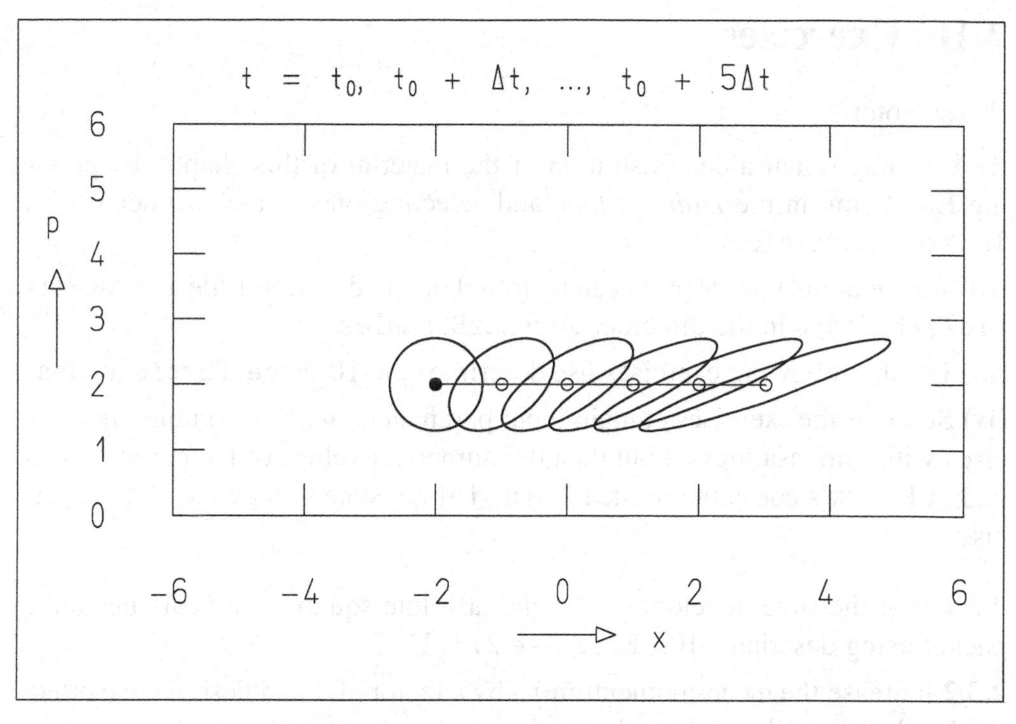

Fig. 2.8. Plot produced with descriptor `Covariance ellipse in classical phase space, free particle` on file `1D_Free_Particle.des`

On the subpanel **Physics—Phase-Space Distr.** you enter the initial parameters x_0, p_0, σ_{p0} as well as a time t_0 (normally 0), a time difference Δt, and an integer number N_t. The covariance ellipse is shown in the x, p plane for the times $t = t_0, t = t_0 + \Delta t, \ldots, t = t_0 + (N_t - 1)\Delta t$.

At the bottom of the subpanel there are 3 check boxes. By enabling them you may show

- the *expectation value* as a circle in the x, p plane for the times $t = t_0, t = t_0 + \Delta t, \ldots$ and the trajectory of the expectation value,
- the *covariance ellipse*,
- a rectangular *frame* enclosing the covariance ellipse.

Movie Capability: Direct. On the subpanel **Movie** (see Sect. A.4) of the parameters panel you may choose to show or not to show the initial and intermediate positions of the covariance ellipse.

Example Descriptor on File `1D_Free_Particle.des`

- `Covariance ellipse in classical phase space, free particle` (see Fig. 2.8)

2.10 Exercises

Please note:

(i) You may watch a demonstration of the material of this chapter by pressing **Run Demo** in the *main toolbar* and selecting one of the two demo files `1D_Free_Particle`.

(ii) More example descriptors can be found on the descriptor file `1DFreeParticle(FE).des` in the directory `FurtherExamples`.

(iii) For the following exercises use descriptor file `1D_Free_Particle.des`.

(iv) Some of the exercises contain input parameters in physical units. In exercises with dimensionless input data the numerical values of the particle mass and of Planck's constant are meant to be 1 if not stated otherwise in the exercise.

2.3.1 Plot the time development of the absolute square of a Gaussian wave packet using descriptor 10: '`Exercise 2.3.1`'.

2.3.2 Increase the mean momentum p_0 by a factor of 2 and describe the effect on the change of the group velocity.

2.3.3 Repeat Exercise 2.3.1; plot **(a)** the real part and **(b)** the imaginary part of the wave function. **(c)** Give a reason why after some time the wavelength of the front part of the real and imaginary parts of the wave packet are shorter than close to the rear end.

2.3.4 Repeat Exercise 2.3.2 for the real part of the wave function. Explain the change of wavelength observed.

2.3.5 **(a)** Repeat Exercise 2.3.1 but increase the momentum width σ_p by a factor of 2. Explain the change in shape of the wave packet as time passes. **(b)** Repeat the exercise, halving the momentum width. **(c)** Study also the real and imaginary parts.

2.3.6 Repeat Exercise 2.3.1. Show quantile positions and quantile trajectories for **(a)** $P = 0.8$, **(b)** $P = 0.5$, **(c)** $P = 0.2$ for the wave packet of Exercise 2.3.1. Use descriptor 11: '`Exercise 2.3.6`'.

2.3.7 Repeat Exercise 2.3.6 for $P = 0.001$. Explain why the quantile trajectory for large t runs backward in x, while $\langle x(t)\rangle$ runs forward.

For the following exercises study Sect. 12.1 'Units and Orders of Magnitude'.

2.3.8 Plot **(a)** the real part, **(b)** the imaginary part, and **(c)** the absolute square of a wave packet of an electron with velocity $1\,\mathrm{m\,s^{-1}}$ and absolute width $\sigma_p = 0.5 \times 10^{-12}\,\mathrm{eV\,s\,m^{-1}}$ moving from an initial position $x_0 = -2\,\mathrm{mm}$ for the instants of time $t_i = 0, 0.5, 1, 1.5, 2, 2.5, 3, 3.5, 4\,\mathrm{ms}$. Start from descriptor 12: '`Exercise 2.3.8`', which already contains the correct time intervals.

(d) Calculate the momentum p_0 of the electron. **(e)** Why is the spreading of the wave packet relatively small in time? **(f)** What is the physical unit along the x axis?

2.3.9 Repeat Exercise 2.3.8 (a–d) with velocity $3\,\mathrm{m\,s^{-1}}$ for the instants of time $t_i = 0, 0.25, 0.5, 0.75, 1, 1.25, 1.5, 1.75, 2$ ms. Start from descriptor 13: '`Exercise 2.3.9`'.

2.3.10 Plot **(a)** the real part, **(b)** the imaginary part, and **(c)** the absolute square of the wave function of an electron of velocity $2.11\,\mathrm{m\,s^{-1}}$ and a relative width of $\sigma_p = 0.75 p_0$ of the corresponding momentum p_0 for a period $0 \leq t \leq 2$ ms in steps of 0.25 ms. The initial position is $x_0 = -2$ mm. Start with descriptor 13: '`Exercise 2.3.9`'. **(d)** Calculate the momentum p_0 of the electron. **(e)** What is the order of the magnitude along the x axis? **(f)** Why do the real and imaginary parts look so different from the earlier exercises and from the picture of the initial descriptor? **(g)** Why are there small wavelengths to either side of the wave packet?

2.4.1 Plot the optical wave packet corresponding to the quantum-mechanical one of Exercise 2.3.1. What is the essential difference between this plot and the one obtained in Exercise 2.3.1? Start from descriptor 14: '`Exercise 2.4.1`'.

2.4.2 Double the mean wave number k_0 of the optical wave packet.

2.4.3 Repeat Exercise 2.4.1; plot **(a)** the real part and **(b)** the imaginary part of the optical wave packet.

2.4.4 Repeat Exercise 2.4.2 for the real part of the optical wave packet.

2.4.5 Adapt Exercise 2.3.5 to the optical wave packet.

2.5.1 For the wave packet of Exercise 2.3.1 draw a set of quantile trajectories for $P = 0.1, 0.2, \ldots, 0.9$ using descriptor 5: '`Free wave packet, quantile trajectories`'.

2.5.2 Repeat Exercise 2.5.1 but for $P = 0.001, 0.5, 0.999$.

2.5.3 Extend Exercise 2.5.2 by creating a 2×2 multiple plot with $\Delta p_0 = 2$, $\Delta\sigma_p = 0.5$. Note the values of p_0 and σ_p for which the quantile trajectory for $P = 0.001$ runs backward.

2.5.4 Determine the asymptotic behavior of $x_P(t)$ for $t \to \infty$ from (2.23) and compare the result with the findings of Exercise 2.5.3.

2.5.5 Determine the asymptotic behavior of the quantum potential $V_Q(x,t)$ for $t \to \infty$ and connect it to the result you obtained in Exercise 2.5.4.

2.6.1 Plot the spectral function corresponding to Exercise 2.3.1 using descriptor 15: '`Exercise 2.6.1`'.

2.6.2 Perform the necessary changes to get the spectral function of the optical wave packet of Exercise 2.4.1.

2.6.3 Plot the spectral functions corresponding to Exercises 2.3.2, 2.4.2, 2.3.5, 2.4.5.

2.6.4 **(a)** Plot the Gaussian spectral function of an electron of velocity $v_0 = 1\,\mathrm{m\,s^{-1}}$ and $v_0 = 3\,\mathrm{m\,s^{-1}}$ for the two widths $\sigma_p = 0.5 \times 10^{-12}\,\mathrm{eV\,s\,m^{-1}}$ and $\sigma_p = 10^{-12}\,\mathrm{eV\,s\,m^{-1}}$ in a multiple plot with four graphs. Start with descriptor 16: '`Exercise 2.6.4`'. **(b)** Calculate the corresponding momenta. **(c)** What is the physical unit at the abscissa? **(d)** Calculate the corresponding kinetic energies.

2.6.5 **(a)** Plot the Gaussian spectral function of an electron of kinetic energy $E = 1\,\mathrm{eV}$ and $E = 3\,\mathrm{eV}$ for the two widths $\sigma_p = 0.5 \times 10^{-6}\,\mathrm{eV\,s\,m^{-1}}$ and $\sigma_p = 10^{-6}\,\mathrm{eV\,s\,m^{-1}}$ in a multiple plot with four graphs. Start with descriptor 17: '`Exercise 2.6.5`'. **(b)** Calculate the corresponding momenta. **(c)** What is the physical unit at the abscissa? **(d)** Calculate the corresponding velocities of the electron. **(e)** To what order is the use of the nonrelativistic formulas still allowed?

2.6.6 Repeat Exercise 2.6.4 for a proton for the two widths $\sigma_p = 0.5 \times 10^{-9}\,\mathrm{eV\,s\,m^{-1}}$ and $\sigma_p = 10^{-9}\,\mathrm{eV\,s\,m^{-1}}$. Start with descriptor 18: '`Exercise 2.6.6`'.

2.6.7 Repeat Exercise 2.6.5 for a proton for the two kinetic energies $E = 1\,\mathrm{keV}$ and $E = 3\,\mathrm{keV}$ and the two widths $\sigma_p = 0.5 \times 10^{-3}\,\mathrm{eV\,s\,m^{-1}}$ and $\sigma_p = 10^{-3}\,\mathrm{eV\,s\,m^{-1}}$. Start with descriptor 19: '`Exercise 2.6.7`'.

2.7.1 Plot a wave packet approximated by a finite sum of harmonic waves using descriptor 7: '`Wave packet as sum of harmonic waves`'.

2.7.2 Study the time development of the harmonic waves and their sum by doing plots for various times (you may do this by using the multiple plot facility). Study the phase velocities of the different harmonic waves and the group velocity of the wave packet.

2.7.3 Repeat Exercises 2.7.1 and 2.7.2 for electromagnetic waves.

2.7.4 Repeat Exercise 2.7.1. Now, gradually, decrease the number of terms in the sum. Why is the resulting sum periodic in x?

2.8.1 Show the time dependence of a classical phase-space distribution using descriptor 8: '`Classical phase-space distribution, free particle`'. **(a)** Display again using $\sigma_{p0} = 0.5, 1, 2$. Study and describe the changes in the distribution $\varrho^{\mathrm{cl}}(x, p, t)$ and in the marginal distributions $\varrho^{\mathrm{cl}}_x(x, t)$ and $\varrho^{\mathrm{cl}}_p(p, t)$. **(b)** By construction, our classical phase-space distribution is of minimum uncertainty and has no correlation at the time $t = 0$. Study it also for negative times by setting $t_0 = -1$. In particular, pay attention to the marginal distributions ϱ^{cl}_x and to the correlation.

2.9.1 **(a)** Repeat the preceding exercise studying only the covariance ellipse; start from descriptor 9: '`Covariance ellipse in classical phase space,` `free particle`'. **(b)** For $\sigma_{p0} = 2$ you will observe the point of smallest p on the ellipse not to change with time. Why?

2.9.2 Study the time dependence of the covariance ellipse for a phase-space distribution that, at average, is at rest (i.e., $p_0 = 0$). Start from descriptor 9: '`Covariance ellipse in classical phase space, free particle`' and adapt the range of the computing coordinates to allow for negative p. Create and watch a movie.

3. Bound States in One Dimension

Contents: Introduction of the time-dependent and stationary Schrödinger equations. Computation of eigenfunctions and eigenvalues in the infinitely deep square-well potential, in the harmonic-oscillator potential, and in the general step and piecewise linear potential. Motion of a wave packet in the deep square-well potential and in the harmonic-oscillator potential.

3.1 Physical Concepts

3.1.1 Schrödinger's Equation with a Potential. Eigenfunctions. Eigenvalues

The motion of a particle under the action of a force given by a *potential* $V(x)$ is governed by the *Schrödinger equation*

$$\mathrm{i}\hbar\frac{\partial}{\partial t}\psi(x,t) = -\frac{\hbar^2}{2m}\frac{\partial^2}{\partial x^2}\psi(x,t) + V(x)\psi(x,t) \quad . \tag{3.1}$$

With the *Hamiltonian*

$$H = T + V \tag{3.2}$$

it reads

$$\mathrm{i}\hbar\frac{\partial}{\partial t}\psi(x,t) = H\psi(x,t) \quad . \tag{3.3}$$

Separation of the variables time t and position x by way of the expression for *stationary wave function*

$$\psi_E(x,t) = \mathrm{e}^{-\mathrm{i}Et/\hbar}\varphi_E(x) \tag{3.4}$$

leads to the *stationary Schrödinger equation*

$$-\frac{\hbar^2}{2m}\frac{\mathrm{d}^2}{\mathrm{d}x^2}\varphi_E(x) + V(x)\varphi_E(x) = E\varphi_E(x) \tag{3.5}$$

or equivalently

$$H\varphi_E(x) = E\varphi_E(x) \tag{3.6}$$

for the *eigenfunction* $\varphi_E(x)$ of the Hamiltonian belonging to the energy eigenvalue E.

S. Brandt et al., *Interactive Quantum Mechanics: Quantum Experiments on the Computer*,
DOI 10.1007/978-1-4419-7424-2_3, © Springer Science+Business Media, LLC 2011

3.1.2 Normalization. Discrete Spectra. Orthonormality

The Hamiltonian H is a *Hermitian* operator for *square-integrable functions* $\varphi(x)$ only. These can be *normalized* to one, i.e.,

$$\int_{-\infty}^{+\infty} \varphi^*(x)\,\varphi(x)\,\mathrm{d}x = 1 \quad . \tag{3.7}$$

For normalizable eigenfunctions the eigenvalues E of H form a set $\{E_1, E_2, \ldots\}$ of discrete real values. This set is the *discrete spectrum* of the Hamiltonian. The corresponding eigenfunctions are called *discrete eigenfunctions* of the Hamiltonian. We will label the eigenfunction belonging to the eigenvalue $E = E_n$ by $\varphi_n(x)$. The eigenfunctions $\varphi_n(x)$, $\varphi_m(x)$ corresponding to different eigenvalues $E_n \neq E_m$ are *orthogonal*:

$$\int \varphi_n^*(x)\,\varphi_m(x)\,\mathrm{d}x = 0 \quad . \tag{3.8}$$

Together with the normalization (3.7) we have the *orthonormality* of the discrete eigenfunctions

$$\int \varphi_n^*(x)\,\varphi_m(x)\,\mathrm{d}x = \delta_{mn} \quad . \tag{3.9}$$

For potentials $V(x)$ bounded below, i.e., $V_0 \le V(x)$ for all x, the eigenvalues lie in the domain $E \ge V_0$. For potentials bounded below, tending to infinity at $x \to -\infty$ as well as $x \to +\infty$, all eigenvalues are discrete.

For potentials bounded below tending to a finite limit $V(-\infty)$ or $V(+\infty)$ at either $x \to +\infty$ or $x \to -\infty$, the discrete eigenvalues can occur in the interval $V_0 \le E_n \le V_c$ with $V_c = \min(V(+\infty), V(-\infty))$.

3.1.3 The Infinitely Deep Square-Well Potential

$$V(x) = \begin{cases} 0 \ , & -d/2 \le x \le d/2 \\ \infty \ , & \text{elsewhere} \end{cases} \quad , \tag{3.10}$$

d: width of potential.

This potential confines the particle to an interval of length d. The eigenfunctions of the Hamiltonian with this potential are

$$\begin{aligned} \varphi_n(x) &= \sqrt{2/d}\,\cos(n\pi x/d) \quad , \quad n = 1, 3, 5, \ldots \quad , \\ \varphi_n(x) &= \sqrt{2/d}\,\sin(n\pi x/d) \quad , \quad n = 2, 4, 6, \ldots \quad , \end{aligned} \tag{3.11}$$

belonging to the eigenvalues

$$E_n = \frac{1}{2m}\left(\frac{\hbar n\pi}{d}\right)^2 \quad , \quad n = 1, 2, 3, \ldots \quad . \tag{3.12}$$

The discrete energies E_n are enumerated by the principal quantum number n.

3.1.4 The Harmonic Oscillator

$$V(x) = \frac{m}{2}\omega^2 x^2 \quad , \tag{3.13}$$

m: mass of particle,
ω: angular frequency.

The eigenfunctions of the Hamiltonian of the harmonic oscillator are

$$\varphi_n(x) = (\sqrt{\pi}2^n n!\sigma_0)^{-1/2} H_n\left(\frac{x}{\sigma_0}\right) \exp\left(-\frac{x^2}{2\sigma_0^2}\right) \quad , \quad n = 0, 1, 2, \ldots \quad , \tag{3.14}$$

belonging to the eigenvalues

$$E_n = \left(n + \tfrac{1}{2}\right) \hbar\omega \quad , \tag{3.15}$$

n: principal quantum number n of the harmonic oscillator,
$H_n(x)$: Hermite polynomial of order n,
$\sigma_0 = \sqrt{\hbar/m\omega}$: width of ground-state wave function.

3.1.5 The Step Potential

$$V(x) = \begin{cases} V_1 \geq 0 \ , \ x < x_1 = 0 & \text{region } 1 \\ V_2 \qquad , \ x_1 \leq x < x_2 & \text{region } 2 \\ \vdots & \\ V_{N-1} \quad , x_{N-2} \leq x < x_{N-1} & \text{region } N-1 \\ V_N = 0 \ , \ x_{N-1} \leq x & \text{region } N \end{cases} \quad . \tag{3.16}$$

The potential possesses discrete eigenvalues E_l for $E < 0$, which can again be enumerated by a principal quantum number l.

For an eigenfunction $\varphi_l(x)$ belonging to the eigenvalue E_l the *de Broglie wave number* in region m is

$$\begin{aligned} k_{lm} &= \left|\sqrt{2m(E_l - V_m)}/\hbar\right| \quad \text{for} \quad E_l > V_m \quad , \\ k_{lm} &= \mathrm{i}\kappa_{lm} \quad , \quad \kappa_{lm} = \left|\sqrt{2m(V_m - E_l)}/\hbar\right| \quad \text{for} \quad E_l < V_m \quad . \end{aligned} \tag{3.17}$$

The wave function $\varphi_l(x)$ is then given for all the N intervals of constant potential by

$$\varphi_l(x) = \begin{cases} \varphi_{l1}(x) \quad , & \text{region } 1 \\ \varphi_{l2}(x) \quad , & \text{region } 2 \\ \vdots & \\ \varphi_{lN-1} \quad , & \text{region } N-1 \\ \varphi_{lN} \quad , & \text{region } N \end{cases} \quad . \tag{3.18}$$

For $E \neq V_m$ the piece φ_{lm} of the wave function,

$$\varphi_{lm}(x) = A_{lm}\mathrm{e}^{\mathrm{i}k_{lm}x} + B_{lm}\mathrm{e}^{-\mathrm{i}k_{lm}x} \quad , \quad x_{m-1} \le x < x_m \quad , \tag{3.19}$$

consists for $E_l > V_m$ of a *right-moving* and a *left-moving harmonic wave* and for $V_m > E_l$ of a *decreasing* and an *increasing exponential function*. For $E_l = V_m$ the piece φ_{lm} is a straight line

$$\varphi_{lm}(x) = A_{lm} + B_{lm}x \quad , \quad x_{m-1} \le x < x_m \quad . \tag{3.20}$$

Because of the normalizability (3.7), the bound-state wave function $\varphi_l(x)$ must decrease exponentially in the regions

$$\begin{aligned} m = 1 : \varphi_{l1} &= B_{l1}\mathrm{e}^{\kappa_{l1}x} \quad , \quad -\infty < x < x_1 = 0 \quad , \\ m = N : \varphi_{lN} &= A_{lN}\mathrm{e}^{-\kappa_{lN}x} \quad , \quad x_{N-1} \le x < \infty \quad , \end{aligned} \tag{3.21}$$

i.e., $A_{l1} = 0$, $B_{lN} = 0$.

The requirement of exponential behavior stipulates $E < 0$. The coefficients A_{lm}, B_{lm} are determined from the requirement of the wave function being *continuous* and *continuously differentiable* at the values x_m, $m = 1, \ldots, N-1$. For $E \neq V_m, V_{m+1}$, these *continuity conditions* read

$$\begin{aligned} A_{lm}\mathrm{e}^{\mathrm{i}k_{lm}x_m} + B_{lm}\mathrm{e}^{\mathrm{i}k_{lm}x_m} &= A_{lm+1}\mathrm{e}^{\mathrm{i}k_{lm+1}x_m} + B_{lm+1}\mathrm{e}^{-\mathrm{i}k_{lm+1}x_m} \quad , \\ k_{lm}(A_{lm}\mathrm{e}^{\mathrm{i}k_{lm}x_m} - B_{lm}\mathrm{e}^{-\mathrm{i}k_{lm}x_m}) & \\ = k_{lm+1}(A_{lm+1}\mathrm{e}^{\mathrm{i}k_{lm+1}x_m} - B_{lm+1}\mathrm{e}^{-\mathrm{i}k_{lm+1}x_m}) & \quad . \end{aligned} \tag{3.22}$$

If $E_l = V_m$ or $E_l = V_{m+1}$, the left-hand or right-hand side has to be replaced by using (3.20). The system represents a set of $2(N-1)$ linear homogeneous equations for $2(N-1)$ unknown coefficients A_{lm}, B_{lm}. It has a non-trivial solution only if its determinant $D(E)$ vanishes. This requirement leads to a transcendental equation for the eigenvalues E_l present in the wave numbers k_{lm}. In general, its solution can be obtained numerically only and is calculated by the computer by finding the zeros of the function

$$D = D(E) \tag{3.23}$$

that coincide with the values E_l at which the determinant vanishes. Once the eigenvalue E_l is determined as a single zero of the transcendental equation, the system of linear equations can be solved yielding the coefficients A_{lm}, B_{lm} as functions of one of them. This undetermined coefficient is then fixed by the normalization condition (3.7). The number of the eigenstates $\varphi_l(x)$ of step potentials is finite; thus, they do not form a complete set. In Chap. 4 we present the continuum eigenfunctions that supplement the $\varphi_l(x)$ to a complete set of functions.

3.1.6 The Piecewise Linear Potential

$$V(x) = \begin{cases} V_1 \geq 0 & , x < x_1 = 0 \quad \text{region } 1 \\ V_{2,a} + (x - x_1)V_2' & , x_1 \leq x < x_2 \quad \text{region } 2 \\ \vdots & \\ V_{N-1,a} + (x - x_{N-2})V_{N-1}' & , x_{N-2} \leq x < x_{N-1} \quad \text{region } N-1 \\ V_N = 0 & , x_{N-1} \leq x \quad \text{region } N \end{cases} ,$$

$$V_j' = \frac{V_{j,b} - V_{j,a}}{x_j - x_{j-1}} \quad , \quad j = 2, \ldots, N-1 \quad . \tag{3.24}$$

The potential resembles the step potential of Sect. 3.1.5, where the constant in each region is replaced by a linear function. In general, the potential is discontinuous at the region boundaries; continuity at x_j, $j = 1, \ldots, N-1$, holds for $V_{j,b} = V_{j+1,a}$.

For a nonzero slope $V_m' = \frac{V_{m,b} - V_{m,a}}{x_m - x_{m-1}}$ of a potential in region m, x_{lm}^{T} defines a reference point for the eigenvalue E_l, the (extrapolated) classical turning point, and α_m represents a scale factor, see also Sect. 4.1.18:

$$x_{lm}^{\mathrm{T}} = x_{m-1} + \frac{E_l - V_{a,m}}{V_m'} \quad , \quad \alpha_m = \left(\frac{\hbar^2}{2m} \frac{1}{V_m'} \right)^{-\frac{1}{3}} \quad . \tag{3.25}$$

The eigenfunction $\varphi_l(x)$ is represented as in (3.18), where for zero potential slope in region m the solutions of Sect. 3.1.5 apply and for a nonzero slope one obtains

$$\varphi_{lm}(x) = A_{lm}\,\mathrm{Ai}(\alpha_m(x - x_{lm}^{\mathrm{T}})) + B_{lm}\,\mathrm{Bi}(\alpha_m(x - x_{lm}^{\mathrm{T}})) \quad , \quad x_{m-1} \leq x < x_m \quad . \tag{3.26}$$

Here Ai and Bi are the Airy functions, representing the solution of the stationary Schrödinger equation for a linear potential, that are decreasing or increasing in the classically forbidden region, respectively, see Sect. 11.1.7.

The outermost regions (1 and N) are governed by constant potentials, thus the restrictions to the exponential behavior ($A_{l1} = 0$, $B_{lN} = 0$) of Sect. 3.1.5 also apply there. The continuity conditions, i.e., the wave function at the boundaries x_m has to be *continuously differentiable*, is also used, now reading

$$\begin{aligned} &A_{lm}\,\mathrm{Ai}(\alpha_m(x_m - x_{lm}^{\mathrm{T}})) + B_{lm}\,\mathrm{Bi}(\alpha_m(x_m - x_{lm}^{\mathrm{T}})) \\ &= A_{lm+1}\,\mathrm{Ai}(\alpha_m(x_m - x_{lm+1}^{\mathrm{T}})) + B_{lm+1}\,\mathrm{Bi}(\alpha_m(x_m - x_{lm+1}^{\mathrm{T}})) \quad , \end{aligned} \tag{3.27}$$

$$\begin{aligned} &A_{lm}\alpha_m\,\mathrm{Ai}'\,(\alpha_m(x_m - x_{lm}^{\mathrm{T}})) + B_{lm}\alpha_m\,\mathrm{Bi}'\,(\alpha_m(x_m - x_{lm}^{\mathrm{T}})) \\ &= A_{lm+1}\alpha_{m+1}\,\mathrm{Ai}'\,(\alpha_{m+1}(x_m - x_{lm+1}^{\mathrm{T}})) \\ &\quad + B_{lm+1}\alpha_{m+1}\,\mathrm{Bi}'\,(\alpha_{m+1}(x_m - x_{lm+1}^{\mathrm{T}})) \quad . \end{aligned}$$

Here Ai′ and Bi′ are the derivatives of the Airy functions. For zero potential slopes the form (3.22) for one or both sides of the equations are to be used.

As for the step potential the system of $2(N-1)$ equations for the $2(N-1)$ unknown coefficients A_{lm}, B_{lm} has a non-trivial solution only for vanishing determinant $D(E)$, (3.23). Again the energy eigenvalues E_l coincide with the zeros of the corresponding transcendental equation. The eigenfunction are finally normalized according to (3.7).

3.1.7 Time-Dependent Solutions

Because the time-dependent Schrödinger equation (3.1) is linear, the time-dependent harmonic waves, (3.4),

$$\psi_n(x,t) = \mathrm{e}^{-\mathrm{i}E_n t/\hbar}\varphi_n(x) \tag{3.28}$$

can be superimposed with time-independent spectral coefficients w_n yielding the solution

$$\psi(x,t) = \sum_n w_n \mathrm{e}^{-\mathrm{i}E_n t/\hbar}\varphi_n(x) \tag{3.29}$$

of the Schrödinger equation. Because the eigenfunctions $\varphi_n(x)$, with n belonging to the discrete spectrum, confine the particle to a bounded region in space, the solutions $\psi(x,t)$ do so for all times.

3.1.8 Harmonic Particle Motion. Coherent States. Squeezed States

For the time $t = 0$ we choose an initial Gaussian wave packet:

$$\psi(x,0) = \frac{1}{(2\pi)^{1/4}\sigma_{x0}} \exp\left\{-\frac{(x-x_0)^2}{4\sigma_{x0}^2}\right\} \quad , \tag{3.30}$$

x_0: initial mean position,
σ_{x0}: initial width of wave packet.

The initial mean momentum is zero. We decompose $\psi(x,0)$ into a sum over the complete set of real eigenfunctions $\varphi_n(x)$ of the harmonic oscillator,

$$\psi(x,0) = \sum_{n=0}^{\infty} w_n\, \varphi(x) \quad . \tag{3.31}$$

The orthonormality condition (3.7) is used to determine the coefficients

$$w_n = \int_{-\infty}^{+\infty} \varphi_n(x)\, \psi(x,0)\, \mathrm{d}x \quad . \tag{3.32}$$

The time-dependent solution $\psi(x,t)$ is then given with these coefficients by (3.29). This expansion can be summed up. For brevity we present only its absolute square explicitly:

$$\varrho(x,t) = |\psi(x,t)|^2 = \frac{1}{\sqrt{2\pi}\sigma_x(t)} \exp\left\{-\frac{(x-\langle x(t)\rangle)^2}{2\sigma_x^2(t)}\right\} \quad . \tag{3.33}$$

It represents a Gaussian wave packet moving with a mean position

$$\langle x(t)\rangle = x_0 \cos\omega t \tag{3.34}$$

oscillating harmonically in time and with an in-general time-dependent width

$$\sigma_x(t) = \frac{\sigma_0}{2\sqrt{2}\sigma_{r0}}(4\sigma_{r0}^4 + 1 + (4\sigma_{r0}^4 - 1)\cos(2\omega t))^{1/2} \quad , \tag{3.35}$$

$\sigma_{r0} = \sigma_{x0}/\sigma_0$: relative initial width of wave packet,
$\sigma_0 = \sqrt{\hbar/m\omega}$: ground-state width.

The time-dependent width itself oscillates with double angular frequency 2ω about an average width

$$\bar{\sigma} = \frac{\sigma_0}{2\sqrt{2}\sigma_{r0}}(4\sigma_{r0}^4 + 1)^{1/2} \quad . \tag{3.36}$$

The *coherent state* is distinguished by a time-independent width $\sigma = \sigma_0/\sqrt{2}$. It is of central importance in quantum optics and quantum electronics, e.g., in lasers and quantum oscillations in electrical circuits. States with oscillating widths are called *squeezed states*.

3.1.9 Quantile Motion in the Harmonic-Oscillator Potential

Equation (3.33) is identical to (2.13). Only the time dependences of mean $\langle x(t)\rangle$ and width $\sigma_x(t)$ are different. Keeping this difference in mind all results of Sects. 2.1.3 and 2.1.4 about the probability-current density $j(x,t)$ and the quantile trajectories $x_P = x_P(t)$ remain valid for a Gaussian wave packet in a harmonic-oscillator potential.

3.1.10 Harmonic Motion of a Classical Phase-Space Distribution

As in Sect. 2.1.7 we consider a classical phase-space probability density which at the initial time $t = 0$ is

$$\varrho_{\mathrm{i}}^{\mathrm{cl}}(x_{\mathrm{i}}, p_{\mathrm{i}}) = \frac{1}{2\pi\sigma_{x0}\sigma_{p0}} \exp\left\{-\frac{1}{2}\left[\frac{(x_{\mathrm{i}}-x_0)^2}{\sigma_{x0}^2} + \frac{(p_{\mathrm{i}}-p_0)^2}{\sigma_{p0}^2}\right]\right\} \quad . \tag{3.37}$$

Here x_0, p_0 are the initial expectation values and σ_{x0}, σ_{p0} are the initial widths of position and momentum, respectively. The covariance ellipse of this bivariate Gaussian is characterized by the exponential being equal to $-1/2$,

$$\frac{(x_\mathrm{i} - x_0)^2}{\sigma_{x0}^2} + \frac{(p_\mathrm{i} - p_0)^2}{\sigma_{p0}^2} = 1 \quad . \tag{3.38}$$

The ellipse is centered around (x_0, q_0) and has the semi-axes σ_{x0} and σ_{q0}, which are parallel to the x axis and the q axis, respectively.

The classical motion of a particle in phase space under the action of a harmonic force is simply

$$\begin{aligned} x &= x_\mathrm{i} \cos \omega t + q_\mathrm{i} \sin \omega t \quad , \\ q &= -x_\mathrm{i} \sin \omega t + q_\mathrm{i} \cos \omega t \quad , \end{aligned} \tag{3.39}$$

if one uses the notation

$$q(t) = \frac{p(t)}{m\omega} \quad , \qquad q_\mathrm{i} = \frac{p_\mathrm{i}}{m\omega} \quad .$$

A classical particle rotates with angular velocity ω on a circle around the origin in the x, q plane. For a given time t and given values $x(t)$, $q(t)$ the initial conditions of a particle are then

$$\begin{aligned} x_\mathrm{i} &= x \cos \omega t - q \sin \omega t \quad , \\ q_\mathrm{i} &= x \sin \omega t + q \cos \omega t \quad . \end{aligned}$$

Introducing this result into the equation (3.38) for the initial covariance ellipse yields

$$\begin{aligned} &\frac{([x - \langle x(t)\rangle] \cos \omega t - [q - \langle q(t)\rangle] \sin \omega t)^2}{\sigma_{x0}^2} \\ &\quad + \frac{([x - \langle x(t)\rangle] \sin \omega t + [q - \langle q(t)\rangle] \cos \omega t)^2}{\sigma_{q0}^2} = 1 \quad . \end{aligned}$$

This is again an equation of an ellipse with principal semi-axes of length σ_{x0} and σ_{q0}. They are, however, no longer parallel to the coordinate directions but rotated by an angle ωt with respect to these. The center of the ellipse is the point $(\langle x(t)\rangle, \langle q(t)\rangle)$ to which the set of initial expectation values (x_0, q_0) has moved at the time t according to (3.39).

The situation is summarized as follows:

- A classical phase-space distribution described by a bivariate Gaussian keeps its Gaussian shape.

- Its center, which is the center of the covariance ellipse, moves on a circle around the center of the x, q plane with angular velocity ω.
- The covariance ellipse keeps its shape but rotates around its center with the same angular velocity ω.

Rotation of the covariance ellipse implies a time dependence of the widths $\sigma_x(t)$ and $\sigma_q(t)$ in x and q as well as a nonvanishing correlation coefficient $c(t)$, which also depends on time. We can rewrite the equation of the covariance ellipse in the form known from Sect. 11.14,

$$\frac{1}{1-c^2(t)}\left\{\frac{(x-\langle x(t)\rangle)^2}{\sigma_x^2(t)} - 2c(t)\frac{(x-\langle x(t)\rangle)(q-\langle q(t)\rangle)}{\sigma_x(t)\sigma_q(t)} + \frac{(q-\langle q(t)\rangle)^2}{\sigma_q^2(t)}\right\} = 1$$

with

$$\sigma_x(t) = \sqrt{\sigma_{x0}^2\cos^2\omega t + \sigma_{q0}^2\sin^2\omega t} \quad ,$$

$$\sigma_q(t) = \sqrt{\sigma_{x0}^2\sin^2\omega t + \sigma_{q0}^2\cos^2\omega t} \quad ,$$

$$c(t) = \frac{(\sigma_{q0}^2 - \sigma_{x0}^2)\sin 2\omega t}{\sqrt{4\sigma_{x0}^2\sigma_{q0}^2 + (\sigma_{x0}^2 - \sigma_{q0}^2)^2\sin^2 2\omega t}} \quad .$$

In the particular case

$$\sigma_{x0} = \sigma_{q0}$$

the covariance ellipse is a circle, σ_x and σ_q are independent of time, and the correlation vanishes for all times. If we require the minimum-uncertainty relation of quantum mechanics,

$$\sigma_{x0}\sigma_{p0} = \frac{\hbar}{2} \quad ,$$

to be fulfilled for our classical phase-space probability density, we have

$$\sigma_{q0} = \frac{\sigma_{p0}}{m\omega} = \frac{\hbar}{2m\omega\sigma_{x0}} \quad .$$

Together with the requirement $\sigma_{x0} = \sigma_{q0}$ we get

$$\sigma_{x0} = \frac{1}{\sqrt{2}}\sqrt{\frac{\hbar}{m\omega}} = \frac{\sigma_0}{\sqrt{2}} \quad , \qquad \sigma_0 = \sqrt{\frac{\hbar}{m\omega}} \quad .$$

For this particular value of the initial width, the width stays constant. For $\sigma_{x0} \neq \sigma_0/\sqrt{2}$ the spatial width of the classical phase-space density oscillates exactly as the quantum-mechanical probability density does.

3.1.11 Particle Motion in a Deep Square Well

For time $t = 0$ we choose an initial wave packet with a bell shape,

$$\psi(x,0) = \sum_{n=N_1}^{N_2} w_n \, \varphi_n(x) \quad , \tag{3.40}$$

$\varphi_n(x)$: eigenfunctions of infinitely deep square well,
w_n: spectral weights.

The spectral weights w_n are taken as the values of a Gaussian spectral distribution in momentum space at the discrete values $k_n = n\pi/d$ of the wave numbers allowed in the infinitely deep square well. The Gaussian is centered at p_0. Its complex phase factor puts at $t = 0$ the initial position expectation value of the wave packet to $x = x_0$. The result is a 'Gaussian' wave packet inside the infinitely deep square well. The explicit formulae for the spectral weights used are

$$\begin{aligned} w_n &= (2\pi)^{1/4}\sqrt{\sigma_\mathrm{r}}[\mathrm{e}^{-\sigma_\mathrm{r}^2(p_0 d/\hbar+n\pi)^2}\mathrm{e}^{\mathrm{i}n\pi x_0/d} \\ &\quad + \mathrm{e}^{-\sigma_\mathrm{r}^2(p_0 d/\hbar-n\pi)^2}\mathrm{e}^{-\mathrm{i}n\pi x_0/d}] \quad , \quad n = 1,3,5,\ldots \quad , \\ w_n &= -\mathrm{i}(2\pi)^{1/4}\sqrt{\sigma_\mathrm{r}}[\mathrm{e}^{-\sigma_\mathrm{r}^2(p_0 d/\hbar+n\pi)^2}\mathrm{e}^{\mathrm{i}n\pi x_0/d} \\ &\quad - \mathrm{e}^{-\sigma_\mathrm{r}^2(p_0 d/\hbar-n\pi)^2}\mathrm{e}^{-\mathrm{i}n\pi x_0/d}] \quad , \quad n = 2,4,6,\ldots \quad , \end{aligned} \tag{3.41}$$

x_0: expectation value of initial position,
p_0: expectation value of initial momentum,
d: width of potential,
σ_{x0}: initial width of wave packet,
$\sigma_\mathrm{r} = \sigma_{x0}/d$: relative initial width,
N_1, N_2: lower and upper limits of summation.

For reasonable localization of the wave packet within the deep well, the relative initial width σ_r must be small compared to one. The moving wave packet is obtained from the time-dependent solution (3.29) with the spectral weights (3.41). In the harmonic oscillator the expectation values of position and of momentum of the wave packet coincide with its classical position and momentum.

For a wave packet in the infinitely deep square well, the motion of the classical particle is after some time drastically different from the motion of the position expectation value of the wave packet. The reason for this phenomenon is the broadening of the wave packet. As soon as its width substantially exceeds the width of the well, the probability density of the particle fills the whole well and its position expectation value just rests at the center of the well. Thus, the original amplitude of the oscillating expectation value within some inner range of the well decreases to zero and the particle rests at the center of the well.

However, the broadening of the wave packet in the infinitely deep square well cannot go on forever as in the case of the free motion of a particle. In the infinitely deep square well, all time-dependent processes are periodic in time with the period T_1. This can be calculated with the following arguments. The energy of the ground state of an infinitely deep square well is

$$E_1 = \frac{\hbar^2}{2m}\frac{\pi^2}{d^2} \quad .$$

The corresponding angular frequency $\omega_1 = E_1/\hbar$ determines a period T_1 for all time-dependent processes in this system:

$$T_1 = 2\pi/\omega_1 = 4md^2/(\pi\hbar) = 8md^2/h \quad .$$

This means, in particular, that after the time T_1 a wave packet moving in the well assumes its initial shape, i.e., the shape it had at $t = 0$. The classical particle of momentum p_0 and mass m bouncing back and forth between the two walls of the well has the bouncing period

$$T_c = 2d/v_0 = 2md/p_0 \quad .$$

At the time T_1, the wave packet has regained its initial width so that its position expectation values show again the bouncing behavior of the initial narrow wave packet. However, its location coincides with the classical particle only if the quantum-mechanical and classical periods T_1 and T_c are compatible. Actually, after half the quantum-mechanical period the wave packet already assumes its original width, however with the opposite of the initial momentum. Because the classical particle position and the expectation value of the wave packet have to coincide at times $0, T_1, 2T_1, \ldots$, the initial momentum p_0 must be chosen so that

$$T_1 = MT_c \quad , \qquad M = 1, 2, \ldots \quad ,$$

i.e.,

$$p_0 = M\frac{\pi}{2d}\hbar = M\frac{h}{4d} \quad .$$

Further Reading

Alonso, Finn: Vol. 3, Chaps. 2, 6
Berkeley Physics Course: Vol. 4, Chaps. 7, 8
Brandt, Dahmen: Chaps. 4, 6, 7
Feynman, Leighton, Sands: Vol. 3, Chaps. 13, 14, 16
Flügge: Vol. 1, Chap. 2A
Gasiorowicz: Chaps. 3, 4
Merzbacher: Chaps. 3, 4, 5, 6
Messiah: Vol. 1, Chaps. 2, 3
Schiff: Chaps. 2, 3, 4

3.2 Eigenstates in the Infinitely Deep Square-Well Potential and in the Harmonic-Oscillator Potential

Aim of this section: Computation and presentation of the eigenfunctions (3.11) and eigenvalue spectrum (3.12) for the deep square-well potential (3.10) and of the eigenfunctions (3.14) and eigenvalues (3.15) for the harmonic-oscillator potential (3.13).

A plot similar to Fig. 3.1 or Fig. 3.2 is produced, which may contain the following items in a plane spanned by the position coordinate x and the energy E:

- the *potential* $V(x)$ shown as a long-stroke dashed line,
- the *eigenvalues* E_n shown as short-stroke horizontal dashed lines,
- the *eigenfunctions* $\varphi_n(x)$ or their squares as 2D function graphs for which the graphical representations of the eigenvalues serve as zero lines,
- the *term scheme* shown as a series of short lines at the positions E_n on the right-hand side of the scale in E.

On the subpanel **Physics—Comp. Coord.** you can select either the **Deep Square Well** or the **Harmonic Oscillator** potential and you can choose to plot

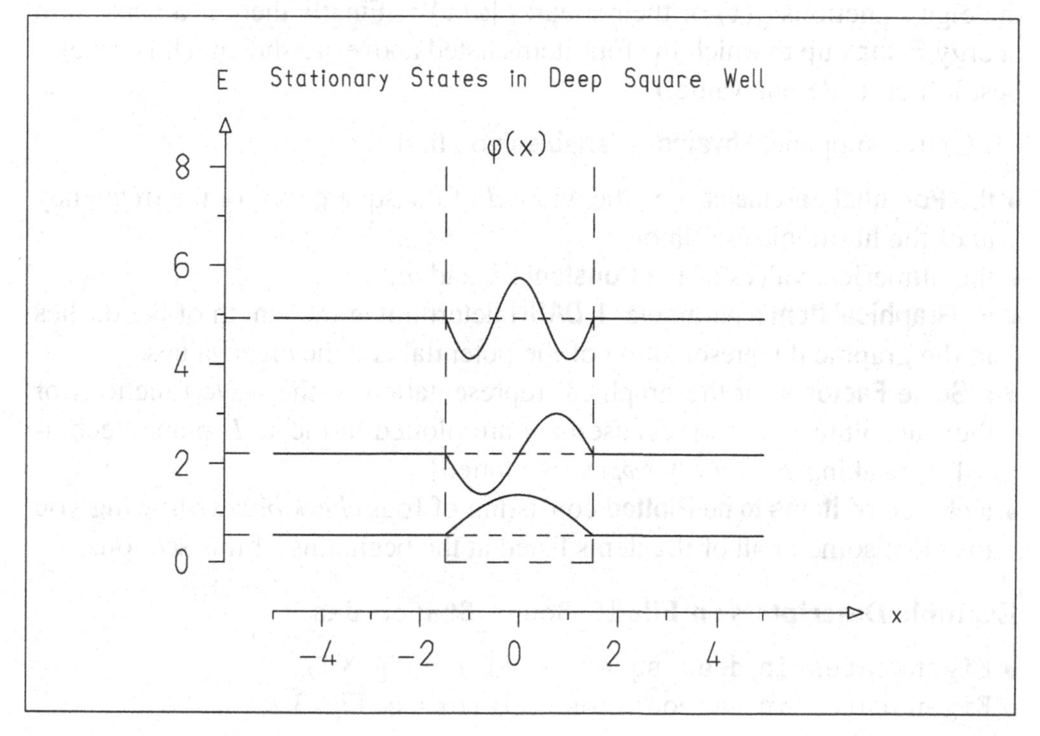

Fig. 3.1. Plot produced with descriptor `Eigenstates in deep square well` on file `1D_Bound_States.des`

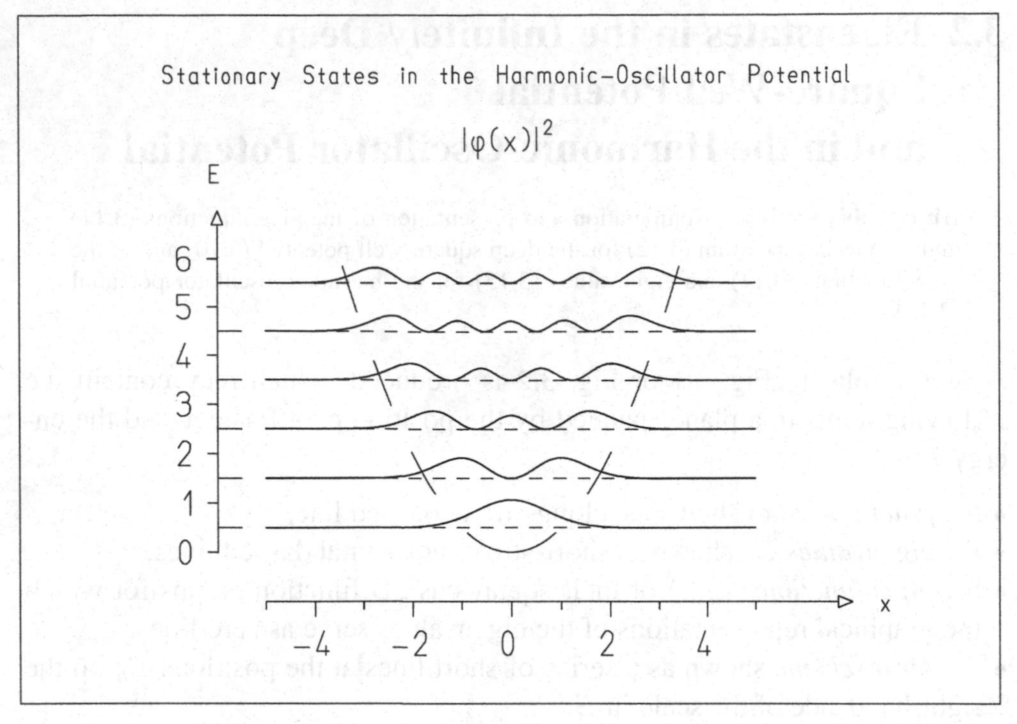

Fig. 3.2. Plot produced with descriptor `Eigenstates in harmonic oscillator` on file `1D_Bound_States.des`

the eigenfunctions $\varphi(x)$ or their squares $|\varphi(x)|^2$. Finally there is a maximum energy **E_max** up to which the four items listed above are shown. (It is usually best left at its default value.)

On the subpanel **Physics—Variables** you find

- the **Potential Parameter**, i.e., the width d of the square well or the frequency ω of the harmonic oscillator,
- the numerical values of the **Constants** $\hbar$ and m,
- as **Graphical Item** a parameter **l_DASH** determining the length of the dashes in the graphical representation of the potential and the eigenvalues,
- a **Scale Factor** s for the graphical representation of the wave functions or their absolute squares [because they are plotted in the x, E plane, technically speaking $E = E_n + s\varphi_n(x)$ is plotted],
- a choice of **Items to be Plotted** consisting of four *check boxes* allowing you to select some or all of the items listed at the beginning of this section.

Example Descriptors on File `1D_Bound_States.des`

- `Eigenstates in deep square well` (see Fig. 3.1)
- `Eigenstates in harmonic oscillator` (see Fig. 3.2)

3.3 Eigenstates in the Step Potential

Aim of this section: Computation and presentation of eigenfunctions $\varphi_l(x)$, (3.18), and the corresponding eigenvalues of bound states in a step potential $V(x)$, (3.16). The value of the determinant $D(E)$, (3.23), is also shown as a function of the energy E.

A plot similar to Fig. 3.3 is produced containing some or all of the items *potential, eigenvalues, eigenfunctions*, and *term scheme.* Because the eigenvalues are found by a numerical search for the zeros in the determinant $D = D(E)$, a graph of the function $D = D(E)$ can be shown as a further item.

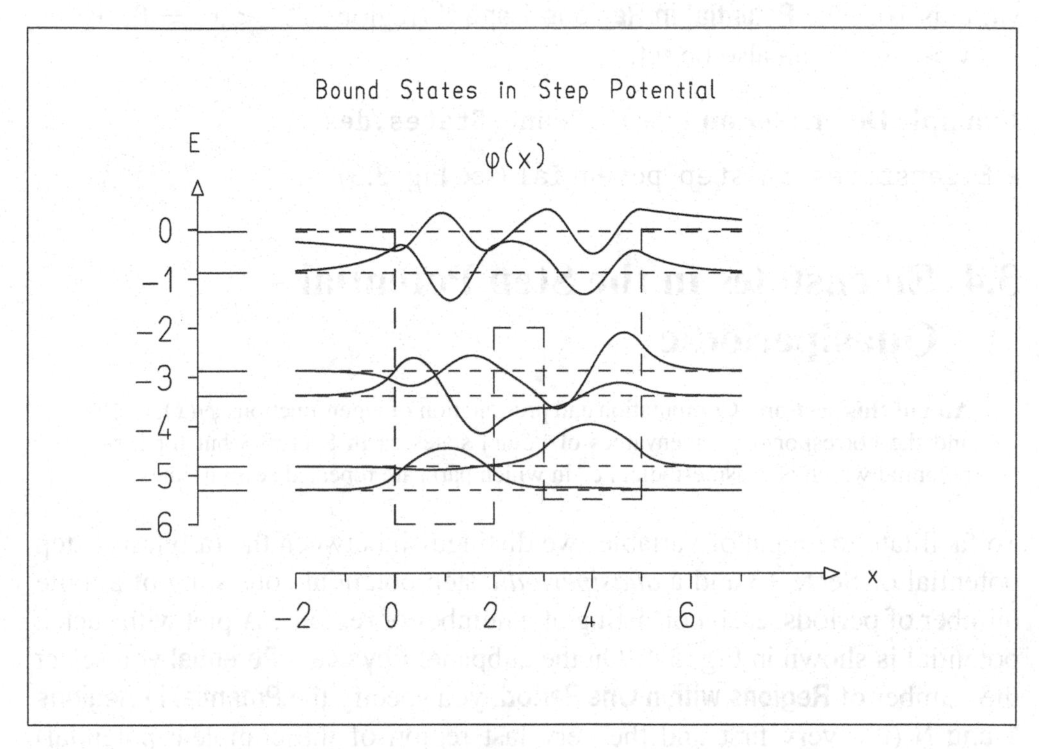

Fig. 3.3. Plot produced with descriptor `Eigenstates in step potential` on file `1D_-Bound_States.des`

On the subpanel **Physics—Comp. Coord.** you can choose to plot the eigenfunctions $\varphi(x)$ or their squares $|\varphi(x)|^2$. There are also four **Search Parameters** needed by the program to perform the numerical search for the eigenvalues. These are the energies **E_min** and **E_max** (defining the range of E in which the search is done), the number of intervals **N_Search** into which this interval is divided for a coarse search, and a parameter **epsilon** that is used for a fine search within intervals in which the coarse search found a zero.

On the subpanel **Physics—Variables** you find (as in Sect. 3.2) the numerical values for the **Constants** $\hbar$ and m, the **Graphical Item** used to determine the dash lengths for eigenvalues and potential, the **Scale Factor** for the eigenfunction, and a choice of **Items to be Plotted**. In addition there is information about the graphical representation of the **Function D=D(E)**. It may be **Shown** or **Not Shown**. Because it is plotted in the x, E plane, technically speaking the function $x = x_D + f_D \sinh^{-1}[s_D D(E)]$ is presented. Here **x_D** is the position in x corresponding to $D = 0$ and **f_D** and **s_D** are scale factors.

On the subpanel **Physics—Potential** the step potential is described. Here the potentials $V_2, \dots, V_{N-1}$ and the region borders $x_2, \dots, x_{N-1}$ are set individually in the field **Regions**. The **Number of Regions** is **N**. Its maximum value is 10. The **Potential in Regions 1 and N** (region 1: $x < x_1 = 0$; region N: $x > x_{N-1}$) can also be set.

Example Descriptor on File `1D_Bound_States.des`

- `Eigenstates in step potential` (see Fig. 3.3)

3.4 Eigenstates in the Step Potential – Quasiperiodic

Aim of this section: Computation and presentation of eigenfunctions $\varphi_l(x)$, (3.18), and the corresponding eigenvalues of bound states as in Sect. 3.3 but for a step potential which is quasiperiodic, i.e., in which parts are repeated several times.

To facilitate the input of variables we distinguish between the (arbitrary) step potential of Sect. 3.3 and a *quasiperiodic* step potential consisting of a finite number of periods, each consisting of a number of regions. A plot with such a potential is shown in Fig. 3.4. On the subpanel **Physics—Potential** you select the number of **Regions within One Period**, you specify the **Potential in Regions 1 and N** (the very first and the very last region of the complete potential) and you define the **Number of Periods** N_P. The different **Regions** within one period are specified by pairs of variables $(\Delta x_2, V_2), \dots, (\Delta x_{N_\mathrm{R}+1}, V_{N_\mathrm{R}+1})$. Here N_R is the number of regions in each period; the regions have the widths $\Delta x_2, \dots$ and the potentials $V_2, \dots$, respectively. **Changes in the Last Period**, i.e., the rightmost period are sometimes helpful in the construction of the overall potential. This can be done by omitting one or more (up to $N_\mathrm{R} - 1$) regions beginning at the far right of the rightmost period.

Example Descriptor on File `1D_Bound_States.des`

- `Eigenstates in quasiperiodic potential` (see Fig. 3.4)

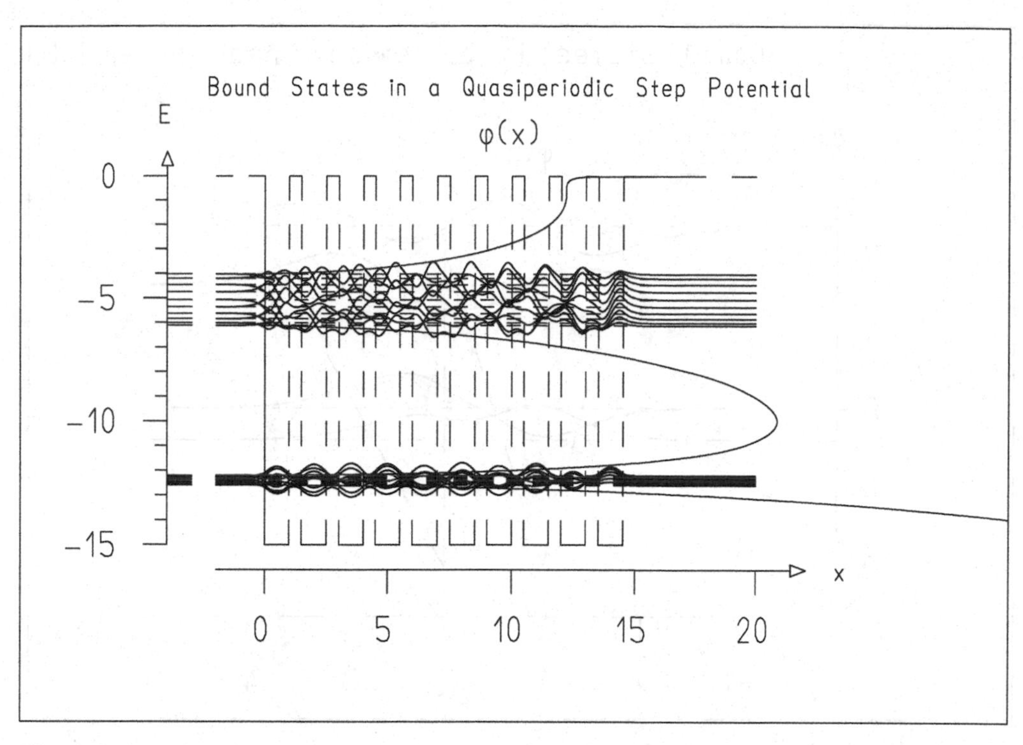

Fig. 3.4. Plot produced with descriptor `Eigenstates in quasiperiodic potential` on file `1D_Bound_States.des`

3.5 Eigenstates in the Piecewise Linear Potential

Aim of this section: Computation and presentation of eigenfunctions $\varphi_l(x)$, ((3.18), (3.26)), and the corresponding eigenvalues of bound states in a piecewise linear potential $V(x)$, (3.24). The value of the determinant $D(E)$ is also shown as a function of the energy E.

A plot similar to Fig. 3.5 is produced containing some or all of the items *potential*, *eigenvalues*, *eigenfunctions*, and *term scheme* for a piecewise linear potential. Since the eigenvalues are found by a numerical search for the zeros in the determinant $D = D(E)$, a graph of the function $D = D(E)$ can be shown as a further item.

For the input of parameters defining the potential you should picture the graph of $V = V(x)$ as a series of straight lines connecting the points $(x_1 = 0, V_1), (x_2, V_2), \ldots, (x_{N-1}, V_{N-1}), (x_{N-1}, V_N)$. In region 1 $(x < x_1 = 0)$ the potential has the constant value V_1; in region N $(x > x_{N-1})$ it has the constant value V_N. The slope in each region is constant, usually zero or finite; but can be made infinite by two identical values of x with two different values of V $(x_3 = 2, V_3 = -2$ and $x_4 = 2, V_3 = -10$ in Fig. 3.5). In this way vertical walls can be included in the potential landscape. The subpanel **Physics—Potential** has the same appearance as for the step potential (Sect. 3.3). Also the subpanels **Physics—Comp. Coord.** and **Physics—Variables** are as in Sect. 3.3.

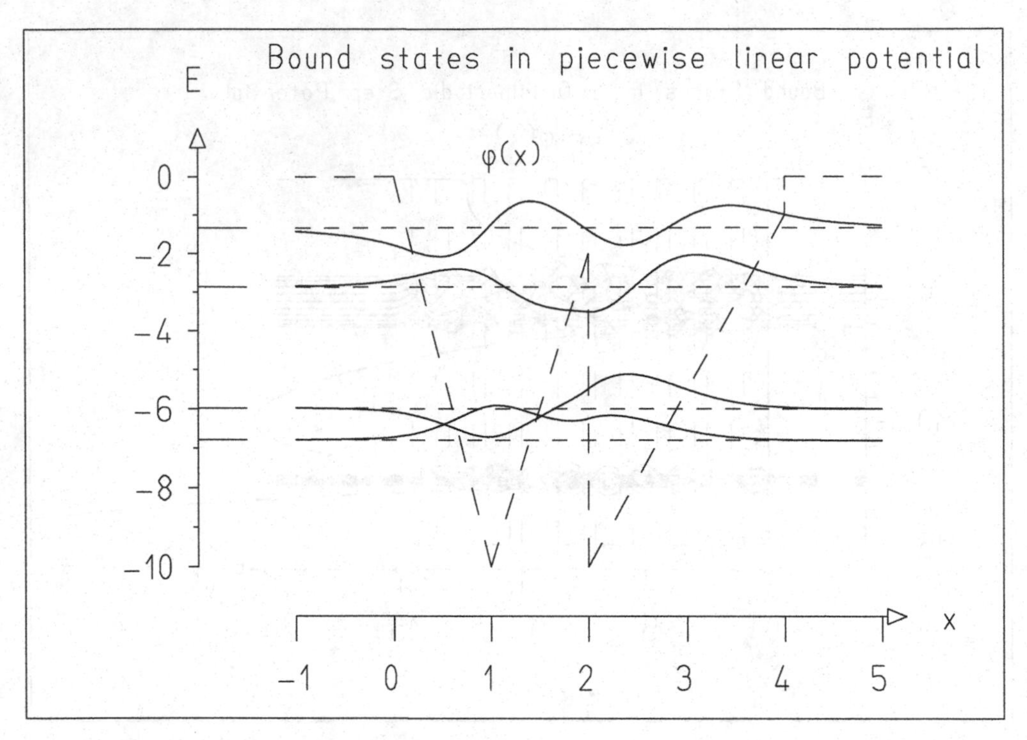

Fig. 3.5. Plot produced with descriptor `Bound states in piecewise linear potential` on file `1D_Bound_States.des`

Example Descriptor on File `1D_Bound_States.des`

- `Bound states in piecewise linear potential` (see Fig. 3.5)

3.6 Eigenstates in the Piecewise Linear Potential – Quasiperiodic

Aim of this section: Computation and presentation of eigenfunctions $\varphi_l(x)$, ((3.18), (3.26)), and the corresponding eigenvalues of bound states as in Sect. 3.5 but for a piecewise linear potential which is quasiperiodic, i.e., in which parts are repeated several times.

We distinguish between the (arbitrary) piecewise linear potential of Sect. 3.5 and a *quasiperiodic* piecewise linear potential consisting of a finite number of periods, each consisting of a number of regions, see, e.g., Fig. 3.6. On the subpanel **Physics—Potential** you select the number of **Regions within One Period**, you specify the **Potential in Regions 1 and N** (the very first and the very last region of the complete potential) and you define the **Number of Periods** N_P. The different **Regions** within one period are specified by pairs of variables $(\Delta x_2, V_2), \ldots, (\Delta x_{N_\mathrm{R}+1}, V_{N_\mathrm{R}+1})$. Here N_R is the number of regions in each period. The regions have the widths $\Delta x_2, \ldots, \Delta x_{N_\mathrm{R}+1}$. Over the width Δx_j

of region j the potential is linear, changing from V_{j-1} to V_j. The leftmost point of the first region Δx_2 of the first period is $x_1 = 0$ with the (outer) potential V_1. The rightmost point of the last region Δx_{N_R+1} of the last period is given the (outer) potential V_N. **Changes in the Last Period** are sometimes helpful in the construction of the overall potential. This can be done by adding one region or by omitting one or more (up to $N_R - 2$) regions beginning at the far right.

Example Descriptor on File `1D_Bound_States.des`

- `Bound states in quasiperiodic, piecewise linear potential` (see Fig. 3.6)

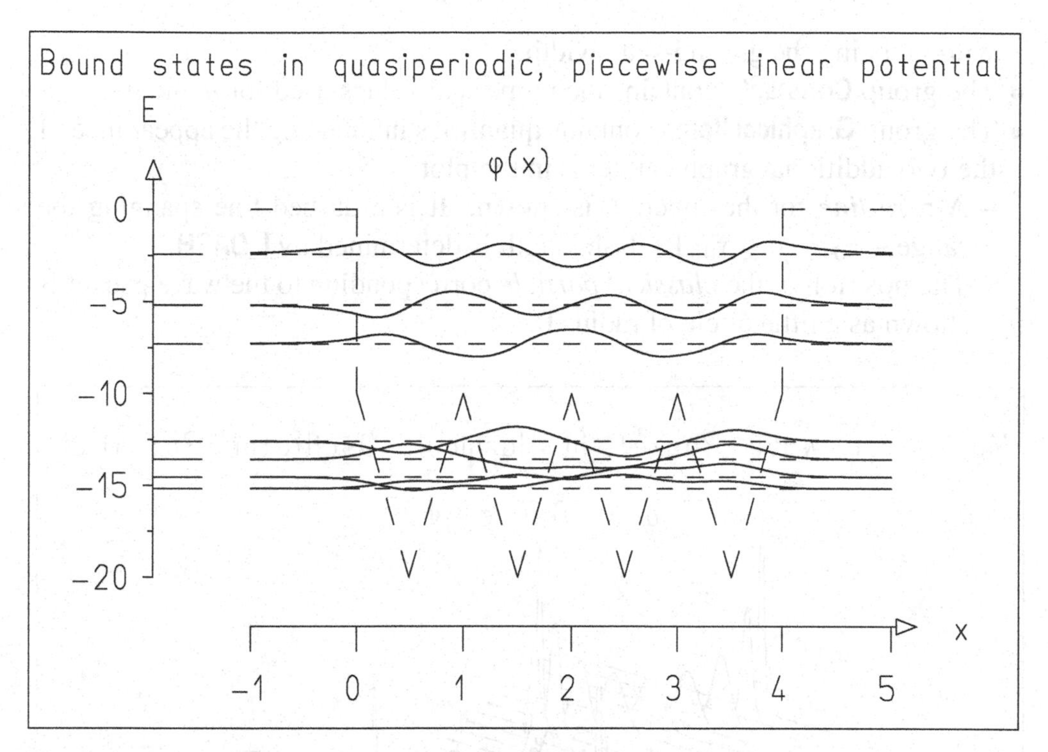

Fig. 3.6. Plot produced with descriptor `Bound states in quasiperiodic, piecewise linear potential` on file `1D_Bound_States.des`

3.7 Harmonic Particle Motion

Aim of this section: Demonstration of the motion of coherent states and squeezed states in the harmonic-oscillator potential. Presentation of absolute square (3.33) of the wave function and also of its real and imaginary part.

On the subpanel **Physics—Comp. Coord.** you can select one of the following functions for plotting:

- **Absolute Square of Wave Function** $|\psi(x,t)|^2$,
- **Real Part of Wave Function** $\mathrm{Re}\,\psi(x,t)$,
- **Imaginary Part of Wave Function** $\mathrm{Im}\,\psi(x,t)$,
- **Probability-Current Density** $j(x,t)$,
- **Bohm's Quantum Potential** $V_{\mathrm{Q}}(x,t)$.

On the subpanel **Physics—Variables** you find four groups of variables:

- The group **Harmonic Oscillator** contains only the circular frequency ω.
- The group **Wave Packet** contains the initial mean position x_0 and a quantity f_σ. It expresses in a convenient way the initial width in x,

$$\sigma_{x0} = f_\sigma \sigma_0/\sqrt{2} \quad ,$$

  with σ_0 being the ground-state width.
- The group **Constants** contains the numerical values used for $\hbar$ and m.
- The group **Graphical Items** contains quantities influencing the appearance of the two additional graphical items in the plot:
  - A *zero line* for the function is shown. It is a dashed line spanning the range $-x_0 \le x \le x_0$. Its dash length is determined by **l_DASH**.
  - The position of the *classical particle* corresponding to the wave packet is shown as a little circle of radius **R**.

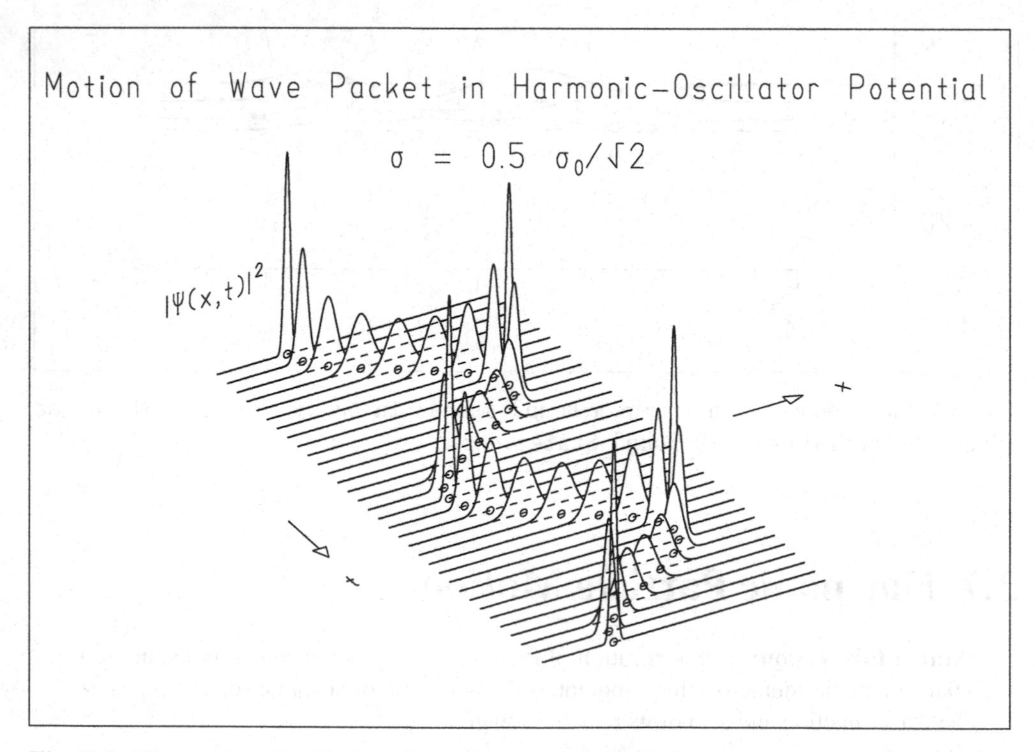

Fig. 3.7. Plot produced with descriptor `Harmonic particle motion` on file `1D_Bound_-States.des`

The contents of the subpanel **Physics—Quantiles** are as described in Sect. 2.3.3.

Movie Capability: Indirect. After conversion to direct movie capability the start and end times can be changed in the subpanel **Movie** (see Sect. A.4) of the parameter panel.

Example Descriptors on File `1D_Bound_States.des`

- `Harmonic particle motion` (see Fig. 3.7)
- `Movie: Harmonic particle motion`
- `Harmonic particle motion, quantile shown`
- `Movie: Harmonic particle motion, quantile shown`

3.8 Harmonic Oscillator: Quantile Trajectories

Aim of this section: Illustration of the quantile trajectories $x_P = x_P(t)$, (2.23), for the motion of a Gaussian wave packet in a harmonic-oscillator potential.

On the subpanel **Physics—Variables** you find the circular frequency ω of the **Harmonic Oscillator**, the initial mean position x_0 of the **Wave Packet**, and the

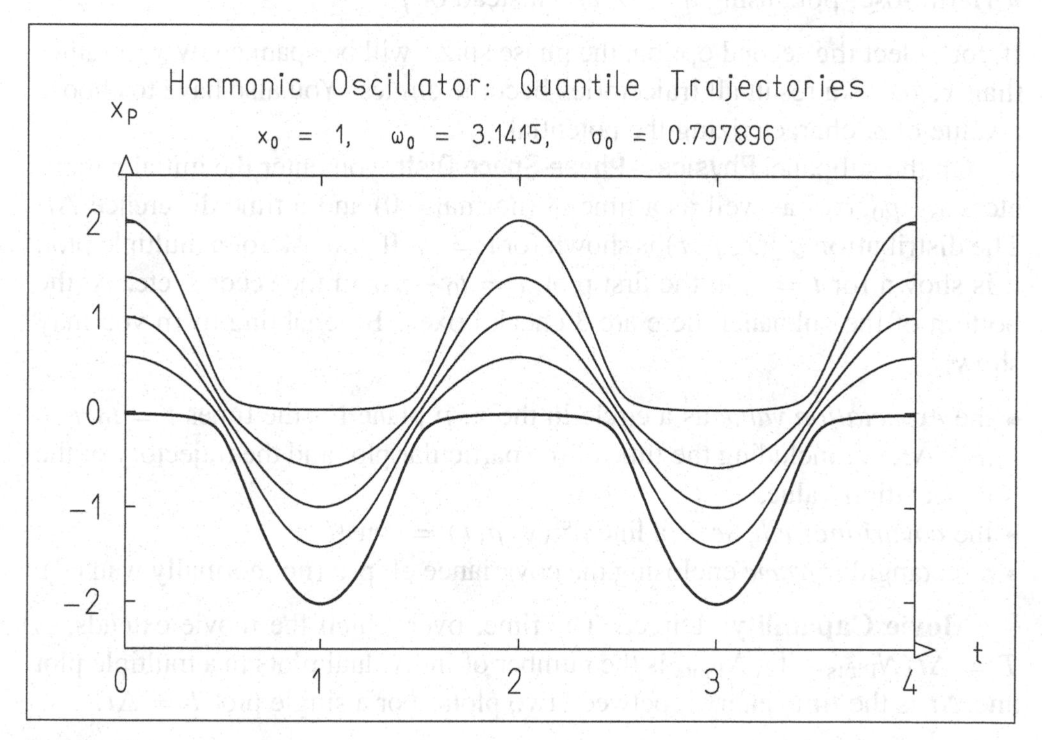

Fig. 3.8. Plot produced with descriptor `Harmonic oscillator, quantile trajectories` on file `1D_Bound_States.des`

factor f_σ (so that the initial width is f_σ times the width of a wave packet formed by the oscillator's ground state).

The content of the subpanel **Physics—Quantiles** is as described in Sect. 2.5.

Example Descriptor on File `1D_Bound_States.des`

- `Harmonic oscillator, quantile trajectories` (see Fig. 3.8)

3.9 Classical Phase-Space Distribution: Harmonic Motion

Aim of this section: Graphical presentation of the phase-space probability density $\varrho^{\rm cl}(x, p, t)$ described in Sect. 3.1.10, which initially (at time $t = 0$) is uncorrelated and fulfills the condition $\sigma_{x0}\sigma_{p0} = \hbar/2$. The marginal distributions $\varrho^{\rm cl}_x(x, t)$ and $\varrho^{\rm cl}_p(p, t)$ are also shown.

On the subpanel **Physics—Comp. Coord.** you have to select one of the following:

- Harmonic-oscillator potential,
- Harm.-osc. pot., using $q = p/m\omega$ instead of p.

If you select the second option, the phase space will be spanned by x, q rather than x, p. As a result all trajectories become circles. You also have to choose a value of ω characterizing the potential.

On the subpanel **Physics—Phase-Space Distr.** you enter the initial parameters x_0, p_0, σ_{p0} as well as a time t_0 (normally 0) and a time difference Δt. The distribution $\varrho^{\rm cl}(x, p, t)$ is shown for $t = t_0$. If you ask for a multiple plot, it is shown for $t = t_0$ in the first plot, $t = t_0 + \Delta t$ in the second, etc. At the bottom of the subpanel there are 3 check boxes. By enabling them you may show

- the *expectation value* as a circle in the x, p plane for the times $t = t_0, t = t_0 + \Delta t, \ldots$ including the time of the particular plot and the trajectory of the expectation value,
- the *covariance ellipse* as a line $\varrho^{\rm cl}(x, p, t) = \text{const}$,
- a rectangular *frame* enclosing the covariance ellipse (not normally wanted).

Movie Capability: Direct. The time, over which the movie extends, is $T = \Delta t(N_{\rm Plots} - 1)$; $N_{\rm Plots}$ is the number of individual plots in a multiple plot and Δt is the time interval between two plots. For a single plot $T = \Delta t$.

Example Descriptor on File `1D_Bound_States.des`

- `Classical phase-sp. density in harm. osc.` (see Fig. 3.9)

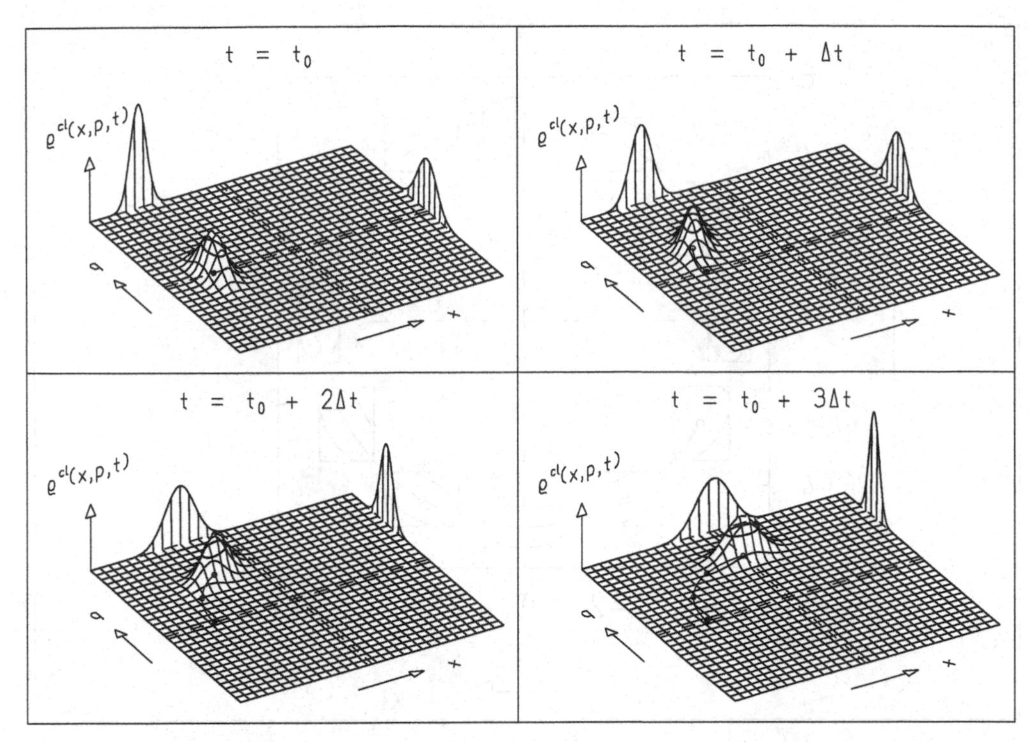

Fig. 3.9. Plot produced with descriptor `Classical phase-sp. density in harm. osc.` on file `1D_Bound_States.des`

3.10 Harmonic Motion of Classical Phase-Space Distribution: Covariance Ellipse

Aim of this section: Graphical presentation of the covariance ellipse, characterizing the phase-space probability density $\varrho^{\mathrm{cl}}(x, p, t)$ of Sect. 3.1.10 which initially (at time $t = 0$) is uncorrelated and fulfills the condition $\sigma_{x0}\sigma_{p0} = \hbar/2$.

On the subpanel **Physics—Comp. Coord.** you have to select one of the following:

- Harmonic-oscillator potential,
- Harm.-osc. pot., using $q = p/m\omega$ instead of p.

If you select the second option, the phase space will be spanned by x, q rather than x, p. As a result all trajectories become circles. You also have to choose a value of ω characterizing the potential.

On the subpanel **Physics—Phase-Space Distr.** you enter the initial parameters x_0, p_0, σ_{p0} as well as a time t_0 (normally 0), a time difference Δt, and an integer number N_t. The covariance ellipse is shown in the x, p plane for the times $t = t_0, t = t_0 + \Delta t, \ldots, t = t_0 + (N_t - 1)\Delta t$.

At the bottom of the subpanel there are 3 check boxes. By enabling them you may show

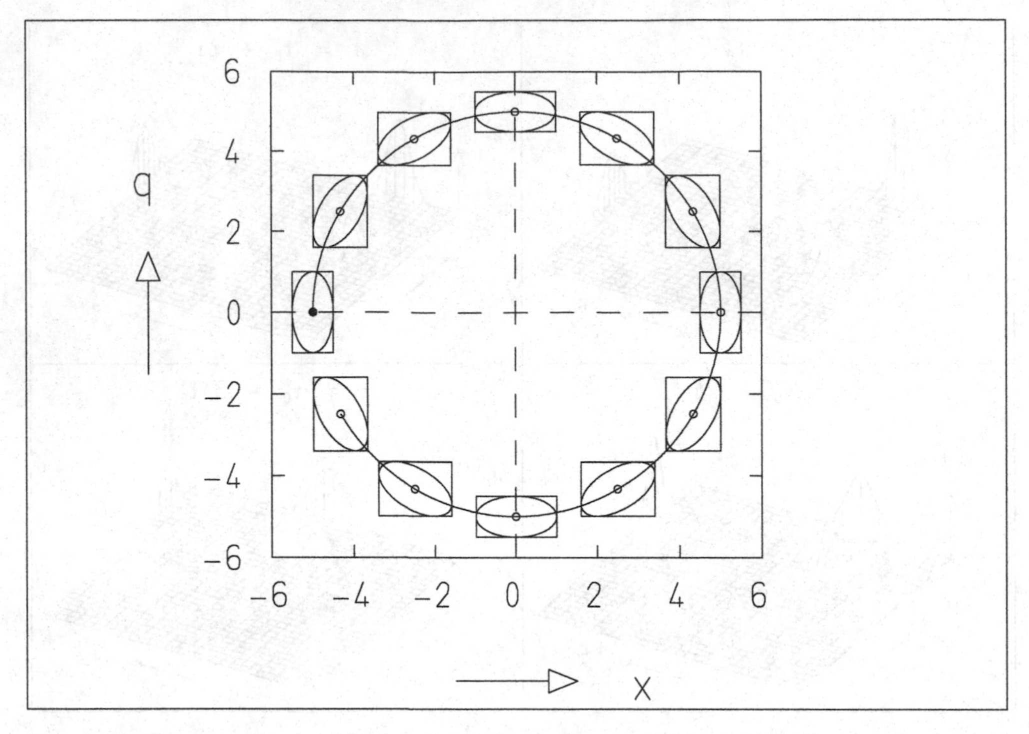

Fig. 3.10. Plot produced with descriptor `Covariance ellipse of class. phase-sp. density in harm. osc.` on file `1D_Bound_States.des`

- the *expectation value* as a circle in the x, p plane for the times $t = t_0, t = t_0 + \Delta t, \ldots$ and the trajectory of the expectation value,
- the *covariance ellipse*,
- a rectangular *frame* enclosing the covariance ellipse.

Movie Capability: Direct. On the subpanel **Movie** (see Sect. A.4) of the parameters panel you may choose to show or not to show the initial and intermediate positions of the covariance ellipse.

Example Descriptor on File `1D_Bound_States.des`

- `Covariance ellipse of class. phase-sp. density in harm. osc.` (see Fig. 3.10)

3.11 Particle Motion in the Infinitely Deep Square-Well Potential

Aim of this section: Study of the motion of a wave packet (3.40) in the deep square-well potential.

The sum (3.40) extends over integer values n ranging from

$$N_1 = N_0 - 5\sigma_0$$

to

$$N_2 = N_0 + 5\sigma_0$$

with N_0 as the nearest integer in the neighborhood of the number $|p_0|d/(\hbar\pi)$. The number σ_0 is the nearest integer of $1/\sigma_r$. If N_1 turns out to be less than one, N_1 is set to one.

On the subpanel **Physics—Comp. Coord.** you can select to compute either the spatial probability density of the classical phase-space distribution or the quantum-mechanical wave function. In the latter case you can choose to show the absolute square of the wave function or its real or its imaginary part.

On the subpanel **Physics—Variables** there are four groups of parameters:

- The group **Square Well** contains the width d of the well.
- The group **Wave Packet** contains the initial mean position x_0, the initial mean momentum p_0, and the initial spatial width σ_{x0} of the packet.
- The group **Constants** contains the numerical values of $\hbar$ and m.
- The group **Graphical Items** refers to the three additional items shown in the plot.
  - The *potential* $V(x)$ is indicated by a dashed horizontal line indicating the bottom and two dashed vertical lines indicating the walls of the potential. The dash length is given by **l_DASH**, the height of the vertical lines by **H**.

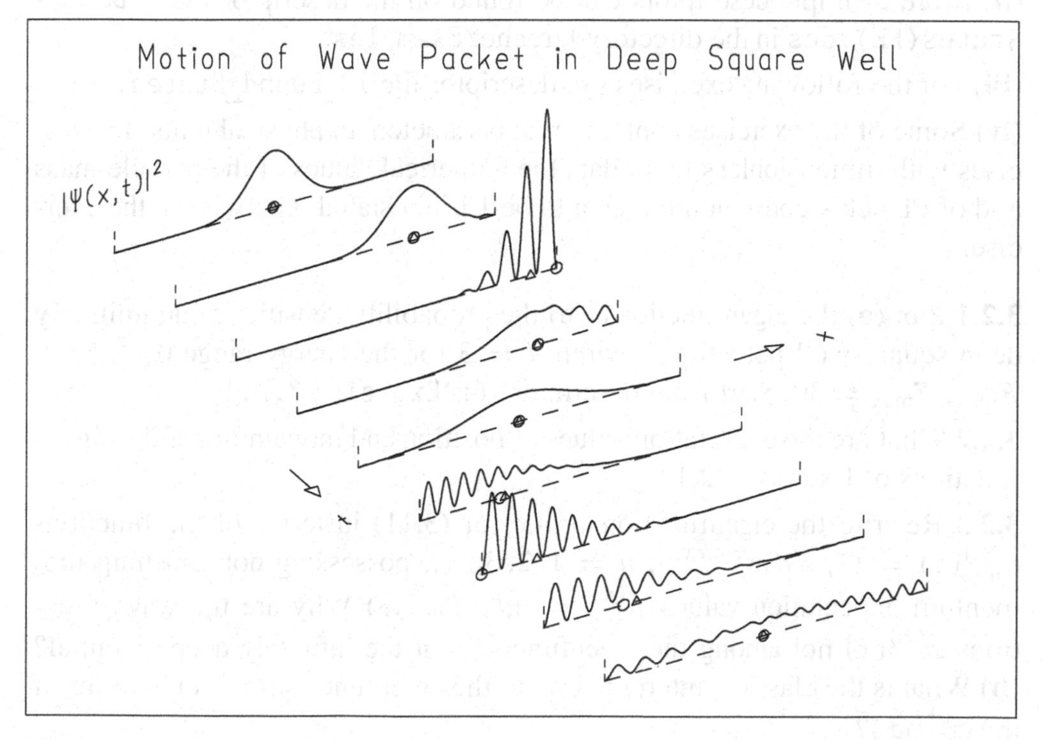

Fig. 3.11. Plot produced with descriptor `Particle motion in deep square well` on file `1D_Bound_States.des`

– The position of the corresponding *classical particle* moving in the well with initial conditions x_0, p_0 indicated by a circle of radius **R**.
– The position of the *quantum-mechanical expectation value* indicated by a triangle. The extension of the triangle is also given by **R**.

Movie Capability: Indirect. After conversion to direct movie capability the start and end times can be changed in the subpanel **Movie** (see Sect. A.4) of the parameter panel.

Example Descriptors on File `1D_Bound_States.des`

- `Particle motion in deep square well` (see Fig. 3.11)
- `Movie: Particle motion in deep square well` (This movie runs over a whole quantum-mechanical period T_1.)

3.12 Exercises

Please note:

(i) You may watch a demonstration of the material of this chapter by pressing **Run Demo** in the *main toolbar* and selecting one of the demo files `1D_Bound_States`.

(ii) More example descriptors can be found on the descriptor file `1DBound-States(FE).des` in the directory `FurtherExamples`.

(iii) For the following exercises use descriptor file `1D_Bound_States.des`.

(iv) Some of the exercises contain input parameters in physical units. In exercises with dimensionless input data the numerical values of the particle mass and of Planck's constant are meant to be 1 if not stated otherwise in the exercise.

3.2.1 Plot **(a)** the eigenfunctions, **(b)** the probability densities of an infinitely deep square-well potential of width $d = 3$ for the energy range $0 \leq E_i \leq E_{\max}$, $E_{\max} = 30$. Start from descriptor 16: '`Exercise 3.2.1`'.

3.2.2 What are the expectation values of position and momentum of the eigenfunctions of Exercise 3.2.1?

3.2.3 Rewrite the eigenfunctions $\varphi_n(x)$ of (3.11) in terms of the functions $\varphi_n^{(\pm)}(x) = (1/\sqrt{d})\mathrm{e}^{\pm \mathrm{i}n\pi x/d}$, $n = 1, 2, 3, \ldots$, possessing nonvanishing momentum expectation values $p_n^{(\pm)} = \pm\hbar\pi/d$. **(a)** Why are the wave functions $\varphi_n^{(\pm)}(x)$ not among the eigenfunctions of the infinitely deep potential? **(b)** What is the classical interpretation of the eigenfunctions $\varphi_n(x)$ in terms of the $\varphi_n^{(\pm)}(x)$?

3.2.4 Calculate the eigenfunctions and eigenvalues in the energy range $0 \leq E_i \leq E_{\max} = 2\,\mathrm{eV}$ for an electron in the infinitely deep square-well potential of width **(a)** 3 nm, **(b)** 5 nm. Start from descriptor 17: '`Exercise 3.2.4`'. **(c)** Calculate a rough estimate of the ground-state energy using Heisenberg's uncertainty relation.

3.2.5 Repeat Exercise 3.2.4 for a proton for the energy range $0 \leq E_i \leq E_{\text{max}}$, $E_{\text{max}} = 1\,\text{meV}$. Start from descriptor 18: '`Exercise 3.2.5`'.

3.2.6 **(a)** Plot the wave functions and the energy levels E_i in the range $0 \leq E_i \leq E_{\text{max}}$, $E_{\text{max}} = 20\,\text{eV}$, for an electron (mass m_e) in an infinitely deep potential of width $d = 3a$. Here, $a = \hbar c/(\alpha m_\text{e} c^2) = 0.05292\,\text{nm}$ is the Bohr radius of the innermost orbit of the hydrogen atom. Read off the energy E_1 of the lowest eigenfunction φ_1. Start with descriptor 19: '`Exercise 3.2.6`'. **(b)** Calculate the potential energy of an electron in Coulomb potential $V(r) = -\alpha\hbar c/r$ at the Bohr radius $r = a$. **(c)** Calculate the sum of the kinetic energy E_1 of the lowest eigenfunction φ_1 as determined in (a) and the potential energy as calculated in (b). Compare it to the binding energy of the electron in the hydrogen ground state.

3.2.7 Plot the eigenfunctions of the harmonic oscillator for an electron with the angular frequencies **(a)** $\omega = 10^{15}\,\text{s}^{-1}$, **(b)** $\omega = 1.5 \times 10^{15}\,\text{s}^{-1}$, **(c)** $\omega = 2 \times 10^{15}\,\text{s}^{-1}$, and **(d–f)** the corresponding probability densities. Start with descriptor 20: '`Exercise 3.2.7`'. **(g)** What is the physical unit of the energy scale? **(h)** Calculate the spring constants $D = m_\text{e}\omega^2$. **(i)** What is the physical unit of the x scale? **(j)** Why is the probability density highest close to the wall of the potential?

3.2.8 Plot **(a,b)** the eigenfunctions, **(c,d)** the probability density of a proton in the harmonic-oscillator potential $V = Dx^2/2$ with the constants $D = 0.1\,\text{eV}\,\text{m}^{-2}$ and $D = 1\,\text{keV}\,\text{m}^{-2}$, respectively. Start from descriptor 21: '`Exercise 3.2.8`'. **(e)** Calculate the angular frequency ω of the oscillator. **(f)** Calculate the width of the ground state of the harmonic oscillator, $\sigma_0 = \sqrt{\hbar m^{-1}\omega^{-1}}$.

3.3.1 Plot **(a)** the eigenfunctions, **(b)** the probability distributions for a square-well potential of width $d = 4$ and depth $V_0 = -6$. Start from descriptor 22: '`Exercise 3.3.1`'. **(c)** Read the energy eigenvalues E_i off the screen and calculate the differences $\Delta_i = E_i - V_0$.

3.3.2 Repeat Exercise 3.3.1 for the widths **(a,b,c)** $d = 2$, **(d,e,f)** $d = 6$. Start from descriptor 22: '`Exercise 3.3.1`'.

3.3.3 **(a)** Calculate the lowest energy eigenvalues E'_i in the infinitely deep square well for the widths $d = 2, 4, 6$. **(b)** Compare the E'_i to the differences Δ_i for the square-well potentials of Exercises 3.3.1, 3.3.2 of corresponding widths. **(c)** Explain why the separation of the eigenvalues E_i is smaller than that of the E'_i.

3.3.4 Plot **(a)** the eigenfunctions, **(b)** the probability densities of a double-well potential

$$V(x) = \begin{cases} 0\,,\, x < 0 \\ -4\,,\, 0 \le x < 1.625 \\ -1.33\,,\, 1.625 \le x < 1.875 \\ -4\,,\, 1.875 \le x < 3.5 \\ 0\,,\, 3.5 \le x \end{cases} .$$

Start with descriptor 23: '`Exercise 3.3.4`'. **(c)** Why is the third eigenfunction a horizontal straight line in the central region? **(d)** Why have the two lowest eigenfunctions energies close to each other? **(e)** Why are they symmetric and antisymmetric?

3.3.5 Plot **(a)** the eigenfunctions, **(b)** the probability densities of a double-well potential

$$V(x) = \begin{cases} 0\,,\, x < 0 \\ -4\,,\, 0 \le x < 1 \\ -1.33\,,\, 1 \le x < 2 \\ -4\,,\, 2 \le x < 3.5 \\ 0\,,\, 3.5 \le x \end{cases} .$$

Start with descriptor 23: '`Exercise 3.3.4`'. **(c)** Why do the wave functions exhibit no symmetric pattern?

3.3.6 Plot **(a)** the eigenfunctions, **(b)** the probability densities of an asymmetric double-well potential

$$V(x) = \begin{cases} 0\,,\, x < 0 \\ -5.72\,,\, 0 \le x < 1 \\ -1\,,\, 1 \le x < 1.5 \\ -4\,,\, 1.5 \le x < 3.5 \\ 0\,,\, 3.5 \le x \end{cases} .$$

Start with descriptor 23: '`Exercise 3.3.4`'. **(c)** Why is the ground-state wave function given by a straight line in the second well?

3.3.7 Plot the eigenfunctions in a set of asymmetric potential wells given by the potentials

$$V(x) = \begin{cases} 0\,,\, x < 0 \\ -4\,,\, 0 \le x < 1 \\ -1.33\,,\, 1 \le x < 2 \\ -4\,,\, 2 \le x < d \\ 0\,,\, d \le x \end{cases} ,$$

where the right edge d is to be set to **(a)** $d = 4$, **(b1)** $d = 4.7$, **(b2)** $d = 4.8$, **(b3)** $d = 4.9$, **(b4)** $d = 5$, **(c)** $d = 6$. Start from descriptor 23: '`Exercise 3.3.4`'. In order to facilitate a direct comparison of the plots (b1–b4) show them in a combined plot; see Appendix A.10. An example of a mother descriptor is descriptor 24: '`Exercise 3.3.7d`', which quotes the descriptors 16: '`Exercise 3.2.1`', 17: '`Exercise 3.2.4`', 18: '`Exercise 3.2.5`', 19: '`Exercise 3.2.6`'. Try it out. Now modify descriptor 23: '`Exercise 3.3.4`' according to question (b1) and append the modified descriptor. Repeat this procedure for (b2), (b3), (b4). By pressing the button **Descriptor** you will get the list of titles of all descriptors and at its end the four descriptors just stored away and their numbers. Enter these four numbers in the subpanel **Table of Descriptors** of descriptor 23: '`Exercise 3.3.4`'. **(d)** Now plot descriptor 23: '`Exercise 3.3.4`' and thus the graphs corresponding to (b1–b4) in a combined plot. **(e)** Explain the behavior of the second and third eigenstate in terms of admixtures of eigenstates in the two single wells.

3.3.8 Repeat Exercise 3.3.7 for the probability densities.

3.4.1 We consider a quasiperiodic potential that consists of r equal potential wells of width 1 and depth -15 with a separation of 0.5. Plot the wave functions in an increasing number of equal wells **(a)** $r = 1$, **(b)** $r = 2$, **(c)** $r = 3$, **(d)** $r = 4$, **(e)** $r = 5$, **(f)** $r = 6$, **(g)** $r = 7$, **(h)** $r = 8$, **(i)** $r = 9$, **(j)** $r = 10$. Start from descriptor 25: '`Exercise 3.4.1`'. **(k)** Give a qualitative reason for the occurrence of two bands of states in the quasiperiodic potential.

3.4.2 Repeat Exercise 3.4.1 for a quasiperiodic potential with 10 wells of width 0.2 and depth -50 and separation of 0.15 between the wells for $\hbar = 0.658$ and $m = 5.685$. **(a)** Plot the term scheme. Start from descriptor 25: '`Exercise 3.4.1`'. **(b)** Plot the wave functions of the lowest band. **(c)** Plot the wave functions of the second-lowest band. **(d)** Plot the wave functions of the highest band. **(e)** Explain the symmetry structure of the wave functions in a band. Start from the form of the wave functions in the wide square-well potential that is obtained by taking all the walls out.

3.4.3 (a–d) Repeat Exercise 3.4.2 (a–d) for a quasiperiodic potential of a depth of -40 and with the value -5 for the potential in the regions of the intermediate walls separating the narrow wells. Start from descriptor 25: '`Exercise 3.4.1`'. **(e)** Show the two highest states of the lowest band in a separate plot. Switch off the plotting of the potential. **(f)** Why are the two highest states separated from the lower eight by a somewhat larger energy gap?

3.5.1 (a) Determine the number of energy eigenvalues in a symmetric triangular potential with the following parameters: $(x_2, V_2) = (20, -2)$, $(x_3, V_3) = (40, 0)$. **(b)** Consider one half of this potential by creating a steep fall at $x_1 = 0$ by setting $x_2 = 0$ and $x_3 = 20$ and determine the number of energy eigenvalues. **(c)** Set now $V_1 = \infty$ and again determine the number of energy eigenvalues. **(d)** Compare the spectrum and the eigenfunctions of (a) and (c). Start from descriptor 26: '`Exercise 3.5.1`' and adjust the x range in computing coordinates appropriately.

3.5.2 Determine **(a)** the number and **(b)** the energy eigenvalues in the following piecewise linear potential (possessing steep rises and falls at x_1 and x_{N-1}, respectively, and representing the potential of Fig. 4.3 for stationary scattering): $(x_2, V_2) = (0, 5)$, $(x_3, V_3) = (2.5, -5)$, $(x_4, V_4) = (5, 5)$, $(x_5, V_5) = (5, 0)$. **(c)** How many bumps are expected for the absolute square of the scattering wave at the first resonance (inside the potential region)? **(d)** What is the approximate minimum depth $|V_3|$ ($V_3 < 0$), for all other potential parameters unchanged, of such a potential to contain at least one bound state? Start from descriptor 26: '`Exercise 3.5.1`' and adjust the x and y ranges in computing coordinates appropriately.

3.6.1 Consider a quasiperiodic piecewise linear potential with a double-well structure and the following parameters containing four and five wells, respectively, of different width and depth: $N_R = 4$, $N_P = 5$, $(\Delta x_2, V_2) = (0.75, 0)$, $(\Delta x_3, V_3) = (0.25, -21.7)$, $(\Delta x_4, V_4) = (0.25, 0)$, $(\Delta x_5, V_5) = (0.75, -10)$. Start from descriptor 27: '`Exercise 3.6.1`'. **(a)** Plot the eigenvalues and eigenfunctions and determine the average energy of the two separated bands. How many states are contained in the lower and upper band, respectively? What band belongs to which set of repeated wells? **(b)** Now change the depths of the potential wells by modifying the values of V_3 and V_5. What values need to be assigned to these potential values in order to exchange the correspondence of the two bands to the repeated wells, where the band centers should stay at the same positions in energy? How many states are now contained in the lower and upper band, respectively? **(c)** Set the outside potentials for both situations of (a) and (b) to infinity. You observe two states separating from the two-band structure, forming so-called surface states that more or less concentrate at the outer boundaries. Determine the numbers of states in the remaining bands and the positions of the surface states in energy for both cases. Explain the resulting structure.

3.7.1 Plot **(a)** the real part, **(b)** the imaginary part, **(c)** the absolute square of a wave packet initially at rest moving in a harmonic-oscillator potential. For $\hbar$ and particle mass use the default values. The oscillator frequency is $\omega = \pi$. The initial data of the wave packet are the initial location $x_0 = -3$ and the relative initial width $f_\sigma = 0.5$. Start from descriptor 28: '`Exercise 3.7.1`'. **(d)** What is the period of the time evolution of the wave function? **(e)** What is the period of the time evolution of the absolute square? **(f)** What is the most general requirement for the periodicity of a wave function describing a physical process periodic in time?

3.7.2 Plot the absolute square of the wave packets initially at rest at $x_0 = -6$ moving in a harmonic-oscillator potential of angular frequency $\omega = \pi$ for the three relative widths **(a)** $f_\sigma = 0.5$, **(b)** $f_\sigma = 2$, **(c)** $f_\sigma = 1$ for two periods of the oscillation. Start from descriptor 29: '`Exercise 3.7.2`'. **(d)** Explain the oscillation of the widths observed in (a) and (b) in terms of a classical particle with inaccurately known initial values of location and momentum.

3.7.3 Plot **(a)** the real part, **(b)** the imaginary part, **(c)** the absolute square of a wave packet initially at rest in the central position of the harmonic-oscillator potential. Choose the relative initial width as $f_\sigma = 1.75$. Start with descriptor

29: 'Exercise 3.7.2'. **(d)** Why does the wave packet periodically change its width?

3.7.4 (a,b,c) Repeat Exercise 3.7.3 (a,b,c) for the relative width $f_\sigma = 1$. **(d)** Why does the wave packet not change its width over time?

3.7.5 Display the quantile motion for $P = 0.5, 0.7, 0.9$. Use descriptor 9: 'Harmonic particle motion, quantile shown'.

3.8.1 Display quantile trajectories using descriptor 11: 'Harmonic oscillator, quantile trajectories'. For the wave-packet parameters used in that descriptor the trajectories with $P = 0.1$ and $P = 0.9$ stay practically constant for half an oscillator period. **(a)** Explain this effect qualitatively. **(b)** Use descriptor 9: 'Harmonic particle motion, quantile shown' with the wave-packet parameters of descriptor 11: 'Harmonic oscillator, quantile trajectories' to check your explanation. **(c)** Plot the quantile trajectories for $P = 0.05, 0.1, 0.15$ to study the behavior for P somewhat smaller and larger than the special case $P = 0.1$.

3.11.1 Plot the motion of a 'Gaussian' wave packet with initial values $x_0 = 0$, $p_0 = 5$, $\sigma_{x0} = 0.75$ in an infinitely deep square-well potential of width $d = 10$ for the time intervals given in time steps $\Delta t = 0.5$. Choose **(a)** $0 \leq t \leq 10$, **(b)** $10 \leq t \leq 20$, **(c)** $20 \leq t \leq 30$. Start from descriptor 30: 'Exercise 3.11.1'. **(d)** Why does the wave packet disperse in time? **(e)** Why does the wave exhibit a wiggly shape when close to the wall?

3.11.2 Plot the motion of a 'Gaussian' wave packet with initial values $x_0 = 0$, $p_0 = 5$, $\sigma_{x0} = 0.5$ in an infinitely deep square-well potential of width $d = 10$ for the time intervals **(a)** $0 \leq t \leq 10$, **(b)** $10 \leq t \leq 20$, **(c)** $20 \leq t \leq 30$, **(d)** $30 \leq t \leq 40$, **(e)** $40 \leq t \leq 50$, **(f)** $50 \leq t \leq 60$, **(g)** $60 \leq t \leq 70$, **(h)** $70 \leq t \leq 80$, **(i)** $80 \leq t \leq 90$, **(j)** $90 \leq t \leq 100$, **(k)** $100 \leq t \leq 110$, **(l)** $110 \leq t \leq 120$, **(m)** $120 \leq t \leq 130$ in time steps $\Delta t = 1$. Start from descriptor 30: 'Exercise 3.11.1'. **(n)** Calculate the time T_1 in which the initial wave packet is re-established. **(o)** Look at the wave packet at time $T_1/2$. **(p)** Why do the classical position of the particle and the position expectation value of the wave packet coincide only at the beginning of the motion. **(q)** Why are there long time intervals in which the position expectation value is almost at rest?

3.11.3 Plot the motion of a 'Gaussian' wave packet in an infinitely deep square well for the initial values $x_0 = 2$, $p_0 = 5.184$, $\sigma_{x0} = 0.5$, for a particle of mass 1 ($\hbar = 1$) in time steps of $\Delta t = 0.5$ for the intervals **(a)** $0 \leq t \leq 10$, **(b)** $58.66 \leq t \leq 68.66$, **(c)** $122.32 \leq t \leq 132.32$. Start from descriptor 30: 'Exercise 3.11.1'. **(d)** Why does the expectation value of the wave packet at $t = 63.66$ coincide with the position of the classical particle?

3.11.4 Plot the motion of a 'Gaussian' wave packet in the infinitely deep square well with initial conditions $x_0 = 2$, $p_0 = 1.09956$, $\sigma_{x0} = 0.6$. **(a)** Start with a plot of the motion of the position expectation value of the wave packet during the time interval $0 \le t \le 127.32$. Subdivide this interval into 40 time steps. To make the wave function disappear set $z_{\rm end} = 1$. For the projection angles choose $\vartheta = 60$ and $\varphi = -90$. Start from descriptor 30: '`Exercise 3.11.1`'. One observes a time interval during which the expectation value of the position of the wave packet is almost at rest. **(b)** What is the expectation value of the energy of the wave packet during this interval? **(c)** What is the approximate expectation value of the momentum of the wave packet during this time interval? Plot the probability density of the 'Gaussian' wave packet (set $z_{\rm end}$ back to 0.001) for **(d)** $0 \le t \le 20$, **(e)** $20 \le t \le 40$, **(f)** $40 \le t \le 60$, **(g)** $60 \le t \le 80$, **(h)** $80 \le t \le 100$, **(i)** $100 \le t \le 120$, **(j)** $120 \le t \le 140$.

3.11.5 (a–g) Repeat Exercise 3.11.4 (a, d, e, ..., j) with $x_0 = 0$, $p_0 = 10.9956$, $\sigma_{x0} = 0.6$, for a particle of mass $m = 10$. Start from descriptor 30: '`Exercise 3.11.1`'. **(h)** Why does the position expectation value of the wave packet in this case follow the motion of the classical particle much longer than in Exercise 3.11.4?

4. Scattering in One Dimension

Contents: Continuum eigenfunctions and continuous spectra. Boundary conditions and stationary solutions of the Schrödinger equation for step and piecewise linear potentials. Continuum normalization. Motion of wave packets in step and piecewise linear potentials. Transmission and reflection coefficients. Unitarity and Argand diagram. Tunnel effect. Resonances. Stationary waves, wave packet, and quantiles for the linear potential. Classical phase-space distribution for linear potential and for reflection by a high potential wall.

4.1 Physical Concepts

4.1.1 Stationary Scattering States. Continuum Eigenstates and Eigenvalues. Continuous Spectra

For a potential with at least one finite limit

$$V(+\infty) = \lim_{x\to\infty} V(x) \quad \text{or} \quad V(-\infty) = \lim_{x\to-\infty} V(x) \tag{4.1}$$

there are normalized eigenstates φ_n only for energies $E < V_c = \min(V(+\infty), V(-\infty))$. In addition to these discrete eigenvalues with normalizable eigenfunctions the Schrödinger equation with a potential satisfying (4.1) also possesses eigenvalues with eigenfunctions that are *not normalizable*. Their fall-off for large values of $|x|$ is not sufficiently fast for the integral of the absolute square $|\varphi|^2$ extended over the whole x axis to have a finite value. Therefore, these eigenfunctions are not normalizable and do not represent actual physical states. The eigenvalues E belonging to the non-normalizable eigenfunctions are no longer discrete points but form continuous sets of values, e.g., intervals or a half axis of energy values. The set of continuous eigenvalues is called the *continuous spectrum*, the corresponding non-normalizable eigenfunctions are called *continuum eigenfunctions* $\varphi(E,x)$. They are solutions of the stationary Schrödinger equation (3.6)

$$H\varphi(E,x) = (T+V)\varphi(E,x) = E\varphi(E,x) \quad . \tag{4.2}$$

S. Brandt et al., *Interactive Quantum Mechanics: Quantum Experiments on the Computer*,
DOI 10.1007/978-1-4419-7424-2_4, © Springer Science+Business Media, LLC 2011

If a nonsingular potential fulfills the relation $V(x) < V_c$ only for a finite number of regions of finite lengths on the x axis, the continuous spectrum is bounded by

$$V_c \leq E \quad . \tag{4.3}$$

Normalizable solutions of the time-dependent Schrödinger equation can be formed as linear superpositions of these continuum eigenfunctions.

4.1.2 Time-Dependent Solutions of the Schrödinger Equation

Because the continuum eigenfunctions $\varphi(E, x)$ are not normalizable, they extend over the x axis to $+\infty$ or $-\infty$, depending on the values of $V(+\infty)$ and $V(-\infty)$. Thus, the continuum eigenfunctions can be used to form moving wave packets far away from the region where the potential actually exerts a force on the particle, i.e., in regions of constant or almost constant potential:

$$\psi(x, 0) = \int_{V_c}^{\infty} w(E)\, \varphi(E, x)\, \mathrm{d}E \quad . \tag{4.4}$$

The time-dependent solution of the Schrödinger equation having $\psi(x, 0)$ as initial state at $t = 0$ then takes the form

$$\psi(x, t) = \int_{V_c}^{\infty} w(E) \mathrm{e}^{-\mathrm{i}Et/\hbar} \varphi(E, x)\, \mathrm{d}E \quad . \tag{4.5}$$

4.1.3 Right-Moving and Left-Moving Stationary Waves of a Free Particle

Equation (2.1) describes the harmonic wave associated with a particle of mass m and momentum p. Because $E = p^2/2m$ is quadratic in p, (2.1) represents two solutions $p = \pm|p|$, with $|p| = \left|\sqrt{2mE}\right|$, for each energy value E. Thus, the wave functions (2.1) can also be interpreted as belonging to one of two sets,

$$\begin{aligned} \psi_+(E, x, t) &= \frac{1}{(2\pi\hbar)^{1/2}} \exp\left\{-\frac{\mathrm{i}}{\hbar}(Et - |p|x)\right\} \quad , \\ \psi_-(E, x, t) &= \frac{1}{(2\pi\hbar)^{1/2}} \exp\left\{-\frac{\mathrm{i}}{\hbar}(Et + |p|x)\right\} \quad . \end{aligned} \tag{4.6}$$

With a spectral function being different from zero for positive values of p only, the superposition (2.4) formed with $\psi_+(E, x, t)$ represents a *right-moving wave packet*, i.e., a wave packet propagating from smaller x values to larger ones. For the same spectral function the wave packet formed with

$\psi_-(E,x,t)$ is *left moving*, i.e., propagating from larger to smaller x values. Actually, the two harmonic waves themselves propagate to the right and to the left, respectively. In analogy to (3.4) they can be factorized:

$$\begin{aligned}\psi_\pm(E,x,t) &= \mathrm{e}^{-\mathrm{i}Et/\hbar}\varphi_\pm(E,x) \quad , \\ \varphi_\pm(E,x) &= (2\pi\hbar)^{-1/2}\mathrm{e}^{\pm\mathrm{i}|p|x/\hbar} \quad ;\end{aligned} \tag{4.7}$$

that is, into a solely time-dependent exponential and eigenfunctions $\varphi_+(E,x)$, $\varphi_-(E,x)$ of the kinetic energy T, i.e., the Hamiltonian $H = T$ of a free particle,

$$H\varphi_\pm(E,x) = E\varphi_\pm(E,x) \quad . \tag{4.8}$$

The time-dependent solutions $\psi_+(x,t)$, superpositions of the stationary waves $\varphi_+(E,x)$,

$$\psi_+(x,t) = \int w(p)\mathrm{e}^{-\mathrm{i}Et/\hbar}\varphi_+(E(p),x)\,\mathrm{d}p \quad , \tag{4.9}$$

only represent right-moving wave packets. Analogously, replacing φ_+ by φ_- leads to left-moving wave packets. For $t = 0$ and real $w(p)$ these wave packets are centered around $x = 0$. If we want to place a right-moving wave packet at $t = t_0$ around $x = x_0$, we have to substitute t with $t - t_0$ and the Gaussian spectral function $f(p)$, (2.5), with

$$w(p) = f(p)\mathrm{e}^{-\mathrm{i}px_0/\hbar} \quad . \tag{4.10}$$

This allows us to construct the wave packets incident on a step potential.

The eigenfunctions belonging to different energy eigenvalues E, E' are orthogonal,

$$\int_{-\infty}^{\infty}\varphi_\pm^*(E',x)\varphi_\pm(E,x)\,\mathrm{d}x = 0 \quad , \tag{4.11}$$

as are those for equal energy eigenvalues but different subscript signs, e.g.,

$$\int_{-\infty}^{\infty}\varphi_+^*(E,x)\varphi_-(E,x)\,\mathrm{d}x = 0 \quad . \tag{4.12}$$

The stationary wave functions $\varphi_{+,-}(E,x)$ are two continuum eigenfunctions to the same energy eigenvalue E, which is therefore called *two-fold degenerate*.

4.1.4 Orthogonality and Continuum Normalization of Stationary Waves of a Free Particle. Completeness

Because the integral over the absolute squares $|\varphi_+|^2$ or $|\varphi_-|^2$ does not exist, a normalization to unity is not possible. The normalization of discrete eigenfunctions is replaced by the *continuum normalization*

$$\int_{-\infty}^{\infty} \varphi_{s'}^*(E(p'), x)\varphi_s(E(p), x)\,\mathrm{d}x = \delta_{ss'}\delta(p - p') \quad , \qquad s = \pm \quad , \quad s' = \pm \quad . \tag{4.13}$$

This ensures that the normalization of the wave packet is equal to one if the spectral function $w(p)$ is correctly normalized to one,

$$\int_{-\infty}^{\infty} w^*(p)w(p)\,\mathrm{d}p = 1 \quad . \tag{4.14}$$

The set of functions $\varphi_s(E, x)$ is *complete*: any absolute-square-integrable function $\varphi(x)$ can be represented by an integral (Fourier's theorem):

$$\varphi(x) = \sum_{s=\pm} \int_0^{\infty} w_s(p)\,\varphi_s(E(p), x)\,\mathrm{d}p \quad . \tag{4.15}$$

4.1.5 Boundary Conditions for Stationary Scattering Solutions in Step Potentials

In the step potential, (3.16), the wave numbers k_j, (3.17), determine the solutions (3.18) for given E. The stationary solutions

$$\varphi_j(E, x) = \varphi_{j+}(E, x) + \varphi_{j-}(E, x) \tag{4.16}$$

are superpositions of two exponentials of opposite exponents in the region j, $x_{j-1} \leq x < x_j$:

$$\varphi_{j+}(E, x) = A'_j \mathrm{e}^{\mathrm{i}k_j x} \quad , \quad \varphi_{j-}(E, x) = B'_j \mathrm{e}^{-\mathrm{i}k_j x} \quad . \tag{4.17}$$

The scattering of a right-moving wave packet incident from $-\infty$ is possible for $E \geq V_1 = 0$. We have to distinguish two cases: $E \geq V_N$ and $E < V_N$.

i) For $E \geq V_N$, $k_N = \left|\sqrt{2m(E - V_N)}/\hbar\right|$, an outgoing wave

$$\varphi_N(E, x) = A'_N \mathrm{e}^{\mathrm{i}k_N x} \tag{4.18}$$

propagates inside the region $x \geq x_{N-1}$, i.e., to the right of the step potential.

ii) For $E < V_N$, $k_N = \mathrm{i}\kappa_N$, $\kappa_N = \left|\sqrt{2m(V_N - E)}/\hbar\right|$, there is only an exponentially decreasing wave function

$$\varphi_N(E, x) = A'_N \mathrm{e}^{-\kappa_N x} \tag{4.19}$$

in the region $x \geq x_{N-1}$.

The scattering of a left-moving wave packet incident from $+\infty$ is possible for $E \geq V_N$. Again, we have to distinguish two cases: $E \geq V_1 = 0$ and $E < V_1 = 0$.

i) For $E \geq V_1 = 0$, $k_1 = \left|\sqrt{2mE}/\hbar\right|$, there exists only an outgoing wave in the region $-\infty < x < x_1 = 0$:

$$\varphi_1(E,x) = B_1' \mathrm{e}^{-\mathrm{i}k_1 x} \quad . \tag{4.20}$$

ii) For $E < V_1 = 0$, $k_1 = \mathrm{i}\kappa_1$, $\kappa_1 = \left|\sqrt{2m|E|}/\hbar\right|$, there exists only a wave function

$$\varphi_1(E,x) = B_1' \mathrm{e}^{\kappa_1 x} \tag{4.21}$$

decreasing exponentially toward $-\infty$ in the region $-\infty < x < x_1 = 0$.

In the following discussion we shall restrict ourselves to right-moving incoming waves. For this scattering situation the boundary condition is given by (4.18) or (4.19) depending either on the relation $E \geq V_N$ or $E < V_N$.

4.1.6 Stationary Scattering Solutions in Step Potentials

The *stationary solutions* of the Schrödinger equation for a right-moving incoming wave incident on a step potential (3.16) with N regions is of the form

$$\begin{aligned}
\varphi_1(E,x) &= A_1' \mathrm{e}^{\mathrm{i}k_1 x} + B_1' \mathrm{e}^{-\mathrm{i}k_1 x} && \text{region 1} \\
\varphi_2(E,x) &= A_2' \mathrm{e}^{\mathrm{i}k_2 x} + B_2' \mathrm{e}^{-\mathrm{i}k_2 x} && \text{region 2} \\
\vdots \quad & \quad \vdots && \vdots \\
\varphi_N(E,x) &= A_N' \mathrm{e}^{\mathrm{i}k_N x} && \text{region } N
\end{aligned} \quad . \tag{4.22}$$

The $(2N-1)$ coefficients A_j', B_j' are again determined from the requirement of the wave function being continuous and continuously differentiable at the values x_m, $m = 1, \ldots, N-1$. This leads once more to the conditions (3.22) for $E \neq V_m, V_{m+1}$. For $E = V_m$ or $E = V_{m+1}$, (3.20) has to be used. Again this yields $2(N-1)$ linear algebraic equations, now, however, for $(2N-1)$ coefficients A_j', B_j'. Thus, for every value $E \geq V_0 = 0$, a number of $2(N-1)$ coefficients can be determined as functions of one of them. We single out the coefficient A_1' as the independent one. Its size determines the amplitude of the right-moving wave coming in from $-\infty$. Thus, it regulates the strength of the incoming current. It will either be fixed in (4.25) below by a normalization, or simply be set to one.

Because for any real value $E \geq V_0 = 0$ of the energy we find a stationary solution in the step potential, all values $E \geq V_0$ form the continuous spectrum of the Hamiltonian. All corresponding stationary solutions $\varphi(E,x)$ are continuum eigenfunctions with right-moving outgoing waves. There is a further set of eigenfunctions of $E \geq V_N$ for scattering processes where the incoming particles move in from $+\infty$ which we do not further discuss.

4.1.7 Constituent Waves

The pieces $\varphi_j(E,x)$ in region j $(x_{j-1} \le x < x_j)$ of the stationary wave function $\varphi(E,x)$ consist of a *right-moving* and a *left-moving constituent wave*

$$\varphi_{j+}(E,x) = A'_j \mathrm{e}^{\mathrm{i}k_j x} \quad \text{and} \quad \varphi_{j-}(E,x) = B'_j \mathrm{e}^{-\mathrm{i}k_j x} \quad , \tag{4.23}$$

if $E > V_j$. For $E < V_j$ the wave number becomes imaginary: $k_j = \mathrm{i}\kappa_j$, κ_j real. In this case,

$$\varphi_{j+}(E,x) = A'_j \mathrm{e}^{-\kappa_j x} \quad \text{and} \quad \varphi_{j-}(E,x) = B'_j \mathrm{e}^{\kappa_j x}$$

represent a decreasing and an increasing exponential, respectively,

x : position variable,
$k_j = \mathrm{i}\kappa_j$, $\kappa_j = \left|\sqrt{2m(V_j - E)}/\hbar\right|$: wave vector for $E < V_j$,
A'_j, B'_j : complex amplitudes.

4.1.8 Normalization of Continuum Eigenstates

As all eigenvectors of Hermitian operators the continuum eigenfunctions belonging to different eigenvalues E, E' are orthogonal,

$$\int_{-\infty}^{+\infty} \varphi^*(E',x)\varphi(E,x)\,\mathrm{d}x = 0 \quad . \tag{4.24}$$

The normalization condition for continuum eigenfunctions for $E = E'$ is in analogy to (4.13) given by $(E = p^2/2m,\ E' = p'^2/2m)$

$$\int_{-\infty}^{+\infty} \varphi^*(E',x)\varphi(E,x)\,\mathrm{d}x = \delta(p'-p) \quad . \tag{4.25}$$

Again this ensures that the normalization of a right-moving wave packet is unity if the spectral function $w(p)$ is normalized as in (4.14). It is the normalization (4.25) that fixes the independent coefficient A'_1 in the stationary scattering solution (4.22).

4.1.9 Harmonic Waves in a Step Potential

The time-dependent waves

$$\psi(E,x,t) = \mathrm{e}^{-\mathrm{i}Et/\hbar}\varphi(E,x) \quad , \tag{4.26}$$

$\varphi(E,x)$: right-moving incident wave function (4.22),
$E = p^2/2m$: energy eigenvalue,
x: position,
t: time,

are solutions of the time-dependent Schrödinger equation. In the regions j with $E > V_j$ they are harmonic waves. Also the time-dependent stationary waves can be decomposed into time-dependent right-moving and left-moving constituent waves,

$$\begin{aligned} \psi_{j+}(E,x,t) &= \mathrm{e}^{-\mathrm{i}Et/\hbar}\varphi_{j+}(E,x) \quad , \\ \psi_{j-}(E,x,t) &= \mathrm{e}^{-\mathrm{i}Et/\hbar}\varphi_{j-}(E,x) \quad . \end{aligned} \tag{4.27}$$

4.1.10 Time-Dependent Scattering Solutions in a Step Potential

If we want to describe a particle coming in from the left by a right-moving Gaussian wave packet of spatial width σ_x, as in classical mechanics, we have to set at the initial time $t = 0$ its position to x_0 and its average momentum to p_0. This is accomplished with the time-dependent superposition

$$\psi(x,t) = \int w(p)\mathrm{e}^{-\mathrm{i}Et/\hbar}\varphi(E(p),x)\,\mathrm{d}p \tag{4.28}$$

of the continuum eigenfunctions $\varphi(E,x)$ with the Gaussian spectral function

$$w(p) = \frac{1}{(2\pi)^{1/4}\sqrt{\sigma_p}}\exp\left\{-\frac{(p-p_0)^2}{4\sigma_p^2} - \mathrm{i}px_0/\hbar\right\} \quad , \tag{4.29}$$

$E = p^2/2m$: energy,
p: momentum,
x: position,
t: time,
p_0: momentum expectation value of incident wave packet,
x_0: position expectation value of incident wave packet,
$\sigma_p = \hbar/2\sigma_{x0}$ momentum width of incident wave packet,
σ_{x0}: spatial width of initial wave packet,
$\varphi(E,x)$: right-moving stationary scattering wave.

The constituent waves $\psi_{j+}(E,x,t)$, $\psi_{j-}(E,x,t)$ of the wave packet can be formed with the stationary constituent waves

$$\psi_{j\pm}(x,t) = \int w(p)\mathrm{e}^{-\mathrm{i}Et/\hbar}\varphi_{j\pm}(E(p),x)\,\mathrm{d}p \quad , \tag{4.30}$$

which are right-moving or left-moving.

4.1.11 Generalization to Piecewise Linear Potentials

In analogy to the step potential in Sect. 4.1.6 the *stationary solutions* of the Schrödinger equation for a right-moving incoming wave incident on a piecewise linear potential (3.24) with N regions is given by

$$
\begin{aligned}
\varphi_1(E,x) &= A_1' e^{ik_1x} + B_1' e^{-ik_1x} && \text{region 1}\\
\varphi_2(E,x) &= A_2' \operatorname{Ai}(\alpha_2(x - x_{E2}^{\mathrm{T}})) + B_2' \operatorname{Bi}(\alpha_2(x - x_{E2}^{\mathrm{T}})) && \text{region 2}\\
&\vdots && \vdots\\
\varphi_{N-1}(E,x) &= A_{N-1}' \operatorname{Ai}(\alpha_{N-1}(x - x_{E\,N-1}^{\mathrm{T}}))\\
&\quad + B_{N-1}' \operatorname{Bi}(\alpha_{N-1}(x - x_{E\,N-1}^{\mathrm{T}})) && \text{region } N-1\\
\varphi_N(E,x) &= A_N' e^{ik_Nx} && \text{region } N
\end{aligned}
\tag{4.31}
$$

Here the solutions in region m are written according to Sect. 3.1.6 with the scale factor α_m and the (extrapolated) classical turning point x_{Em}^{T}, in this case referring to the continuous energy E. The $(2N-1)$ coefficients A_j', B_j' are determined in analogy to Sect. 4.1.6. (The solutions in region m assume the special from of that section for a zero potential slope.) As in Sect. 4.1.9 the time-dependent solution is

$$
\psi(E,x,t) = e^{-iEt/\hbar}\varphi(E,x) \quad . \tag{4.32}
$$

Such solutions can be superimposed to form a wave packet as in Sect. 4.1.10.

For the case of piecewise linear potentials we restrict ourself to discuss the full solution (4.31) only.

The continuum normalization is done according to Sect. 4.1.8; it is determined by the solution for the potentials in the outermost regions, which are the same for both cases. Harmonic waves (see Sect. 4.1.9) and their superposition to a wave packet (see Sect. 4.1.10) for the full solution can be constructed in exactly the same way as described there.

4.1.12 Transmission and Reflection. Unitarity. The Argand Diagram

For $E \geq V_N$, i.e., $k_N = \left|\sqrt{2m(E-V_N)}/\hbar\right|$, the solution (4.22) is interpreted in the following way:

i) $A_1' e^{ik_1x}$ is the right-moving harmonic wave coming in from $-\infty$.
ii) $A_N' e^{ik_Nx}$ is the transmitted wave. It is right moving, going out to $+\infty$.
iii) $B_1' e^{-ik_1x}$ is the reflected wave. It is left moving, going out to $-\infty$.

For $E < V_N$, i.e., $k_N = i\kappa_N$, $\kappa_N = \left|\sqrt{2m(V_N-E)}/\hbar\right|$, the solution (4.22) contains the term

$$
A_N' e^{ik_Nx} = A_N' e^{-\kappa_Nx} \quad , \tag{4.33}
$$

which represents an exponentially decreasing function in the region N. It approaches zero for $x \to +\infty$. Thus, there is no transmission for $E < V_N$. The incoming wave $A_1' e^{ik_1x}$ is *totally reflected* to produce the left-moving reflected wave $B_1' e^{-ik_1x}$, which goes out to $-\infty$.

For $E > V_N$ the complex functions $A_N = \sqrt{k_1/k_N} A'_N(E)$ and $B_1 = B'_1(E)$ are called the *transmission* and *reflection coefficients*, respectively. Their normalization is fixed by setting the independent coefficient $A_1 = A'_1 = 1$. They depend on the energy E of the incoming wave and fulfill the *unitarity relation*

$$|A_N|^2 + |B_1|^2 = |A_1|^2 = 1 \quad . \tag{4.34}$$

This relation states that, for varying energy E, the complex quantities $A_N(E)$ and $B_1(E)$ move inside a circle of radius 1 around the origin in the complex plane. This representation of the coefficients A_N, B_1 in the complex plane is known as the *Argand diagram.* The coefficients A_N and B_1 are also called the *scattering-matrix elements* or *S-matrix elements* of transmission and reflection, respectively. Accordingly, (4.34) is called a unitarity relation of the S matrix. A detailed discussion of the physical interpretation of the prominent features of the Argand diagram, e.g., in relation to resonances, is given in Sect. 12.2.

For $E < V_N$ the complex function $B_1(E)$ again is the reflection coefficient. There is no transmission of a wave that goes out to infinity. For the normalization $A'_1 = 1$ the reflection coefficient fulfills for $E < V_N$ the unitarity relation

$$|B_1|^2 = |A_1|^2 = 1 \quad . \tag{4.35}$$

Thus, for $E < V_N$, the complex reflection coefficient $B_1(E)$ moves for varying energy E on the unit circle in the complex plane.

Related quantities are the transition-matrix elements or *T-matrix elements* T_T, T_R of transmission and reflection,

$$T_\mathrm{T}(E) = (A_N(E) - 1)/2\mathrm{i} \quad \text{and} \quad T_\mathrm{R}(E) = B_1(E)/2\mathrm{i} \quad . \tag{4.36}$$

The T-matrix elements fulfill the T-matrix unitarity relation

$$\operatorname{Im} T_\mathrm{T} = |T_\mathrm{T}|^2 + |T_\mathrm{R}|^2 \quad . \tag{4.37}$$

This states that the element $T_\mathrm{T}(E)$ moves for varying energy E inside a circle of radius $1/2$ with its center at the point $\mathrm{i}/2$ in the complex plane. The element $T_\mathrm{R}(E)$ varies inside the circle of radius $1/2$ around the origin of the complex plane.

4.1.13 The Tunnel Effect

We consider a simple potential with three regions,

$$V(x) = \begin{cases} 0 \ , x < x_1 = 0 & \text{region 1} \\ V_0 \ , 0 \le x < d & \text{region 2} \\ 0 \ , d \le x & \text{region 3} \end{cases} \quad , \tag{4.38}$$

x: position coordinate,
d: width of potential,
$V_0 > 0$: potential height.

For energies $0 < E < V_0$ a classical particle will be reflected. Quantum mechanics allows a nonvanishing transmission coefficient A_3,

$$|A_3|^2 = \frac{4E(V_0 - E)}{4E(V_0 - E) + V_0^2 \sinh^2 \kappa d} \quad , \quad \kappa = \left| \sqrt{2m(V_0 - E)}/\hbar \right| \quad . \tag{4.39}$$

Thus, there is a nonvanishing probability of the particle being transmitted from region 1 into the classically forbidden region 3, if $E < V_0$. This phenomenon is called the *tunnel effect*.

For general potentials, the tunnel effect means that penetration through a repulsive wall is possible if the incident energy is larger than the potential on the other side of the wall.

4.1.14 Resonances

In a step potential or in a piecewise linear potential with N regions and $V_1 = 0$, transmission is possible for positive energies if $E > V_N$. The transmission coefficient A_N varies with the energy E of the incident particle.

The maxima of the absolute square $|A_N|^2$ of the transmission coefficient are called *transmission resonances*. The energies at which these maxima occur are the *resonance energies*. Because of the unitarity relation (4.34) the absolute square $|B_1|^2$ of the reflection coefficient exhibits a minimum at the resonance energy of transmission. Therefore, in a plot of the energy dependence of the absolute square of the wave function, transmission resonances can be recognized at energies where the interference pattern of the incoming and reflected wave in the region 1 is least prominent or absent.

4.1.15 Phase Shifts upon Reflection at a Steep Rise or Deep Fall of the Potential

We study the reflection and transmission in two adjacent regions l and $l + 1$ with large differences in the values V_l, V_{l+1} of the potentials:

$$V(x) = \begin{cases} V_l & , \quad x_{l-1} \le x < x_l \text{ region } l \\ V_{l+1} , & x_l \le x < x_{l+1} \text{ region } l+1 \end{cases} \quad . \tag{4.40}$$

A particle with the kinetic energy E is incident on the potential step at $x = x_l$. The wave function in the regions l and $l + 1$ is given by

$$\begin{aligned} \varphi_l(E, x) &= A_l' \mathrm{e}^{\mathrm{i}k_l x} + B_l' \mathrm{e}^{-\mathrm{i}k_l x} \quad , \\ \varphi_{l+1}(E, x) &= A_{l+1}' \mathrm{e}^{\mathrm{i}k_{l+1} x} + B_{l+1}' \mathrm{e}^{-\mathrm{i}k_{l+1} x} \quad . \end{aligned} \tag{4.41}$$

The continuity conditions to be satisfied at $x = x_l$ are

$$A'_l e^{ik_l x_l} + B'_l e^{-ik_l x_l} = A'_{l+1} e^{ik_{l+1} x_l} + B'_{l+1} e^{-ik_{l+1} x_l} \quad ,$$
$$A'_l e^{ik_l x_l} - B'_l e^{-ik_l x_l} = \frac{k_{l+1}}{k_l} (A'_{l+1} e^{ik_{l+1} x_l} - B'_{l+1} e^{-ik_{l+1} x_l}) \quad . \qquad (4.42)$$

This leads to the solutions

$$A'_l e^{ik_l x_l} = \frac{1}{2}\left(1 + \frac{k_{l+1}}{k_l}\right) A'_{l+1} e^{ik_{l+1} x_l} + \frac{1}{2}\left(1 - \frac{k_{l+1}}{k_l}\right) B'_{l+1} e^{-ik_{l+1} x_l} \quad ,$$
$$B'_l e^{-ik_l x_l} = \frac{1}{2}\left(1 - \frac{k_{l+1}}{k_l}\right) A'_{l+1} e^{ik_{l+1} x_l} + \frac{1}{2}\left(1 + \frac{k_{l+1}}{k_l}\right) B'_{l+1} e^{-ik_{l+1} x_l} \quad .$$
$$(4.43)$$

i) Reflection and transmission at a sudden increase in potential energy ($V_l \ll V_{l+1}$).
For a particle with kinetic energy $E \geq V_{l+1}$ closely above the potential value V_{l+1} in the region $(l+1)$, the quotient of the wave numbers satisfies $k_{l+1}/k_l \ll 1$. In this case (4.43) yields

$$B'_l e^{-ik_l x_l} \approx A'_l e^{ik_l x_l} \quad \text{for} \quad E \geq V_{l+1} \quad . \qquad (4.44)$$

We conclude that the reflected wave, i.e., the left-moving constituent wave in region l,

$$\varphi_{l-}(E, x) = B'_l e^{-ik_l x} \quad , \qquad (4.45)$$

does not show a *phase shift* compared to the incident wave, i.e., to the right-moving constituent wave in this region,

$$\varphi_{l+}(E, x) = A'_l e^{ik_l x} \quad . \qquad (4.46)$$

The analogous situation in optics is the reflection of light on an optically thinner medium, which does not exhibit a phase shift either. The analogy rests on the relation of the wave numbers k_l and k_{l+1} in the two adjacent regions. In optics and in quantum mechanics reflection on a 'thinner medium' requires $k_l > k_{l+1}$. Actually, to obtain a vanishing phase shift in quantum mechanics the relation has to be stronger, i.e., $k_l \gg k_{l+1}$.

ii) Reflection and transmission at a sudden decrease in potential energy ($V_l \gg V_{l+1}$).
For a particle of kinetic energy $E \geq V_l$ close above the potential value V_l in the region l, the quotient of wave numbers satisfies $k_{l+1}/k_l \gg 1$. For kinetic energies E slightly larger than V_l, (4.43) then leads to the relation

$$B'_l e^{-ik_l x_l} \approx -A'_l e^{ik_l x_l} \quad \text{for} \quad E \geq V_l \quad , \qquad (4.47)$$

which is tantamount to a *phase shift* of π between the reflected wave, i.e., the left-moving constituent wave in region l,

$$\varphi_{l-}(E,x) = B'_l \mathrm{e}^{-\mathrm{i}k_l x} \approx -A'_l \mathrm{e}^{2\mathrm{i}k_l x_l - \mathrm{i}k_l x} = A'_l \mathrm{e}^{-\mathrm{i}(k_l x - 2k_l x_l + \pi)} \quad , \tag{4.48}$$

and the incident or right-moving constituent wave in this region,

$$\varphi_{l+}(E,x) = A'_l \mathrm{e}^{\mathrm{i}k_l x} \quad . \tag{4.49}$$

This corresponds to the reflection of light on an optically denser medium (Sect. 4.12, 'Analogies in Optics'). Both the quantum-mechanical and the optical situation are characterized by $k_l < k_{l+1}$. In quantum mechanics the phase shift upon reflection on a 'denser medium' approaches the value π for the limiting case $k_l \ll k_{l+1}$ only.

iii) Reflection at a high potential step.
A particle with a kinetic energy E satisfying $V_l < E \ll V_{l+1}$ is only reflected at $x = x_l$; there is vanishing transmission, i.e., $A'_{l+1} = B'_{l+1} = 0$. In region $(l+1)$ the wave number is imaginary, $k_{l+1} = \mathrm{i}\kappa_{l+1}$, and, furthermore, $\kappa_{l+1}/k_l \gg 1$. This leads to the relation

$$B'_l \mathrm{e}^{-\mathrm{i}k_l x_l} \approx -A'_l \mathrm{e}^{\mathrm{i}k_l x_l} \quad \text{for} \quad V_l < E \ll V_{l+1} \quad . \tag{4.50}$$

As under ii), we conclude that the reflected wave in region l suffers a phase shift of π. This situation is analogous to the reflection at a fixed end.

4.1.16 Transmission Resonances upon Reflection at 'More- and Less-Dense Media'

We investigate a simple repulsive potential of three regions,

$$V(x) = \begin{cases} 0 & x < x_1 = 0 \quad \text{region 1} \\ V_0 > 0 & 0 \le x < x_2 = d \quad \text{region 2} \\ 0 & x_2 \le x \quad \text{region 3} \end{cases} \quad . \tag{4.51}$$

A particle of kinetic energy E slightly larger than V_0 is incident on this potential from the left. Reflection occurs at $x = 0$ and $x = d$. At $x = 0$ the reflection occurs as in optics on a 'thinner medium'; thus, the reflected wave in region 1 suffers no phase shift. At $x = d$ reflection occurs on a 'denser medium' and thus with a phase shift of π for the reflected wave in region 2. The left-moving constituent wave $\varphi_{1-}(E,x)$ in region 1 can be thought of as consisting of two parts interfering with each other: the one reflected at $x = 0$ on a thinner medium and the other reflected at $x = d$ on a denser medium and transmitted into region 1 at $x = 0$. The phase difference of the two parts consists of the phase shift π upon reflection on the denser medium at $x = d$ and the phase shift due to the longer path $k_2(2d)$ of the wave in region 2. The *total phase shift* amounts to

$$\delta = 2k_2 d + \pi \quad . \tag{4.52}$$

For destructive interference of the two parts making up the reflected, i.e., left-moving, constituent wave $\varphi_{1-}(E, x)$ in region 1, this phase difference has to be equal to an odd multiple of π. Thus, a transmission resonance for the potential (4.51) under the condition $E - V_0 \ll V_0$ occurs if

$$2k_2 d + \pi = (2l + 1)\pi \quad \text{or} \quad k_2 = l\pi/d \quad \text{for} \quad l = 1, 2, 3, \ldots \quad . \tag{4.53}$$

For the corresponding wavelength we find

$$\lambda_2 = 2\pi/k_2 = 2d/l \quad , \quad l = 1, 2, 3, \ldots \quad , \tag{4.54}$$

i.e., whenever the wavelength in region 2 is an integer fraction of twice the width of the step potential, transmission is at a maximum. The largest wavelength for which this happens is just twice the width of the potential region. It should be remembered, however, that the validity of the simple formula (4.53) hinges on the condition $k_2 \ll k_1$ at resonance energy E_l, i.e.,

$$l\hbar\frac{\pi}{d} \ll \sqrt{2mE_l} \quad . \tag{4.55}$$

Under this condition the resonance energies of the kinetic energy of the incident particles are

$$E_l = V_0 + l^2 \frac{1}{2m}\left(\frac{\hbar\pi}{d}\right)^2 \quad . \tag{4.56}$$

The spacing of the resonances increases like l^2 for not too large values of the integer l.

4.1.17 The Quantum-Well Device and the Quantum-Effect Device

Two developments in circuit elements based on the tunnel effect are the quantum-well device and the quantum-effect device. For an introductory article we refer the reader to R. T. Bates "Quantum-Effect Device: Tomorrow's Transistor?" in Scientific American Vol. 258, No. 3, p. 78 (March 1988).

A *quantum-well device* (QWD) with one-dimensional confinement is an arrangement of five layers of material, Fig. 4.1a. The two outer layers are n-doped gallium arsenide, GaAs. The two slices to the left and right of the middle layer are made of aluminum gallium arsenide, AlGaAs. The middle slice is gallium arsenide, GaAs. The band structure of AlGaAs is such that no classical electron current flowing in the outer n-doped GaAs can pass it. The middle layer acts like a potential well between the two AlGaAs layers, which act like two barriers. Thus, the one-dimensional potential representing the quantum-well device possesses five regions with $V_1 = 0,\ V_2 > 0,\ V_3 < V_2,\ V_4 = V_2,\ V_5 = 0$, Fig. 4.1b.

The electrons in the first region, usually called the emitter, can be transmitted into the fifth region, the collector, only if their initial energy in region 1 matches a resonance energy in the well. In this case the tunnel effect through the barrier (region 2) into the well (region 3) and from here through the second barrier (region 4) into region 5 leads to a sizable transmission coefficient. The adaptation of the resonance energy in the well can be facilitated by connecting the material in regions 1 and 5 to a battery. By varying the voltage between emitter and collector, Fig. 4.1c, the potential can be changed and thus the resonance energy. This effect can be used to steer the current through the quantum-well device.

Another possible way to influence the current is to connect a third electrical contact (base) to the middle layer (region 3) of the quantum-well device. This contact can be used to change the potential V_3 in the well for a fixed voltage between emitter and collector. A circuit element of this kind is called a *quantum-effect device*.

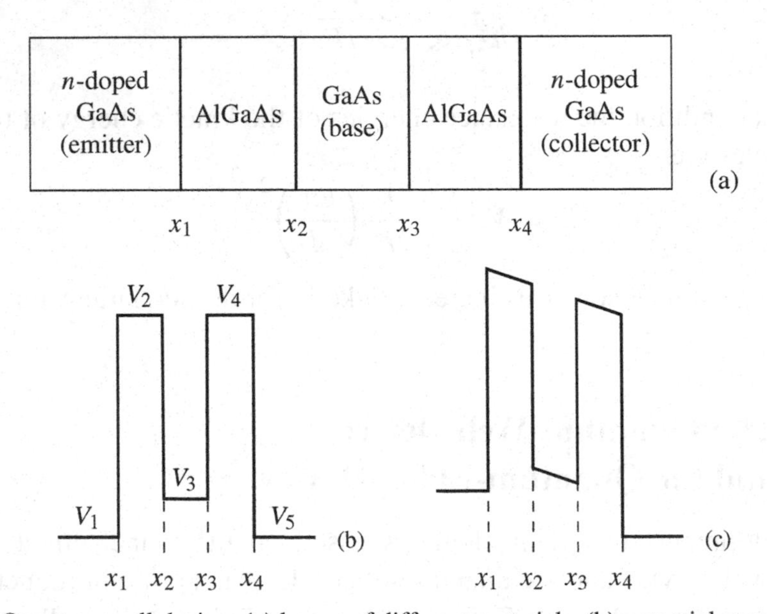

Fig. 4.1. Quantum-well device: (a) layers of different materials, (b) potential at zero voltage, (c) potential with voltage between emitter and collector

4.1.18 Stationary States in a Linear Potential

We consider the potential

$$V(x) = -mgx \tag{4.57}$$

corresponding to a *constant force* $F = mg$. The stationary Schrödinger equation reads

$$\left(-\frac{\hbar^2}{2m}\frac{\mathrm{d}^2}{\mathrm{d}x^2} - mgx\right)\varphi(x) = E\varphi(x) \quad . \tag{4.58}$$

Introducing the *classical turning point*

$$x_{\mathrm{T}} = -E/mg$$

of a particle with total energy E and the dimensionless variable

$$\xi = \frac{x - x_{\mathrm{T}}}{\ell_0} \quad , \quad \ell_0 = \left(\frac{\hbar^2}{2m^2 g}\right)^{1/3} \quad ,$$

we give the Schrödinger equation the form

$$\left(\frac{\mathrm{d}^2}{\mathrm{d}\xi^2} + \xi\right)\phi(\xi) = 0 \quad , \quad \phi(\xi) = \varphi(\ell_0\xi + x_{\mathrm{T}}) \quad .$$

It is solved by the *Airy function* $\mathrm{Ai}(\xi)$, Sect. 11.1.7, multiplied by a normalization constant:

$$\phi(\xi) = N\mathrm{Ai}(-\xi) \quad , \quad N = \left(\frac{2m^{1/2}}{g^{1/2}\hbar^2}\right)^{1/3} \quad .$$

Returning to the stationary wave function we find

$$\varphi(x) = N\mathrm{Ai}\left(-\frac{x - x_{\mathrm{T}}}{\ell_0}\right) \quad . \tag{4.59}$$

Note that the wave function $\varphi(x)$ is a real function of x.

4.1.19 Wave Packet in a Linear Potential

Also in the linear potential one can have a Gaussian wave packet with the probability density

$$\varrho(x,t) = |\psi(x,t)|^2 = \frac{1}{\sqrt{2\pi}\sigma_x(t)}\exp\left[-\frac{(x - \langle x(t)\rangle)^2}{2\sigma_x^2(t)}\right] \quad , \tag{4.60}$$

familiar from the free Gaussian wave packet, Sect. 2.1.2, and from the Gaussian wave packet in the harmonic-oscillator potential, Sect. 3.1.8. We denote by $x_0 = \langle x(0)\rangle$ the *initial mean position*, by $p_0 = \langle p(0)\rangle$ the *initial mean momentum*, and by $\sigma_{x0} = \sigma_x(0)$ the *initial spatial width*. The time-dependent mean

$$\langle x(t)\rangle = x_0 + \frac{p_0}{m}t + \frac{g}{2}t^2 \tag{4.61}$$

is identical to the position $x(t)$ of a classical particle with the initial conditions x_0, p_0. The time dependence of the spatial width $\sigma_x(t)$ is the same as for the free wave packet,

$$\sigma_x^2(t) = \sigma_{x0}^2\left(1 + \frac{\hbar^2}{4\sigma_{x0}^4}\frac{t^2}{m^2}\right) = \sigma_{x0}^2\left(1 + \frac{4\sigma_{p0}^4}{\hbar^2}\frac{t^2}{m^2}\right) \quad . \tag{4.62}$$

Remember that for $t = 0$ the widths in position and momentum are related by $\sigma_{x0}\sigma_{p0} = \hbar/2$.

4.1.20 Quantile Motion in a Linear Potential

Equation (4.60) is identical to (2.13). Only the time dependence of the mean $\langle x(t)\rangle$ is different. Keeping this difference in mind all results of Sects. 2.1.3 and 2.1.4 about the probability-current density $j(x,t)$ and the quantile trajectories $x_P = x_P(t)$ remain valid for a Gaussian wave packet in a linear potential.

4.1.21 Classical Phase-Space Density in a Linear Potential

In Sects. 2.1.7 and 3.1.10 we saw that, for a free particle and a particle under the influence of a harmonic force, there is a close analogy between the quantum-mechanical treatment as a wave packet and the classical description as a phase-space probability density. Here we show that this analogy also holds for a constant force, i.e., the linear potential (4.57). In this case the classical equations of motion for a point (x, p) in phase space are $x = x_\mathrm{i} + (p_\mathrm{i}/m)t + (g/2)t^2$, $p = p_\mathrm{i} + gt$. We proceed as in Sect. 2.1.7 by solving the equations of motion for the initial position and momentum x_i, p_i at $t = 0$,

$$x_\mathrm{i} = x - (p_\mathrm{i}/m)t - (g/2)t^2 \quad , \qquad p_\mathrm{i} = p - gt \quad , \tag{4.63}$$

and by inserting them into (2.38) describing an uncorrelated Gaussian phase-space probability density at $t = 0$. The resulting time-dependent density is also Gaussian. Its expectation values are

$$\langle x(t)\rangle = x_0 + \frac{p_0}{m}t + \frac{g}{2}t^2 \quad , \qquad \langle p(t)\rangle = p_0 + gt \quad . \tag{4.64}$$

The widths and the correlation coefficients are identical to those of the force-free case, they are given by (2.42) and (2.43), respectively. As in that case we conclude: The classical considerations performed here and the quantum-mechanical ones of Sect. 4.1.19 yield identical results, provided the initial classical phase-space probability density fulfills the minimum uncertainty condition $\sigma_{x0}\sigma_{p0} = \hbar/2$ of quantum mechanics.

4.1.22 Classical Phase-Space Density Reflected by a High Potential Wall

The close analogy between the behaviors of a quantum-mechanical wave packet and a classical phase-space probability density holds for potentials that are constant, linear, or quadratic in x. As a drastic counterexample we study the behavior of a classical phase-space density under the influence of a very high potential step at $x = 0$. We assume that initially (at $t = 0$) the distribution is concentrated far left of the step and describe it by a Gaussian as given in Sect. 2.1.7, characterized by the initial position and momentum expectation values x_0, p_0 and their uncertainties σ_{x0}, σ_p,

$$\varrho_+^{\mathrm{cl}}(x,p,t) = \frac{1}{2\pi\sigma_{x0}\sigma_p}\exp\left\{-\frac{1}{2(1-c^2)}\left[\frac{(x-[x_0+p_0t/m])^2}{\sigma_x^2(t)}\right.\right.$$
$$\left.\left.-2c\frac{(x-[x_0+p_0t/m])}{\sigma_x(t)}\frac{(p-p_0)}{\sigma_p}+\frac{(p-p_0)^2}{\sigma_p^2}\right]\right\} .$$

Its spatial width σ_x and correlation coefficient c,

$$\sigma_x(t) = \sqrt{\sigma_{x0}^2+\sigma_p^2t^2/m^2} \quad , \qquad c = \frac{\sigma_p t}{\sigma_x(t)m} \quad ,$$

are time dependent. This distribution, traveling to the right, is a valid description of our problem as long as it stays well left of the step, i.e., as long as $x_0+p_0t/m \ll -\sigma_x(t)$. For much larger times, for which $x_0+p_0t/m \gg \sigma_x(t)$, complete reflection has taken place: the phase-space density moves towards the left and behaves just as if it would have started at $t = 0$ with the expectation values $(-x_0, -p_0)$, i.e., it is described by

$$\varrho_-^{\mathrm{cl}}(x,p,t) = \frac{1}{2\pi\sigma_{x0}\sigma_p}\exp\left\{-\frac{1}{2(1-c^2)}\left[\frac{(x-[-x_0-p_0t/m])^2}{\sigma_x^2(t)}\right.\right.$$
$$\left.\left.-2c\frac{(x-[-x_0-p_0t/m])}{\sigma_x(t)}\frac{(p+p_0)}{\sigma_p}+\frac{(p+p_0)^2}{\sigma_p^2}\right]\right\} .$$

For all times we can then describe the phase-space distribution under the action of a reflecting force at $x = 0$ by the sum of $\varrho_+^{\rm cl}$ and $\varrho_-^{\rm cl}$ left of the step (vanishing, of course, to the right of it),

$$\varrho^{\rm cl}(x, p, t) = \begin{cases} \varrho_+^{\rm cl}(x, p, t) + \varrho_-^{\rm cl}(x, p, t) \quad , & x < 0 \quad , \\ 0 \quad , & x > 0 \quad . \end{cases}$$

The marginal distribution in x,

$$\begin{aligned} \varrho_x^{\rm cl}(x, t) &= \varrho_{x+}^{\rm cl}(x, t) + \varrho_{x-}^{\rm cl}(x, t) \\ &= \frac{1}{\sqrt{2\pi}\sigma_x(t)} \exp\left\{-\frac{(x - [x_0 + p_0 t/m])^2}{2\sigma_x^2(t)}\right\} \\ &\quad + \frac{1}{\sqrt{2\pi}\sigma_x(t)} \exp\left\{-\frac{(x - [-x_0 - p_0 t/m])^2}{2\sigma_x^2(t)}\right\} \quad , \end{aligned}$$

is simply the sum of the marginal distributions of $\varrho_+^{\rm cl}$ and $\varrho_-^{\rm cl}$ for $x < 0$ and vanishes for $x > 0$. It is the classical spatial probability density. For times long before or long after the reflection process it is identical to the quantum-mechanical probability density. During the period of reflection, however, $t \approx m|x_0|/p_0$, the quantum-mechanical probability density $\varrho(x, t) = |\psi(x, t)|^2$ shows the typical interference pattern, whereas the classical density $\varrho_x^{\rm cl}(x, t)$ is smooth. This striking difference is due to the fact that in the quantum-mechanical calculation the wave functions $\psi_{1+}(x, t)$ and $\psi_{1-}(x, t)$ are added and the absolute square of the sum is taken to form $\varrho(x, t)$ whereas in the classical calculation the marginal densities $\varrho_{x+}^{\rm cl}(x, t)$ and $\varrho_{x-}^{\rm cl}(x, t)$ of the constituent phase-space distributions are added directly.

Further Reading

Alonso, Finn: Vol. 3, Chap. 2
Berkeley Physics Course: Vol. 4, Chaps. 7, 8
Brandt, Dahmen: Chaps. 4, 5, 7
Feynman, Leighton, Sands: Vol. 3, Chaps. 9, 16
Flügge: Vol. 1, Chap. 2A
Gasiorowicz: Chap. 4
Merzbacher: Chaps. 3, 4, 5, 6
Messiah: Vol. 1, Chaps. 2, 3
Schiff: Chaps. 2, 3, 4

4.2 Stationary Scattering States in the Step Potential and in the Piecewise Linear Potential

Aim of this section: Computation and demonstration of the stationary solution $\varphi(E, x)$ of the Schrödinger equation for a right-moving incoming wave in a step potential, (4.22), or in a piecewise linear potential, (4.31), as a function of position x and energy E or momentum p. For the step potential the stationary constituent solutions $\varphi_{j\pm}(E, x)$, (4.23), can also be shown.

A plot similar to Fig. 4.2 or Fig. 4.3 is produced. It contains graphical representations of

- the step or piecewise linear potential $V(x)$ as long-stroke dashed line,
- the total energy E as short-stroke dashed line,
- the stationary wave function $\varphi(E, x)$, displayed as $|\varphi(E, x)|^2$, $\mathrm{Re}\,\varphi(E, x)$, or $\mathrm{Im}\,\varphi(E, x)$ for which the short-stroke line serves as zero line.

Technically, the plot is of the type *surface over Cartesian grid in 3D*. The number of energies and wave functions shown is given by **n_y** on the subpanel **Graphics—Accuracy**.

On the bottom of the subpanel **Physics—Comp. Coord.** you can choose to show either the **Absolute Square** or the **Real Part** or the **Imaginary Part** of the wave function.

On the subpanel **Physics—Wave**, for the case of step potentials, you can choose to show the solution valid in the full x range or the right-moving or the left-moving constituent wave valid only in one of the potential regions. For all cases you can choose whether the y coordinate represents momentum or energy and you can set the position x_0 for which the phase of the incident wave vanishes. Here you may also choose whether the **y Coordinate represents** energy (as in Fig. 4.2) or momentum.

On the subpanel **Physics—Misc.** you find under **Constants** the numerical values of $\hbar$ and m, under **Graphical Item** a variable ℓ_{DASH} that determines the dash lengths used for the graphical representations of potential and energy, and under **Scale Factor** the factor s used to scale the graphical representation of the wave function.

On the subpanel **Physics—Potential** you can choose as **Type of Potential** either the **Step Potential** or the **Piecewise Linear Potential**. On that panel you also find all the parameters determining the potential, the number N of regions, the boundaries x_i between region i and region $i + 1$, and the potentials V_i. Note:

- For the step potential V_i is the potential in region i.

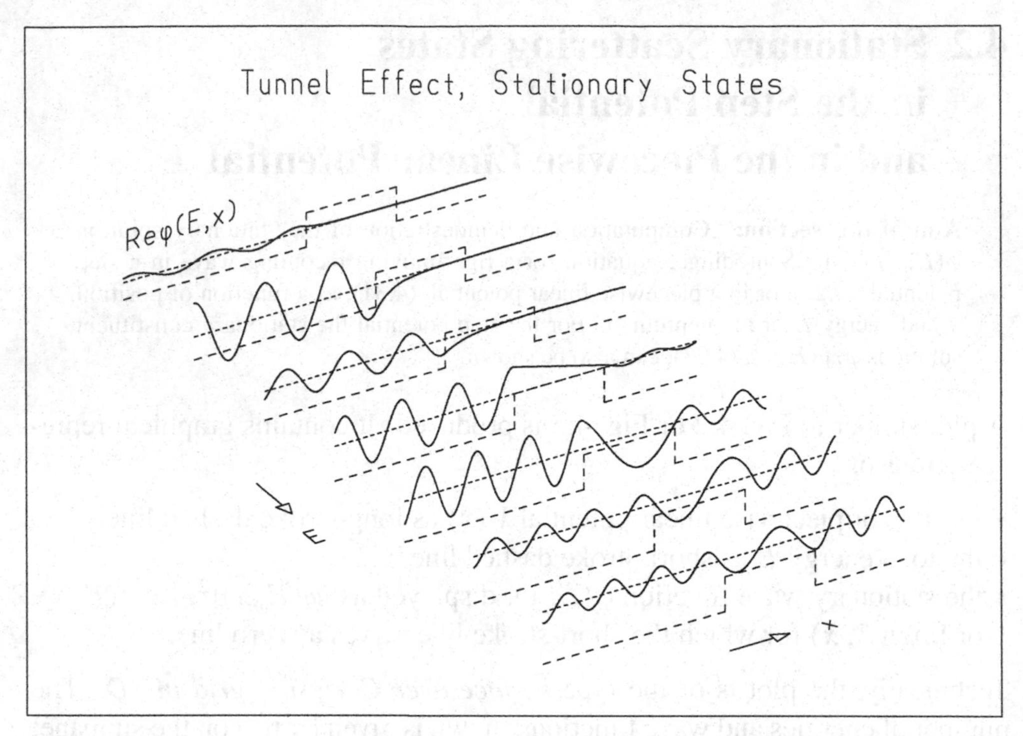

Fig. 4.2. Plot produced with descriptor `Step pot.: stationary scattering` on file `1D_Scattering.des`

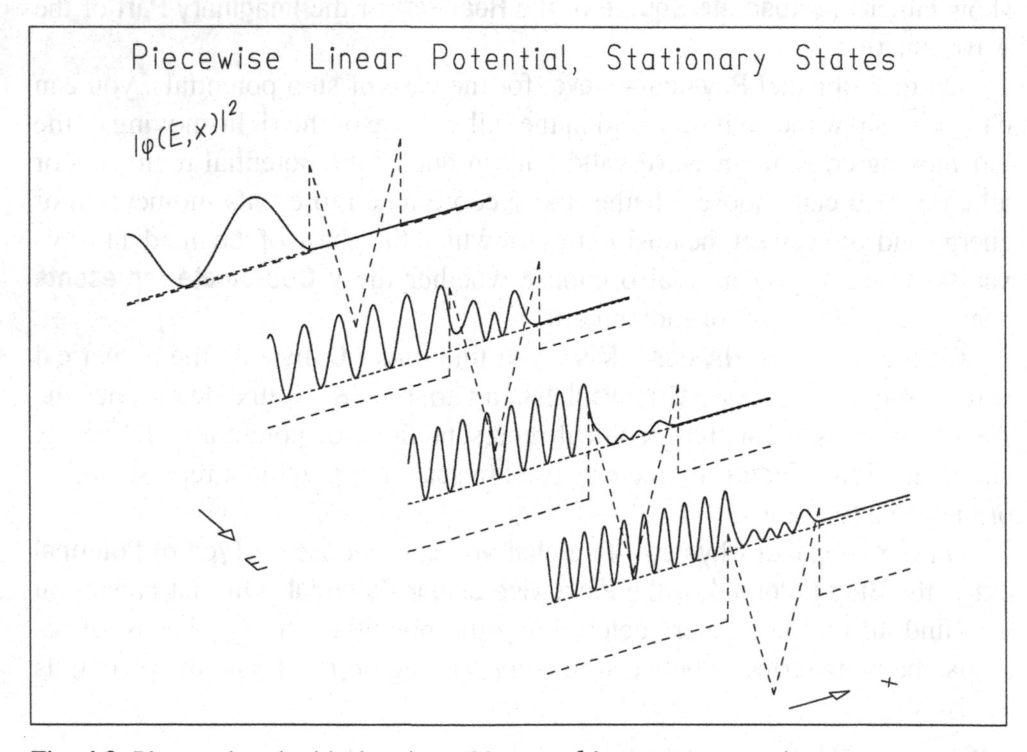

Fig. 4.3. Plot produced with descriptor `Piecew. lin. pot.: stationary scattering` on file `1D_Scattering.des`

- For the piecewise linear potential $V(x) = V_i$ at x_i and $V(x) = V_{i+1}$ at x_{i+1} with a linear variation in between. Vertical potential edges can be constructed by defining two points with identical positions but different potentials. To give an explicit example: $(x_3 = 5, V_3 = 0)$ and $(x_4 = 5, V_4 = 10)$ indicates a sudden rise from $V = 0$ to $V = 10$ at the point $x = 5$.

Movie Capability: Indirect. After conversion to direct movie capability the start and end values of energy (or momentum) can be changed in the subpanel **Movie** (see Sect. A.4) of the parameter panel. In the resulting movie energy (or momentum) of the stationary state changes with time. Such a movie, in particular, is useful for the demonstration of resonances.

Example Descriptors on File `1D_Scattering.des`

- `Step pot.: stationary scattering` (see Fig. 4.2)
- `Piecew. lin. pot.: stationary scattering` (see Fig. 4.3)

4.3 Time-Dependent Scattering by the Step Potential and by the Piecewise Linear Potential

Aim of this section: Computation and demonstration of a time-dependent harmonic wave $\psi(E, x, t)$ coming in from the left and scattered by a step potential or by a piecewise linear potential. For the step potential the right-moving and left-moving constituent waves $\psi_{j\pm}(E, x, t)$, (4.27), can also be displayed. Study of the time development of a Gaussian wave packet $\psi(x, t)$ scattered by a step potential or by a piecewise linear potential.

A plot is produced showing for various instances in time either a harmonic wave scattered by a step potential, Fig. 4.4, or by a piecewise linear potential, Fig. 4.5, or a wave packet, Figs. 4.6 and 4.7, scattered by one of these potentials. For the step potential also the constituent solutions $\psi_{j\pm}$ can be displayed. Note that the constituent solutions $\psi_{j\pm}$ have physical significance only in region j although they are drawn in the complete x range determined by the C3 window, unless you restrict the range to region j. Note also that only the sum ψ_j is a solution of the Schrödinger equation.

As in Sect. 2.7, for the computation of the Gaussian wave packet, the integration over p has to be performed numerically and is thus approximated by a sum (f_σ is a reasonably large positive number, e.g., $f_\sigma = 3$):

$$\Psi(x, t) = \int_{-\infty}^{\infty} f(p)\psi_p(x, t)\,\mathrm{d}p \;\rightarrow\; \Delta p \sum_{n=0}^{N_{\mathrm{sum}}} f(p_n)\psi_{p_n}(x, t) \quad ,$$

$$p_n = p_0 - f_\sigma\sigma_p + n\Delta p \quad , \quad \Delta p = \frac{2f_\sigma\sigma_p}{N_{\mathrm{sum}} - 1} \quad . \qquad (4.65)$$

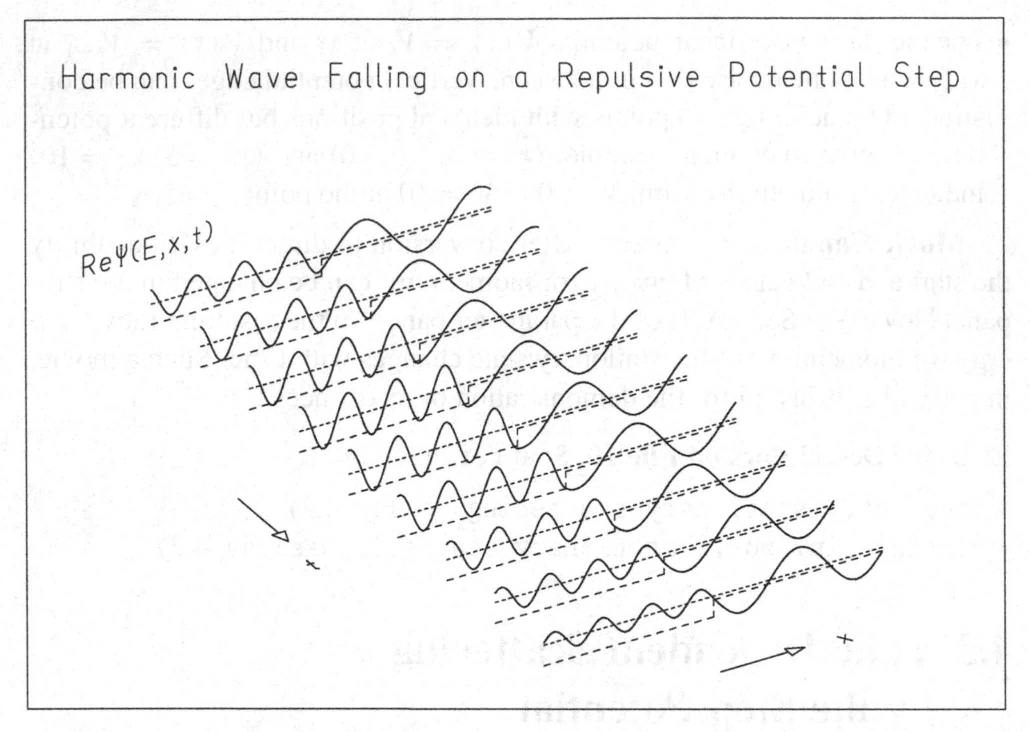

Fig. 4.4. Plot produced with descriptor `Step pot.: time-dependent scattering: harmonic wave` on file `1D_Scattering.des`

Again the solution obtained numerically as a sum with a finite number of terms will be periodic in x. That is, only in a limited x region you will get a good approximation to the true solution. The patterns for $\psi(x)$ [or $\psi_{j+}(x)$ and $\psi_{j-}(x)$] will repeat themselves (quasi-)periodically along the x direction, the period Δx becoming larger as N_{sum} increases. You will have to make sure that the x interval of your C3 window is small compared to Δx.

On the bottom of the subpanel **Physics—Comp. Coord.** you can choose to show either the **Absolute Square** or the **Real Part** or the **Imaginary Part** of the wave function.

On the subpanel **Physics—Wave (Packet)** you find the following items:

- **Full/Constituent Solution** – Here you can choose to plot ψ or ψ_{j+} or ψ_{j-}. In the latter two cases, of course, the **Region Number** j has to be given. (This field is available only for step potentials.)
- **Incoming Wave** – Here you can choose between **Harmonic Wave** and **Wave Packet**. In the latter case the number N_{int}, $1 \le N_{\mathrm{int}} \le 200$, determines the number $N_{\mathrm{sum}} = 2N_{\mathrm{int}}+1$ of terms in the sum approximating the momentum integral for the wave packet and f_σ is the other parameter needed for the numerical approximation of the wave packet, see (4.65).

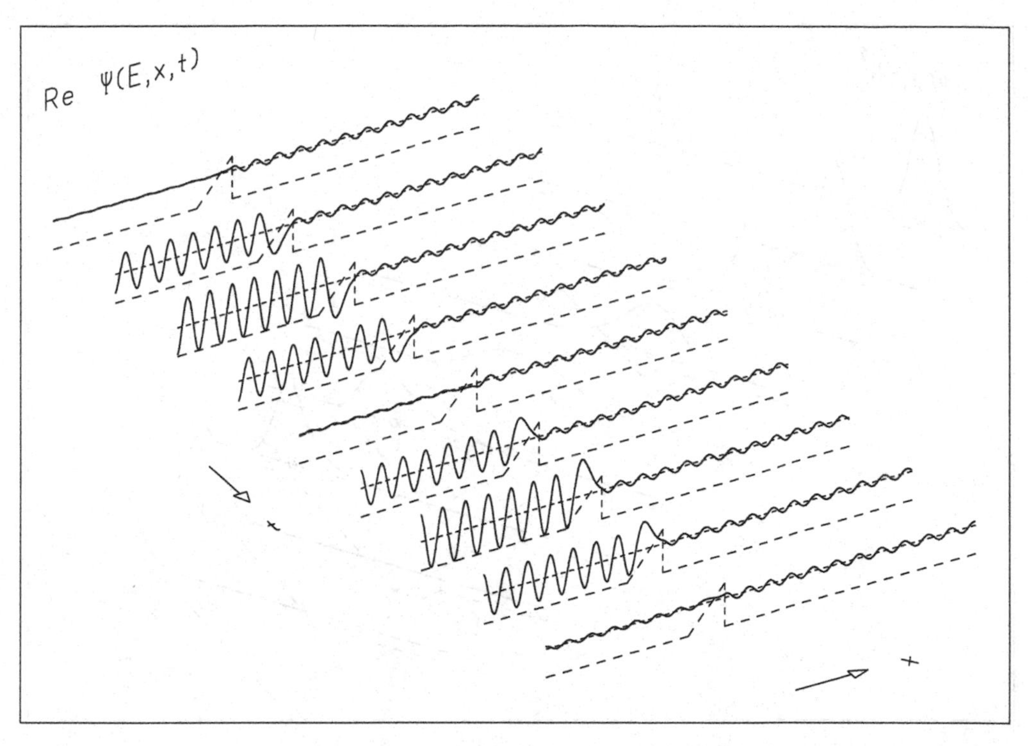

Fig. 4.5. Plot produced with descriptor `Piecew. lin. pot.: time-dependent scattering - harmonic wave` on file `1D_Scattering.des`

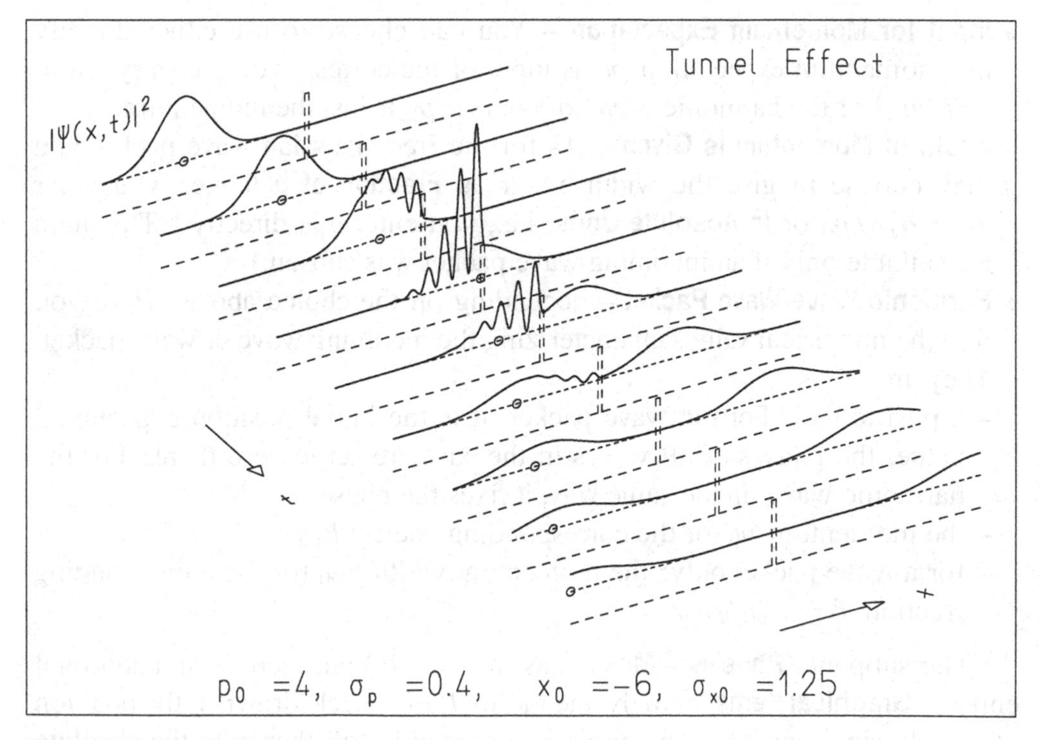

Fig. 4.6. Plot produced with descriptor `Step pot.: time-dependent scattering: wave packet` on file `1D_Scattering.des`

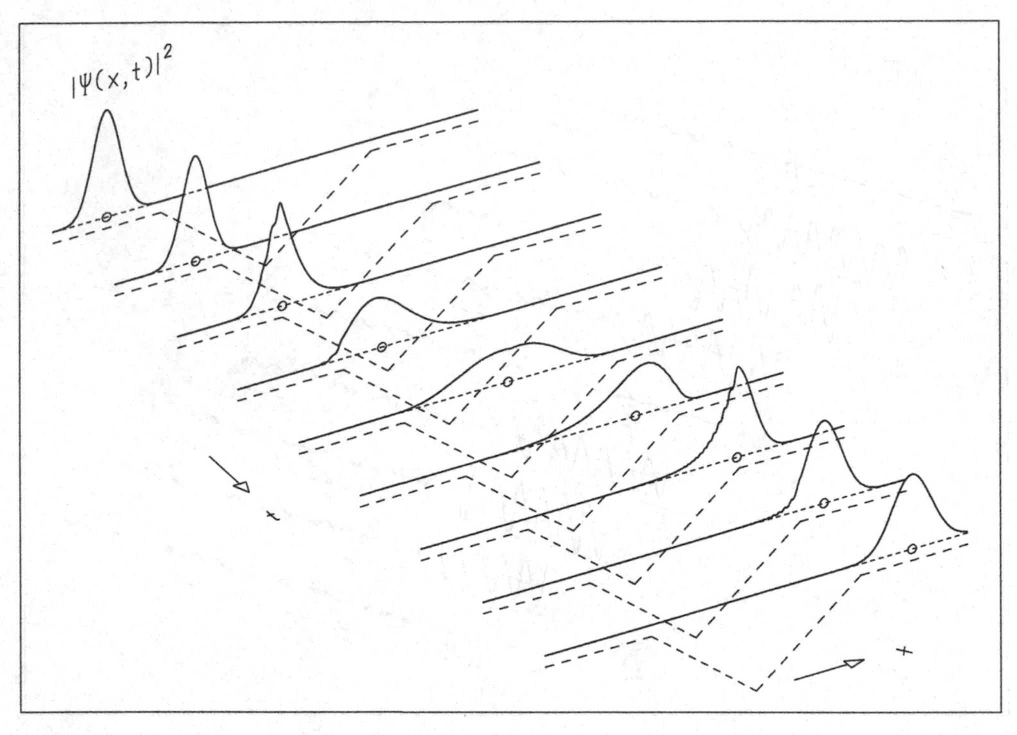

Fig. 4.7. Plot produced with descriptor `Piecew. lin. pot.: time-dependent scattering - wave packet` on file `1D_Scattering.des`

- **Input for Momentum Expectation** – You can choose to use either directly the momentum expectation p_0 as input or the corresponding energy $E_0 = p_0^2/2m$. For the harmonic wave, of course, p_0 is just the momentum.
- **Width in Momentum is Given** – As for the free Gaussian wave packet you may choose to give the width either **as Fraction of p_0**, i.e., you enter $f = \sigma_{p0}/p_0$ or **in Absolute Units**, i.e., you enter σ_{p0} directly. (This item is available only if an incoming wave packet was chosen.)
- **Harmonic Wave/Wave Packet** – depending on the choice above. Here you find the numerical values characterizing the incoming wave or wave packet. They are
  - a position x_0 (For the wave packet, it is the initial position expectation value, the phases of all waves in the sum are set to zero there. For the harmonic wave, in the same way, it fixes the phase.),
  - the momentum p_0 (or the corresponding energy E_0),
  - for a wave packet only: the momentum width σ_{p0} (or the corresponding fraction $f = \sigma_{p0}/p_0$).

The subpanel **Physics—Misc.** is as in Sect. 4.2 but there is an additional entry in **Graphical Items**, namely, the radius R of a circle drawn at the position of the classical particle. The circle is drawn only together with the absolute square of the wave function of a packet.

The subpanel **Physics—Potential** is as in Sect. 4.2.

Movie Capability: Indirect. After conversion to direct movie capability the start and end times can be changed in the subpanel **Movie** (see Sect. A.4) of the parameter panel.

Example Descriptors on File `1D_Scattering.des`

- `Step pot.: time-dependent scattering: harmonic wave` (see Fig. 4.4)
- `Piecew. lin. pot.: time-dependent scattering - harmonic wave` (see Fig. 4.5)
- `Step pot.: time-dependent scattering: wave packet` (see Fig. 4.6)
- `Piecew. lin. pot.: time-dependent scattering - wave packet` (see Fig. 4.7)
- `Movie: wave packet in step potential`
- `Movie: wave packet in piecew. lin. potential`

4.4 Transmission and Reflection. The Argand Diagram

Aim of this section: Presentation of the complex transmission coefficient $A_N(E)$ and the complex reflection coefficient $B_1(E)$ and of the complex T-matrix elements of transmission $T_T(E)$ and of reflection $T_R(E)$, see (4.36).

If $C(E)$ is one of these quantities, we want to illustrate its energy dependence by four different graphs,

- the Argand diagram $\mathrm{Im}\{C(E)\}$ vs. $\mathrm{Re}\{C(E)\}$,
- the real part $\mathrm{Re}\{C(E)\}$ as a function of E,
- the imaginary part $\mathrm{Im}\{C(E)\}$ as a function of E,
- the absolute square $|C(E)|^2$ as a function of E.

Alternatively, these functions can also be plotted with momentum p (instead of energy E) as independent variable.

It is customary to draw an Argand diagram ($\mathrm{Im}\{C(E)\}$ vs. $\mathrm{Re}\{C(E)\}$) and graphs $\mathrm{Im}\{C(E)\}$ and $\mathrm{Re}\{C(E)\}$ in such a way that the graphs appear to be projections to the right and below the Argand diagram, respectively. You can do that by producing a *combined plot* using a mother descriptor, which in turn quotes several individual descriptors (see Appendix A.10) as in the example plots, Figs. 4.8 and 4.9.

All four plots in Fig. 4.8 and also those in Fig. 4.9 are of the type *2D function graph*. The Argand diagram (top left) is a *parameter representation* $x = x(\alpha)$, $y = y(\alpha)$. The two plots on the right-hand side are *Cartesian*

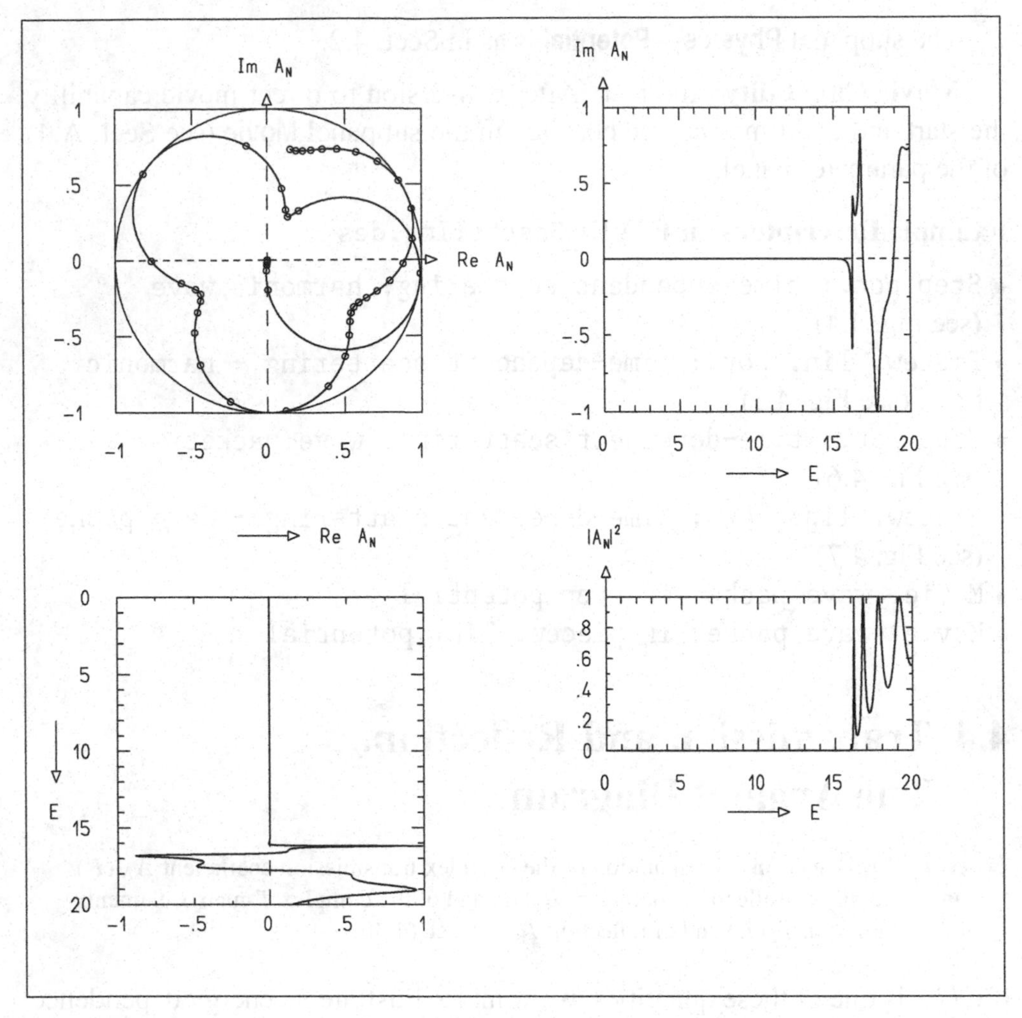

Fig. 4.8. Combined plot produced with descriptor `Step pot.: Argand diagram - combined plot` on file `1D_Scattering.des`. This descriptor quotes four other descriptors to generate the individual plots

plots $y = y(x)$. The plot in the bottom-left corner is an *inverse Cartesian plot* $x = x(y)$.

On the subpanel **Physics—Comp. Coord.** you can select to use as **Independent Variable** the energy E (as in the formulae above) or the momentum p and to compute one of the four functions

$$T_{\mathrm{T}},\ T_{\mathrm{R}},\ A_N,\ B_1 \quad .$$

You can further choose the type of 2D function graph you want to produce:

- Argand diagram,
- $y = y(x)$,
- $x = x(y)$.

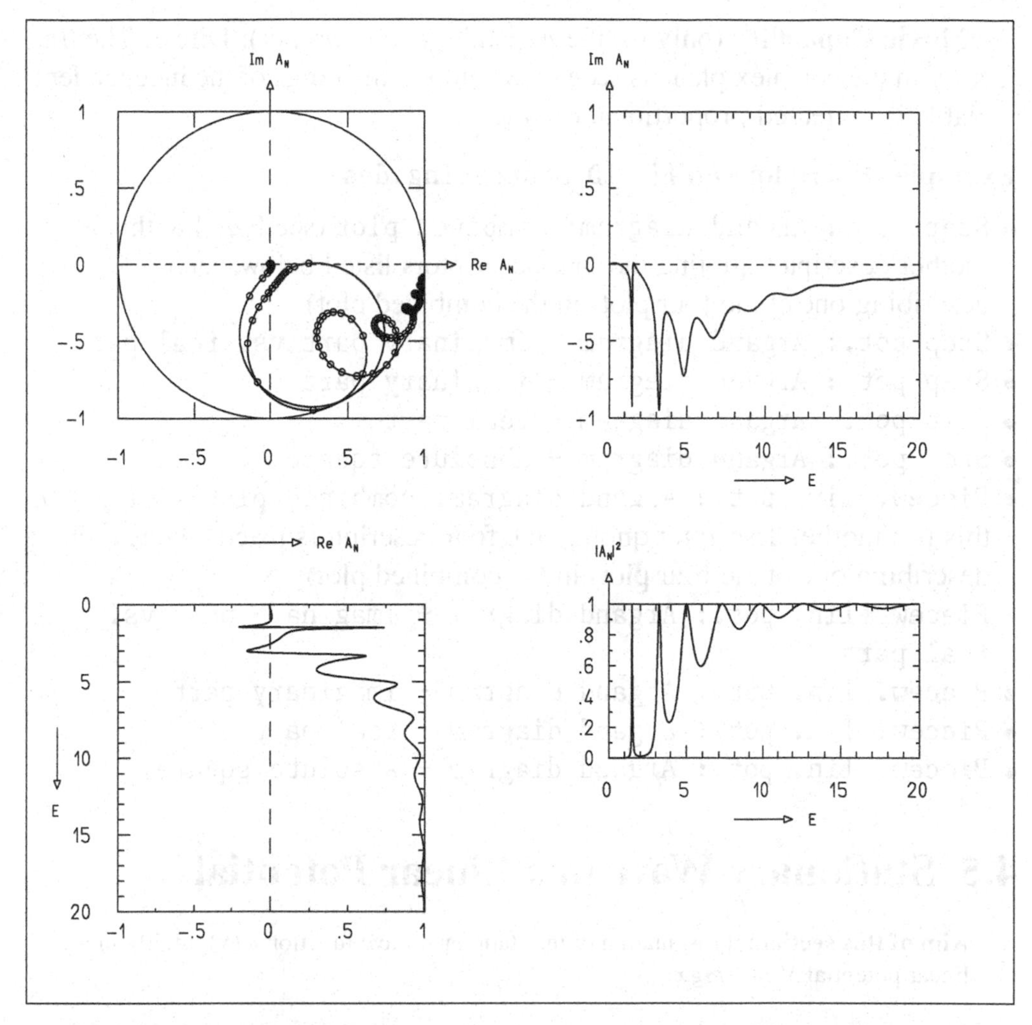

Fig. 4.9. Combined plot produced with descriptor `Piecew. lin. pot.: Argand diagram - combined plot` on file `1D_Scattering.des`. This descriptor quotes four other descriptors to generate the individual plots. The potential is the same as in Fig. 4.3.

For the latter two types of function graph you still can choose to present as dependent variable the **Absolute Square**, the **Real Part**, or the **Imaginary Part** of the function selected.

On the subpanel **Physics—Variables** you find three items:

- **Phase-Fixing Position** – At x_0 the phase of the incoming wave is zero.
- **Constants** – the numerical values of $\hbar$ and m.
- **Range of Independent Variable in Argand Diagram** – The variable is the energy E (or the momentum p). It is varied between the boundaries E_{beg} and E_{end} (or p_{beg} and p_{end}).

The subpanel **Physics—Potential** is as in Sect. 4.2.

Movie Capability (only for the Argand diagram proper): Direct. The trajectory in the complex plane is seen to develop as the range of the independent variable is increased proportional to time.

Example Descriptors on File `1D_Scattering.des`

- `Step pot.: Argand diagram: combined plot` (see Fig. 4.8, this is a mother descriptor quoting the four descriptors listed below, each describing one of the four plots in the combined plot)
- `Step pot.: Argand diagram - imaginary part vs. real part`
- `Step pot.: Argand diagram - imaginary part`
- `Step pot.: Argand diagram - real part`
- `Step pot.: Argand diagram - absolute square`
- `Piecew. lin. pot.: Argand diagram: combined plot` (see Fig. 4.9, this is a mother descriptor quoting the four descriptors listed below, each describing one of the four plots in the combined plot)
- `Piecew. lin. pot.: Argand diagram - imaginary part vs. real part`
- `Piecew. lin. pot.: Argand diagram - imaginary part`
- `Piecew. lin. pot.: Argand diagram - real part`
- `Piecew. lin. pot.: Argand diagram - absolute square`

4.5 Stationary Wave in a Linear Potential

Aim of this section: Presentation of the stationary wave function $\varphi(x)$, (4.59), in a linear potential $V = -mgx$.

A plot similar to Fig. 4.10 is produced showing the linear potential $V = -mgx$ as a long-stroke dashed line and, for various values of the energy E, the wave function $\varphi(x)$. In each case the energy E is shown as a short-stroke dashed line, which also serves as a zero line for the wave function.

On the subpanel **Physics—Comp. Coord.** you can choose to show the absolute square, the real part, or the imaginary part of $\varphi(x)$. (Note that the latter is always zero.)

On the subpanel **Physics—Variables** there are four items:

- **Acceleration** – Here you find the numerical value of the acceleration.
- **Linear Potential** – You can choose whether the potential is
  - **Shown** (as in Fig. 4.10) or
  - **Not Shown.**

  In the former case the wave function is shown as $z = E + s\varphi(x)$ where s is a scale factor, see below.
- **Scale Factor** – contains the scale factor s just mentioned.
- **Constants** – contains the numerical values of the constants $\hbar$ and m.

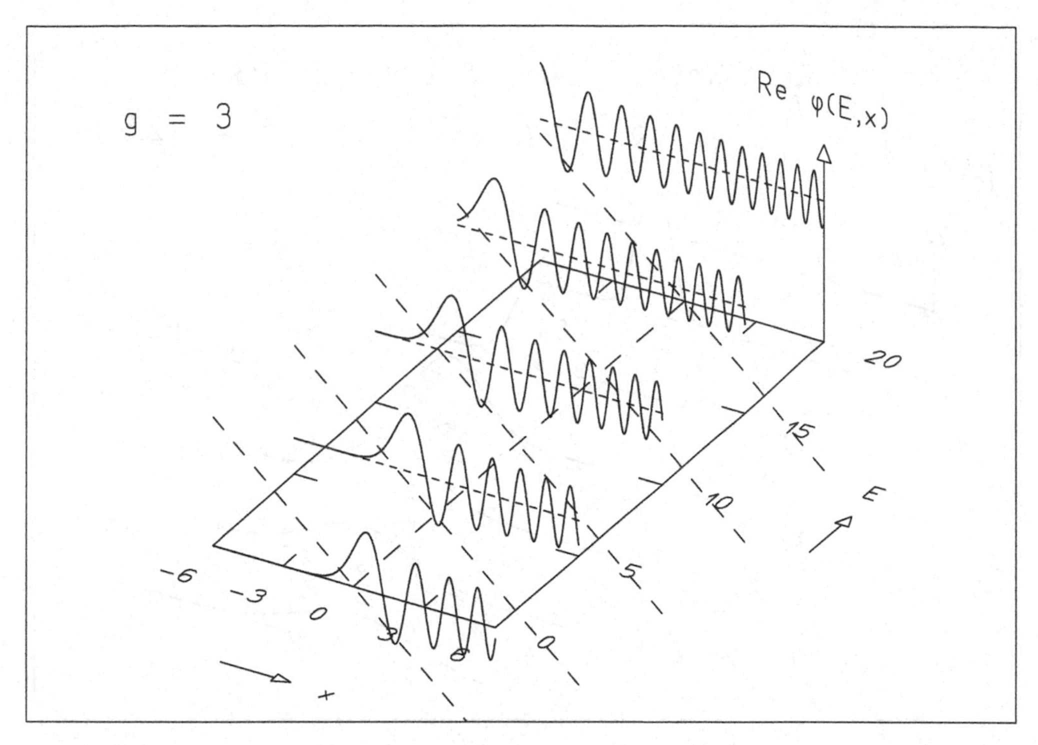

Fig. 4.10. Plot produced with descriptor `Stationary wave in linear potential` on file `1D_Scattering.des`

Example Descriptor on file `1D_Scattering.des`

- `Stationary wave in linear potential` (see Fig. 4.10)

4.6 Gaussian Wave Packet in a Linear Potential

Aim of this section: Illustration of the motion of a wave packet (4.60) in a linear potential including quantile motion.

On the subpanel **Physics—Comp. Coord.** you can choose to plot one of five different quantities as in Sect. 2.3.1.

The subpanel **Physics—Wave Packet** is as described in Sect. 2.3.2.

The subpanel **Physics—Pot.** contains three of the four items described in Sect. 4.5 (for the subpanel **Physics—Variables**).

The subpanel **Physics—Quantile** is as described in Sect. 2.3.3.

Movie Capability: Indirect. After conversion to direct movie capability the start and end times can be changed in the subpanel **Movie** (see Sect. A.4) of the parameter panel.

Example Descriptor on file `1D_Scattering.des`

- `Wave packet in linear potential` (see Fig. 4.11)

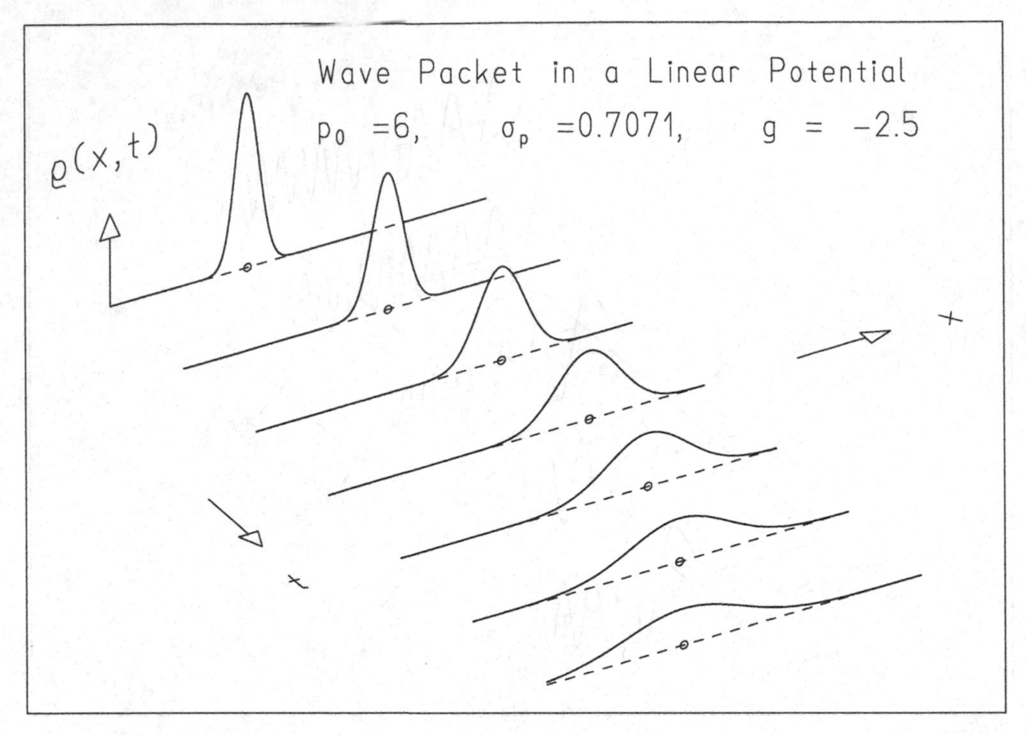

Fig. 4.11. Plot produced with descriptor `Wave packet in linear potential` on file `1D_Scattering.des`

4.7 Quantile Trajectories in a Linear Potential

Aim of this section: Presentation of quantile trajectories $x_P = x_P(t)$, Sect. 4.1.20, for the motion of a Gaussian wave packet in a linear potential.

On the bottom of the subpanel **Physics—Comp. Coord.** you find the **Acceleration** g, the parameter of the linear potential $V(x) = -mgx$.

The subpanels **Physics—Quantiles** and **Physics—Wave Packet** are as described in Sect. 2.5.

Example Descriptor on file `1D_Scattering.des`

- `Linear potential: quantile trajectories` (see Fig. 4.12)

4.8 Classical Phase-Space Density in a Linear Potential

Aim of this section: Graphical presentation of the phase-space probability density $\varrho^{\text{cl}}(x, p, t)$ of Sect. 4.1.21 under the influence of a linear potential $V = -mgx$. The density initially (at time $t = 0$) is uncorrelated and fulfills the condition $\sigma_{x0}\sigma_{p0} = \hbar/2$. The marginal distributions $\varrho^{\text{cl}}_x(x, t)$ and $\varrho^{\text{cl}}_p(p, t)$ are also shown.

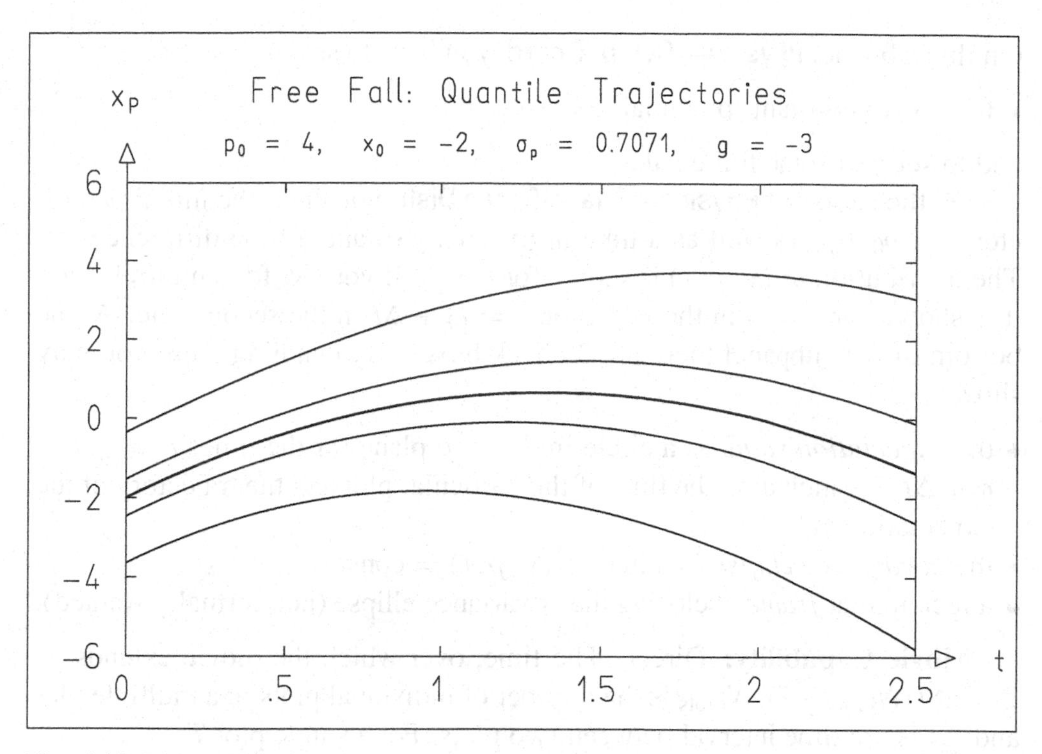

Fig. 4.12. Plot produced with descriptor `Linear potential: quantile trajectories` on file `1D_Scattering.des`

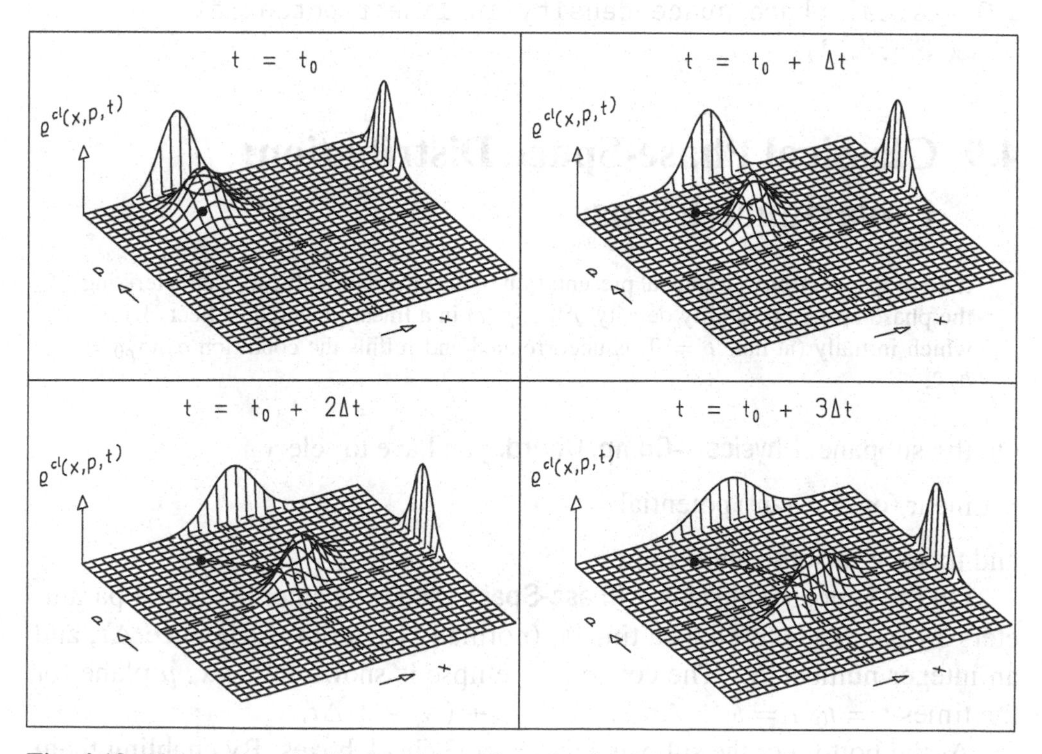

Fig. 4.13. Plot produced with descriptor `Classical phase-space density in linear potential` on file `1D_Scattering.des`

On the subpanel **Physics—Comp. Coord.** you have to select

- Linear (or constant) potential

and to set g to some finite value.

On the subpanel **Physics—Phase-Space Distr.** you enter the initial parameters x_0, p_0, σ_{p0} as well as a time t_0 (normally 0) and a time difference Δt. The distribution $\varrho^{\mathrm{cl}}(x, p, t)$ is shown for $t = t_0$. If you ask for a multiple plot, it is shown for $t = t_0$ in the first plot, $t = t_0 + \Delta t$ in the second, etc. At the bottom of the subpanel there are 3 check boxes. By enabling them you may show

- the *expectation value* as a circle in the x, p plane for the times $t = t_0, t = t_0 + \Delta t, \ldots$ including the time of the particular plot and the trajectory of the expectation value,
- the *covariance ellipse* as a line $\varrho^{\mathrm{cl}}(x, p, t) = \mathrm{const}$,
- a rectangular *frame* enclosing the covariance ellipse (not normally wanted).

Movie Capability: Direct. The time, over which the movie extends, is $T = \Delta t(N_{\mathrm{Plots}} - 1)$; N_{Plots} is the number of individual plots in a multiple plot and Δt is the time interval between two plots. For a single plot $T = \Delta t$.

Example Descriptor on File `1D_Scattering.des`

- `Classical phase-space density in linear potential` (see Fig. 4.13)

4.9 Classical Phase-Space Distribution: Covariance Ellipse

Aim of this section: Graphical presentation of the covariance ellipse, characterizing the phase-space probability density $\varrho^{\mathrm{cl}}(x, p, t)$ in a linear potential of Sect. 4.1.21, which initially (at time $t = 0$) is uncorrelated and fulfills the condition $\sigma_{x0}\sigma_{p0} = \hbar/2$.

On the subpanel **Physics—Comp. Coord.** you have to select

- Linear (or constant) potential

and to set g to some finite value.

On the subpanel **Physics—Phase-Space Distr.** you enter the initial parameters x_0, p_0, σ_{p0} as well as a time t_0 (normally 0), a time difference Δt, and an integer number N_t. The covariance ellipse is shown in the x, p plane for the times $t = t_0, t = t_0 + \Delta t, \ldots, t = t_0 + (N_t - 1)\Delta t$.

At the bottom of the subpanel there are 3 check boxes. By enabling them you may show

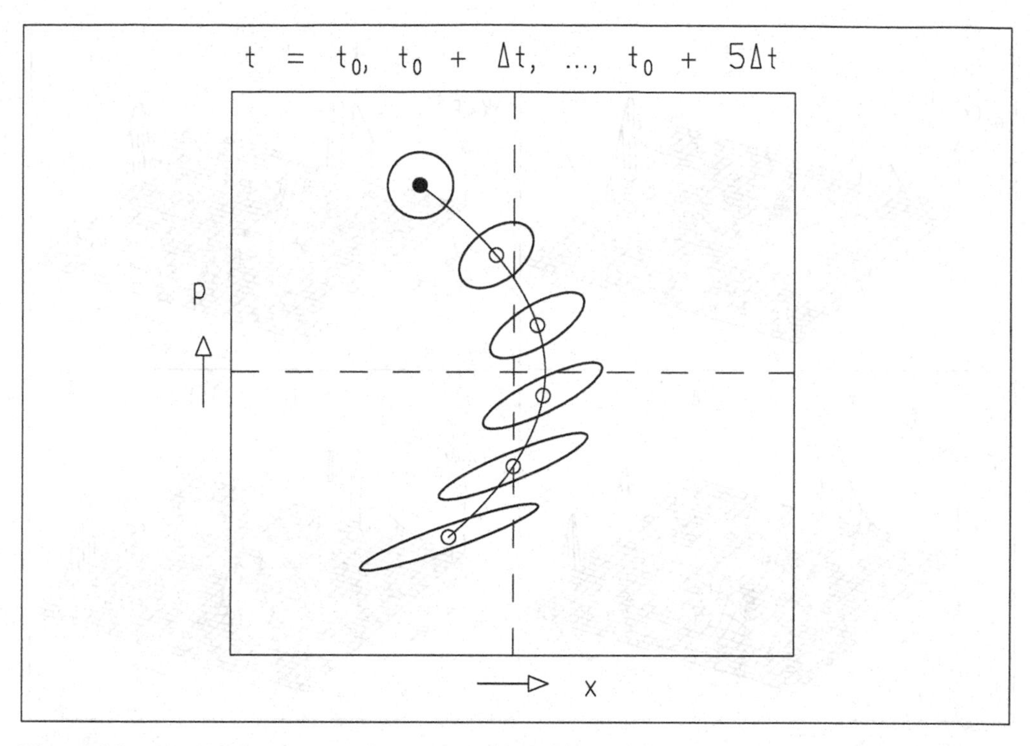

Fig. 4.14. Plot produced with descriptor `Cl. phase-sp. density in linear potential: covariance ellipse` on file `1D_Free_Particle.des`

- the *expectation value* as a circle in the x, p plane for the times $t = t_0, t = t_0 + \Delta t, \ldots$ and the trajectory of the expectation value,
- the *covariance ellipse*,
- a rectangular *frame* enclosing the covariance ellipse.

Movie Capability: Direct. On the subpanel **Movie** (see Sect. A.4) of the parameters panel you may choose to show or not to show the initial and intermediate positions of the covariance ellipse.

Example Descriptor on File `1D_Scattering.des`

- `Cl. phase-sp. density in linear potential: covariance ellipse` (see Fig. 4.14)

4.10 Classical Phase-Space Density Reflected by a High Potential Wall

Aim of this section: Graphical presentation of the phase-space probability density $\varrho^{\mathrm{cl}}(x, p, t)$ of Sect. 4.1.22, which is reflected at a high potential wall at $x = 0$. The density initially (at time $t = 0$) is uncorrelated and fulfills the condition $\sigma_{x0}\sigma_{p0} = \hbar/2$. The marginal distributions $\varrho_x^{\mathrm{cl}}(x, t)$ and $\varrho_p^{\mathrm{cl}}(p, t)$ are also shown.

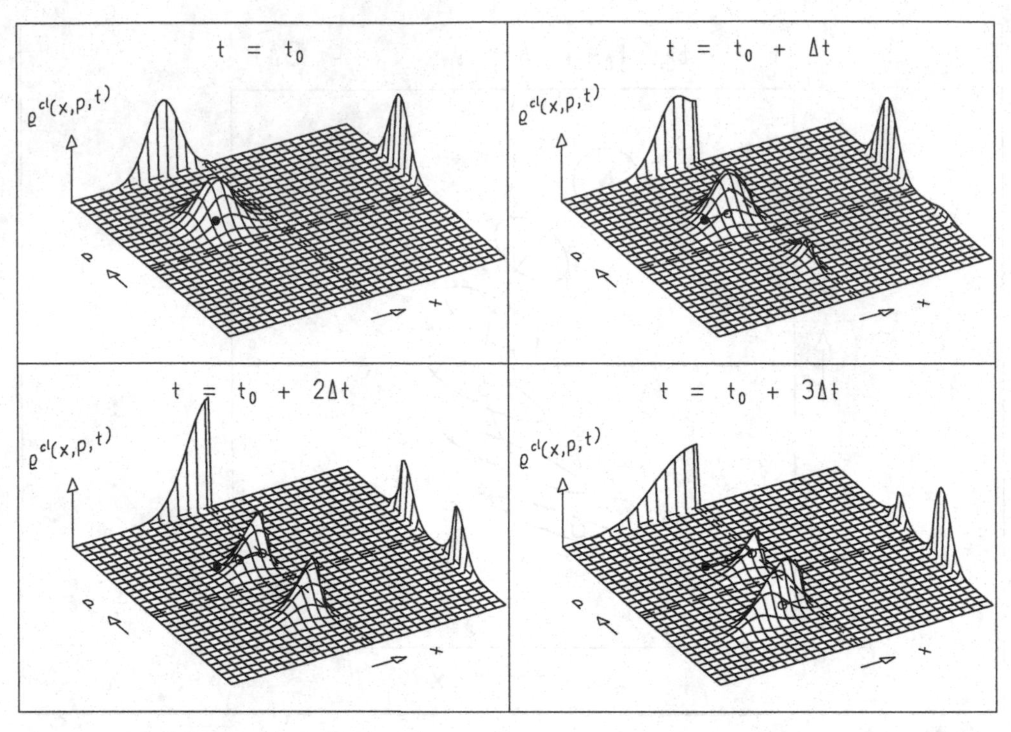

Fig. 4.15. Plot produced with descriptor `Cl. phase-sp. density, reflected at high potential wall` on file `1D_Scattering.des`

On the subpanel **Physics—Comp. Coord.** you have to select

- High wall at $x = 0$.

On the subpanel **Physics—Phase-Space Distr.** you enter the initial parameters x_0, p_0, σ_{p0} as well as a time t_0 (normally 0) and a time difference Δt. The distribution $\varrho^{\mathrm{cl}}(x, p, t)$ is shown for $t = t_0$. If you ask for a multiple plot, it is shown for $t = t_0$ in the first plot, $t = t_0 + \Delta t$ in the second, etc. At the bottom of the subpanel there are 3 check boxes. By enabling the first one you may show

- the *expectation value* as a circle in the x, p plane for the times $t = t_0, t = t_0 + \Delta t, \ldots$ including the time of the particular plot and the trajectory of the expectation value.

(The other two check boxes have no meaning in the present case.)

Movie Capability: Direct. The time, over which the movie extends, is $T = \Delta t(N_{\mathrm{Plots}} - 1)$; N_{Plots} is the number of individual plots in a multiple plot and Δt is the time interval between two plots. For a single plot $T = \Delta t$.

Example Descriptor on File `1D_Scattering.des`

- `Cl. phase-sp. density, reflected at high potential wall` (see Fig. 4.15)

4.11 Exercises

Please note:

(i) You may watch a demonstration of the material of this chapter by pressing the button **Run Demo** in the *main toolbar* and selecting one of the demo files `1D_Scattering`.

(ii) More example descriptors can be found on the descriptor file `1DScatt-(FE).des` in the directory `FurtherExamples`.

(iii) For the following exercises use descriptor file `1D_Scattering.des`.

(iv) If not stated otherwise in the exercises the numerical values of the mass of the particle and of Planck's constant are put to 1.

4.2.1 Plot **(a)** the real part, **(b)** the imaginary part, **(c)** the absolute square of the scattering wave function for the potential

$$V(x) = \begin{cases} 0\,,\ x < 0 \\ 2\,,\ 0 \le x < 2 \\ 0\,,\ 2 \le x \end{cases} .$$

Start from descriptor 36: '`Exercise 4.2.1`'. **(d)** Why is the wave function a linear function in the region of the potential for one of the energies? **(e)** Explain the trend of the transmission coefficient for increasing energies in the plot.

4.2.2 (a,b,c) Repeat Exercise 4.2.1 (a,b,c) for the potential

$$V(x) = \begin{cases} 0\,,\ x < 0 \\ 2\,,\ 0 \le x < 0.5 \\ 0\,,\ 0.5 \le x \end{cases} .$$

(d) Why does the tunnel probability increase in comparison with Exercise 4.2.1? **(e)** Why is the absolute square constant in the region beyond the potential? **(f)** Why does the absolute square show a wave pattern in the region to the left of the potential wall?

4.2.3 Plot **(a)** the real part, **(b)** the imaginary part, **(c)** the absolute square of the scattering wave function for the potential

$$V(x) = \begin{cases} 0\,,\ x < 0 \\ 2\,,\ 0 \le x < 0.5 \\ 0\,,\ 0.5 \le x < 1.5 \\ 2\,,\ 1.5 \le x < 2 \\ 0\,,\ 2 \le x \end{cases} .$$

Start from descriptor 36: '`Exercise 4.2.1`'. **(d)** Why is the amplitude of the wave pattern of the absolute square to the left of the double well very small for one of the energies? At which energy does this phenomenon occur?

4.2.4 Study **(a)** the real part, **(b)** the imaginary part, **(c)** the absolute square of the wave function in the potential of Exercise 4.2.3 in the neighborhood of the particular energy determined in 4.2.3 (d). Choose in particular the energy interval $1.08 \le E \le 1.14$. Start from descriptor 37: ‘Exercise 4.2.4’. **(d)** Why does the resonance phenomenon occur at the value 1.12 for the energy? **(e)** Is there another resonance at lower energies?

4.2.5 Study the resonances of a repulsive double-well potential:

$$V(x) = \begin{cases} 0\,, x < 0 \\ 2.5\,, 0 \le x < 0.5 \\ 0\,, 0.5 \le x < 2.5 \\ 2.5\,, 2.5 \le x < 3 \\ 0\,, 3 \le x \end{cases} .$$

(a) Plot the absolute square of the wave function in the energy region $0.4 \le E \le 2.4$ in 10 intervals. In which energy regions do you see indications for the occurrence of resonances? Determine their energies roughly by changing the limits of the energy interval. Start from descriptor 37: ‘Exercise 4.2.4’. **(b)** Plot the absolute square of the wave function in the neighborhood of the first resonance as an interval of 0.04 in steps of 0.008 energy units. Place the interval in such a way that the resonating wave function comes out best. **(c)** Plot the real part of the wave function in the interval of (b). **(d)** Repeat (b) for the neighborhood of the second resonance. Choose an interval of 0.5 energy units in steps of 0.1. **(e)** Plot the real part of the wave function in the interval of (d). **(f)** What distinguishes the two resonances from each other? How is the difference in the two wave functions correlated to their difference in energy?

4.2.6 Study the resonances of an asymmetric triple-well potential:

$$V(x) = \begin{cases} 0\,, x < 0 \\ 7\,, 0 \le x < 0.5 \\ 0\,, 0.5 \le x < 2.5 \\ 7\,, 2.5 \le x < 3 \\ 0\,, 3 \le x < 4.5 \\ 7\,, 4.5 \le x < 5 \\ 0\,, 5 \le x \end{cases} .$$

Plot the absolute square of the wave function in the energy regions **(a)** $0.01 \le E \le 1.01$, **(b)** $1.01 \le E \le 2.01$, **(c)** $2.01 \le E \le 3.01$, **(d)** $3.01 \le E \le 4.01$, **(e)** $4.01 \le E \le 5.01$, **(f)** $5.01 \le E \le 6.01$, **(g)** $6.01 \le E \le 7.01$, in steps of $\Delta E = 0.1$. Find the energies of the resonances in these regions. Start from descriptor 37: ‘Exercise 4.2.4’. For (h–l) start from descriptors 38: ‘Exercise 4.2.6h’, 39: ‘Exercise 4.2.6i’, ..., 42: ‘Exercise 4.2.6l’, respectively. **(h)** Plot the first resonance in the left-hand well as one of the center lines in a set of 10 lines with an energy resolution 0.01. **(i)** Plot the second resonance in the left-hand well as the center line in a set of 10 lines with an energy resolution 0.01. **(j)** Plot the third resonance in the left-hand well as the center line in a set of 10 lines with an energy resolution 0.05. **(k)** Plot the first resonance in the right-hand well in a set of 10 lines with an energy resolution of 0.02. **(l)** Plot the second resonance in the right-hand well in a set of 10 lines with an energy resolution of 0.05.

4.2.7 Plot the transmission probability $|A_N|^2$ for the potential of Exercise 4.2.6. Start from descriptor 43: '`Exercise 4.2.7`'. Determine the energies of the maxima of $|A_N|^2$ and compare them with the resonance energies of Exercise 4.2.6.

4.2.8 Study the resonance behavior in a double-well potential corresponding to the left-hand potential of Exercise 4.2.6:

$$V(x) = \begin{cases} 0\,, x < 0 \\ 7\,, 0 \le x < 0.5 \\ 0\,, 0.5 \le x < 2.5 \\ 7\,, 2.5 \le x < 3 \\ 0\,, 3 \le x \end{cases} .$$

Plot the absolute square of the wave function in the energy regions **(a)** $0.01 \le E \le 1.01$, **(b)** $1.01 \le E \le 2.01$, **(c)** $2.01 \le E \le 3.01$, **(d)** $3.01 \le E \le 4.01$, **(e)** $4.01 \le E \le 5.01$, **(f)** $5.01 \le E \le 6.01$, **(g)** $6.01 \le E \le 7.01$, in steps of $\Delta E = 0.1$. Find the energies of the resonances in these regions. Start from descriptor 37: '`Exercise 4.2.4`'. **(h)** Plot the first resonance as one of the center lines in a set of 10 lines with an energy resolution 0.01. **(i)** Plot the second resonance as one of the center lines in a set of 10 lines with an energy resolution 0.01. **(j)** Plot the third resonance as one of the center lines in a set of 10 lines with an energy resolution 0.05.

4.2.9 Plot the transmission probability $|A_N|^2$ for the potential of Exercise 4.2.8. Start from descriptor 43: '`Exercise 4.2.7`'. Determine the energies of the maxima of $|A_N|^2$ and compare them with the resonance energies of Exercise 4.2.8.

4.2.10 (a–i) Repeat Exercise 4.2.8 (a–i) for the double-well potential corresponding to the right-hand potential of Exercise 4.2.6:

$$V(x) = \begin{cases} 0\,, x < 2.5 \\ 7\,, 2.5 \le x < 3 \\ 0\,, 3 \le x < 4.5 \\ 7\,, 4.5 \le x < 5 \\ 0\,, 5 \le x \end{cases} .$$

Start from descriptor 37: '`Exercise 4.2.4`'. Choose as energy resolution in (a–g) $\Delta E = 0.1$, (h) $\Delta E = 0.02$, (i) $\Delta E = 0.05$.

4.2.11 Plot the transmission probability $|A_N|^2$ for the potential of Exercise 4.2.10. Start from descriptor 43: '`Exercise 4.2.7`'. Determine the energies of the maxima of $|A_N|^2$ and compare them to the resonance energies of Exercise 4.2.10.

4.2.12 We consider a potential with five steps (a model of the quantum-well device)

$$V(x) = \begin{cases} V_1\,, & x < 0 \quad \text{region 1} \\ V_2\,, & 0 \le x < 0.5 \ \text{region 2} \\ V_3\,, & 0.5 \le x < 1 \ \text{region 3} \\ V_4\,, & 1 \le x < 1.5 \ \text{region 4} \\ V_5\,, & 1.5 \le x \quad \text{region 5} \end{cases} .$$

For the potential values $V_1 = 0$, $V_2 = 7$, $V_3 = -0.5$, $V_4 = 6.5$, $V_5 = -1$, the energy $E_{\rm res}$ for a transmission resonance is $E_{\rm res} = 3.67$. Plot the absolute square of the wave function for **(a)** 'zero voltage', $V_1 = 0$, $V_2 = 7$, $V_3 = 0$, $V_4 = 7$, $V_5 = 0$, **(b)** a voltage below resonance, $V_1 = 0$, $V_2 = 7$, $V_3 = -0.25$, $V_4 = 6.75$, $V_5 = -0.5$, **(c)** resonance voltage, $V_1 = 0$, $V_2 = 7$, $V_3 = -0.5$, $V_4 = 6.5$, $V_5 = -1$, **(d)** a voltage above resonance, $V_1 = 0$, $V_2 = 7$, $V_3 = -0.75$, $V_4 = 6.25$, $V_5 = -1.5$. Start from descriptor 36: '`Exercise 4.2.1`'. **(e)** For comparison put the descriptors (a–d) into one combined plot. Start from the mother descriptor 44: '`Exercise 4.2.12`'. **(f)** The voltage U at the quantum well is given by the difference $U = V_1 - V_5$. The variation of the voltage is $\Delta U = \Delta V_1 - \Delta V_5 = -\Delta V_5$ because the potential V_1 is kept fixed in the cases (a–d). The electric current I through the quantum-well device is proportional to the absolute square $|A_5|^2$ of the coefficient A_5: $I = \alpha |A_5|^2$. Thus, the variation ΔI of the electric current is proportional to the variation $\Delta |A_5|^2$ of the absolute square of the coefficient A_5: $\Delta I = \alpha \Delta |A_5|^2$. For a given variation $\Delta U = -\Delta V_5$ of the voltage the quotient $R = \Delta U / \Delta I = \Delta V_5 / \Delta I$ of ΔU and the corresponding variation ΔI of the electric current is called the differential resistance. It is given by $R = (1/\alpha) \Delta U / \Delta |A_5|^2$. What is the trend of the differential resistance that can be learned for $\Delta U = 0.5$ from the comparison of the situations (ab), (bc), (cd)?

4.2.13 Use the potential of Exercise 4.2.12 with the potential parameters at resonance $V_1 = 0$, $V_2 = 60$, $V_3 = -2$, $V_4 = 58$, $V_5 = -4$. There is a lowest transmission resonance in this step potential at $E_1 = 8.464$. Plot the absolute square of the wave function in the neighborhood of the lowest resonance for **(a)** a voltage below resonance, $V_1 = 0$, $V_2 = 60$, $V_3 = -1.95$, $V_4 = 58.05$, $V_5 = -3.9$, **(b)** a voltage close below resonance, $V_1 = 0$, $V_2 = 60$, $V_3 = -1.993$, $V_4 = 58.007$, $V_5 = -3.986$, **(c)** resonance voltage, $V_1 = 0$, $V_2 = 60$, $V_3 = -2$, $V_4 = 58$, $V_5 = -4$, **(d)** a voltage above resonance, $V_1 = 0$, $V_2 = 60$, $V_3 = -2.007$, $V_4 = 57.993$, $V_5 = -4.014$. Start from descriptor 36: '`Exercise 4.2.1`'. **(e)** For comparison put the graphs (a–d) into one combined plot. Start from the mother descriptor 44: '`Exercise 4.2.12`'. **(f)** Why is the variation of the differential resistance much faster than in Exercise 4.2.12?

4.2.14 Repeat Exercise 4.2.13 for a much higher voltage between emitter and collector so that the step potential has at $E_{\rm res} = 8.464$ the parameters $V_1 = 0$, $V_2 = 60$, $V_3 = -33.25$, $V_4 = 26.75$, $V_5 = -66.5$. Plot the absolute square of the wave function in the neighborhood of **(a)** the voltage below resonance, $V_1 = 0$, $V_2 = 60$, $V_3 = -33.1$, $V_4 = 26.9$, $V_5 = -66.2$, **(b)** the voltage below resonance, $V_1 = 0$, $V_2 = 60$, $V_3 = -33.2$, $V_4 = 26.8$, $V_5 = -66.4$, **(c)** the resonance voltage, $V_1 = 0$, $V_2 = 60$, $V_3 = -33.25$, $V_4 = 26.75$, $V_5 = -66.5$, **(d)** the voltage above resonance, $V_1 = 0$, $V_2 = 60$, $V_3 = -33.3$, $V_4 = 26.7$, $V_5 = -66.6$. Start from descriptor 36: '`Exercise 4.2.1`'. **(e)** For comparison put the plots (a–d) into one combined plot. Start from the mother descriptor 44: '`Exercise 4.2.12`'. **(f)** Why is the variation of the differential resistance for this second resonance slower than for the lowest resonance as studied in Exercise 4.2.13?

4.2.15 This exercise models a quantum-effect device. Make use of the potential of five regions of Exercise 4.2.12. Plot the absolute square of the wave function for different voltages at the base (region 3), i.e., for the different potential values V_3 **(a)** below resonance, $V_3 = -1.9$, **(b)** slightly below resonance, $V_3 = -1.98$, **(c)** at resonance, $V_3 = -2$, **(d)** above resonance, $V_3 = -2.03$, whereas the potential values in the other regions remain unchanged at $V_1 = 0$, $V_2 = 40$, $V_4 = 38$, $V_5 = -4$. Start from descriptor 36: '`Exercise 4.2.1`'. **(e)** Put the four plots (a–d) into one combined plot. Start from the mother descriptor 44: '`Exercise 4.2.12`'. **(f)** For the case of a quantum-effect device the variation of the potential V_3 leads to a change $\Delta U = -\Delta V_3$ in the voltage $U = V_1 - V_3$. Thus, the differential resistance [see Exercise 4.2.12 (f)] is accordingly defined as $R = \Delta U/\Delta I = -\Delta V_3/\Delta I$. What is the trend of the differential resistance that can be learned from the comparison of the situations (ab), (bc), (cd)?

4.2.16 Make use of the potential of five regions as in Exercise 4.2.12. The potential in four of the five regions is kept fixed to the values $V_1 = 0$, $V_2 = 10$, $V_4 = 9.5$, $V_5 = -1$. In region 3 the potential values are changed. Plot the absolute square of the wave function **(a)** below resonance, $V_3 = -0.2$, **(b)** at resonance, $V_3 = -0.5$, **(c)** slightly above resonance, $V_3 = -0.8$, **(d)** above resonance, $V_3 = -1$. Start from descriptor 36: '`Exercise 4.2.1`'. **(e)** Put the four plots (a–d) into one combined plot. Start from mother descriptor 44: '`Exercise 4.2.12`'.

4.2.17 Use a potential of seven regions

$$V(x) = \begin{cases} V_1\,, & x < 0 \quad \text{region 1} \\ V_2\,, & 0 \le x < 0.5 \ \text{region 2} \\ V_3\,, & 0.5 \le x < 1 \ \text{region 3} \\ V_4\,, & 1 \le x < 1.5 \ \text{region 4} \\ V_5\,, & 1.5 \le x < 2 \ \text{region 5} \\ V_6\,, & 2 \le x < 2.5 \ \text{region 6} \\ V_7\,, & 2.5 \le x \quad \text{region 7} \end{cases} .$$

The values $V_1 = 0$, $V_2 = 10$, $V_4 = 9.5$, $V_6 = 9$, $V_7 = -1$ are kept fixed. The values in the regions 3 and 5 are varied: **(a)** below resonance $V_3 = V_5 =$

-0.35, **(b)** at resonance $V_3 = V_5 = -0.5$, **(c)** slightly above resonance $V_3 = V_5 = -0.65$, **(d)** above resonance $V_3 = V_5 = -0.75$. Start from descriptor 36: '`Exercise 4.2.1`'. **(e)** Put the four plots (a–d) for the absolute square of the wave function into one combined plot. Start from mother descriptor 44: '`Exercise 4.2.12`'. **(f)** Why is the variation of the differential resistance in this case much faster than in Exercise 4.2.16?

4.2.18 Consider the piecewise linear potential $(x_2, V_2) = (0, 5)$, $(x_3, V_3) = (2.5, -5)$, $(x_4, V_4) = (5, 5)$, $(x_5, V_5) = (5, 0)$ of Fig. 4.3. Switch the function to either real or imaginary part. **(a)** How can one determine the number of bound states contained in a potential by watching the wave function at small scattering energies or momenta? **(b)** Determine the number of bound states for this potential. It is customary to restrict the x interval boundaries of the plot to values of or slightly outside the potential region and to adjust the scale factor of the wave function. **(c)** Determine the number of bound states contained in the potential $(x_2, V_2) = (20, -15)$, $(x_3, V_3) = (40, 0)$ (i.e., the potential parameters of Fig. 4.7) in the same way.

4.2.19 **(a)** Plot the absolute square of the scattering wave function of the attractive triangular potential $(x_2, V_2) = (2, -1)$, $(x_3, V_3) = (4, 0)$ in the energy region $0.1 \le E \le 1$ and keep the plot on the screen. Start from descriptor 45: '`Exercise 4.2.19`'. **(b)** Consider now the following attractive potential with 10 regions: $(x_2, V_2) = (1.5, -0.07)$, $(x_3, V_3) = (2.5, -0.4)$, $(x_4, V_4) = (3, -0.8)$, $(x_5, V_5) = (3.5, -1)$, $(x_6, V_6) = (4, -0.8)$, $(x_7, V_7) = (4.5, -0.4)$, $(x_8, V_8) = (5.5, -0.07)$, $(x_9, V_9) = (7, 0)$. Plot again the absolute square of the scattering wave function. Start also from descriptor 45: '`Exercise 4.2.19`'. **(c)** What is the striking difference between these two results? What is the physical interpretation?

4.2.20 **(a)** Repeat Exercise 4.2.19 (b) for the (approximately linearly scaled) potential $(x_2, V_2) = (1.5, -0.15)$, $(x_3, V_3) = (2.5, -1.06)$, $(x_4, V_4) = (3, -2.4)$, $(x_5, V_5) = (3.5, -3)$, $(x_6, V_6) = (4, -2.4)$, $(x_7, V_7) = (4.5, -1.06)$, $(x_8, V_8) = (5.5, -0.15)$, $(x_9, V_9) = (7, 0)$. **(b)** Repeat Exercise 4.2.19 (b) for the (approximately linearly scaled) potential $(x_2, V_2) = (1.5, -0.3)$, $(x_3, V_3) = (2.5, -2.2)$, $(x_4, V_4) = (3, -4.4)$, $(x_5, V_5) = (3.5, -6)$, $(x_6, V_6) = (4, -4.4)$, $(x_7, V_7) = (4.5, -2.2)$, $(x_8, V_8) = (5.5, -0.3)$, $(x_9, V_9) = (7, 0)$.

4.3.1 Plot **(a)** the real part, **(b)** the imaginary part, **(c)** the absolute square of a harmonic wave of momentum $p_0 = 2.2$ for the time range $0 \le t \le 6$ in the potential

$$V(x) = \begin{cases} 0\,, & x < 0 \\ 32\,, & 0 \le x \end{cases} .$$

Start from descriptor 46: '`Exercise 4.3.1`'. **(d)** Why does the wave function look similar to the reflection of a wave on a fixed end?

4.3.2 (a,b,c) Repeat Exercise 4.3.1 (a,b,c) for the potential

$$V(x) = \begin{cases} 0\,,\ x < 0 \\ -3000\,,\ 0 \le x \end{cases}.$$

(d) Why does the wave function look similar to the reflection of a light wave on a denser medium?

4.3.3 (a,b,c) Repeat Exercise 4.3.1 (a,b,c) for a potential

$$V(x) = \begin{cases} 0\,,\ x < 0 \\ 32\,,\ 0 \le x \end{cases}$$

for a harmonic wave of momentum $p_0 = 8$ for the time range $0 \le t \le 6$. Start from descriptor 46: ‘`Exercise 4.3.1`’. **(d)** Why does the reflection pattern now look like one at a thinner medium?

4.3.4 Plot **(a)** the real part, **(b)** the imaginary part, **(c)** the absolute square of a harmonic wave of momentum $p_0 = 1.2$ for the time range $0 \le t \le 6$ in the potential of Exercise 4.2.2. Start from descriptor 46: ‘`Exercise 4.3.1`’. **(d)** Why do the plots of Exercise 4.2.2 (c) at $p_0 = 1.2$ and 4.3.4 (c) look alike?

4.3.5 Plot **(a)** the real part, **(b)** the imaginary part, **(c)** the absolute square of a harmonic wave of momentum $p_0 = 1.2$ for the time range $0 \le t \le 6$ in the potential of Exercise 4.2.3. Start from descriptor 46: ‘`Exercise 4.3.1`’. **(d)** Why is the amplitude of the real part of wave function to the right of the potential barrier time independent? **(e)** Why is the amplitude of the real part of the wave function to the left of the potential barrier time dependent?

4.3.6 Plot the motion of the wave packet incident on a step potential

$$V(x) = \begin{cases} 0\,,\ x < 0 \\ 6\,,\ x \ge 0 \end{cases}$$

for the width $\sigma_p = 0.3$ and the energies **(a)** $E = 2$, **(b)** $E = 6.5$, **(c)** $E = 8$. Start from descriptor 47: ‘`Exercise 4.3.6`’. **(d)** Describe the trend of the reflection probability. **(e)** Why is the transmitted wave packet in (b) faster than the classical particle?

4.3.7 Plot the motion of a wave packet incident on a down-step potential

$$V(x) = \begin{cases} 0\,,\ x < 0 \\ -6\,,\ x \ge 0 \end{cases}$$

for the width $\sigma_p = 0.3$ and the energies **(a)** $E = 1$, **(b)** $E = 2$, **(c)** $E = 4$. The initial position expectation value of the wave packet is $x_0 = -6$. Start from descriptor 47: ‘`Exercise 4.3.6`’.

4.3.8 (a–c) Repeat Exercise 4.3.7 for the potential

$$V(x) = \begin{cases} 0\,,\ x < 0 \\ -12\,,\ x \ge 0 \end{cases}.$$

(d) Why does the reflection probability increase with the height of the step (see Exercise 4.3.7)?

4.3.9 Plot the transmission probability $|A_N|^2$ for the potential

$$V(x) = \begin{cases} 0\,, & x < 0 \\ 16\,, & 0 \leq x < 4 \\ 0\,, & 4 \leq x \end{cases}$$

for the energy range $15 \leq E \leq 22$. Determine the energies of the four transmission resonances ($|A_N| = 1$) in this energy range. Start from descriptor 48: '`Exercise 4.3.9`'.

4.3.10 (a,b) Plot the scattering of the wave packet of width $\sigma_p = 0.05$ at the repulsive square-well potential of Exercise 4.3.9 for the lowest two resonance energies determined in Exercise 4.3.9. Choose eight intervals in the time range $0 \leq t \leq 12$. Start from descriptor 47: '`Exercise 4.3.6`'. **(c)** Explain the occurrence of the resonance phenomena. Give a qualitative argument for the energies at which they occur. **(d)** For the last four consecutive time instants determine the ratios of the maxima of the wave functions within the range of the barrier. Explain why these ratios are approximately equal.

4.3.11 (a) Plot the scattering of the wave packet of width $\sigma_p = 0.05$ at the repulsive square-well potential of Exercise 4.3.9 for the third resonance energy determined in Exercise 4.3.9. Choose eight intervals in the time range $0 \leq t \leq 12$. **(b)** Study the time range $8 \leq t \leq 12$ in eight intervals. Start from descriptor 47: '`Exercise 4.3.6`'.

4.3.12 (a) Repeat Exercise 4.3.11 (a) for the fourth resonance energy determined in Exercise 4.3.9. Choose eight intervals in the time range $0 \leq t \leq 8$. **(b)** Study the time range $6 \leq t \leq 8$ in eight intervals. Start from descriptor 47: '`Exercise 4.3.6`'. **(c)** Why is the exponential decay faster with higher resonance energy? Compare with the result of Exercises 4.3.7 and 4.3.8 and look at the Argand diagram of Exercise 4.3.9.

4.3.13 In the energy range $0 \leq E \leq 40$ study the energy dependence of the complex transmission amplitude **(a)** $|A_N|^2$ (descriptor 49: '`Exercise 4.3.13a`'), **(b)** $\mathrm{Re}\, A_N$ (descriptor 50: '`Exercise 4.3.13b`'), **(c)** $\mathrm{Im}\, A_N$ (descriptor 51: '`Exercise 4.3.13c`'), **(d)** $\mathrm{Im}\, A_N$ vs. $\mathrm{Re}\, A_N$ in an Argand plot (descriptor 52: '`Exercise 4.3.13d`') for the potential

$$V(x) = \begin{cases} 0\,, & x < 0 \\ -16\,, & 0 \leq x < 4 \\ 0\,, & 4 \leq x \end{cases} .$$

(e) Put the above plots into one combined plot with the positioning (d) upper-left field, (c) upper-right field, (b) lower-left field, (a) lower-right field. Start from descriptor 53: '`Exercise 4.3.13e`'. **(f)** Relate the prominent features of the absolute square and of the real and imaginary parts to the ones of the Argand plot.

4.3.14 Study the energy dependence of the quantities **(a)** $|T_\mathrm{T}|^2$, **(b)** $\mathrm{Re}\, T_\mathrm{T}$, **(c)** $\mathrm{Im}\, T_\mathrm{T}$, **(d)** $\mathrm{Im}\, T_\mathrm{T}$ vs. $\mathrm{Re}\, T_\mathrm{T}$ in an Argand plot for the potential of Exercise 4.3.13. Use the same descriptors as in Exercise 4.3.13. **(e)** Put the above plots into one combined plot with the same positioning as in Exercise 4.3.13 (e). Start from descriptor 53: ‘`Exercise 4.3.13e`’. **(f)** Relate the behavior of this quantity to the complex transmission amplitude A_N.

4.3.15 Repeat Exercise 4.3.14 for the complex reflection amplitude B_1.

4.3.16 Repeat Exercise 4.3.14 for the quantity T_{R}.

4.3.17 Repeat Exercise 4.3.13 with the energy range $0 \le E \le 1$ for the potential

$$V(x) = \begin{cases} 0\,, \; x < 0 \\ -5\,, \; 0 \le x < 1 \\ 0\,, \; 1 \le x \end{cases} .$$

Determine the energy of the lowest resonance.

4.3.18 Plot the bound states in the potential of Exercise 4.3.17 and determine their energy eigenvalues, see Sect. 3.3. Start from descriptor 54: ‘`Exercise 4.3.18`’.

4.3.19 Plot the time development of a wave packet under the action of the potential of Exercise 4.3.17. The wave packet comes in from the initial position $x_0 = -6$ at $t = 0$ with the energy $E = 0.1$ and the absolute width $\sigma_p = 0.05$. The time ranges are **(a)** $0 \le t \le 40$, **(b)** $40 \le t \le 80$, **(c)** $80 \le t \le 120$. They should be subdivided into eight intervals each. Start from descriptor 55: ‘`Exercise 4.3.19`’. **(d)** Using also the result of Exercise 4.3.18 interpret the behavior of the wave function inside the potential region.

4.3.20 **(a–e)** Repeat Exercise 4.3.13 (a–e) for the potential

$$V(x) = \begin{cases} 0\,, \; x < 0 \\ 16\,, \; 0 \le x < 0.2 \\ 0\,, \; 0.2 \le x < 2.2 \\ 16\,, \; 2.2 \le x < 2.4 \\ 0\,, \; 2.4 \le x \end{cases} .$$

(f) Determine the four lowest resonance energies.

4.3.21 Plot the time development of the wave packet under the action of the potential of Exercise 4.3.20 starting at $t = 0$ at $x_0 = -6$ with the lowest resonance energy and the relative width $\sigma_p/p_0 = 0.3$. **(a)** Plot nine instants in the time range $0 \le t \le 16$. Start from descriptor 56: ‘`Exercise 4.3.21a`’. **(b)** Plot 11 instants in the time range $8 \le t \le 24$. Change the range in x to $-3 \le x \le 3$. Start from descriptor 57: ‘`Exercise 4.3.21b`’.

4.3.22 Repeat Exercise 4.3.21 for the second resonance energy for the relative width $\sigma = 0.15$. **(a)** $0 \le t \le 6$, **(b)** $5 \le t \le 9$.

4.3.23 Repeat Exercise 4.3.21 for the third resonance energy for the relative width $\sigma_p/p_0 = 0.1$. **(a)** $0 \le t \le 4$, **(b)** $2 \le t \le 4$.

4.3.24 Repeat Exercise 4.3.21 for the fourth resonance energy for the relative width $\sigma_p/p_0 = 0.05$. **(a)** $0 \le t \le 3$, **(b)** $2.5 \le t \le 3$.

4.3.25 Study the behavior of a wave packet in a five-step potential modeling a quantum-effect device

$$V(x) = \begin{cases} V_1\,, & x < 0 \quad \text{region 1} \\ V_2\,, & 0 \le x < 0.5 \text{ region 2} \\ V_3\,, & 0.5 \le x < 1 \text{ region 3} \\ V_4\,, & 1 \le x < 1.5 \text{ region 4} \\ V_5\,, & 1.5 \le x < 2 \text{ region 5} \end{cases} .$$

The potential values $V_1 = 0$, $V_2 = 10$, $V_4 = 10.5$, $V_5 = 1$, remain unchanged. The value in region 3 varies **(a)** below resonance, $V_3 = 1.5$, **(b)** slightly below resonance, $V_3 = 1$, **(c)** at resonance, $V_3 = 0.5$, **(d)** above resonance, $V_3 = -0.5$. For the above cases (a–d) plot the absolute square of the wave function of a wave packet with the initial data at $t = 0$: $E_0 = 5.410$, $\sigma_p = 0.01 p_0$, $x_0 = -6$, moving in the interval $-50 \le x \le 50$ during the time interval $0 \le t \le 20$ in time steps of 1. Start from descriptor 47: '`Exercise 4.3.6`'. **(e)** Put the four plots (a–d) into a combined plot. Start from mother descriptor 53: '`Exercise 4.3.13e`'.

4.3.26 **(a)** Plot the time development of the absolute square of a wave packet ($x_0 = -20$, $E_0 = 0.25$, $\sigma_p = 0.1\,p_0$) incident on a symmetric attractive triangular potential $x_2 = 2$, $(x_3, V_3) = (4, 0)$ for values of $2.4 \le V_2 \le 3.0$ in steps of 0.1. Start from descriptor 58: '`Exercise 4.3.26`'. For what value of V_2 is the wiggly pattern in the leftmost region (or the reflected part of the wave packet) smallest? **(b)** For the resulting value of V_2 in (a) plot the time development of wave packets with increased energy values $E_0 = 0.35$ and $E_0 = 0.5$. Interpret the result.

4.3.27 Consider the piecewise linear potential $(x_2, V_2) = (1.5, -0.15)$, $(x_3, V_3) = (2.5, -1.06)$, $(x_4, V_4) = (3, -2.4)$, $(x_5, V_5) = (3.5, -3)$, $(x_6, V_6) = (4, -2.4)$, $(x_7, V_7) = (4.5, -1.06)$, $(x_8, V_8) = (5.5, -0.15)$, $(x_9, V_9) = (7, 0)$ and plot the time development of the scattering process for the wave packet of Exercise 4.3.26 (a) in this potential. Start from descriptor 58: '`Exercise 4.3.26`'. Plot also for the energies of Exercise 4.3.26 (b). Interpret the result. You may want to watch a movie for any of these situations.

4.3.28 Repeat Exercise 4.3.27 for the potential $(x_2, V_2) = (1.5, -0.3)$, $(x_3, V_3) = (2.5, -2.2)$, $(x_4, V_4) = (3, -4.4)$, $(x_5, V_5) = (3.5, -6)$, $(x_6, V_6) = (4, -4.4)$, $(x_7, V_7) = (4.5, -2.2)$, $(x_8, V_8) = (5.5, -0.3)$, $(x_9, V_9) = (7, 0)$.

4.4.1 In the range $0.1 \le E \le 40$ study the energy dependence of the complex transmission amplitude. Plot **(a)** $|A_N|^2$ (descriptor 49: '`Exercise 4.3.13a`') and read off the resonance energies, **(b)** Re A_N (descriptor 50: '`Exercise 4.3.13b`'), **(c)** Im A_N (descriptor 51: '`Exercise 4.3.13c`'), **(d)** Im A_N vs. Re A_N in an Argand plot (descriptor 52: '`Exercise 4.3.13d`') for the potential

$$V(x) = \begin{cases} 0\,, x < 0 \\ 16\,, 0 \le x < 1 \\ 0\,, 1 \le x \end{cases}.$$

(e) Put the above plots into one combined plot with the positioning (d) upper-left field, (c) upper-right field, (b) lower-left field, (a) lower-right field. Start from descriptor 53: '`Exercise 4.3.13e`'. **(f)** Relate the prominent features of the absolute square and of the real and imaginary parts to the ones of the Argand plot. **(g)** Calculate the resonance energies according to (4.56) and compare them to the values read off $|A_N|^2$ in (a).

4.4.2 In the range $0.1 \le E \le 40$ study the energy dependence of the quantities **(a)** $|T_\mathrm{T}|^2$, **(b)** $\mathrm{Re}\, T_\mathrm{T}$, **(c)** $\mathrm{Im}\, T_\mathrm{T}$, **(d)** $\mathrm{Im}\, T_\mathrm{T}$ vs. $\mathrm{Re}\, T_\mathrm{T}$ in an Argand plot for the potential of Exercise 4.4.1. Use the same descriptors as in Exercise 4.4.1. **(e)** Put the above plots into one combined plot with the same positioning as in Exercise 4.4.1 (e). Start from descriptor 53: '`Exercise 4.3.13e`'. **(f)** Relate the behavior of this quantity to the complex transmission amplitude A_N.

4.4.3 Repeat Exercise 4.4.2 for the complex reflection amplitude B_1.

4.4.4 Repeat Exercise 4.4.2 for the quantity T_R.

4.4.5 Repeat Exercise 4.4.1 for the potential

$$V(x) = \begin{cases} 0\,, x < 0 \\ 16\,, 0 \le x < 2 \\ 0\,, 2 \le x \end{cases}.$$

4.4.6 Repeat Exercise 4.4.2 for the potential of Exercise 4.4.5.

4.4.7 Repeat Exercise 4.4.6 for the complex reflection amplitude B_1.

4.4.8 Repeat Exercise 4.4.6 for the quantity T_R.

4.4.9 Explain qualitatively the differences in the behavior of the quantities A_N, T_T, B_1, T_R for the two potentials used in Exercises 4.4.1–4.4.4 and Exercises 4.4.5–4.4.8, respectively. In particular, argue why the distances of the energies at which $|A_N|^2 = 1$ vary with the potential the way it is observed in Exercises 4.4.1 and 4.4.5.

4.4.10 Repeat Exercise 4.4.1 for the potential

$$V(x) = \begin{cases} 0\,, x < 0 \\ 16\,, 0 \le x < 4 \\ 0\,, 4 \le x \end{cases}.$$

4.4.11 Repeat Exercise 4.4.2 for the potential of Exercise 4.4.10.

4.4.12 Repeat Exercise 4.4.11 for the complex reflection amplitude B_1.

4.4.13 Repeat Exercise 4.4.11 for the quantity T_R.

4.4.14 In the ranges **(a)** $0 \le E \le 60$, **(b)** $0 \le E \le 15$ plot the energy dependence of the absolute square $|A_N|^2$ of the transmission amplitude for a quasiperiodic potential of nine regions with four square wells of width 1 and depth -44 and three separating walls of width 0.2 and depth 0. **(c)** Why do the lowest transmission resonances form a band of four?

4.4.15 (a) In the energy range $0.001 \leq E \leq 5$ plot **(a1)** Im A_N vs. Re A_N, **(a2)** Im A_N, **(a3)** Re A_N, **(a4)** $|A_N|^2$ in an Argand plot for the symmetric attractive triangular potential $(x_2, V_2) = (2, -3)$, $(x_3, V_3) = (4, 0)$. Start from descriptors 15: '`Piecew. lin. pot.: Argand diagram - imaginary part vs. real part`', 16: '`Piecew. lin. pot.: Argand diagram - imaginary part`', 17: '`Piecew. lin. pot.: Argand diagram - real part`', 18: '`Piecew. lin. pot.: Argand diagram - absolute square`' for the individual plots and from 14: '`Piecew. lin. pot.: Argand diagram - combined plot`' for the mother descriptor. (The latter can be kept unchanged if the single descriptors are placed directly behind it in the given order.) Describe the behavior of A_N. **(b)** Repeat part (a) for the piecewise linear potential $(x_2, V_2) = (1.5, -0.15)$, $(x_3, V_3) = (2.5, -1.06)$, $(x_4, V_4) = (3, -2.4)$, $(x_5, V_5) = (3.5, -3)$, $(x_6, V_6) = (4, -2.4)$, $(x_7, V_7) = (4.5, -1.06)$, $(x_8, V_8) = (5.5, -0.15)$, $(x_9, V_9) = (7, 0)$. **(c)** Describe the differences between the cases (a) and (b). Plot also for both cases the quantity $|A_N|^2$ by setting the lower bound of the y interval (ordinate) to 0.9.

4.5.1 Produce a plot with descriptor 19: '`Stationary wave in linear potential`'. **(a)** Describe the difference between the curves of $\varphi(E, x)$ shown for different fixed values of E. **(b)** Leaving the plot on the screen produce a second plot with descriptor 19: '`Stationary wave in linear potential`' and, for this plot, change the sign of the acceleration g. Explain the symmetry between the two plots.

4.5.2 Produce side by side (using descriptor 19: '`Stationary wave in linear potential`') two plots of $\varphi(E, x)$ for $g = 3$ and $g = 1.5$. Explain the difference between them. [Compute the momentum $p(E, x)$ of a classical particle with total energy E at a point x and translate it into a de Broglie wavelength.]

4.6.1 Produce a plot with descriptor 20: '`Wave packet in linear potential`'. **(a)** Show Re $\psi(x, t)$. **(b)** Show the probability-current density $j(x, t)$.

4.6.2 (a) Use descriptor 20: '`Wave packet in linear potential`' to display quantile positions for $P = 0.8$. **(b)** Turn the plot into a multiple plot with two rows and two columns.

4.7.1 Produce a plot of quantile trajectories using descriptor 22: '`Linear potential: quantile trajectories`'. The middle trajectory (for $P = 0.5$) corresponds to the motion of a classical particle in a linear potential. Is that true also for the other trajectories?

4.12 Analogies in Optics

For a right-moving plane wave of light vertically incident on glass or other dielectrics we study reflection and refraction. We choose the x axis normal to the plane surface of the glass. The dielectric may consist of layers $1, 2, \dots, N$ of different refractive indices $n'_1, \dots, n'_N$. This divides the x axis into N regions:

$$n(x) = \begin{cases} n_1 = 1 \,, x < x_1 = 0 & \text{region } 1 \\ n_2 \qquad , x_1 \leq x < x_2 & \text{region } 2 \\ \vdots & \\ n_{N-1} \quad , x_{N-2} \leq x < x_{N-1} & \text{region } N-1 \\ n_N \qquad , x_{N-1} \leq x & \text{region } N \end{cases} \quad . \tag{4.66}$$

For simplicity we have set $n_1 = 1$. Hereby, all the different n_i are relative refractive indices $n_\ell = n'_\ell / n'_1$, with n'_ℓ ($\ell = 1, \dots, N$) being the absolute refractive index. For all further considerations in this system we may suppress the coordinates y and z so that we deal with a one-dimensional problem. The complex electric field strength (2.29) can be factorized into time-dependent and purely x-dependent factors

$$E_\mathrm{c}(\omega, x, t) = \mathrm{e}^{-\mathrm{i}\omega t} E_\mathrm{s}(\omega, x) \quad , \qquad E_\mathrm{s}(\omega, x) = A\mathrm{e}^{\mathrm{i}kx} \quad , \qquad k = \omega/c \quad . \tag{4.67}$$

The time-independent factor $E_\mathrm{s}(\omega, x)$ is called the *stationary electric field strength*. For $k > 0$, the real part of $E_\mathrm{c}(\omega, x, t)$,

$$\mathrm{Re}\, E_\mathrm{c}(\omega, x, t) = |A| \cos(\omega t - kx - \alpha) \quad , \tag{4.68}$$

represents a right-moving harmonic wave. Here we have decomposed the complex amplitude A into modulus $|A|$ and phase α:

$$A = |A| \mathrm{e}^{\mathrm{i}\alpha} \quad . \tag{4.69}$$

Therefore, for $k > 0$ we call

$$A\mathrm{e}^{\mathrm{i}kx} = E_{\mathrm{s}+}(\omega, x) \tag{4.70}$$

a 'right-moving' stationary electric field strength. By the same token, for $k > 0$,

$$A\mathrm{e}^{-\mathrm{i}kx} = E_{\mathrm{s}-}(\omega, x) \tag{4.71}$$

is called a 'left-moving' stationary field strength.

If a right-moving incoming monochromatic light wave of angular frequency ω and wave number $k = \omega/c$ falls onto the arrangement of dielectrics (4.66), we have an outgoing wave in region N,

$$E_N(\omega, x, t) = A'_N e^{ik_N x} \quad , \quad k_N = n_N k_1 \quad , \quad \text{region } N \quad , \tag{4.72}$$

only. For all other regions there is a reflected left-moving wave in addition to the right-moving one. Thus, the stationary electric field in any region ℓ, $1 \le \ell \le N-1$, is a superposition:

$$E_\ell(\omega, x) = A'_\ell e^{ik_\ell x} + B'_\ell e^{-ik_\ell x} \quad , \quad k_\ell = n_\ell k_1 \quad , \quad \text{region } \ell \quad . \tag{4.73}$$

The complex electric field strength is given by

$$E_c(\omega, x, t) = e^{-i\omega t} E_s(\omega, x) \tag{4.74}$$

with

$$E_s(\omega, x) = \begin{cases} E_1(\omega, x) \ , x < x_1 = 0 \\ E_2(\omega, x) \ , 0 \le x < x_2 \\ \vdots \\ E_N(\omega, x) \ , x_{N-1} \le x \end{cases} , \tag{4.75}$$

$\omega = c_\ell k_\ell$: angular frequency ($\ell = 1, \ldots, N$),
$k_\ell = n_\ell k_1 = n'_\ell k$: wave number in region ℓ,
$c_\ell = c/n'_\ell = c_1/n_\ell$: speed of light in region ℓ,
c : speed of light in vacuum,
n'_ℓ : absolute refractive index,
$n_\ell = n'_\ell / n'_1$: relative refractive index.

The expression (4.74) solves Maxwell's equations if the coefficients A'_ℓ and B'_ℓ in (4.73) are determined such that the function $E_s(\omega, x)$ is continuous and continuously differentiable at the end points of the regions $1, \ldots, N-1$, i.e.,

$$\begin{aligned} E_\ell(\omega, x_\ell) &= E_{\ell+1}(\omega, x_\ell) \\ \frac{dE_\ell}{dx}(\omega, x_\ell) &= \frac{dE_{\ell+1}}{dx}(\omega, x_\ell) \end{aligned} \quad , \quad \ell = 1, \ldots, N-1 \quad . \tag{4.76}$$

This yields a system of equations

$$\begin{aligned} A'_\ell e^{ik_\ell x_\ell} + B'_\ell e^{-ik_\ell x_\ell} &= A'_{\ell+1} e^{ik_{\ell+1} x_\ell} + B'_{\ell+1} e^{-ik_{\ell+1} x_\ell} \quad , \\ k_\ell (A'_\ell e^{ik_\ell x_\ell} - B'_\ell e^{-ik_\ell x_\ell}) &= k_{\ell+1} (A'_{\ell+1} e^{ik_{\ell+1} x_\ell} - B'_{\ell+1} e^{-ik_{\ell+1} x_\ell}) \end{aligned} \tag{4.77}$$

for $\ell = 1, \ldots, N-1$. The condition (4.72) in region N is implemented by setting $B_N = 0$. The $(2N-1)$ coefficients A'_ℓ, B'_ℓ are determined by the system (4.77) of $(2N-2)$ equations. Choosing again A'_1 as the independent variable, determining the incoming flux of light, (4.77) constitutes a system of $(2N-2)$ inhomogeneous linear equations, the term with A'_1 being the inhomogeneity. Its solution yields the coefficients $A'_2, \ldots, A'_N$ and $B'_1, \ldots, B'_{N-1}$ as functions of the wave number $k = k_1$ of the incident wave.

The energy density of the electromagnetic field of light in vacuum averaged[1] over one period $T = 2\pi/\omega$ is given by (2.37). In glass with refractive index n_ℓ it is

$$w_\ell = n_\ell^2 \frac{\varepsilon_0}{2} E_\ell^* E_\ell \quad . \tag{4.78}$$

The *average density of the energy flux* in the wave in glass is

$$S_\ell = w_\ell c_\ell = n_\ell c \frac{\varepsilon_0}{2} E_\ell^* E_\ell \quad , \tag{4.79}$$

where $c_\ell = c/n_\ell$ is the speed of light in glass. Because of the discontinuity of n when passing from one material to the other, neither of the two quantities is continuous. Therefore, we plot in addition to Re E_c and Im E_c the absolute square $E_c^* E_c$.

For stationary waves the current densities S of the electromagnetic energy of right-moving and left-moving waves are proportional to the squares of the transmission and reflection coefficients

$$A_\ell = \sqrt{n_\ell} A'_\ell \quad , \qquad B_\ell = \sqrt{n_\ell} B'_\ell \quad . \tag{4.80}$$

Current conservation of the electromagnetic energy simply states that the sum of the transmitted and reflected currents is equal to the incoming current. As a conventional normalization we shall set $A_1 = 1$. Then the conservation of the current of electromagnetic energy leads to the *unitarity relation*

$$|A_N|^2 + |B_1|^2 = |A_1|^2 = 1 \quad , \tag{4.81}$$

which is of the same form as (4.34) for the transmission and reflection coefficients in one-dimensional quantum mechanics.

A *transmission resonance* occurs for those values of the wave number k for which $|A_N|$ is at a maximum and $|B_1|$ therefore at a minimum. For three regions with refractive indices n_1, n_2, n_3, the resonant wave numbers can be determined by simple arguments. In this arrangement there are two surfaces at $x_1 (= 0)$ and $x_2 (= d)$ where reflection occurs. The coefficient $|B_1|$ is at a minimum if the two waves reflected at x_1 and x_2 interfere destructively in the region $x < x_1 = 0$ to the left of $x_1 = 0$. Because there is a phase shift of π for reflection on an optically denser medium, we have to distinguish two cases:

i)

$$1 = n_1 < n_2 < n_3 \quad \text{or} \quad 1 = n_1 > n_2 > n_3 \quad . \tag{4.82}$$

In these cases the relative phase shift of the two waves reflected at x_1 and x_2 is simply given by $\delta = 2k_2 d$, $k_2 = n_2 k$. For maximally destructive interference the phase shift δ has to be equal to odd multiples of π,

[1] See footnote on page 11.

$$2k_2d = (2m+1)\pi \quad \text{or} \quad k_2 = \frac{(2m+1)\pi}{2d} \quad , \quad m = 0, 1, 2, \dots \quad . \tag{4.83}$$

For the wavelength λ_2 in region 2 we find in terms of the thickness of the material

$$\lambda_2 = 4d/(2m+1) \quad . \tag{4.84}$$

The resonant wavelengths in region 2 are odd fractions of $4d$. The longest resonant wavelength ($m = 0$) is then four times the thickness of the middle layer of material. This is the well-known $d = \lambda/4$ condition that ensures a minimization of reflection for light of this wavelength in an optical system with three different refractive indices. It is used to produce antireflex lenses, etc., by coating the surface of the glass with a material transparent for visible light and a thickness of $d = \lambda/4$ for an average wavelength of the visible spectrum. In order to achieve the absence of reflection for this wavelength – and therefore, little reflection for neighboring wavelengths – the refractive indices have to be chosen suitably according to

$$n_2 = \sqrt{n_1 n_3} \quad . \tag{4.85}$$

For coated lenses in air n_1 is the refractive index of air, n_2 that of the coating, and n_3 that of the glass.

ii)

$$1 = n_1 < n_2 > n_3 \quad \text{or} \quad 1 = n_1 > n_2 < n_3 \quad . \tag{4.86}$$

For these cases in addition to the relative phase shift $\delta = 2k_2d$ caused by the difference $2d$ in the length of the light path there is the phase shift of π from the reflection at the denser medium. For maximally destructive interference we find the condition

$$2k_2d + \pi = (2m+1)\pi \quad \text{or} \quad k_2 = m\pi/d \quad , \quad m = 1, 2, \dots \quad . \tag{4.87}$$

For the wavelength λ_2 in region 2 we find

$$\lambda_2 = 2d/m \quad . \tag{4.88}$$

The longest resonant wavelength is $2d$. All others are integer fractions of $2d$.

The stationary waves can be superimposed to form a wave packet of finite energy content in analogy to (2.32). Using the harmonic electric field strength (4.74), with the stationary field strength (4.75), we find that

$$E_{\mathrm{c}}(x,t) = E_0 \int f(k) \mathrm{e}^{-\mathrm{i}kx_0} E_{\mathrm{c}}(\omega, x, t)\,\mathrm{d}k \tag{4.89}$$

is a Gaussian wave packet centered at $t = 0$ around the initial position $x = x_0$, if we choose the Gaussian spectral function

$$f(k) = \frac{1}{\sqrt{2\pi}\sigma_k} \exp\left[-\frac{(k-k_0)^2}{2\sigma_k^2}\right] \quad . \tag{4.90}$$

Further Reading

Alonso, Finn: Vol. 3, Chaps. 19, 20
Berkeley Physics Course: Vol. 3, Chaps. 4, 5, 6
Brandt, Dahmen: Chap. 2
Feynman, Leighton, Sands: Vol. 2, Chaps. 32, 33
Hecht, Zajac: Chaps. 4, 7, 8, 9

4.13 Reflection and Refraction of Stationary Electromagnetic Waves

Aim of this section: Computation and demonstration of the stationary electric field E_s, (4.75), for a right-moving incoming wave in a system (4.66) of dielectrics as a function of position x and wave number k.

This section is the analog in optics of Sect. 4.2. A plot similar to Fig. 4.16 is produced showing the stationary electric field strength $E_s(k, x)$ as a function of x for various values of the wave number k in the x, k plane. Zero lines are shown as dashed lines. The vertical lines crossing them indicate the boundaries between media with different refractive index.

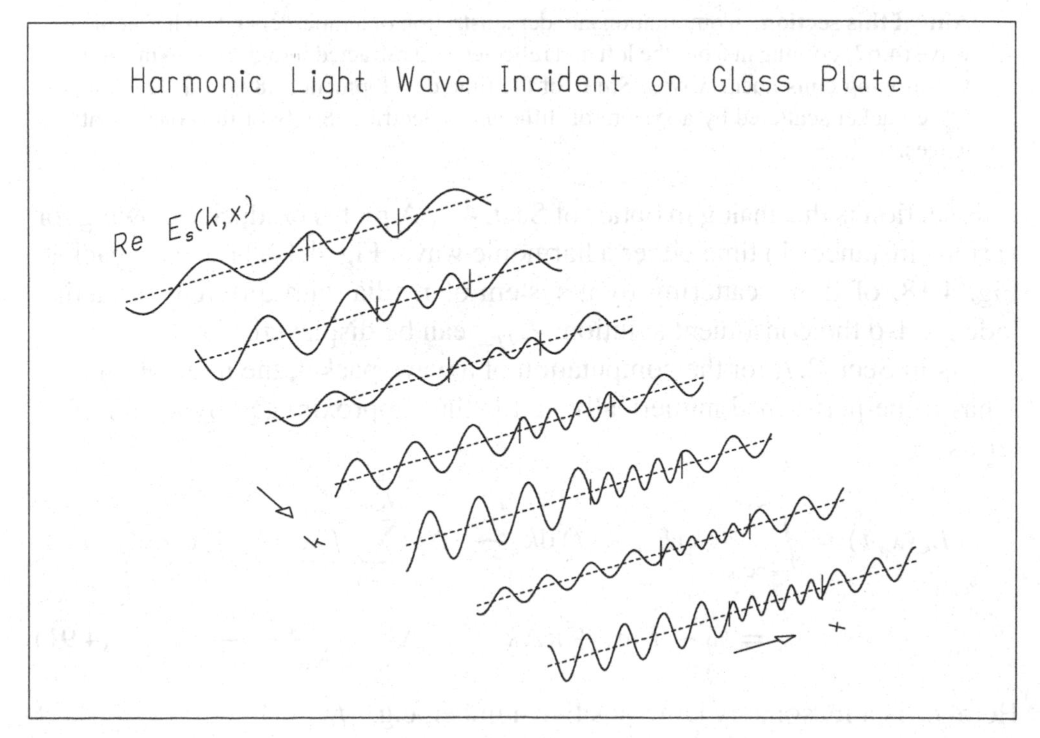

Fig. 4.16. Plot produced with descriptor `Stationary 1D scattering (optics)` on file `1D_Scattering.des`

On the subpanel **Physics—Wave** you can choose to show the full solution or the right-moving or left-moving constituent wave in one of the media. You can also set the position x_0 for which the phase of the incident wave vanishes.

On the subpanel **Physics—Misc.** there are two **Graphical Items**, the dash length of the zero lines and the height of the vertical lines separating the regions with different media.

On the subpanel **Physics—Media** you find the parameters defining the systems of media. These are the number N of regions, the boundaries x_i between region i and region $i+1$, and the refractive indices n_i.

Movie Capability: Indirect. After conversion to direct movie capability the start end end values of the wave number can be changed in the subpanel **Movie** (see Sect. A.4) of the parameter panel. In the resulting movie the wave number of the stationary state changes with time. Such a movie, in particular, is useful for the demonstration of transmission resonances.

Example Descriptor on File `1D_Scattering.des`

- `Stationary 1D scattering (optics)` (see Fig. 4.16)

4.14 Time-Dependent Scattering of Light

Aim of this section: Computation and demonstration of a time-dependent harmonic wave (4.67) coming in from the left and reflected and refracted into right-moving and left-moving constituent waves. Study of the time development (4.89) of a Gaussian wave packet scattered by a system of different dielectrics. Study of the constituent waves.

This section is the analog in optics of Sect. 4.3. A plot is produced showing for various instances in time either a harmonic wave, Fig. 4.17, or a wave packet, Fig. 4.18, of light scattering by a system of media with different refractive index. Also the constituent solutions $E_{cj\pm}$ can be displayed.

As in Sect. 2.7, for the computation of a wave packet, the integration over k has to be performed numerically and is thus approximated by a sum of N terms,

$$E_c(x,t) = \int_{-\infty}^{\infty} f(k) E_{ck}(x,t)\,\mathrm{d}k \;\rightarrow\; \Delta k \sum_{n=0}^{N_{\text{sum}}} f(k_n) E_{ck_n}(x,t) \quad ,$$

$$k_n = k_0 - f_\sigma \sigma_k + n\Delta k \quad , \quad \Delta k = \frac{2 f_\sigma \sigma_k}{N_{\text{sum}} - 1} \quad . \tag{4.91}$$

Here f_σ is a reasonably large positive number, e.g., $f_\sigma = 3$.

Again the solution will be periodic in x. That is, only in a limited x region you will get a good approximation to the true solution. The patterns for $E_c(x,t)$ [or $E_{cj+}(x,t)$ and $E_{cj-}(x,t)$] will repeat themselves periodically

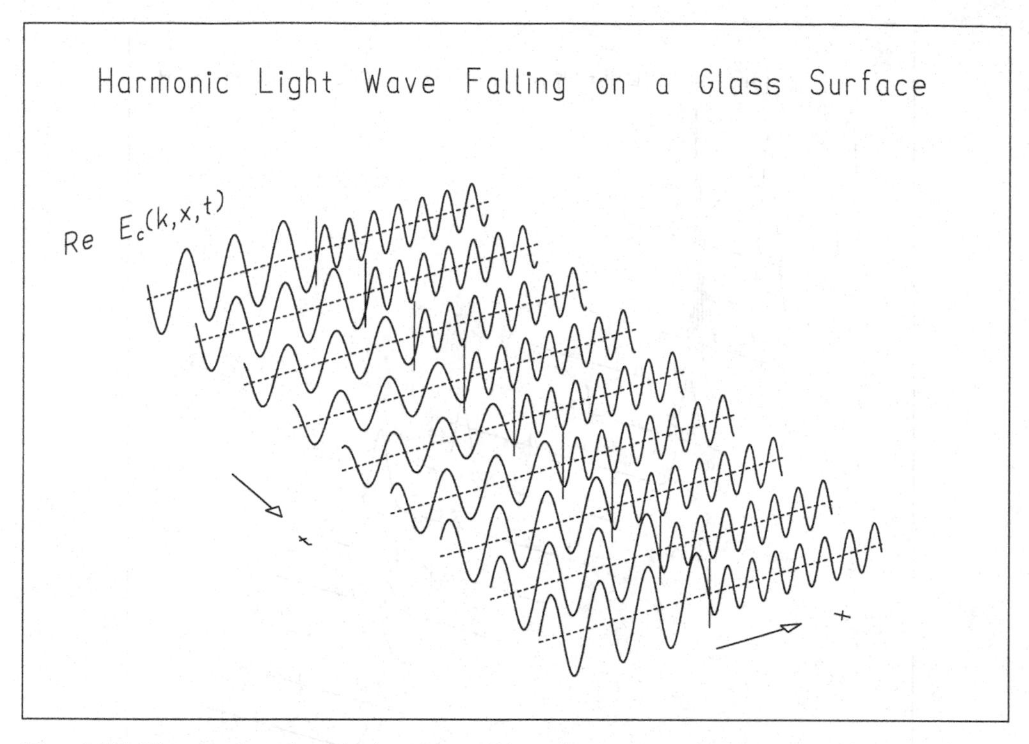

Fig. 4.17. Plot produced with descriptor `Time-dependent 1D scattering (optics): harmonic wave` on file `1D_Scattering.des`

along the x direction, the period Δx becoming larger as $N_{\rm sum}$ increases. You will have to make sure that the x interval of your C3 window is small compared to Δx.

On the bottom of the subpanel **Physics—Comp. Coord.** you can choose to show either the **Absolute Square** or the **Real Part** or the **Imaginary Part** of the complex electric field strength.

The subpanel **Physics—Wave (Packet)** carries the following items:

- **Full/Constituent Solution** – You can choose to plot $E_{\rm c}$ or $E_{{\rm c}j+}$ or $E_{{\rm c}j-}$. If you want to plot a constituent wave, you have to give the region j for which it is valid.
- **Incoming Wave** – You can choose either a **Harmonic Wave** or a **Wave Packet**. The number $N_{\rm int}$, $1 \le N_{\rm int} \le 200$, determines the number $N_{\rm sum} = 2N_{\rm int}+1$ of terms and f_σ is the other parameter needed in the numerical approximation of the wave packet, see (4.91).
- **Width in Wave Number is Given** – You may choose to give the width either **as Fraction of k_0**, i.e., you enter $f = \sigma_{k0}/k_0$, or **in Absolute Units**, i.e., you enter σ_{k0} directly. (This item is available only if an incoming wave packet was chosen.)
- **Harmonic Wave/Wave Packet** – depending on the choice above. Here you find the numerical values for the incoming wave or wave packet:

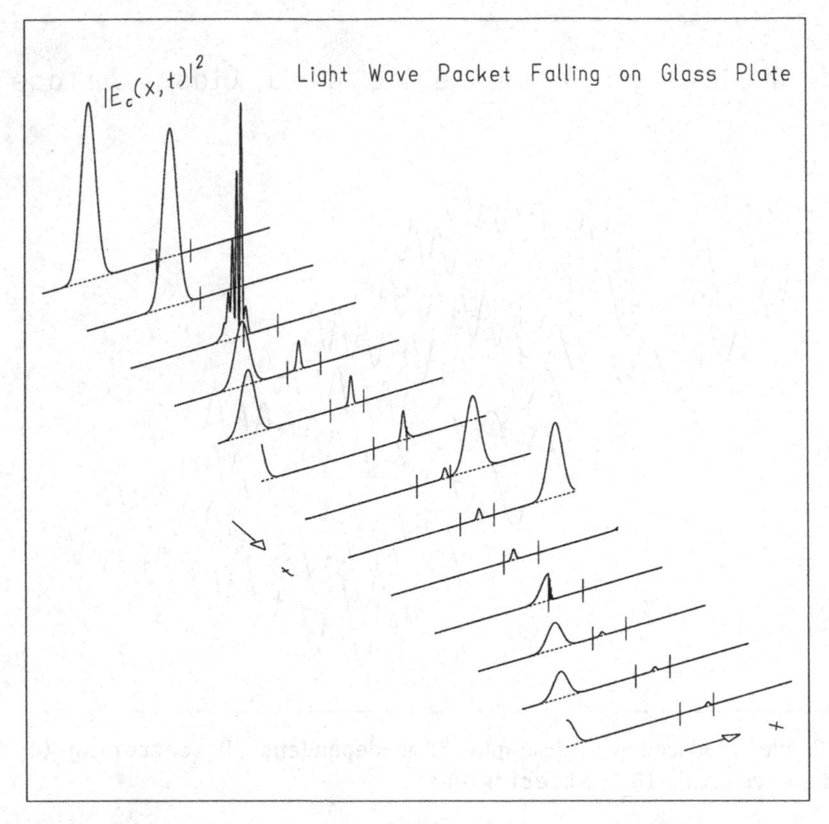

Fig. 4.18. Plot produced with descriptor `Time-dependent 1D scattering (optics): wave packet` on file `1D_Scattering.des`

– a position x_0 for which the phase of the incoming wave (for a packet: of all waves in the sum) vanishes,
– the wave number k_0,
– for a wave packet only: the width σ_{k0} in wave number (or the corresponding fraction $f = \sigma_{k0}/k_0$).

The subpanel **Physics—Misc.** contains two **Graphical Items**, the dash length of the zero lines and the height of the vertical lines separating different media.

The subpanel **Physics—Media** is as in Sect. 4.13.

Movie Capability: Indirect. After conversion to direct movie capability the start and end times can be changed in the subpanel **Movie** (see Sect. A.4) of the parameter panel.

Example Descriptors on File `1D_Scattering.des`

- `Time-dependent 1D scattering (optics): harmonic wave` (see Fig. 4.17)
- `Time-dependent 1D scattering (optics): wave packet` (see Fig. 4.18)

4.15 Transmission, Reflection, and Argand Diagram for a Light Wave

Aim of this section: Presentation of the complex transmission coefficient $A_N(k)$ and the complex reflection coefficient $B_1(k)$, see (4.80).

If $C(k)$ is one of these quantities, we want to illustrate its wave-number dependence by four different graphs,

- the Argand diagram $\mathrm{Im}\{C(k)\}$ vs. $\mathrm{Re}\{C(k)\}$,
- the real part $\mathrm{Re}\{C(k)\}$ as a function of k,
- the imaginary part $\mathrm{Im}\{C(k)\}$ as a function of k,
- the absolute square $|C(k)|^2$ as a function of k.

It is customary to draw an Argand diagram ($\mathrm{Im}\{C(k)\}$ vs. $\mathrm{Re}\{C(k)\}$) and graphs $\mathrm{Im}\{C(k)\}$ and $\mathrm{Re}\{C(k)\}$ in such a way that the graphs appear to be projections to the right and below the Argand diagram, respectively. You can do that by using a mother descriptor, which in turn quotes several individual descriptors (see Appendix A.10) as in the example plot, Fig. 4.19.

All four plots in Fig. 4.19 are of the type *2D function graph*. The Argand diagram (top left) is a *parameter representation* $x = x(p)$, $y = y(p)$. The two plots on the right-hand side are *Cartesian plots* $y = y(x)$. The plot in the bottom-left corner is an *inverse Cartesian plot* $x = x(y)$.

On the subpanel **Physics—Comp. Coord.** you can select to compute one of the two functions

$$A_N(k),\ B_1(k) \quad .$$

You can further choose the type of 2D function graph you want to produce:

- Argand diagram,
- $y = y(x)$,
- $x = x(y)$.

For the latter two types of function graphs you still can choose to present as dependent variable the **Absolute Square**, the **Real Part**, or the **Imaginary Part** of the function selected.

On the subpanel **Physics—Variables** you find two items,

- **Phase-Fixing Position** – At x_0 the phase of the incoming wave is zero.
- **Range of Independent Variable in Argand Diagram** – The variable is the wave number k. It is varied between the boundaries k_{beg} and k_{end}.

On the subpanel **Physics—Media** you find all parameters defining the system of media, namely, the number N of regions, the boundaries x_j between the regions, and the constant refractive indices n_j in the regions.

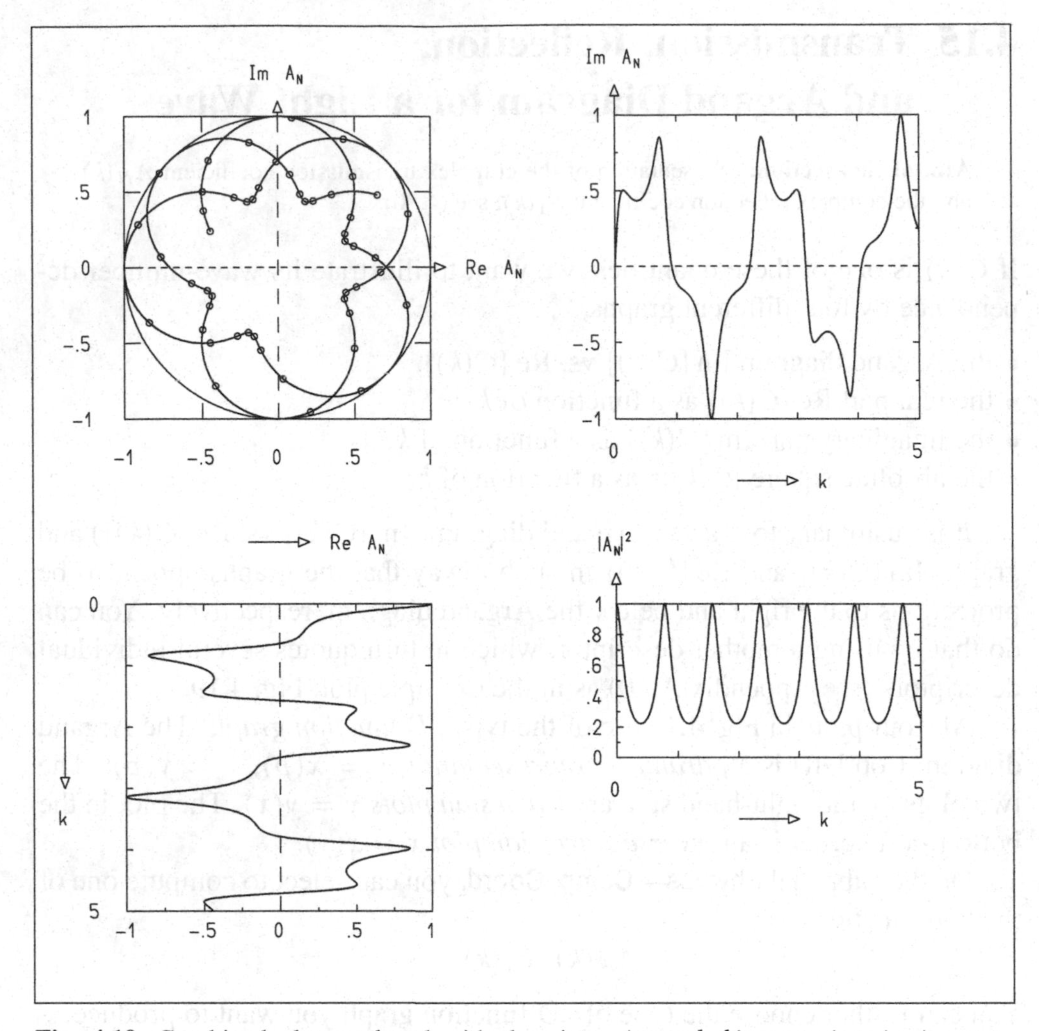

Fig. 4.19. Combined plot produced with descriptor `Argand diagram (optics): combined plot` on file `1D_Scattering.des`. This descriptor quotes four other descriptors to generate the individual plots situated in the top-left, top-right, bottom-left, and bottom-right corners, respectively

Movie Capability (only for the Argand diagram proper): Direct. The trajectory in the complex plane is seen to develop as the range of the independent variable is increased proportional to time.

Example Descriptors on File `1D_Scattering.des`

- `Argand diagram (optics): combined plot` (see Fig. 4.19, this is a mother descriptor quoting the four descriptors listed below, each describing one of the four plots in the combined plot)
- `Argand diagram (optics): imaginary part vs. real part`
- `Argand diagram (optics): imaginary part`
- `Argand diagram (optics): real part`
- `Argand diagram (optics): absolute square`

4.16 Exercises

Please note:

(i) For the following exercises use descriptor file `1D_Scattering.des`.

(ii) In many exercises we use refractive indices with numerical values much larger than available in ordinary dielectrics like glass.

4.13.1 Plot **(a)** the real part, **(b)** the imaginary part, **(c)** the absolute square of the stationary electric field strength in vacuum vertically incident on glass of refractive index $n_2 = 2$ extending from $x = 0$ to infinity. As the range for the wave numbers choose $0.1 \leq k \leq 5$. Start from descriptor 59: '`Exercise 4.13.1`'. **(d)** Why is the wavelength in region 2, $x \geq 0$, shorter than in region 1, $x < 0$? **(e)** Why is the absolute square constant in region 2? **(f)** What is the origin of the wiggly pattern in region 1?

4.13.2 (a,b,c) Repeat Exercise 4.13.1 (a,b,c) for $n_2 = 10$. Start from descriptor 59: '`Exercise 4.13.1`'. **(d)** Why is the transmission of light into region 2 close to zero?

4.13.3 (a,b,c) Repeat Exercise 4.13.1 (a,b,c) for $n_2 = 0.1$. Start from descriptor 59: '`Exercise 4.13.1`'. This case corresponds to the reflection and refraction of a wave propagating inside an optically denser medium ($n_1 = 1$) incident on a ten-times-thinner medium ($n_2 = 0.1$). **(d)** Why is the transmission into the thinner medium large?

4.13.4 Determine the phase shift between the incident and reflected wave for **(a)** reflection at an optically denser medium $n_2 = 10$, **(b)** reflection at an optically thinner medium $n_2 = 0.1$. To this end compare the phases of the incoming and reflected constituent waves in region 1 with each other. Choose a wave-number range $0.01 \leq k \leq 5$ in six wave-number intervals of equal length. Start from descriptor 59: '`Exercise 4.13.1`'.

4.13.5 Plot **(a)** the real part, **(b)** the imaginary part, **(c)** the absolute square of the stationary electric field strength for a sheet of glass of thickness $d = 1$ and refractive index 2. In the domain behind the glass choose $n = 1$. Divide the wave-number range $0.001 \leq k \leq 3\pi/2$ into six intervals of length $\pi/4$. Start from descriptor 59: '`Exercise 4.13.1`'. **(d)** What is the phenomenon behind the absence of wiggles of the absolute square in region 1 for the wave number $k = \pi/2, \pi, 3\pi/2$? **(e)** Why do the resonances in this case occur at $\lambda = 2d/m, m = 1, 2, 3, \ldots$?

4.13.6 (a,b,c) Repeat Exercise 4.13.5 (a,b,c), however, with a denser medium with $n = 4$ behind the glass. Start from descriptor 59: '`Exercise 4.13.1`'. **(d)** Why do the transmission resonances occur at values $\lambda = 4d/(2m + 1)$? **(e)** Why is the square of the electric field in region 3 different from region 1?

4.14.1 Plot the motion of **(a)** the real part, **(b)** the imaginary part, **(c)** the absolute square of a harmonic light wave for different refractive indices in three regions: $n_1 = 1, x < 0$; $n_2 = 2, 0 \le x < 1$; $n_3 = 1, 1 \le x$ for the wave number $k = 5\pi/4$. Choose the time range $0.001 \le t \le 1$ and the x range $-4 \le x \le 4$. Start from descriptor 60: '`Exercise 4.14.1`'. **(d)** Why are the amplitudes of the electric field strength time independent in region 3? **(e)** Why is this not so in region 1?

4.14.2 Plot the real parts of the constituent waves for Exercise 4.14.1: **(a)** E_{1+}, **(b)** E_{1-}, **(c)** E_{2+}, **(d)** E_{2-}, **(e)** E_{3+}, **(f)** E_{3-}. Start from descriptor 60: '`Exercise 4.14.1`'. **(g)** Why is there no constituent wave E_{3-}?

4.14.3 Repeat Exercise 4.14.1 (a,b,c) for $k = 3\pi/2$.

4.14.4 (a–f) Repeat Exercise 4.14.2 (a–f) for $k = 3\pi/2$. **(g)** Why is there no constituent wave E_{1-}?

4.14.5 A light wave packet of $k_0 = 7.854$ and relative width $\sigma_k/k_0 = 0.01$ is incident on glass of refractive index $n = 4$ and thickness $d = 9.9$ mounted between $x = 0.1$ and $x = 10$. Its initial position is $x_0 = -15$. Plot the time dependence of the absolute square of the electric field strength of the wave packet for 10 intervals between **(a)** $0 \le t \le 60$ and **(b)** $60 \le t \le 120$. The interference pattern upon reflection is resolved only for sufficiently high accuracy. Start from descriptor 61: '`Exercise 4.14.5`'. **(c)** By which factor is the speed of the wave packet in the glass slower than in vacuum? **(d)** Why is the amplitude of $|E_c|^2$ inside the glass plate so much smaller than in vacuum? **(e)** Why does the width of the wave packet shrink upon entering the glass?

4.14.6 (a,b) Repeat Exercise 4.14.5 (a,b), however, with two additional layers of refractive index $n = 2$ for $0 \le x < 0.1$ and the other for $10 \le x < 10.1$. Start from descriptor 61: '`Exercise 4.14.5`'. **(c)** Why does practically no reflection occur at any of the surfaces of the regions of different refractive index?

4.14.7 (a) Repeat Exercise 4.14.5 (a) for coated glass layer 1 (coating): $0 \le x < 0.1, n_1 = 1.2247$; layer 2 (glass): $0.1 \le x < 10, n_2 = 1.5$; layer 3 (coating): $10 \le x < 10.1, n_3 = 1.2247$. As the average wave number k_0 of the wave packet, choose 12.826. Start from descriptor 61: '`Exercise 4.14.5`'. **(b)** Calculate the thickness of a coating of the above refractive index for visible light of a vacuum wavelength $\lambda = 550\,\text{nm}$.

4.14.8 (a) Repeat Exercise 4.14.7 (a) without the coating. Start from descriptor 61: '`Exercise 4.14.5`'. **(b)** Why is the reflected part of the wave packet so much smaller than in Exercise 4.14.5?

4.14.9 (a) Repeat Exercise 4.14.7 (a) for a wave number 1.5 times larger. **(b)** Explain why coatings of actual optical lenses often reflect bluish light.

4.15.1 Study the transmission and reflection coefficients of an arrangement of dielectrics of three regions for a range in wave number, $0.001 \le k \le 30$:

region 1: $n_1 = 1$; region 2: $0 \le x < 0.1$, $n_2 = 2$; region 3: $0.1 \le x$, $n_3 = 4$. Start from descriptor 62: '`Exercise 4.15.1`'. Plot **(a)** the absolute square $|A_N|^2$ of the transmission coefficient, **(b)** the absolute square $|B_1|^2$ of the reflection coefficient, **(c)** the Argand diagram of A_N, **(d)** the Argand diagram of B_1. For (c) and (d) start from descriptor 63: '`Exercise 4.15.1d`'. **(e)** Read the wave numbers of resonant transmission off the graph and compare them with the values given by the $(2m + 1)\lambda/4 = d$ condition.

4.15.2 Repeat Exercise 4.15.1 for the choice of refractive indices $n_2 = 1.2247$, $n_3 = 1.5$.

5. A Two-Particle System: Coupled Harmonic Oscillators

Contents: Wave function of two distinguishable particles. Hamiltonian of two coupled oscillators. Separation of center-of-mass and relative coordinates. Stationary two-particle wave functions and eigenvalues. Initial Gaussian wave packet for distinguishable particles. Time evolution of the Gaussian wave packet. Marginal distributions. Wave functions for distinguishable particles. Symmetrization and antisymmetrization. Pauli principle. Bosons and Fermions. Normal oscillations.

5.1 Physical Concepts

5.1.1 The Two-Particle System

The *wave function of a two-particle system* in one spatial dimension, $\psi(x_1, x_2, t)$, is a function of two coordinates x_1, x_2 and the time t. The Hamiltonian

$$H = T_1 + T_2 + V(x_1, x_2) \tag{5.1}$$

consists of the two kinetic energies

$$T_i = -\frac{\hbar^2}{2m_i}\frac{\partial^2}{\partial x_i^2} \quad , \quad i = 1, 2 \quad , \tag{5.2}$$

and the potential energy $V(x_1, x_2)$ of the two particles. The time-dependent Schrödinger equation has the usual form

$$\mathrm{i}\hbar\frac{\partial}{\partial t}\psi(x_1, x_2, t) = H\psi(x_1, x_2, t) \quad . \tag{5.3}$$

With the separation of time and spatial coordinates

$$\psi(x_1, x_2, t) = \mathrm{e}^{-\mathrm{i}Et/\hbar}\varphi_E(x_1, x_2) \tag{5.4}$$

we obtain the stationary Schrödinger equation

S. Brandt et al., *Interactive Quantum Mechanics: Quantum Experiments on the Computer*,
DOI 10.1007/978-1-4419-7424-2_5, © Springer Science+Business Media, LLC 2011

$$H\varphi_E(x_1, x_2) = E\varphi_E(x_1, x_2) \tag{5.5}$$

for the stationary wave function $\varphi_E(x_1, x_2)$.

In the following we shall deal with two particles of equal mass $m_1 = m_2 = m$ throughout. These particles are bound to the origin by harmonic-oscillator potentials

$$V_1(x_1) = \frac{k}{2}x_1^2 \quad , \quad V_2(x_2) = \frac{k}{2}x_2^2 \tag{5.6}$$

with the same *spring constants* $k > 0$. In addition they are coupled to each other through the harmonic two-particle potential

$$V_c(x_1 - x_2) = \frac{\kappa}{2}(x_1 - x_2)^2 \tag{5.7}$$

with the coupling constant κ. Thus the Hamiltonian reads

$$H = T_1 + T_2 + V_1(x_1) + V_2(x_2) + V_c(x_1 - x_2) \quad . \tag{5.8}$$

With the *total mass* M and the *reduced mass* μ,

$$M = 2m \quad , \quad \mu = m/2 \quad , \tag{5.9}$$

and *center-of-mass coordinate* R and *relative coordinate* r,

$$R = (x_1 + x_2)/2 \quad , \quad r = x_2 - x_1 \quad , \tag{5.10}$$

the Hamiltonian can be separated into two terms,

$$H = H_R + H_r \quad , \tag{5.11}$$

each depending on one coordinate only:

$$H_R = -\frac{\hbar^2}{2M}\frac{\mathrm{d}^2}{\mathrm{d}R^2} + kR^2 \quad , \quad H_r = -\frac{\hbar^2}{2\mu}\frac{\mathrm{d}^2}{\mathrm{d}r^2} + \frac{1}{2}\left(\frac{k}{2} + \kappa\right)r^2 \quad . \tag{5.12}$$

Both the *center-of-mass motion* and the *relative motion* are harmonic oscillations. They can be separated by factorizing the stationary wave function

$$\varphi_E(x_1, x_2) = U_N(R)u_n(r) \quad , \tag{5.13}$$

where N and n are the quantum numbers of the center-of-mass and relative motion, respectively. The factors fulfill the stationary Schrödinger equations

$$\begin{aligned} H_R U_N(R) &= \left(N + \tfrac{1}{2}\right)\hbar\omega_R\, U_N(R) \quad , \\ H_r u_n(r) &= \left(n + \tfrac{1}{2}\right)\hbar\omega_r\, u_n(r) \end{aligned} \tag{5.14}$$

with the angular frequencies

$$\omega_R^2 = k/m \quad , \quad \omega_r^2 = (k + 2\kappa)/m \quad . \tag{5.15}$$

The eigenvalue in (5.5),

$$E = E_N + E_n \quad , \tag{5.16}$$

is the sum of the eigenvalues

$$E_N = \left(N + \tfrac{1}{2}\right) \hbar\omega_R \quad \text{and} \quad E_n = \left(n + \tfrac{1}{2}\right) \hbar\omega_r \tag{5.17}$$

of the center-of-mass and relative motion. The eigenfunctions $U_N(R)$ and $u_n(r)$ are the eigenfunctions (3.14) of harmonic oscillators of single particles with the angular frequencies ω_R and ω_r, respectively.

5.1.1.1 Entanglement If there is no coupling, i.e., for $\kappa = 0$, the eigenstates (5.13) of the system of two oscillators are just products of two single-particle oscillator eigenstates,

$$\varphi_E(x_1, x_2) = \varphi_{E_1}(x_1)\varphi_{E_2}(x_2) \quad , \quad E = E_1 + E_2 \quad .$$

Such a simple factorization does not hold, however, for the eigenfunctions of the system if the two oscillators are coupled, i.e., if they interact with each other. The appearance of a two-particle wave function, which is not simply the product of the single-particle wave functions of the two particles forming the system, was called *entanglement* by Schrödinger who first discussed the situation. For $\kappa \neq 0$ the wave function (5.13) are *entangled states.* Such states could be used in possible future *quantum computers.*

5.1.2 Initial Condition for Distinguishable Particles

For the moment we assume that the two particles are *distinguishable*, e.g., that one is a proton and the other one a neutron. As the generalization of the initial condition of the single-particle oscillator we take a Gaussian two-particle wave packet

$$\psi(x_1, x_2, 0) = \frac{1}{\sqrt{2\pi\sigma_1\sigma_2}(1-c^2)^{1/4}} \exp\left\{-\frac{1}{4(1-c^2)}\right.$$
$$\left. \times \left[\frac{(x_1 - \langle x_1\rangle)^2}{\sigma_1^2} - 2c\frac{(x_1 - \langle x_1\rangle)}{\sigma_1}\frac{(x_2 - \langle x_2\rangle)}{\sigma_2} + \frac{(x_2 - \langle x_2\rangle)^2}{\sigma_2^2}\right]\right\} \quad , \tag{5.18}$$

x_1, x_2: coordinates of particles 1 and 2,
σ_1, σ_2: widths in x_1, x_2 of Gaussian wave packet,
c: correlation between x_1 and x_2, $-1 < c < 1$,
$\langle x_1\rangle, \langle x_2\rangle$: position expectation values.

5.1.3 Time-Dependent Wave Functions and Probability Distributions for Distinguishable Particles

The time evolution of the above initial wave function can be calculated by expanding $\psi(x_1, x_2, 0)$ into a sum over the complete set of eigenfunctions $U_N(R)u_n(r)$,

$$\psi(x_1, x_2, 0) = \sum_{N=0}^{\infty}\sum_{n=0}^{\infty} w_{Nn} U_N(R) u_n(r) \quad , \tag{5.19}$$

which determines the coefficients w_{Nn}. The time-dependent solution of the Schrödinger equation is given by

$$\psi(x_1, x_2, t) = \sum_{N=0}^{\infty}\sum_{n=0}^{\infty} w_{Nn} \mathrm{e}^{-\mathrm{i}(E_N+E_n)t/\hbar} U_N(R) u_n(r) \quad . \tag{5.20}$$

For brevity we discuss only the absolute square of the time-dependent wave function:

$$\begin{aligned} \varrho_{\mathrm{D}}(x_1, x_2, t) &= |\psi(x_1, x_2, t)|^2 \\ &= \frac{1}{2\pi\sigma_1(t)\sigma_2(t)(1-c^2(t))^{1/2}} \exp\left\{ -\frac{1}{2(1-c^2(t))} \left[\frac{x_1 - \langle x_1(t)\rangle)^2}{\sigma_1^2(t)} \right.\right. \\ &\quad \left.\left. - 2c(t)\frac{(x_1 - \langle x_1(t)\rangle)}{\sigma_1(t)}\frac{(x_2 - \langle x_2(t)\rangle)}{\sigma_2(t)} + \frac{(x_2 - \langle x_2(t)\rangle)^2}{\sigma_2^2(t)} \right] \right\} \quad . \end{aligned} \tag{5.21}$$

It differs from the absolute square of $\psi(x_1, x_2, 0)$ only through the time dependence of the expectation values $\langle x_1(t)\rangle$, $\langle x_2(t)\rangle$, the widths $\sigma_1(t)$, $\sigma_2(t)$, and the correlation $c(t)$. For $t = 0$ they assume the values of the initial wave packet (5.18). The quantity $\varrho_{\mathrm{D}}(x_1, x_2, t)$ is the *joint probability density* for finding at time t the distinguishable particles 1 and 2 at the locations x_1 and x_2, respectively.

5.1.4 Marginal Distributions for Distinguishable Particles

The probability distribution of particle 1, independent of the position of particle 2, is given by

$$\varrho_{\mathrm{D}1}(x_1, t) = \int_{-\infty}^{+\infty} \varrho_{\mathrm{D}}(x_1, x_2, t)\,\mathrm{d}x_2 \quad . \tag{5.22}$$

Consequently, the probability density $\varrho_{\mathrm{D}2}(x_2, t)$ of particle 2, independent of the position of particle 1, is given by integrating $\varrho_{\mathrm{D}}(x_1, x_2, t)$ over x_1:

$$\varrho_{\mathrm{D}2}(x_2, t) = \int_{-\infty}^{+\infty} \varrho_{\mathrm{D}}(x_1, x_2, t)\,\mathrm{d}x_1 \quad . \tag{5.23}$$

5.1.5 Wave Functions for Indistinguishable Particles. Symmetrization for Bosons. Antisymmetrization for Fermions

For *indistinguishable* particles (e.g., two protons) the Hamiltonian is symmetric under permutations of the coordinates x_1 and x_2 of the particles 1 and 2:

$$H(x_1, x_2) = H(x_2, x_1) \quad . \tag{5.24}$$

Because the particles cannot be distinguished by measurement, all measurable quantities are symmetric in the particles 1 and 2. To ensure the symmetry of the expectation values, the two-particle wave functions of indistinguishable particles are *symmetric* for *bosons*:

$$\psi_{\mathrm{B}}(x_1, x_2, t) = \psi_{\mathrm{B}}(x_2, x_1, t) \quad ; \tag{5.25}$$

or *antisymmetric* for *fermions*:

$$\psi_{\mathrm{F}}(x_1, x_2, t) = -\psi_{\mathrm{F}}(x_2, x_1, t) \quad . \tag{5.26}$$

The requirement of antisymmetrization is the *Pauli principle*. Its most important physical implication is that two indistinguishable fermions cannot occupy the same state or be at the same position:

$$\psi_{\mathrm{F}}(x, x, t) = 0 \quad . \tag{5.27}$$

Because of the symmetry of the Hamiltonian (5.24), time-dependent wave functions for bosons or fermions can be obtained by symmetrization or antisymmetrization of the time-dependent solution (5.20):

$$\psi_{\mathrm{B,F}}(x_1, x_2, t) = N_{\mathrm{B,F}} \frac{1}{\sqrt{2}} [\psi(x_1, x_2, t) \pm \psi(x_2, x_1, t)] \quad . \tag{5.28}$$

The factor $N_{\mathrm{B,F}}$ ensures the normalization of the boson or fermion wave function. The probability density for bosons and fermions is given by

$$\varrho_{\mathrm{B}}(x_1, x_2, t) = |\psi_{\mathrm{B}}(x_1, x_2, t)|^2 = |N_{\mathrm{B}}|^2 [\varrho_{\mathrm{S}}(x_1, x_2, t) + \varrho_{\mathrm{I}}(x_1, x_2, t)] \tag{5.29}$$

and

$$\varrho_{\mathrm{F}}(x_1, x_2, t) = |\psi_{\mathrm{F}}(x_1, x_2, t)|^2 = |N_{\mathrm{F}}|^2 [\varrho_{\mathrm{S}}(x_1, x_2, t) - \varrho_{\mathrm{I}}(x_1, x_2, t)] \quad , \tag{5.30}$$

where

$$\varrho_{\mathrm{S}}(x_1, x_2, t) = \tfrac{1}{2} [\varrho_{\mathrm{D}}(x_1, x_2, t) + \varrho_{\mathrm{D}}(x_2, x_1, t)] \tag{5.31}$$

is the *symmetrized probability density* of distinguishable particles. The term ϱ_{I} is the *interference term*

$$\varrho_{\mathrm{I}}(x_1, x_2, t) = \tfrac{1}{2} [\psi^*(x_1, x_2, t)\psi(x_2, x_1, t) + \psi^*(x_2, x_1, t)\psi(x_1, x_2, t)] \quad . \tag{5.32}$$

Whereas $\varrho_B, \varrho_F, \varrho_S \geq 0$, the interference term ϱ_I can assume positive and negative values. The joint probability densities $\varrho_B(x_1, x_2, t)$ and $\varrho_F(x_1, x_2, t)$ are both symmetric under permutation of x_1 and x_2.

5.1.6 Marginal Distributions of the Probability Densities of Bosons and Fermions

Because the joint probability densities ϱ_B and ϱ_F are symmetric in x_1 and x_2, there is only one marginal distribution for bosons and one for fermions:

$$\varrho_B(x, t) = \int_{-\infty}^{\infty} \varrho_B(x, x_2, t)\,\mathrm{d}x_2 \tag{5.33}$$

and

$$\varrho_F(x, t) = \int_{-\infty}^{\infty} \varrho_F(x, x_2, t)\,\mathrm{d}x_2 \quad . \tag{5.34}$$

Their physical significance is that they give the probability for finding one of the two particles at the position x independent of the position of the other one. Of course it is possible to compute marginal distributions also for the densities ϱ_S and ϱ_I although these have no direct physical significance:

$$\varrho_{S,I}(x, t) = \int_{-\infty}^{\infty} \varrho_{S,I}(x, x_2, t)\,\mathrm{d}x_2 \quad . \tag{5.35}$$

5.1.7 Normal Oscillations

In our system of two identical oscillators in one dimension, with a harmonic coupling, the two *normal oscillations* of classical mechanics are:

i) the oscillations of the center of mass with a time-independent relative coordinate, and
ii) the oscillation in the relative motion with the center of mass at rest.

The initial conditions (with the two particles initially at rest) that correspond to the two normal oscillations are:

i) identical initial positions $x_{10} = x_{20}$ for the two particles, and
ii) opposite initial positions $x_{10} = -x_{20}$ for the two particles.

In quantum mechanics also, these initial positions lead to the corresponding normal oscillations for the expectation values of the positions $\langle x_1(t)\rangle$, $\langle x_2(t)\rangle$ of the two particles.

Further Reading

Alonso, Finn: Vol. 3, Chap. 4
Brandt, Dahmen: Chaps. 8, 9

Feynman, Leighton, Sands: Vol. 3, Chap. 4
Gasiorowicz: Chap. 13
Merzbacher: Chap. 20
Messiah: Vol. 2, Chap. 14
Schiff: Chap. 10

5.2 Stationary States

Aim of this section: Computation and presentation of the stationary wave function $\varphi_E(x_1, x_2) = U_N(R)u_n(r)$, (5.13).

A plot like Fig. 5.1 is produced showing the eigenfunction of two coupled harmonic oscillators as surface over the x_1, x_2 plane. Also shown as a dashed line is an ellipse given by $E = V(x_1, x_2)$. It is the boundary of the region accessible in classical mechanics.

On the bottom of the subpanel **Physics—Comp. Coord.** you can choose to plot either φ or $|\varphi|^2$.

On the subpanel **Physics—Variables** you find the three variables characterizing the system of **Coupled Oscillators**:

- the spring constant k of the individual oscillators,

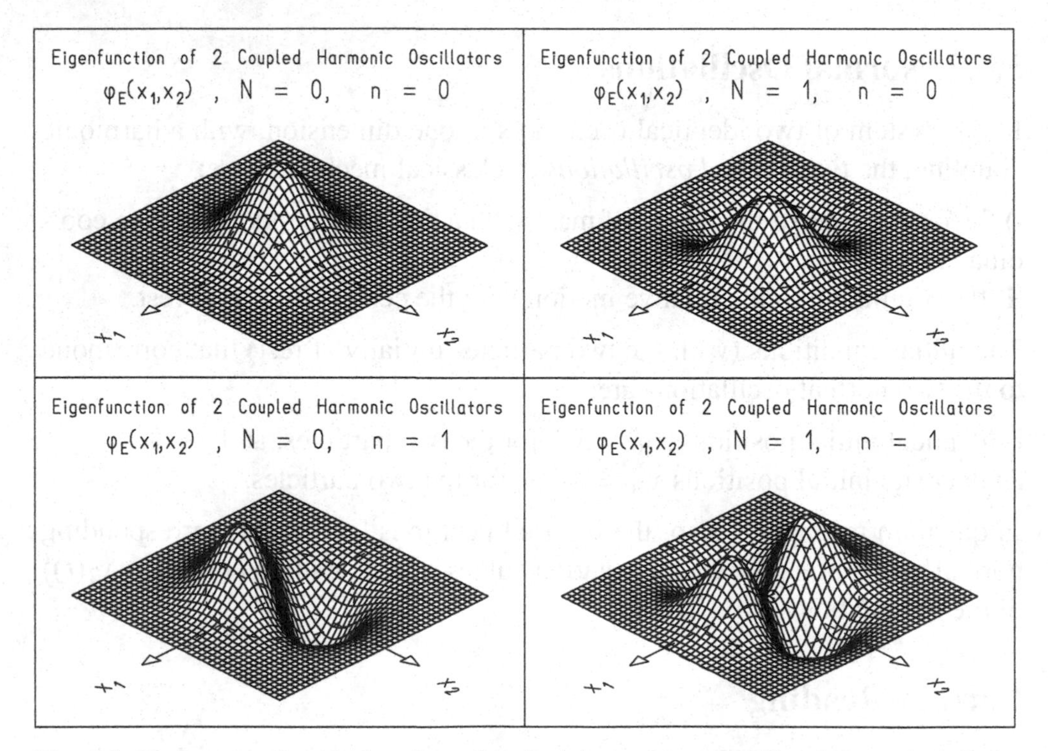

Fig. 5.1. Plot produced with descriptor `Coupled harmonic oscillators: stationary states` on file `Two_Particles.des`

• the spring constant κ of coupling,
• the mass m of the oscillators.

The quantum numbers N and n are determined by the position of an individual plot within a multiple plot. The column index within a multiple plot is N, the row index is n.

Example Descriptor on File `Two_Particles.des`

• `Coupled harmonic oscillators: stationary states` (see Fig. 5.1)

5.3 Time Dependence of Global Variables

Aim of this section: Illustration of the time dependence $\langle x_1(t)\rangle$ and $\langle x_2(t)\rangle$ of the position expectation values, the time dependence $\sigma_1(t)$ and $\sigma_2(t)$ of the widths in x_1 and x_2, and of the time dependence $c(t)$ of the correlation coefficient.

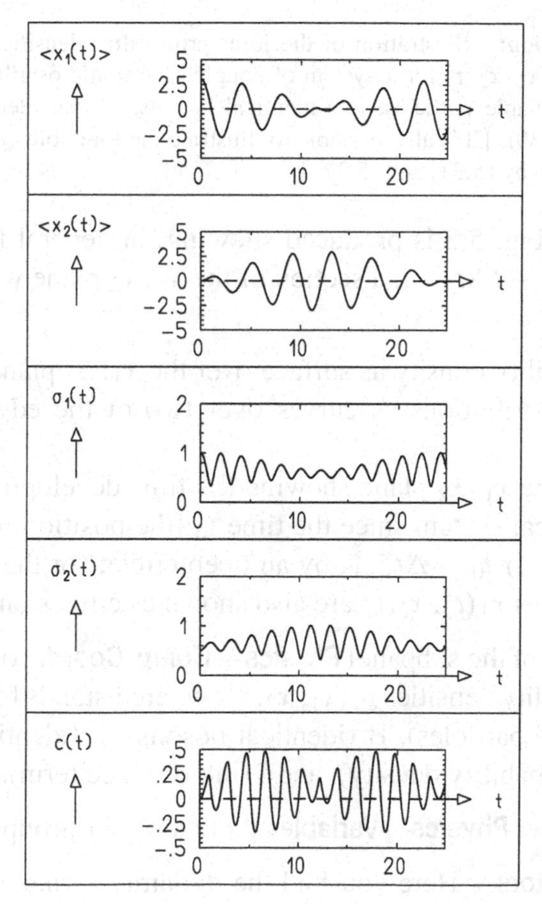

Fig. 5.2. Plot produced with descriptor `Coupled harmonic oscillators: global variables` on file `Two_Particles.des`

The time dependence of the selected global variables is presented as a *2D function graph*. On the bottom of the subpanel **Physics—Comp. Coord.** you can select to plot one of the five global variables. You may also choose to plot all five in a multiple plot.

On the subpanel **Physics—Variables** you find eight parameters:

- the spring constant k of the individual oscillators, the spring constant κ of coupling, and the mass m of the oscillators,
- the initial values (for $t = 0$) of the five global variables, i.e., $\langle x_{10} \rangle$, $\langle x_{20} \rangle$, σ_{10}, σ_{20}, c_0.

Example Descriptor on File `Two_Particles.des`

- `Coupled harmonic oscillators: global variables` (see Fig. 5.2)

5.4 Joint Probability Densities

Aim of this section: Illustration of the joint probability densities $\varrho_D(x_1, x_2, t)$, $\varrho_B(x_1, x_2, t)$, $\varrho_F(x_1, x_2, t)$ for a system of coupled harmonic oscillators composed of two distinguishable particles, two identical bosons, or two identical fermions, (5.21), (5.29), (5.30). [It is also possible to illustrate the functions $\varrho_S(x_1, x_2, t)$ and $\varrho_I(x_1, x_2, t)$ given by (5.31) and (5.32).]

A plot similar to Fig. 5.3 is produced showing, in general for several times $t = t_0, t_0 + \Delta t, t_0 + 2\Delta t, \ldots$, a section of the x_1, x_2 plane with the following items:

- the joint probability density as surface over the x_1, x_2 plane,
- the marginal distributions as curves over two of the edges of the x_1, x_2 plane,
- a trajectory in the x_1, x_2 plane showing the time development of the corresponding classical system since the time t_0 [the position for t_0 is indicated by a full circle, for $t_0 + \Delta t, \ldots$ by an open circle; for the current time the classical positions $x_1(t)$, $x_2(t)$ are also shown as circles on the margins].

On the bottom of the subpanel **Physics—Comp. Coord.** you can choose one of the five probability densities $\varrho_i(x_1, x_2, t)$, where i stands for the five indices D (distinguishable particles), B (identical bosons), F (identical fermions), S (symmetrized probability density), and I (interference term).

On the subpanel **Physics—Variables** there are four groups of variables:

- **Coupled Oscillators** – Here you find the dynamic variables k, κ, and m of the system and the initial conditions $\langle x_{10} \rangle$, $\langle x_{20} \rangle$, σ_{10}, σ_{20}, and c_0.
- **Time** – You can enter the time t_0 for the first plot and the time interval Δt between successive plots.

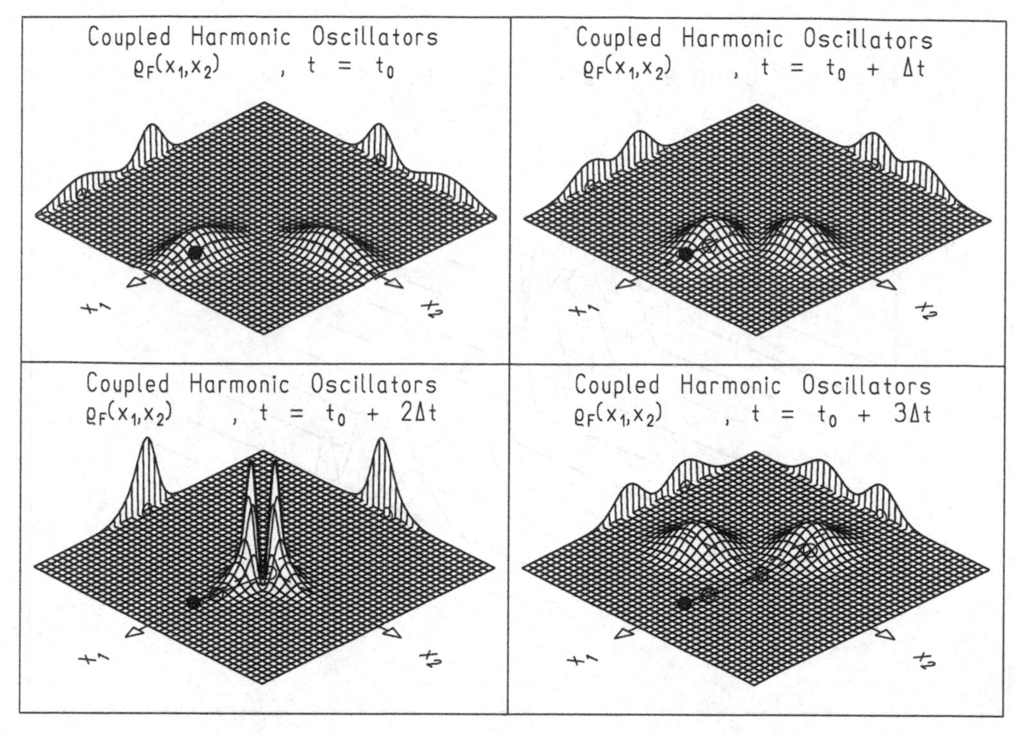

Fig. 5.3. Plot produced with descriptor `Coupled harmonic oscillators: joint probability density` on file `Two_Particles.des`

- **Graphical Item** – This is the radius R of the circle indicating the classical position of the system.
- **For Distinguishable Particles Only** – You can choose to show the covariance ellipse because in that case the probability density is a Gaussian distribution of x_1 and x_2.

Movie Capability: Direct. The time, over which the movie extends, is $T = \Delta t(N_{\text{Plots}} - 1)$; N_{Plots} is the number of individual plots in a multiple plot and Δt is the time interval between two plots. For a single plot $T = \Delta t$.

Example Descriptor on File `Two_Particles.des`

- `Coupled harmonic oscillators: joint probability density` (see Fig. 5.3)

5.5 Marginal Distributions

Aim of this section: Illustration of the marginal distributions $\varrho_{\text{D1}}(x_1, t)$, $\varrho_{\text{D2}}(x_2, t)$, $\varrho_{\text{B}}(x, t)$, $\varrho_{\text{F}}(x, t)$, $\varrho_{\text{S}}(x, t)$, $\varrho_{\text{I}}(x, t)$ discussed in Sect. 5.1, (5.22), (5.23), (5.33), (5.34), (5.35).

A plot similar to Fig. 5.4 is produced, showing for various times t one of the marginal distribution of a system of coupled harmonic oscillators. Also

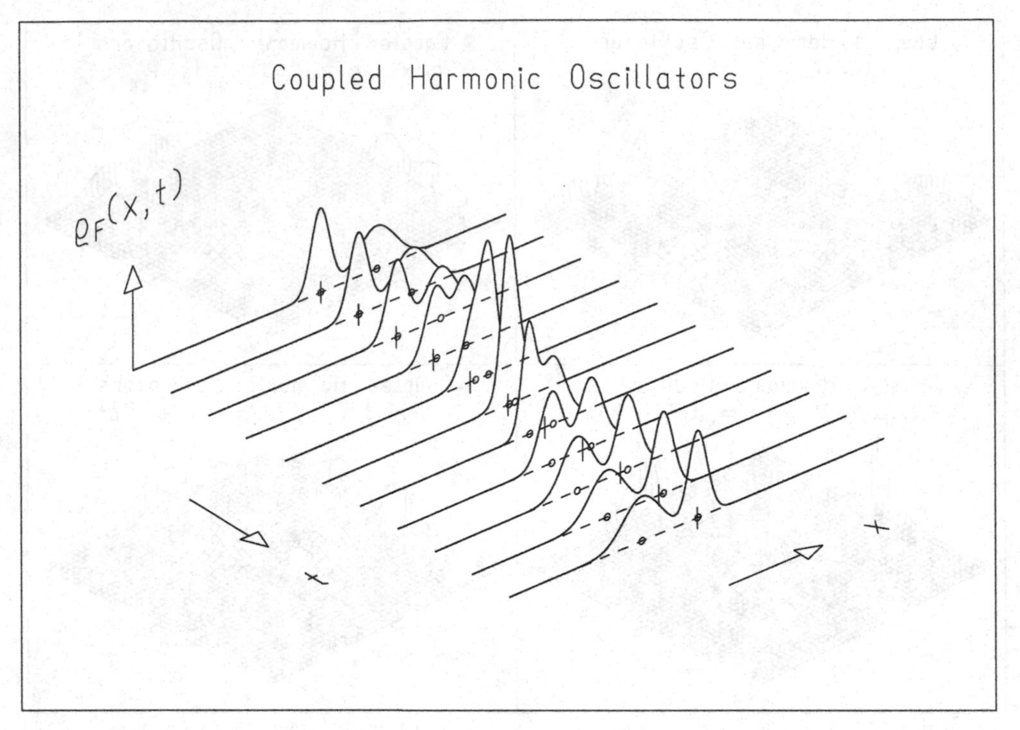

Fig. 5.4. Plot produced with descriptor `Coupled harmonic oscillators: marginal distribution` on file `Two_Particles.des`

shown for every time are a dashed zero line, the positions of the corresponding two classical particles as small circles, and a short vertical line indicating the equilibrium position $x = 0$.

On the bottom of the subpanel **Physics—Comp. Coord.** you can choose between the following six types of marginal distribution:

$$\varrho_{\mathrm{D1}}(x_1, t),\ \varrho_{\mathrm{D2}}(x_2, t),\ \varrho_{\mathrm{B}}(x, t),\ \varrho_{\mathrm{F}}(x, t),\ \varrho_{\mathrm{S}}(x, t),\ \varrho_{\mathrm{I}}(x, t) \quad .$$

On the subpanel **Physics—Variables** there are two groups of variables:

- **Coupled Oscillators** – Here you find the dynamic variables k, κ, and m of the system and the initial conditions $\langle x_{10}\rangle$, $\langle x_{20}\rangle$, σ_{10}, σ_{20}, and c_0.
- **Graphical Items** – These are the dash length ℓ_{DASH} of the zero line (which is also the length of the short vertical line) and the radius R of the circles indicating the positions of the classical oscillators.

Movie Capability: Indirect. After conversion to direct movie capability the start and end times can be changed in the subpanel **Movie** (see Sect. A.4) of the parameter panel.

Example Descriptor on File `Two_Particles.des`

- `Coupled harmonic oscillators: marginal distribution` (see Fig. 5.4)

5.6 Exercises

Please note:

(i) You may watch a demonstration of the material of this chapter by pressing the button **Run Demo** in the *main toolbar* and selecting one of the demo files `Two_Particles`.

(ii) More example descriptors can be found on the descriptor file `TwoParticles(FE).des` in the directory `FurtherExamples`.

(iii) For the following exercises use descriptor file `Two_Particles.des`.

(iv) The numerical value of Planck's constant is put to 1.

5.2.1 Plot the stationary wave functions of two uncoupled oscillators with $k = 2$, $m = 1$ for the quantum numbers $N = 0, 1$, $n = 0, 1$ in a multiple plot. Start from descriptor 6: '`Exercise 5.2.1`'.

5.2.2 Plot the stationary wave functions of two uncoupled oscillators with $k = 2$, $m = 1$, in four multiple plots for the quantum numbers **(a)** $N = 0, 1$, $n = 0, 1$, **(b)** $N = 0, 1$, $n = 2, 3$, **(c)** $N = 2, 3$, $n = 0, 1$, **(d)** $N = 2, 3$, $n = 2, 3$. Start from descriptor 6: '`Exercise 5.2.1`'. **(e)** What determines the number and location of the node lines of the wave functions?

5.2.3 (a–d) Repeat Exercise 5.2.2 (a–d) with $m = 2$. **(e)** Which effect on the wave functions does the doubling of the mass have?

5.2.4 (a–d) Repeat Exercise 5.2.2 (a–d) with nonvanishing coupling $\kappa = 2$. **(e)** How does the coupling affect the wave function?

5.2.5 (a–d) Repeat Exercise 5.2.2 (a–d) with nonvanishing attractive coupling $\kappa = 5$. **(e)** How does the coupling affect the wave function? **(f)** Which correlation in the particle coordinates x_1, x_2 do you observe?

5.2.6 (a–d) Repeat Exercise 5.2.2 (a–d) with nonvanishing repulsive coupling $\kappa = -0.6$. **(e)** How does the repulsive coupling affect the wave function? **(f)** Which correlation in the particle coordinates x_1, x_2 do you observe?

5.2.7 (a) Plot in a multiple plot the probability density of two coupled oscillators with $k = 2$, $m = 1$ for the quantum numbers $N = 1, 2$, $n = 1, 2$ for the spring constant $\kappa = 5$ of the coupling. Start from descriptor 6: '`Exercise 5.2.1`'. **(b)** Why are the outer maxima in the plot for $N = 2$, $n = 2$ higher than the inner ones? **(c)** Why are the widths of the inner maxima smaller than the ones of the outer maxima? **(d)** Why does the region where the probability density is essentially different from zero not in all plots evenly fill the region of possible classical orbits given by the dashed ellipse?

5.2.8 (a) Repeat Exercise 5.2.7 (a) with repulsive coupling $\kappa = -0.6$. **(b)** Why do the locations of the maxima of the probability density form a rectangular grid in the x_1, x_2 plane?

5.2.9 Plot in multiple plots the probability density of two coupled oscillators with $k = 1$, $m = 1$, $\kappa = 4$ for the quantum numbers **(a)** $N = 2, 3$, $n = 0, 1$, **(b)** $N = 2, 3$, $n = 2, 3$, **(c)** $N = 4, 5$, $n = 0, 1$, **(d)** $N = 4, 5$, $n = 2, 3$. Start from descriptor 6: '`Exercise 5.2.1`'. **(e)** Calculate the total energies of the coupled oscillators for the above quantum numbers. **(f)** Compare the plots for $N = 5$, $n = 0$ with $N = 2$, $n = 2$. Explain why the graphs for high N and small n extend mainly along one of the principal axes of the dashed ellipse.

5.3.1 Consider a system of two uncoupled harmonic oscillators with mass $m = 1$ and spring constants $k = 2$, $\kappa = 0$. **(a)** Plot the time dependence of **i)** the position expectation values $\langle x_1(t)\rangle$, $\langle x_2(t)\rangle$, **ii)** the widths $\sigma_1(t)$, $\sigma_2(t)$ in x_1 and x_2, **iii)** the correlation coefficient $c(t)$ for the initial values $\langle x_{10}\rangle = 3$, $\langle x_{20}\rangle = 0$, $\sigma_{10} = 1.01$, $\sigma_{20} = 0.5$, $c_0 = 0$. Start from descriptor 7: '`Exercise 5.3.1`'. **(b)** Why is the expectation value $\langle x_2(t)\rangle$ equal to zero? **(c)** Why do the widths $\sigma_1(t)$, $\sigma_2(t)$ vary with time? **(d)** Why is the correlation $c(t)$ equal to zero?

5.3.2 **(a)** Repeat Exercise 5.3.1 for nonvanishing correlation $c_0 = 0.2$. Start from descriptor 7: '`Exercise 5.3.1`'. **(b)** Explain the time dependence of the correlation.

5.3.3 Repeat Exercise 5.3.1 for nonvanishing $\langle x_{20}\rangle = 3$ and $\sigma_{10} = \sigma_{20} = \sigma_0/\sqrt{2}$, where $\sigma_0 = \sqrt{\hbar/m\omega}$ is the ground-state width of the uncoupled oscillators. Start from descriptor 7: '`Exercise 5.3.1`'. **(a)** Calculate $\sigma_0/\sqrt{2}$. **(b)** Plot the global quantities $\langle x_1(t)\rangle$, $\langle x_2(t)\rangle$, $\sigma_1(t)$, $\sigma_2(t)$, $c(t)$. **(c)** Why are the widths $\sigma_1(t)$, $\sigma_2(t)$ time independent? **(d)** Do you expect time-independent widths also for nonvanishing correlation?

5.3.4 **(a)** Repeat Exercise 5.3.3 for nonvanishing correlation $c_0 = 0.4$. Start from descriptor 7: '`Exercise 5.3.1`'. **(b)** Why do the widths $\sigma_1(t)$, $\sigma_2(t)$ no longer remain time independent? **(c)** Why does the correlation $c(t)$ change periodically?

5.3.5 **(a)** Repeat Exercise 5.3.3 for the anticorrelation $c_0 = -0.4$. Start from descriptor 7: '`Exercise 5.3.1`'. **(b)** Explain the difference between the correlation $c(t)$ of this exercise and of Exercise 5.3.4.

5.3.6 **(a)** Repeat Exercise 5.3.1 for nonvanishing $\langle x_{20}\rangle = 1.5$ and $\sigma_{10} = \sigma_{20} = \sigma_0/\sqrt{2}$, where $\sigma_0 = \sqrt{\hbar/m\omega}$ is the ground-state width of the uncoupled oscillators. Start from descriptor 7: '`Exercise 5.3.1`'. **(b)** Why is the result qualitatively similar to that in Exercise 5.3.3?

5.3.7 Consider the system of two coupled oscillators. Choose the same initial data as in Exercise 5.3.1, however, take a nonvanishing spring constant $\kappa = 0.5$ for the coupling of the two oscillators. Start from descriptor 7: '`Exercise 5.3.1`'. **(a)** Plot the global quantities. **(b)** Explain the behavior of the time-dependent expectation values $\langle x_1(t)\rangle$, $\langle x_2(t)\rangle$. **(c)** Why does the correlation $c(t)$ become different from zero as time increases?

5.3.8 Repeat Exercise 5.3.7, however, with equal initial widths $\sigma_{10} = \sigma_{20} = 0.5$. Start from descriptor 7: '`Exercise 5.3.1`'.

5.3.9 (a) Repeat Exercise 5.3.8, however, with initial positions $\langle x_{10} \rangle = 3$, $\langle x_{20} \rangle = -3$. Start from descriptor 7: '`Exercise 5.3.1`'. **(b)** What oscillation do the expectation values $\langle x_1(t) \rangle$, $\langle x_2(t) \rangle$ perform?

5.3.10 (a,b) Repeat Exercise 5.3.9 (a,b), however, with strong initial anticorrelation $c_0 = -0.95$. Start from descriptor 7: '`Exercise 5.3.1`'.

5.3.11 Repeat Exercise 5.3.9 with strong positive correlation $c_0 = 0.95$. Start from descriptor 7: '`Exercise 5.3.1`'.

5.3.12 (a) Repeat Exercise 5.3.9 with initial position expectation values $\langle x_{10} \rangle = 3$, $\langle x_{20} \rangle = 3$ and vanishing initial correlation $c_0 = 0$. Start from descriptor 7: '`Exercise 5.3.1`'. **(b)** What kind of oscillation do the two position expectation values perform?

5.4.1 Plot the joint probability density $\varrho_D(x_1, x_2)$ for two uncoupled harmonic oscillators of distinguishable particles of mass $m = 1$ and spring constant $k = 2$ for the initial values $\langle x_{10} \rangle = 3$, $\langle x_{20} \rangle = 0$, $\sigma_1 = 1.0$, $\sigma_2 = 0.5$, $c_0 = 0$ in two 2×2 multiple plots **(a)** for the times $t_n = n\Delta t$, $n = 0, 1, 2, 3$, $\Delta t = 0.501$, and **(b)** for the times $t_n = n\Delta t$, $n = 4, 5, 6, 7$, $\Delta t = 0.501$. Start from descriptor 8: '`Exercise 5.4.1`'. **(c)** Why are the axes of the two-dimensional Gauss distributions parallel to the coordinate axes?

5.4.2 (a,b) Repeat Exercise 5.4.1 (a,b) for $\langle x_{20} \rangle = 3$, $c_0 = 0.8$, and the step width $\Delta t = 0.55536$. Start from descriptor 8: '`Exercise 5.4.1`'. **(c)** Why are the axes of the two-dimensional Gauss distribution no longer parallel to the coordinate axes? **(d)** How does the positivity of the initial correlation c_0 show in the plots?

5.4.3 Plot the joint probability density $\varrho_D(x_1, x_2)$ for two uncoupled harmonic oscillators of distinguishable particles of mass $m = 1$ and spring constant $k = 2$ for the initial values $\langle x_{10} \rangle = -2$, $\langle x_{20} \rangle = 3$, $\sigma_{10} = 1$, $\sigma_{20} = 0.5$, $c_0 = -0.8$ in two 2×2 multiple plots **(a)** for the times $t_n = n\Delta t$, $n = 0, 1, 2, 3$, $\Delta t = 0.501$, and **(b)** for the times $t_n = n\Delta t$, $n = 4, 5, 6, 7$, $\Delta t = 0.501$. Start from descriptor 8: '`Exercise 5.4.1`'. **(c)** How does the negative initial correlation show in the plots?

5.4.4 (a,b) Repeat Exercise 5.4.1 (a,b) for a coupling spring constant $\kappa = 0.8$. Start from descriptor 8: '`Exercise 5.4.1`'. **(c)** Which effect showing in the plots is due to the coupling of the two oscillators?

5.4.5 (a,b) Repeat Exercise 5.4.1 (a,b) for a coupling spring constant $\kappa = 1.5$. Start from descriptor 8: '`Exercise 5.4.1`'. **(c)** Why does the amplitude in the variable x_2 grow faster than in the plots of Exercise 5.4.4?

5.4.6 (a,b) Repeat Exercise 5.4.1 (a,b) for a coupling spring constant $\kappa = 3$. Start from descriptor 8: '`Exercise 5.4.1`'. **(c)** What causes the change of the initially uncorrelated Gauss distribution to a correlated one?

5.4.7 (a,b) Repeat Exercise 5.4.1 (a,b) for a coupling spring constant $\kappa = 5$. Start from descriptor 8: 'Exercise 5.4.1'. **(c)** Why does the expectation value (circle) in the x_1, x_2 plane oscillate more often than in the former exercises?

5.4.8 (a,b) Repeat Exercise 5.4.1 (a,b) for a coupling spring constant $\kappa = 10$. Start from descriptor 8: 'Exercise 5.4.1'. **(c)** Why is the Gaussian distribution most narrow when its position expectation values $\langle x_1 \rangle, \langle x_2 \rangle$ are close to zero?

5.4.9 Plot the joint probability density $\varrho_D(x_1, x_2)$ for two coupled harmonic oscillators of distinguishable particles of mass $m = 1$, spring constant $k = 2$, and coupling spring constant $\kappa = 20$ for the initial conditions $\langle x_{10} \rangle = 3$, $\langle x_{20} \rangle = -3$, $\sigma_1 = 1.0$, $\sigma_2 = 0.5$, $c_0 = 0$ in two 2×2 multiple plots **(a)** for $t_n = n\Delta t$, $n = 0, 1, 2, 3$, $\Delta t = 0.2$, and **(b)** for $t_n = n\Delta t$, $n = 4, 5, 6, 7$, $\Delta t = 0.2$. Start from descriptor 8: 'Exercise 5.4.1'. **(c)** What particular kind of oscillation do you observe?

5.4.10 (a,b) Repeat Exercise 5.4.9 (a,b) for a coupling spring constant $\kappa = 20$ and a step width of $\Delta t = 0.2$. Take as initial values for the position expectation values $\langle x_{10} \rangle = 3$, $\langle x_{20} \rangle = 3$. Start from descriptor 8: 'Exercise 5.4.1'. **(c)** What particular kind of oscillation do you observe? **(d)** Why is the motion so much slower than in Exercise 5.4.9?

5.4.11 (a,b) Repeat Exercise 5.4.1 (a,b) for a repulsive coupling spring constant $\kappa = -0.8$ and step width $\Delta t = 0.5$. Start from descriptor 8: 'Exercise 5.4.1'. **(c)** How does the repulsive coupling between the two oscillators make itself felt in the plots?

5.4.12 (a,b) Repeat Exercise 5.4.1 (a,b) for a repulsive spring constant $\kappa = -0.95$ and step width $\Delta t = 2$. Start from descriptor 8: 'Exercise 5.4.1'. **(c)** Why does the initially uncorrelated Gauss distribution develop a correlation of the kind observed?

5.4.13 (a,b) Repeat Exercise 5.4.1 (a,b) for bosons. **(c)** Why do you initially observe two humps? **(d)** What creates the very high peak in the plots where the two bosons are close together?

5.4.14 (a,b) Repeat Exercise 5.4.2 (a,b) for bosons. **(c)** Why do you observe only one hump in the plots?

5.4.15 (a,b) Repeat Exercise 5.4.3 (a,b) for bosons. **(c)** How does the correlation show in the initial double humps?

5.4.16 (a,b) Repeat Exercise 5.4.4 for bosons.

5.4.17 (a,b) Repeat Exercise 5.4.8 (a,b) for bosons. **(c)** How does the strong attractive coupling show in the plots?

5.4.18 (a,b) Repeat Exercise 5.4.9 (a,b) for bosons. **(c)** How is the difference from the graphs for distinguishable particles explained?

5.4.19 (a,b) Repeat Exercise 5.4.10 (a,b) for bosons. **(c)** Why do you observe quick oscillations of the width in $(x_2 - x_1)$ of the Gaussian hump?

5.4.20 Repeat Exercise 5.4.11 for bosons.

5.4.21 (a,b) Repeat Exercise 5.4.1 (a,b) for fermions. **(c)** Why is the joint probability always exactly zero along the line $x_1 = x_2$?

5.4.22 Repeat Exercise 5.4.4 for fermions.

5.4.23 (a,b) Repeat Exercise 5.4.8 (a,b) for fermions.

5.4.24 (a,b) Repeat Exercise 5.4.9 (a,b) for fermions.

5.4.25 (a,b) Repeat Exercise 5.4.10 (a,b) for fermions, however, for $\sigma_{10} = \sigma_{20} = 1$. **(c)** Why does the joint probability distribution vanish?

5.4.26 (a,b) Repeat Exercise 5.4.10 (a,b) for fermions and nonvanishing initial correlation. **(c)** Why does the joint probability density still vanish?

5.4.27 (a,b) Repeat Exercise 5.4.10 (a,b) for fermions and vanishing correlation $c_0 = 0$ and different initial widths $\sigma_1 = 1$, $\sigma_2 = 0.2$. **(c)** Why does the existence of a fermion wave function with identical expectation values for the two fermions not contradict the Pauli principle?

5.5.1 Study the marginal distributions of two distinguishable particles of equal mass $m = 1$ in two coupled oscillators ($k = 2$, $\kappa = 0.8$) with the initial conditions $\langle x_{10} \rangle = 3$, $\langle x_{20} \rangle = 0$, $\sigma_{10} = 1$, $\sigma_{20} = 0.5$, $c_0 = 0$. **(a)** Plot the marginal distribution $\varrho_{D1}(x_1, t)$ of particle 1 for the time interval $0 \le t \le 4$ in ten steps. Start from descriptor 9: '`Exercise 5.5.1a`'. **(b)** Plot the marginal distribution $\varrho_{D2}(x_2, t)$ of particle 2 for the same interval. Start from descriptor 10: '`Exercise 5.5.1b`'. **(c)** For simpler comparison plot both distributions (a), (b) as a combined plot (a) above (b). Start from mother descriptor 11: '`Exercise 5.5.1c`'.

5.5.2 (a–c) Repeat Exercise 5.5.1 (a–c) for a longer time interval $0 \le t \le 10$.

5.5.3 Study the marginal distributions of indistinguishable particles with the same parameters as in Exercise 5.5.1. **(a)** Plot the marginal distribution for bosons. **(b)** Plot the marginal distributions for fermions. **(c)** For comparison, plot both distributions above each other in a double plot. **(d)** What are the differences in the two plots? **(e)** Why are these differences so marginal?

6. Free Particle Motion in Three Dimensions

Contents: Description of the three-dimensional motion of a free particle of sharp momentum by a harmonic plane wave. Schrödinger equation of free motion in three dimensions. Gaussian wave packet. Angular momentum. Spherical harmonics as eigenfunctions of angular momentum. Radial Schrödinger equation of free motion. Spherical Bessel functions. Partial-wave decomposition of plane wave and Gaussian wave packet.

6.1 Physical Concepts

6.1.1 The Schrödinger Equation of a Free Particle in Three Dimensions. The Momentum Operator

Classical free motion in three dimensions can be viewed as three simultaneous one-dimensional motions in the coordinates x, y, z. In quantum mechanics the situation is the same. Three-dimensional free motion is viewed as three one-dimensional harmonic waves (2.1) propagating simultaneously in the coordinates x, y, z of the position vector $\mathbf{r}$:

$$\psi_{\mathbf{p}}(\mathbf{r}, t) = \frac{1}{(2\pi\hbar)^{3/2}} \exp\left[-\frac{\mathrm{i}}{\hbar}\left(\frac{p_x^2}{2M}t - p_x x\right)\right] \times \exp\left[-\frac{\mathrm{i}}{\hbar}\left(\frac{p_y^2}{2M}t - p_y y\right)\right] \exp\left[-\frac{\mathrm{i}}{\hbar}\left(\frac{p_z^2}{2M}t - p_z z\right)\right] \quad . \tag{6.1}$$

Because the kinetic energy E of the particle is given by

$$E = (p_x^2 + p_y^2 + p_z^2)/2M = \mathbf{p}^2/2M \quad , \tag{6.2}$$

M: mass of particle,
$\mathbf{p} = (p_x, p_y, p_z)$: momentum of particle.

The expression (6.1) can be rewritten in the form

$$\psi_{\mathbf{p}}(\mathbf{r}, t) = \frac{1}{(2\pi\hbar)^{3/2}} \mathrm{e}^{-\mathrm{i}(Et - \mathbf{p}\cdot\mathbf{r})/\hbar} = \frac{1}{(2\pi\hbar)^{3/2}} \mathrm{e}^{-\mathrm{i}(\omega t - \mathbf{k}\cdot\mathbf{r})} \quad , \tag{6.3}$$

S. Brandt et al., *Interactive Quantum Mechanics: Quantum Experiments on the Computer*,
DOI 10.1007/978-1-4419-7424-2_6, © Springer Science+Business Media, LLC 2011

$\omega = E/\hbar$: angular frequency of the wave function,
$\mathbf{k} = \mathbf{p}/\hbar$: wave-number vector.

The phase velocity of this wave is

$$\mathbf{v} = \mathbf{p}/2M \quad . \tag{6.4}$$

The wave function (6.3) is the solution of the *free Schrödinger equation in three dimensions*

$$\mathrm{i}\hbar \frac{\partial}{\partial t}\psi(\mathbf{r}, t) = T\psi(\mathbf{r}, t) \tag{6.5}$$

having the same formal appearance as (2.2). However, the operator T of the kinetic energy is now that of a particle in three dimensions,

$$T = -\frac{\hbar^2}{2M}\left(\frac{\partial^2}{\partial x^2} + \frac{\partial^2}{\partial y^2} + \frac{\partial^2}{\partial z^2}\right) \quad . \tag{6.6}$$

With the help of the *gradient operator*

$$\nabla = \left(\frac{\partial}{\partial x}, \frac{\partial}{\partial y}, \frac{\partial}{\partial z}\right) \tag{6.7}$$

it takes the form

$$T = -\frac{\hbar^2}{2M}\nabla^2 = -\frac{\hbar^2}{2M}\Delta \quad , \tag{6.8}$$

i.e., a multiple of the *Laplacian* Δ.

The wave function (6.3) lends itself to the factorization

$$\psi_{\mathbf{p}}(\mathbf{r}, t) = \mathrm{e}^{-\mathrm{i}Et/\hbar}\varphi_{\mathbf{p}}(\mathbf{r}) \tag{6.9}$$

into a time-dependent exponential and a *stationary wave function*

$$\varphi_{\mathbf{p}}(\mathbf{r}) = \frac{1}{(2\pi\hbar)^{3/2}}\mathrm{e}^{\mathrm{i}\mathbf{p}\cdot\mathbf{r}/\hbar} = \frac{1}{(2\pi\hbar)^{3/2}}\mathrm{e}^{\mathrm{i}\mathbf{k}\cdot\mathbf{r}} \quad . \tag{6.10}$$

If we choose the unit vector $\mathbf{e}_z$ in the z direction parallel to the *wave-number vector* $\mathbf{k} = \mathbf{p}/\hbar$, we have $\mathbf{k} = k\mathbf{e}_z$, and the stationary wave assumes the simple form

$$\varphi_{\mathbf{p}}(\mathbf{r}) = \frac{1}{(2\pi\hbar)^{3/2}}\mathrm{e}^{ikz} \quad . \tag{6.11}$$

This is a complex function of the coordinate z only. It can be decomposed into real and imaginary parts,

$$\varphi_{\mathbf{p}}(\mathbf{r}) = \mathrm{Re}\,\varphi_{\mathbf{p}}(\mathbf{r}) + \mathrm{i}\,\mathrm{Im}\,\varphi_{\mathbf{p}}(\mathbf{r}) \quad , \tag{6.12}$$

with

$$\mathrm{Re}\,\varphi_{\mathbf{p}}(\mathbf{r}) = \frac{1}{(2\pi\hbar)^{3/2}}\cos kz \quad , \quad \mathrm{Im}\,\varphi_{\mathbf{p}}(\mathbf{r}) = \frac{1}{(2\pi\hbar)^{3/2}}\sin kz \quad . \tag{6.13}$$

On inserting (6.9) into the time-dependent equation (6.5) we obtain

$$T\varphi_{\mathbf{p}}(\mathbf{r}) = E\varphi_{\mathbf{p}}(\mathbf{r}) \quad \text{or} \quad -\frac{\hbar^2}{2M}\Delta\varphi_{\mathbf{p}}(\mathbf{r}) = E\varphi_{\mathbf{p}}(\mathbf{r}) \tag{6.14}$$

as the *stationary* (time-independent) *Schrödinger equation* for the stationary wave function $\varphi_{\mathbf{p}}(\mathbf{r})$. Equation (6.14) is also viewed as an eigenvalue equation, where E is the continuous eigenvalue of the kinetic energy and $\varphi_{\mathbf{p}}(\mathbf{r})$ a continuum eigenfunction of the operator T of the kinetic energy.

In accordance with the classical relation for the kinetic energy, $T = \mathbf{p}^2/2M$, of a single particle with momentum $\mathbf{p}$ we conclude from (6.8) that

$$\hat{\mathbf{p}} = \frac{\hbar}{\mathrm{i}}\nabla \quad , \quad \text{i.e.,} \quad \hat{p}_x = \frac{\hbar}{\mathrm{i}}\frac{\partial}{\partial x} \quad , \quad \hat{p}_y = \frac{\hbar}{\mathrm{i}}\frac{\partial}{\partial y} \quad , \quad \hat{p}_z = \frac{\hbar}{\mathrm{i}}\frac{\partial}{\partial z} \quad , \tag{6.15}$$

is the *momentum operator*. The stationary wave function (6.10) is a continuum eigenfunction of the momentum operator

$$\hat{\mathbf{p}}\varphi_{\mathbf{p}}(\mathbf{r}) = \frac{\hbar}{\mathrm{i}}\nabla\varphi_{\mathbf{p}}(\mathbf{r}) = \mathbf{p}\varphi_{\mathbf{p}}(\mathbf{r}) \tag{6.16}$$

or in components

$$\hat{p}_x\varphi_{\mathbf{p}}(\mathbf{r}) = \frac{\hbar}{\mathrm{i}}\frac{\partial}{\partial x}\varphi_{\mathbf{p}}(\mathbf{r}) = p_x\varphi_{\mathbf{p}}(\mathbf{r}) \quad , \quad \text{etc.} \tag{6.17}$$

6.1.2 The Wave Packet. Group Velocity. Normalization. The Probability Ellipsoid

The wave function (6.3) does not correspond to an actual physical situation, because the norm of a plane wave diverges. A physical particle corresponds to a wave packet formed with a *spectral function* as in (2.4),

$$\psi(\mathbf{r}, t) = \int f(\mathbf{p})\mathrm{e}^{-\mathrm{i}Et/\hbar}\varphi_{\mathbf{p}}(\mathbf{r} - \mathbf{r}_0)\,\mathrm{d}^3\mathbf{p} \quad , \tag{6.18}$$

$\mathbf{r} = (x, y, z)$: position vector,
$\mathbf{r}_0 = (x_0, y_0, z_0)$: initial position expectation value of wave packet at $t = 0$,
$\mathbf{p} = (p_x, p_y, p_z)$: momentum vector,
$f(\mathbf{p})$: spectral function of wave packet.

We choose again a *Gaussian spectral function* in three dimensions as a product

$$f(\mathbf{p}) = f_x(p_x) f_y(p_y) f_z(p_z) \tag{6.19}$$

of three Gaussians, one for every coordinate $a = x, y, z$,

$$f_a(p_a) = \frac{1}{(2\pi)^{1/4}\sigma_{p_a}^{1/2}} \exp\left\{-\frac{(p_a - p_{a0})^2}{4\sigma_{p_a}^2}\right\} \quad , \tag{6.20}$$

p_a: a coordinate of momentum, $a = x, y, z$,
p_{a0}: expectation value of momentum,
σ_{p_a}: width of Gaussian spectral function f_a.

The factors in front of the exponential of (6.20) *normalize* the spectral function $f(\mathbf{p})$ to one,

$$\int f^2(\mathbf{p})\, \mathrm{d}^3\mathbf{p} = 1 \quad , \tag{6.21}$$

and thus the wave packet (6.18):

$$\int |\psi(\mathbf{r}, t)|^2\, \mathrm{d}^3\mathbf{r} = 1 \quad . \tag{6.22}$$

Because of the factorization of the time-dependent exponential and of the stationary wave function $\varphi_\mathbf{p}$ into factors depending on one momentum component only, the integral in (6.18) yields the wave function of the three-dimensional Gaussian wave packet,

$$\psi(\mathbf{r}, t) = M_x(x, t)\mathrm{e}^{\mathrm{i}\phi_x(x,t)} M_y(y, t)\mathrm{e}^{\mathrm{i}\phi_y(y,t)} M_z(z, t)\mathrm{e}^{\mathrm{i}\phi_z(z,t)} \quad , \tag{6.23}$$

where the explicit expressions for the *modulus* M_a and the *phase* ϕ_a can be derived easily from (2.7) and (2.8).

The absolute square of $\psi(\mathbf{r}, t)$ yields the *probability density*

$$\varrho(\mathbf{r}, t) = \frac{1}{(2\pi)^{3/2}\sigma_x\sigma_y\sigma_z} \exp\left[-\frac{(x - \langle x\rangle)^2}{2\sigma_x^2} - \frac{(y - \langle y\rangle)^2}{2\sigma_y^2} - \frac{(z - \langle z\rangle)^2}{2\sigma_z^2}\right] \tag{6.24}$$

for a particle at the position $\mathbf{r}$ at time t. The *position expectation value*

$$\langle \mathbf{r}(t)\rangle = (\langle x(t)\rangle, \langle y(t)\rangle, \langle z(t)\rangle)$$

is given by

$$\langle \mathbf{r}(t)\rangle = \mathbf{r}_0 + \mathbf{v}t \quad . \tag{6.25}$$

This represents the motion of a particle with constant velocity

$$\mathbf{v} = \mathbf{p}_0/M \tag{6.26}$$

along a straight line, starting at $t = 0$ with the initial position $\mathbf{r}_0$. The velocity $\mathbf{v}$ is called the *group velocity* because it determines the propagation of a wave packet or wave group. It is different from the phase velocity (6.4).

The *width* of the Gaussian is time dependent:

$$\sigma_a^2(t) = \sigma_{a0}^2 + \left(\frac{\hbar t}{2M}\frac{1}{\sigma_{a0}}\right)^2 \quad , \quad a = x, y, z \quad , \tag{6.27}$$

$\sigma_{a0} = \hbar/2\sigma_{p_a}$: initial width of wave packet in the coordinate $a = x, y, z$.

The plots produced with **IQ** refer to the two-dimensional distribution

$$\varrho(x, y, t) = \int_{-\infty}^{+\infty} \varrho(x, y, z, t)\,\mathrm{d}z \quad , \tag{6.28}$$

which represents the probability density for a particle having at time t the coordinates x and y irrespective of z. The explicit result for $\varrho(x, y, t)$ is

$$\varrho(x, y, t) = \frac{1}{2\pi\sigma_x\sigma_y}\exp\left\{-\frac{(x - \langle x\rangle)^2}{2\sigma_x^2} - \frac{(y - \langle y\rangle)^2}{2\sigma_y^2}\right\} \quad . \tag{6.29}$$

6.1.3 Angular Momentum. Spherical Harmonics

Angular momentum is a vector $\mathbf{L} = (L_x, L_y, L_z)$ of the form

$$\mathbf{L} = \mathbf{r} \times \mathbf{p} \quad ; \tag{6.30}$$

in components

$$L_x = yp_z - zp_y \quad , \quad L_y = zp_x - xp_z \quad , \quad L_z = xp_y - yp_x \quad . \tag{6.31}$$

In quantum mechanics the momentum $\mathbf{p}$ is a multiple of the del or nabla operator, see (6.15), so that the *operator of angular momentum* $\hat{\mathbf{L}}$ is given by

$$\hat{\mathbf{L}} = \hat{\mathbf{r}} \times \hat{\mathbf{p}} = \frac{\hbar}{\mathrm{i}}\mathbf{r} \times \nabla \quad . \tag{6.32}$$

Its three components

$$\hat{L}_x = \frac{\hbar}{\mathrm{i}}\left(y\frac{\partial}{\partial z} - z\frac{\partial}{\partial y}\right) \quad , \quad \hat{L}_y = \frac{\hbar}{\mathrm{i}}\left(z\frac{\partial}{\partial x} - x\frac{\partial}{\partial z}\right) \quad ,$$

$$\hat{L}_z = \frac{\hbar}{\mathrm{i}}\left(x\frac{\partial}{\partial y} - y\frac{\partial}{\partial x}\right) \tag{6.33}$$

do not commute with each other. Instead, they satisfy the *commutation relations*

$$[\hat{L}_x, \hat{L}_y] = \mathrm{i}\hbar \hat{L}_z \quad , \quad [\hat{L}_y, \hat{L}_z] = \mathrm{i}\hbar \hat{L}_x \quad , \quad [\hat{L}_z, \hat{L}_x] = \mathrm{i}\hbar \hat{L}_y \quad . \tag{6.34}$$

Each of these components does commute, however, with the square

$$\hat{\mathbf{L}}^2 = \hat{L}_x^2 + \hat{L}_y^2 + \hat{L}_z^2 \tag{6.35}$$

of the angular-momentum vector $\hat{\mathbf{L}}$,

$$[\hat{\mathbf{L}}^2, \hat{L}_a] = 0 \quad , \quad a = x, y, z \quad . \tag{6.36}$$

Polar coordinates (radius r, polar angle ϑ, azimuth φ) are related to Cartesian coordinates (x, y, z) by

$$x = r \sin\vartheta \cos\varphi \quad , \quad y = r \sin\vartheta \sin\varphi \quad , \quad z = r \cos\vartheta \quad . \tag{6.37}$$

Using these polar coordinates the components and the square of $\hat{\mathbf{L}}$ have the representations

$$\hat{L}_x = \mathrm{i}\hbar \left(\sin\varphi \frac{\partial}{\partial\vartheta} + \operatorname{cotan}\vartheta \cos\varphi \frac{\partial}{\partial\varphi} \right) \quad ,$$

$$\hat{L}_y = \mathrm{i}\hbar \left(\cos\varphi \frac{\partial}{\partial\vartheta} - \operatorname{cotan}\vartheta \sin\varphi \frac{\partial}{\partial\varphi} \right) \quad ,$$

$$\hat{L}_z = -\mathrm{i}\hbar \frac{\partial}{\partial\varphi} \quad , \tag{6.38}$$

$$\hat{\mathbf{L}}^2 = -\hbar^2 \left[\frac{1}{\sin\vartheta} \frac{\partial}{\partial\vartheta} \left(\sin\vartheta \frac{\partial}{\partial\vartheta} \right) + \frac{1}{\sin^2\vartheta} \frac{\partial^2}{\partial\varphi^2} \right] \quad . \tag{6.39}$$

Thus, the eigenfunctions of $\hat{\mathbf{L}}^2$ and $\hat{L}_z$ are the *spherical harmonics* $Y_{\ell m}(\vartheta, \varphi)$ depending on ϑ and φ only, with the two indices relating to the eigenvalues of the square and the z component of angular momentum,

$$\hat{\mathbf{L}}^2 Y_{\ell m} = \ell(\ell+1)\hbar^2 Y_{\ell m} \quad , \quad \ell = 0, 1, 2, \ldots \quad , \tag{6.40}$$

$$\hat{L}_z Y_{\ell m} = m\hbar Y_{\ell m} \quad , \quad -\ell \le m \le \ell \quad . \tag{6.41}$$

The *angular-momentum quantum number* ℓ is interpreted as the modulus of angular momentum and m – usually called *magnetic quantum number* – as its z component. Together with (6.39), (6.40) is up to a factor $\hbar^2$ identical with (11.15) of Chap. 11, which deals with mathematical functions. The details of the spherical harmonics $Y_{\ell m}$ are given there.

6.1.4 The Stationary Schrödinger Equation in Polar Coordinates. Separation of Variables. Spherical Bessel Functions. Continuum Normalization. Completeness

The stationary Schrödinger equation (6.14) of a free particle can be expressed in polar coordinates where the kinetic energy is

$$T\varphi(\mathbf{r}) = \left[-\frac{\hbar^2}{2M}\frac{1}{r}\frac{\partial^2}{\partial r^2}r + \frac{1}{2Mr^2}\hat{\mathbf{L}}^2 \right] \varphi(\mathbf{r}) = E\varphi(\mathbf{r}) \tag{6.42}$$

and where the square $\hat{\mathbf{L}}^2$ of angular momentum is given by (6.39). Separation of the radial variable r and the angles ϑ and φ is achieved by factorization:

$$\varphi(\mathbf{r}) = R_\ell(r)Y_{\ell m}(\vartheta, \varphi) \quad . \tag{6.43}$$

Then, the radial wave function $R_\ell(r)$ satisfies the *free radial Schrödinger equation*

$$-\frac{\hbar^2}{2M}\left[\frac{1}{r}\frac{\mathrm{d}^2}{\mathrm{d}r^2}r - \frac{\ell(\ell+1)}{r^2}\right] R_\ell = ER_\ell \quad , \quad r > 0 \quad . \tag{6.44}$$

It is equivalent to, $k^2 = 2ME/\hbar^2$,

$$r^2\frac{\mathrm{d}^2}{\mathrm{d}r^2}R_\ell(k,r) + 2r\frac{\mathrm{d}}{\mathrm{d}r}R_\ell(k,r) + [k^2r^2 - \ell(\ell+1)]R_\ell(k,r) = 0 \quad . \tag{6.45}$$

Choosing $x = kr$ as a dimensionless variable and setting $z_\ell(x) \sim R_\ell(k,r)$, we arrive at the differential equation (11.30) of Chap. 11 for the spherical Bessel functions $z_\ell(x)$.

The kinetic energy of radial motion, $T^{\mathrm{rad}} = -(\hbar^2/2M)r^{-1}(\mathrm{d}^2/\mathrm{d}r^2)r$, is a Hermitian operator only for wave functions that are not singular at $r = 0$. This requirement restricts the $R_\ell(k,r)$ to be proportional to the *spherical Bessel functions of the first kind* j_ℓ:

$$R_\ell(k,r) = \sqrt{\frac{2}{\pi}}kj_\ell(kr) \quad . \tag{6.46}$$

The factor in front of j_ℓ in (6.46) ensures a continuum normalization of the kind

$$\int_0^\infty R_\ell(k,r)R_\ell(k',r)r^2\,\mathrm{d}r = \delta(k'-k) \quad . \tag{6.47}$$

The eigenfunctions $\varphi_{\ell m}(k,\mathbf{r})$ of the kinetic-energy operator T belonging to the energy eigenvalue $E = \hbar^2k^2/2M$ and to the angular-momentum quantum numbers ℓ, m are called *free partial waves*,

$$\varphi_{\ell m}(k, \mathbf{r}) = R_\ell(k, r) Y_{\ell m}(\vartheta, \varphi) \quad . \tag{6.48}$$

These eigenfunctions exhibit a continuum normalization in k,

$$\int \varphi^*_{\ell' m'}(k', \mathbf{r}) \varphi_{\ell m}(k, \mathbf{r}) \, \mathrm{d}V = \delta(k' - k) \delta_{\ell' \ell} \delta_{m' m} \quad . \tag{6.49}$$

Their completeness relation (11.18) reads

$$\sum_{\ell, m} \int_0^\infty \varphi^*_{\ell m}(k, \mathbf{r}') \varphi_{\ell m}(k, \mathbf{r}) k^2 \, \mathrm{d}k = \delta^3(\mathbf{r}' - \mathbf{r}) \quad . \tag{6.50}$$

It allows a decomposition of wave functions into free partial waves.

6.1.5 Partial-Wave Decomposition of the Plane Wave

The stationary plane wave of momentum $\mathbf{p} = \hbar \mathbf{k}$ has the form

$$\mathrm{e}^{\mathrm{i}\mathbf{p}\cdot\mathbf{r}/\hbar} = \mathrm{e}^{\mathrm{i}\mathbf{k}\cdot\mathbf{r}} = \mathrm{e}^{\mathrm{i}kr\cos\vartheta} \quad . \tag{6.51}$$

In a system of polar coordinates with the z axis in the direction of $\mathbf{k}$, the above formula shows that there is no φ dependence. Thus, a decomposition into free partial waves $\varphi_{\ell 0}(k, \mathbf{r})$, (6.48), containing only the spherical harmonics (11.16) with $m = 0$,

$$Y_{\ell 0}(\vartheta, \varphi) = \sqrt{\frac{2\ell + 1}{4\pi}} P_\ell(\cos\vartheta) \quad , \tag{6.52}$$

is possible. Here P_ℓ is a Legendre polynomial. One obtains

$$\mathrm{e}^{\mathrm{i}\mathbf{k}\cdot\mathbf{r}} = \mathrm{e}^{\mathrm{i}kr\cos\vartheta} = \sum_{\ell=0}^{\infty} \varphi_\ell = \sum_{\ell=0}^{\infty} \mathrm{i}^\ell (2\ell + 1) j_\ell(kr) P_\ell(\cos\vartheta) \quad , \tag{6.53}$$

$\mathbf{r}$: radius vector,
$\mathbf{k}$: wave vector of plane wave,
$\cos\vartheta = \mathbf{k}\cdot\mathbf{r}/kr$: cosine of polar angle,
ℓ: angular-momentum quantum number,
$P_\ell(\cos\vartheta)$: Legendre polynomial of order ℓ,
$j_\ell(kr)$: spherical Bessel function of first kind of order ℓ, (11.31).

6.1.6 Partial-Wave Decomposition of the Gaussian Wave Packet

The Gaussian wave packet (6.18) at time $t = 0$ is decomposed into free partial waves starting from the completeness relation (6.50):

$$\psi(r,0)=\frac{2}{\pi}\sum_{\ell=0}^{\infty}\sum_{m=-\ell}^{\ell}\int b_{\ell m}(k)j_\ell(kr)Y_{\ell m}(\vartheta,\varphi)k^2\,\mathrm{d}k \quad . \tag{6.54}$$

The probability $W_{\ell m}$ of finding a contribution of angular momentum ℓ, m irrespective of the wave number k is given by

$$W_{\ell m}=\frac{2}{\pi}\int b^*_{\ell m}(k)b_{\ell m}(k)k^2\,\mathrm{d}k \quad . \tag{6.55}$$

For a Gaussian wave packet with a probability sphere, i.e., $\sigma_x=\sigma_y=\sigma_z=\sigma(t)$, the probability $W_{\ell m}$ is given by

$$W_{\ell m}=\exp\left\{-\sigma_0^2\lambda^2\right\}\sqrt{\frac{\pi}{2\sigma_0^2k^2}}I_{\ell+\frac{1}{2}}(\sigma_0^2k^2)(2\ell+1) \times\frac{(\ell-|m|)!}{(\ell+|m|)!}\left|P_\ell^m(\cos\varphi'')\right|^2|\zeta|^{-2m} \quad , \tag{6.56}$$

$\mathbf{r}_0=a\hat{\mathbf{k}}_0+b\hat{\mathbf{b}}$: initial position expectation value,
$\hat{\mathbf{k}}_0$: unit vector in the direction of wave-vector expectation value,
$\hat{\mathbf{b}}$: unit vector, perpendicular to $\hat{\mathbf{k}}_0$, $\hat{\mathbf{k}}_0\cdot\hat{\mathbf{b}}=0$,
$k^4=\left(k_0^2-r_0^2/4\sigma_0^4\right)^2+(\mathbf{k}_0\cdot\mathbf{r}_0)^2/\sigma_0^4$,
$\lambda^2=k_0^2+r_0^2/4\sigma_0^4$,
P_ℓ^m: associated Legendre function with complex argument,
$I_{\ell+\frac{1}{2}}$: modified Bessel function.

The complex vector

$$\boldsymbol{\kappa}=\mathbf{k}_0-\mathrm{i}\frac{1}{2\sigma_0^2}\mathbf{r}_0 \tag{6.57}$$

is decomposed into

$$\boldsymbol{\kappa}=\kappa\{\mathbf{e}_1\cos\varphi'+\mathbf{e}_2\sin\varphi'\} \quad . \tag{6.58}$$

The quantity κ is the complex square root

$$\kappa=\left[\left(k_0-\frac{\mathrm{i}}{2\sigma_0^2}a\right)^2-\frac{1}{4\sigma_0^4}b^2\right]^{1/2} \quad . \tag{6.59}$$

The complex angle φ' is defined by

$$\cos\varphi'=\frac{1}{\kappa}(\mathbf{e}_1\cdot\boldsymbol{\kappa}) \quad , \quad \sin\varphi'=\frac{1}{\kappa}(\mathbf{e}_2\cdot\boldsymbol{\kappa}) \quad . \tag{6.60}$$

So far m was the quantum number of the z component of angular momentum, i.e., the *quantization axis* was the z axis. For an arbitrary quantization axis we rotate the coordinate system

$$\mathbf{e}_1 = -\hat{\mathbf{k}}_0 \quad , \quad \mathbf{e}_2 = \hat{\mathbf{b}} \quad , \quad \mathbf{e}_3 = \mathbf{e}_1 \times \mathbf{e}_2$$

into the system η_1, η_2, η_3 with the transformation

$$\eta_j = \sum_{i=1}^{3} \mathbf{e}_i R_{ij} \quad .$$

The matrix R_{ij} represents a rotation with the angle ϑ about the axis

$$\hat{\boldsymbol{\alpha}} = -\mathbf{e}_1 \sin\varphi + \mathbf{e}_2 \cos\varphi \quad , \tag{6.61}$$

i.e., ϑ and φ are the polar and azimuthal angles of the unit vector η_3 in the original system $\mathbf{e}_1, \mathbf{e}_2, \mathbf{e}_3$. With this we define the complex angle φ'' through its cosine,

$$\cos\varphi'' = R_{13}\cos\varphi' + R_{23}\sin\varphi' \quad ,$$

and the quantity

$$\zeta = \frac{(R_{11} + \mathrm{i}R_{12})\cos\varphi' + (R_{21} + \mathrm{i}R_{22})\sin\varphi'}{\sin\varphi''} \quad .$$

Here

$$\begin{aligned}
R_{11} &= \cos\vartheta + (1 - \cos\vartheta)\sin^2\varphi \quad , \\
R_{12} = R_{21} &= -(1 - \cos\vartheta)\cos\varphi\sin\varphi \quad , \\
R_{22} &= \cos\vartheta + (1 - \cos\vartheta)\cos^2\varphi \quad , \\
R_{13} = -R_{31} &= \sin\vartheta\cos\varphi \quad , \\
R_{23} = -R_{32} &= \sin\vartheta\sin\varphi \quad , \\
R_{33} &= \cos\vartheta \quad .
\end{aligned}$$

The quantization axis with respect to which m is defined is η_3. The marginal distribution

$$W_\ell = \sum_{m=-\ell}^{\ell} W_{\ell m} = \mathrm{e}^{-\sigma_0^2\lambda^2} \sqrt{\frac{\pi}{2\sigma_0^2 k^2}}\, I_{\ell+1/2}(\sigma_0^2 k^2)(2\ell + 1) P_\ell\left(\frac{\lambda^2}{k^2}\right) \tag{6.62}$$

describes the weight of the contribution of the angular momentum ℓ in the wave packet.

Further Reading

Alonso, Finn: Vol. 3, Chap. 3
Berkeley Physics Course: Chaps. 7, 8
Brandt, Dahmen: Chaps. 10, 11
Feynman, Leighton, Sands: Vol. 3, Chap. 18
Flügge: Vol. 1, Chaps. 2B, 2C
Gasiorowicz: Chaps. 6, 7, 8
Merzbacher: Chap. 9
Messiah: Vol. 1, Chap. 9, Vol. 2, Chap. 13
Schiff: Chaps. 2, 4

6.2 The 3D Harmonic Plane Wave

Aim of this section: Illustration of the stationary harmonic wave (6.11). The illustration can be presented as a surface over a Cartesian grid or as a surface over a polar grid.

A plot similar to Fig. 6.1 or Fig. 6.2 is produced. The type of plot (Cartesian or polar grid) depends on the type of descriptor that you select. You cannot change the plot type without changing the descriptor.

On the bottom of the subpanel **Physics—Comp. Coord.** you can select for plotting one of the three functions

$$|\mathrm{e}^{ikz}|^2 , \ \mathrm{Re}\,\mathrm{e}^{ikz} , \ \mathrm{Im}\,\mathrm{e}^{ikz} \quad .$$

On the subpanel **Physics—Variables** you can choose whether the **Input is Taken** directly as the **Wave Number** k or as the **Energy** E from which k is then computed. The relation between the two quantities is

$$k = \sqrt{2ME}/\hbar \quad .$$

The numerical value for k (or E) is found in a field labeled **Wave Number / Energy**.

Please note: In all computations of quantum mechanics in 3D we set

$$\hbar = 1 \quad , \quad M = 1 \quad .$$

For the present case that implies $k = \sqrt{2E}$.

Example Descriptors on File `3D_Free_Particle.des`

- `3D harmonic wave: surface over Cartesian grid` (see Fig. 6.1)
- `3D harmonic wave: surface over polar grid` (see Fig. 6.2)

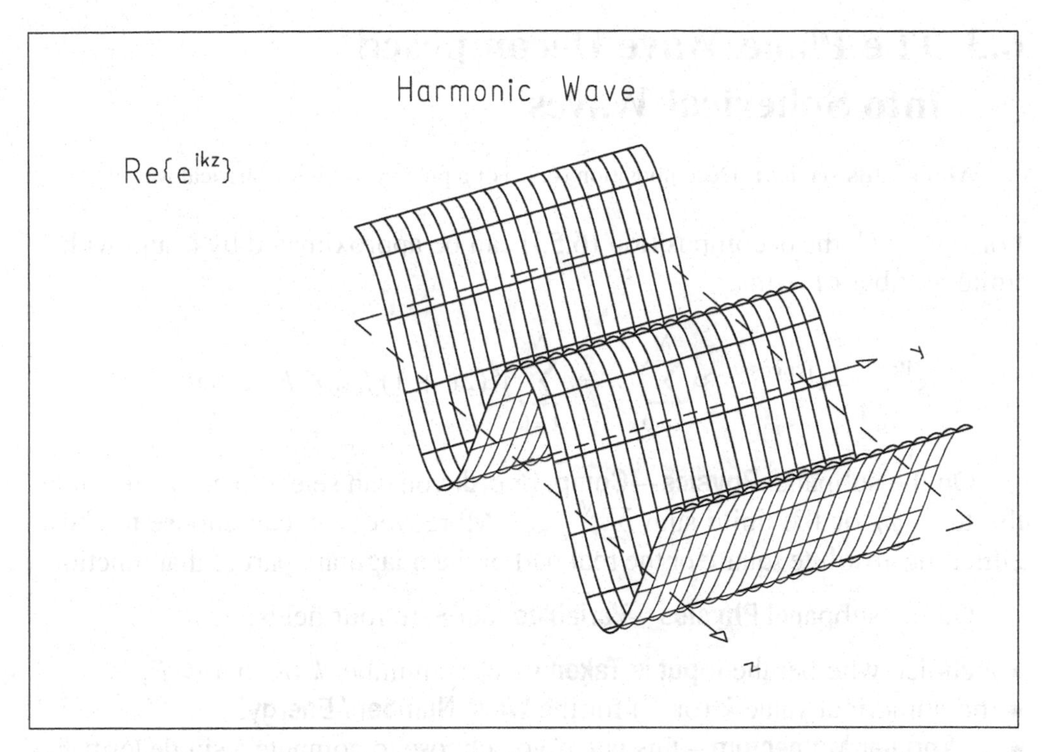

Fig. 6.1. Plot produced with descriptor `3D harmonic wave: surface over Cartesian grid` on file `3D_Free_Particle.des`

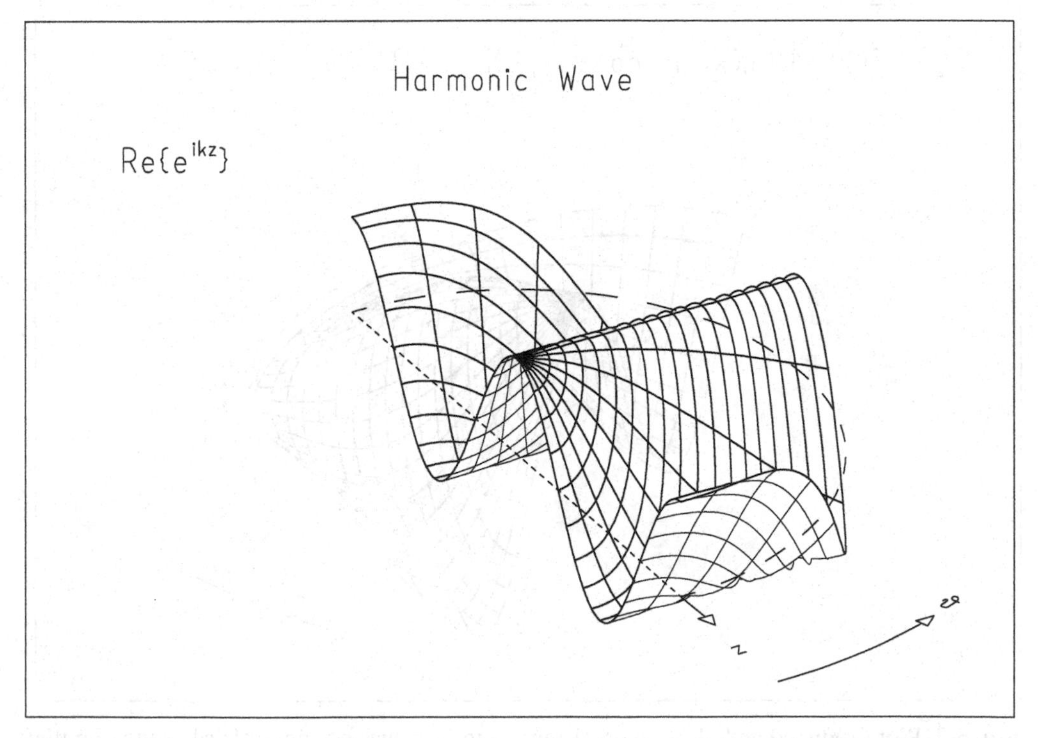

Fig. 6.2. Plot produced with descriptor `3D harmonic wave: surface over polar grid` on file `3D_Free_Particle.des`

6.3 The Plane Wave Decomposed into Spherical Waves

Aim of this section: Decomposition (6.53) of a plane wave into spherical waves.

For $r \le N/k$ the decomposition (6.53) can be approximated by a sum with a finite number of terms,

$$\mathrm{e}^{\mathrm{i}\mathbf{k}\cdot\mathbf{r}} = \mathrm{e}^{ikr\cos\vartheta} \approx \sum_{\ell=0}^{N} \varphi_\ell = \sum_{\ell=0}^{N} \mathrm{i}^\ell (2\ell+1) j_\ell(kr) P_\ell(\cos\vartheta) \quad .$$

On the subpanel **Physics—Comp. Coord.** you can select to compute either the term φ_ℓ or the finite sum $\sum_{\ell=0}^{N} \varphi_\ell$. Moreover, you can choose to show either the absolute square or the real part or the imaginary part of that function.

On the subpanel **Physics—Variables** there are four fields:

- a choice whether the **Input is Taken as** wave number k or energy E,
- the numerical value k (or E) for the **Wave Number / Energy**,
- an **Angular Momentum** – this is ℓ if you choose to compute a single term φ_ℓ. It is N if you choose to plot the sum $\sum_{\ell=0}^{N} \varphi_\ell$,

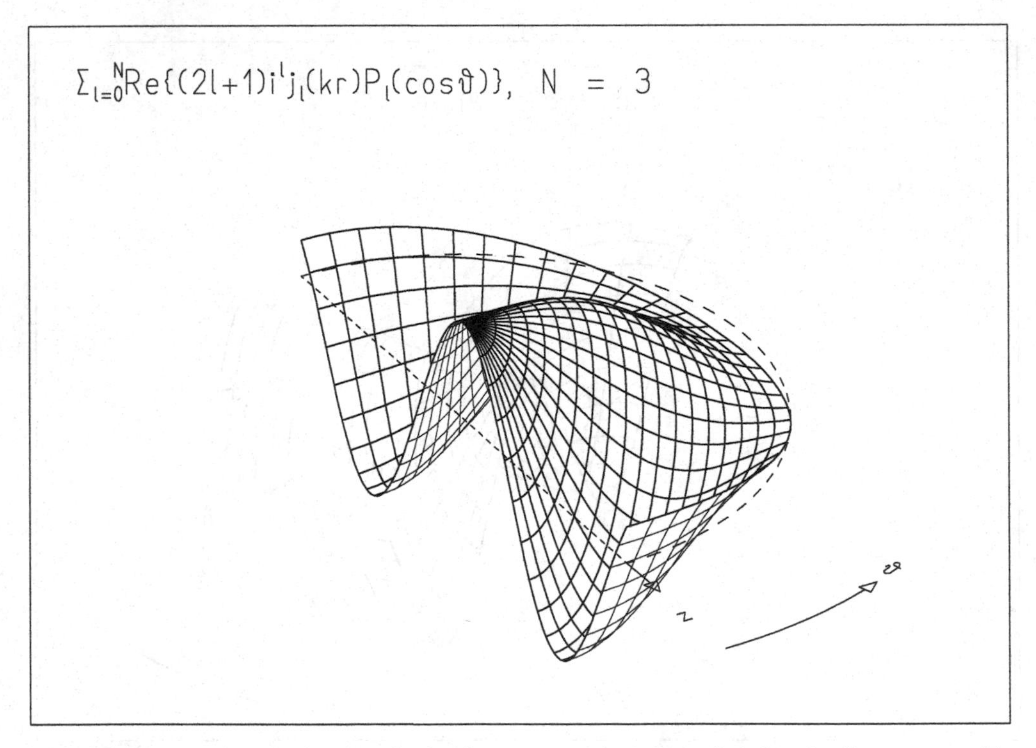

Fig. 6.3. Plot produced with descriptor `Plane wave as sum of spherical waves` on file `3D_Free_Particle.des`

- an explanation **For Multiple Plot:** The angular momentum (ℓ or N) is taken for the first plot in a multiple plot. It is successively increased by one for each subsequent plot.

Example Descriptor on File `3D_Free_Particle.des`

- `Plane wave as sum of spherical waves` (see Fig. 6.3)

6.4 The 3D Gaussian Wave Packet

Aim of this section: Illustration of the probability density $\varrho(x, y, t)$, (6.29), and the corresponding wave function $\psi(x, y, t)$.

A plot similar to Fig. 6.4 is produced depicting various aspects of the wave function ($|\psi|^2$, Re ψ, or Im ψ) in the x, y plane for a Gaussian wave packet. The expectation value of the packet moves freely in the x, y plane. In a *multiple plot* the situation is shown for times $t = 0, \Delta t, 2\Delta t, \ldots$. If the display of $|\psi|^2$ is chosen, additional graphical items are displayed. These are

- the position of the classical particle (as full circle for $t = 0$, as open circle for $t = \Delta t, 2\Delta t, \ldots$),

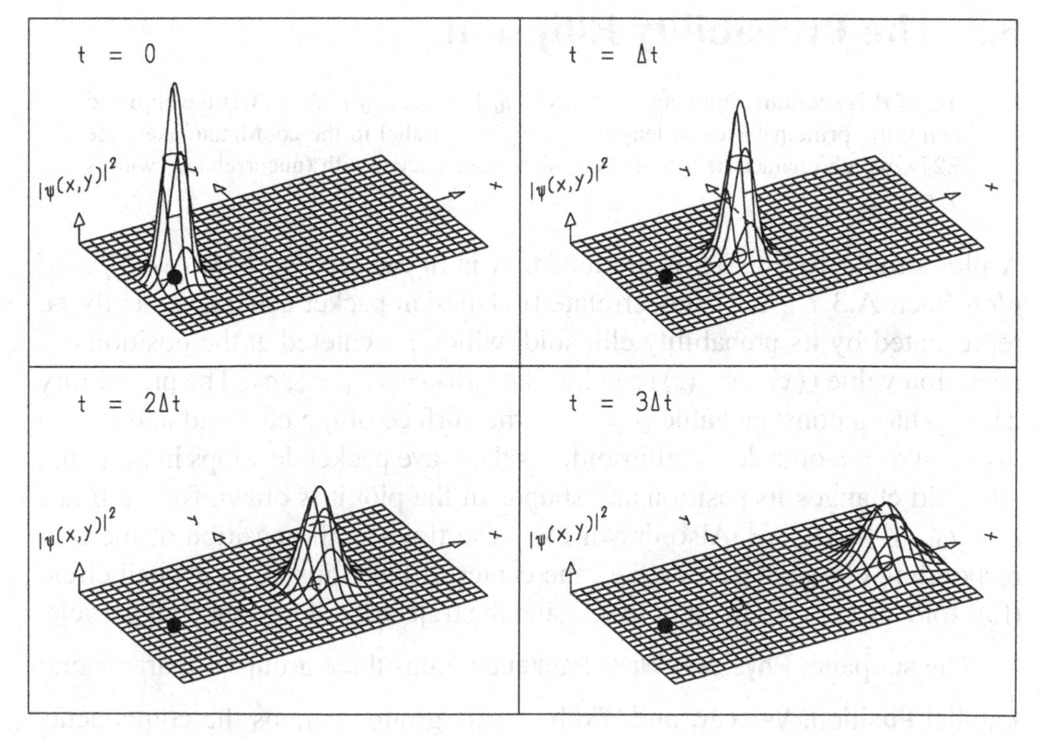

Fig. 6.4. Plot produced with descriptor `Free 3D wave packet: surface over Cartesian grid` on file `3D_Free_Particle.des`

- the trajectory of the classical particle,
- the covariance ellipse of the Gaussian probability distribution.

On the subpanel **Physics—Wave Packet** you find five items:

- **Initial Position** – the coordinates x_0, y_0 of the position expectation value at $t = 0$,
- **Initial Velocity** – the components v_{x0}, v_{y0} of mean velocity at $t = 0$,
- **Initial Width** – the widths σ_{x0}, σ_{y0} in x and y at $t = 0$,
- **Graphical Item** – the radius R of the circle indicating the position of the classical particle,
- **Time Step** – the time difference Δt between the situation shown in successive plots.

Movie Capability: Direct. The time, over which the movie extends, is $T = \Delta t\,(N_{\mathrm{Plots}} - 1)$; N_{Plots} is the number of individual plots in a multiple plot and Δt is the time interval between two plots. For a single plot $T = \Delta t$.

Example Descriptors on File `3D_Free_Particle.des`

- `Free 3D wave packet: surface over Cartesian grid` (see Fig. 6.4)
- `Dispersion of wave packet at rest, real part`

6.5 The Probability Ellipsoid

Aim of this section: Drawing for times $t_0, t_0 + \Delta t, \ldots, t_0 + (N-1)\Delta t$ the ellipsoid having the principal axes of lengths σ_x, σ_y, σ_z parallel to the coordinate axes, see (6.27), which characterizes a 3D Gaussian wave packet with (uncorrelated) widths $\sigma_x, \sigma_y, \sigma_z$.

A plot similar to Fig. 6.5 is produced. It is of the type *probability-ellipsoid plot*, Sect. A.3.7. A 3D (uncorrelated) Gaussian packet can conveniently be represented by its probability ellipsoid, which is centered at the position expectation value $(\langle x\rangle, \langle y\rangle, \langle z\rangle)$ and has the half-axes $\sigma_x, \sigma_y, \sigma_z$. The probability density has a constant value $\varrho = c$ on the surface of the ellipsoid and is $> c$ inside and $< c$ outside the ellipsoid. As the wave packet develops in time, the ellipsoid changes its position and shape. In the plot it is drawn for the times $t = t_0, t_0 + \Delta t, \ldots$. Also drawn for these times is the position of the corresponding classical particle, i.e., the center of the ellipsoid as a small circle (full for $t = t_0$, open for later times) and the trajectory of the classical particle.

The subpanel **Physics—Wave Packet** contains three groups of parameters:

- **Initial Position, Velocity, and Width** – This group contains the components x_0, y_0, z_0 of the initial (time $t = 0$) position expectation value and the components v_{x0}, v_{y0}, v_{z0} of the initial velocity expectation value as well as the initial widths σ_{x0}, σ_{y0}, σ_{z0}.

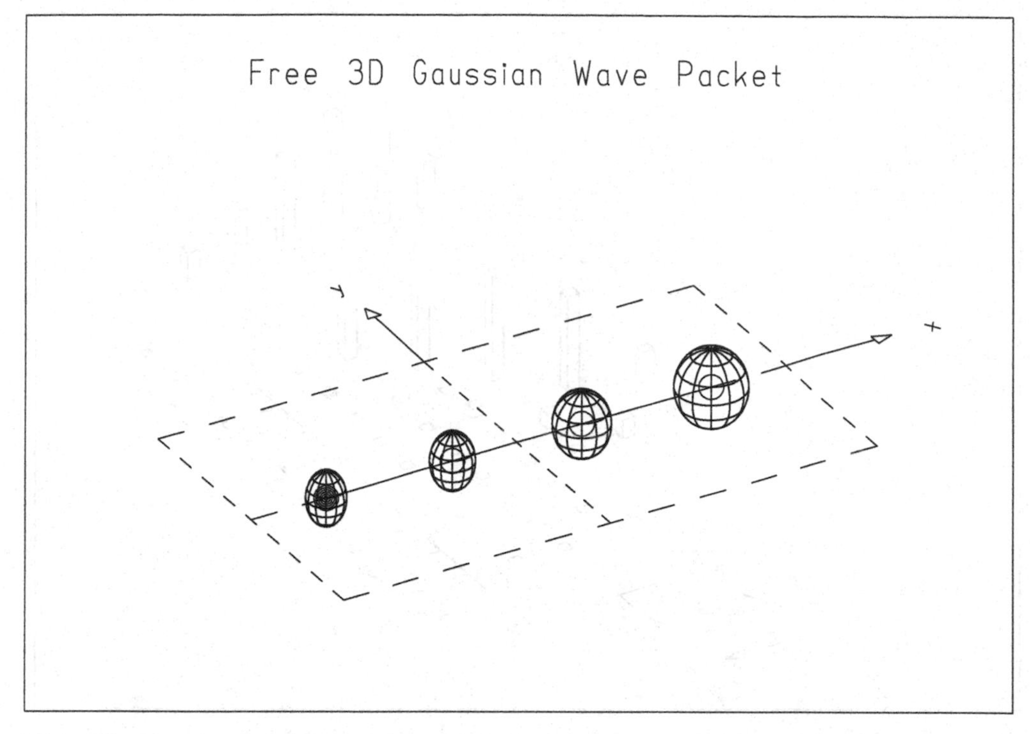

Fig. 6.5. Plot produced with descriptor `Free 3D wave packet: ellipsoids` on file `3D_Free_Particle.des`

- **Time** – Here you find the time t_0 for which the first ellipsoid is shown at the time Δt, i.e., the time difference between ellipsoids adjacent to each other. In total N ellipsoids are drawn.
- **Graphical Item** – This is the radius R of the circle indicating the position of the classical particle.

Movie Capability: Direct. On the subpanel **Movie** (see Sect. A.4) of the parameters panel you may choose to show or not to show the initial and intermediate positions of the ellipsoid.

Example Descriptor on File `3D_Free_Particle.des`

- `Free 3D wave packet: ellipsoids` (see Fig. 6.5)

6.6 Angular-Momentum Decomposition of a Wave Packet

Aim of this section: Computation and presentation of the probabilities $W_{\ell m}$ and W_ℓ that a particle represented by a wave packet of spherical symmetry has the angular-momentum quantum numbers ℓ and m, respectively; see (6.55) and (6.62).

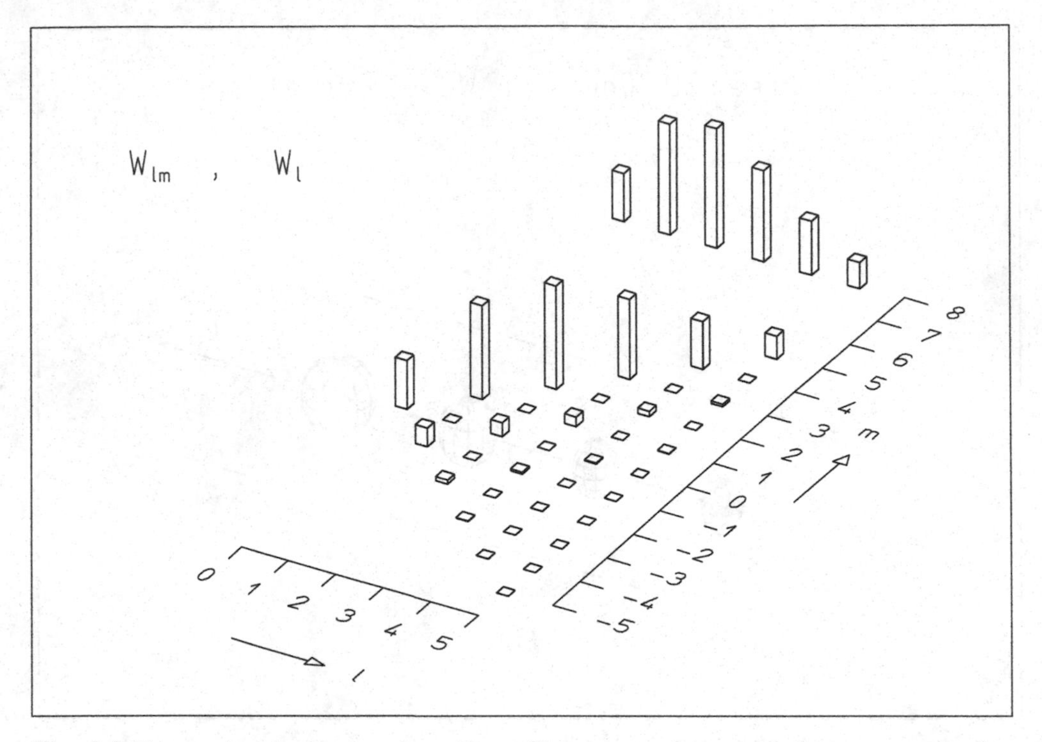

Fig. 6.6. Plot produced with descriptor `Free 3D wave packet: angular-momentum decomposition` on file `3D_Free_Particle.des`

A plot similar to Fig. 6.6 is produced showing as a *3D column plot*, Sect. A.3.8. In an ℓ, m plane the probabilities $W_{\ell m}$ and – at the high m margin – the probabilities W_ℓ are represented as columns. The wave packet has its initial position expectation value ($x_0 = a, y_0 = b, z_0 = 0$) in the x, y plane. The momentum expectation value ($p_x = -p_0, p_y = 0, p_z = 0$) has only an x component. This configuration ensures that for $b > 0, p_0 > 0$ the classical angular momentum ($L_x = 0, L_y = 0, L_z = bp_0$) points in z direction. The orientation of the quantization axes is given by its polar angle ϑ and its azimuth φ.

On the subpanel **Physics—Wave Packet** there are three groups of parameters:

- **Wave Packet** – Here you find the four parameters p_0 (absolute value of the momentum expectation value), σ_0 (initial width of the spherically symmetric packet), a (x component of initial position), b (impact parameter).
- **Quantization Axis** – This is given by its polar angle ϑ and its azimuth φ.
- **Maximum Angular-Momentum Quantum Number** – The $W_{\ell m}$ and W_ℓ are shown in the range $0 \le \ell \le \ell_{\max}$.

Example Descriptor on File `3D_Free_Particle.des`

- `Free 3D wave packet: angular-momentum decomposition` (see Fig. 6.6)

6.7 Exercises

Please note:

(i) You may watch a demonstration of the material of this chapter by pressing the button **Run Demo** in the *main toolbar* and selecting one of the demo files `3D_Free_Particle`.

(ii) More example descriptors can be found on the descriptor file `3DFreeParticle(FE).des` in the directory `FurtherExamples`.

(iii) For the following exercises use descriptor file `3D_Free_Particle.des`.

(iv) The numerical values of the particle mass and of Planck's constant are put to 1.

6.2.1 Plot the three-dimensional plane wave in a Cartesian plot for the wave number $k = 2$ for the interval $0 \leq x \leq 2\pi$: **(a)** real part, **(b)** imaginary part, **(c)** absolute square. Start from descriptor 1: '`3D harmonic wave: surface over Cartesian grid`'.

6.2.2 Plot the three-dimensional plane wave in a polar plot for the wave number $k = 2$ for the radial interval $0 \leq r \leq 2\pi$: **(a)** real part, **(b)** imaginary part, **(c)** absolute square. Start from descriptor 2: '`3D harmonic wave: surface over polar grid`'.

6.3.1 Study the partial waves for different ℓ appearing in the decomposition of the plane harmonic wave for **(a)** $\ell = 0$, **(b)** $\ell = 1$, **(c)** $\ell = 2, \ldots,$ **(l)** $\ell = 13$, **(m)** $\ell = 20$. Start from descriptor 3: '`Plane wave as sum of spherical waves`'. For even (odd) values of ℓ the real (imaginary) part of the partial waves is nonvanishing. **(n)** Calculate the minimal classical angular momentum in units of $\hbar$ for a particle of wave number $k = 2$ that does not enter the radial domain $0 \leq r \leq 2\pi$. **(o)** Plot the partial wave for $\ell = 13$. **(p)** Why does the partial wave become more and more suppressed in the central region (r close to zero) for increasing ℓ?

6.3.2 Study the sums of partial waves for different N approximating the plane harmonic wave for **(a)** $N = 0$, **(b)** $N = 1, \ldots,$ **(k)** $N = 10$, **(l)** $N = 13$, **(m)** $N = 20$. Start from descriptor 3: '`Plane wave as sum of spherical waves`'. **(n)** How fast does the region in which the plotted partial sum resembles a plane harmonic wave grow in radius with increasing N?

6.4.1 A three-dimensional Gaussian wave packet in the x, y plane has the initial position $x_0 = -2$, $y_0 = -2$, initial velocity $v_{x0} = 4$, $v_{y0} = 4$, initial width $\sigma_{x0} = 0.5$, $\sigma_{y0} = 0.5$. Plot for the times $t = 0$ and $t = 1$ **(a)** the absolute square, **(b)** the real part, **(c)** the imaginary part of the wave function. Start from descriptor 4: '`Free 3D wave packet: surface over Cartesian grid`'. **(d)** Calculate the spatial widening of the wave packet with time ($M = 1$, $\hbar = 1$).

6.4.2 (a–d) Repeat Exercise 6.4.1 for the initial position $x_0 = -2$, $y_0 = -2$, initial velocity $v_{x0} = 4$, $v_{y0} = 2$, initial width $\sigma_{x0} = 0.25$, $\sigma_{y0} = 0.8$. **(e)** Calculate the vector of classical angular momentum of a particle with the initial expectation values of the wave packet ($M = 1$, $\hbar = 1$).

6.4.3 (a,b,c) Repeat Exercise 6.4.1 (a,b,c) for the initial position $x_0 = -4$, $y_0 = -3$, initial velocity $v_{x0} = 4$, $v_{y0} = 3$, initial width $\sigma_{x0} = 0.1$, $\sigma_{y0} = 0.3$ for $\Delta t = 1.5$. **(d)** Explain the different speeds by which the main axes of the covariance ellipse grow. **(e)** Which direction do the ripples in the real and imaginary part of the wave function possess?

6.4.4 A wave packet at rest is to be plotted with the initial parameters $x_0 = 0$, $y_0 = 0$, $v_{x0} = 0$, $v_{y0} = 0$, $\sigma_{x0} = 1$, $\sigma_{y0} = 1$ for $\Delta t = 2.5$. Plot **(a)** the absolute square, **(b)** the real part, **(c)** the imaginary part. Start from descriptor 4: '`Free 3D wave packet: surface over Cartesian grid`'. **(d)** Why does this wave packet exhibit a rotationally invariant structure?

6.5.1 (a) Plot the probability ellipsoids of a particle with the initial conditions $x_0 = -3$, $y_0 = -2$, $z_0 = -1$, $v_{x0} = 5$, $v_{y0} = 1.5$, $v_{z0} = 1$, $\sigma_{x0} = 1.0$, $\sigma_{y0} = 1.05$, $\sigma_{z0} = 2$ for three instants in time, $t = 0, 1, 2$. Start from descriptor 6: '`Free 3D wave packet: ellipsoids`'. **(b)** Calculate the angular-momentum vector of the classical particle with the above initial data ($M = 1$, $\hbar = 1$).

6.5.2 Plot the probability ellipsoids of a particle with the initial conditions $x_0 = -3$, $y_0 = -2$, $z_0 = -1$, $v_{x0} = 5$, $v_{y0} = 1.5$, $v_{z0} = 1$, $\sigma_{x0} = 0.25$, $\sigma_{y0} = 0.35$, $\sigma_{z0} = 0.5$ for three instants in time, $t = 0, 0.8, 1.6$. Start from descriptor 6: '`Free 3D wave packet: ellipsoids`'.

6.5.3 (a) Plot the probability ellipsoids of a particle with the initial conditions $x_0 = -5$, $y_0 = -2$, $z_0 = -1$, $v_{x0} = 5$, $v_{y0} = 1.5$, $v_{z0} = 1$, $\sigma_{x0} = 1.8$, $\sigma_{y0} = 1.5$, $\sigma_{z0} = 1$ for the three instants in time, $t = 0, 1, 2$. Start from descriptor 6: '`Free 3D wave packet: ellipsoids`'. **(b)** Why does the ellipsoid enlarge much more slowly in time relative to its initial size than in Exercise 6.5.2?

6.5.4 (a) Plot the probability ellipsoids for a spherical wave packet at rest with the widths $\sigma_{x0} = \sigma_{y0} = \sigma_{z0} = 0.2$ for four instants in time, $t = 0, 0.2, 0.4, 0.6$. Start from descriptor 6: '`Free 3D wave packet: ellipsoids`'. **(b)** Why does the radius of the sphere grow almost linearly over time only for later times?

6.6.1 (a) Plot the probabilities $W_{\ell m}$, W_ℓ of the partial-wave decomposition for the quantization axis $\hat{\mathbf{n}}$ pointing in the z direction of a wave packet with initial conditions $p_0 = 1.5$, $\sigma_0 = 1$, $a = 2$, $b = 0.6667$. Start from descriptor 7: '`Free 3D wave packet: angular-momentum decomposition`'. **(b)** Calculate the angular-momentum vector of a classical particle possessing the above initial data ($\hbar = 1$). **(c)** Calculate the angular momentum for a particle with the impact parameters $b' = b + \sigma_0$ and $b'' = b - \sigma_0$. **(d)** Interpret the results of (b) and (c) in terms of the partial probabilities $W_{\ell m}$ as plotted in (a).

6.6.2 **(a)** Plot the probabilities $W_{\ell m}$, W_ℓ of the partial-wave decomposition for the quantization axis $\hat{\mathbf{n}}$ pointing in the z direction for a wave packet with initial conditions $p_0 = 3$, $\sigma_0 = 0.5$, $a = 2$, $b = 0.3333$. Start from descriptor 7: '`Free 3D wave packet: angular-momentum decomposition`'. **(b)** Why does the plot look the same as in Exercise 6.6.1 (a)?

6.6.3 **(a)** Plot the probabilities $W_{\ell m}$, W_ℓ of the partial-wave decomposition for the quantization axis $\hat{\mathbf{n}}$ pointing in the z direction for a wave packet with initial conditions $p_0 = 3$, $\sigma_0 = 1$, $a = 2$, $b = 0.3333$. Start from descriptor 7: '`Free 3D wave packet: angular-momentum decomposition`'. **(b)** Why are the $W_{\ell m} = 0$ for $m = \ell-1, \ell-3, \ldots, -\ell+1$ for the wave packets investigated so far? **(c)** Why has the distribution of the $W_{\ell m}$ and the W_ℓ widened compared to Exercise 6.6.2?

7. Bound States in Three Dimensions

Contents: Introduction of the Schrödinger equation with potential. Partial-wave decomposition. Spherical harmonics as eigenfunctions of angular momentum. Radial Schrödinger equation. Centrifugal barrier. Normalization and orthogonality of bound-state wave functions. Infinitely deep square-well potential. Spherical step potential. Harmonic oscillator. Coulomb potential. Harmonic particle motion.

7.1 Physical Concepts

7.1.1 The Schrödinger Equation for a Particle under the Action of a Force. The Centrifugal Barrier. The Effective Potential

The Schrödinger equation (6.5) of a free particle of mass M introduced in Sect. 6.1 contains the kinetic energy T as the only term in the Hamiltonian operator. Under the action of a conservative force $\mathbf{F}(\mathbf{r}) = -\nabla V(\mathbf{r})$ the Hamiltonian H contains both the kinetic energy T and the potential energy $V(\mathbf{r})$,

$$H = T + V(\mathbf{r}) = -\frac{\hbar^2}{2M}\Delta + V(\mathbf{r}) \quad . \tag{7.1}$$

The Schrödinger equation for the three-dimensional motion under the action of a force then reads

$$\mathrm{i}\hbar\frac{\partial}{\partial t}\psi(\mathbf{r},t) = \left(-\frac{\hbar^2}{2M}\Delta + V(\mathbf{r})\right)\psi(\mathbf{r},t) \quad . \tag{7.2}$$

With a separation of time and space coordinates,

$$\psi(\mathbf{r},t) = \mathrm{e}^{-\mathrm{i}Et/\hbar}\varphi_E(\mathbf{r}) \quad , \tag{7.3}$$

the stationary Schrödinger equation is an eigenvalue equation,

$$H\varphi_E(\mathbf{r}) = E\varphi_E(\mathbf{r})$$

DOI 10.1007/978-1-4419-7424-2_7,

or

$$\left(-\frac{\hbar^2}{2M}\Delta + V(\mathbf{r})\right)\varphi_E(\mathbf{r}) = E\varphi_E(\mathbf{r}) \quad . \tag{7.4}$$

Again E is the energy eigenvalue, $\varphi_E(\mathbf{r})$ the corresponding eigenfunction.

For a *spherically symmetric potential*

$$V(\mathbf{r}) = V(r)$$

a further separation of radial and angular coordinates by means of an eigenfunction corresponding to the energy eigenvalue E and the angular-momentum quantum numbers ℓ, m,

$$\varphi_{E\ell m}(\mathbf{r}) = R_{E\ell}(r)\, Y_{\ell m}(\vartheta, \varphi) \quad , \tag{7.5}$$

is carried out along the same lines as in Sect. 6.1. We arrive at the *radial Schrödinger equation* for the *radial wave function* $R_{E\ell}(r)$,

$$\left(-\frac{\hbar^2}{2M}\frac{1}{r}\frac{\mathrm{d}^2}{\mathrm{d}r^2}r + V_\ell^{\mathrm{eff}}(r)\right) R_{E\ell}(r) = E R_{E\ell}(r) \quad , \quad r > 0 \quad , \tag{7.6}$$

with the *effective potential*

$$V_\ell^{\mathrm{eff}}(r) = \frac{\hbar^2}{2M}\frac{\ell(\ell+1)}{r^2} + V(r) \quad . \tag{7.7}$$

The first term of the left-hand side of (7.6) represents the *kinetic energy of the radial motion*

$$T^{\mathrm{rad}} = \hat{p}_r^2/2M \quad , \tag{7.8}$$

with the *operator of radial momentum*

$$\hat{p}_r = \frac{\hbar}{\mathrm{i}}\left(\frac{1}{r} + \frac{\partial}{\partial r}\right) \quad . \tag{7.9}$$

The first term of (7.7) is the *centrifugal barrier.* It corresponds to the *rotational energy* relative to the origin of the coordinate frame,

$$T^{\mathrm{rot}} = \hat{\mathbf{L}}^2/2\Theta \quad , \tag{7.10}$$

of a particle with squared angular momentum $\hat{\mathbf{L}}^2 Y_{\ell m} = \hbar^2\ell(\ell+1)Y_{\ell m}$ and a *moment of inertia* $\Theta = Mr^2$ with respect to the origin. The second term is the spherically symmetric potential $V(r)$ of the force $\mathbf{F}(\mathbf{r}) = -\nabla V(r) = -\mathbf{e}_r\, \mathrm{d}V(r)/\mathrm{d}r$ acting on the particle. Because the centrifugal barrier is a repulsive potential (for $\ell \geq 1$), it tends to push the particle away from the origin $r = 0$.

The solutions of (7.6) are physical for $r > 0$. The radial kinetic energy is a Hermitian operator, i.e., a physical observable, only for wave functions free of singularities at $r = 0$. This requirement represents the *boundary condition* for solutions of the Schrödinger equation at $r = 0$.

7.1.2 Bound States. Scattering States. Discrete and Continuous Spectra

We denote by V_∞ the value of the spherically symmetric potential far out,

$$V_\infty = \lim_{r\to\infty} V(r) \quad . \tag{7.11}$$

We consider potentials only for which the intervals in r, for which $V(r) \le E < V_\infty$ for any given energy value have a finite total length. Then there are two types of solution:

i) *bound states* for $E < V_\infty$: there exist only solutions at *discrete energy eigenvalues* $E_{n\ell}$; the integer n is the principal quantum number used to enumerate the eigenvalue, ℓ is the quantum number of angular momentum;
ii) *scattering states* for $E \ge V_\infty$: there is a *continuous spectrum of eigenvalues* $E_\ell(k)$; it fills the domain $V_\infty \le E_\ell(k)$.

In this chapter we deal with bound states only.

The radial wave function for a bound state with energy eigenvalue $E_{n\ell}$ will be denoted by $R_{n\ell}(r)$. It satisfies the radial Schrödinger equation

$$\left(-\frac{\hbar^2}{2M}\frac{1}{r}\frac{\mathrm{d}^2}{\mathrm{d}r^2}r + V_\ell^{\mathrm{eff}}(r)\right) R_{n\ell}(r) = E_{n\ell} R_{n\ell}(r) \quad . \tag{7.12}$$

In the domains in r where $V_\ell^{\mathrm{eff}}(r) > E_{n\ell}$ the integral over the absolute square of the wave function must be finite, otherwise the contributions to the potential energy coming from these domains would diverge. Thus, the integral over the absolute square of the bound-state wave function over the range $0 \le r < \infty$ must be finite so that the integral over the absolute square of the radial bound-state function can be normalized to one:

$$\int_0^\infty R_{n\ell}^*(r) R_{n\ell}(r) r^2 \,\mathrm{d}r = 1 \quad . \tag{7.13}$$

Because the radial kinetic energy and the effective potential energy in (7.12) are Hermitian operators, the eigenfunctions $R_{n\ell}$ are orthogonal for different principal quantum numbers, so that together with (7.13) we have the *orthonormality relation* for the radial bound-state wave functions:

$$\int_0^\infty R_{n'\ell}^*(r) R_{n\ell}(r) r^2 \,\mathrm{d}r = \delta_{n'n} \quad . \tag{7.14}$$

The total bound-state wave functions

$$\varphi_{n\ell m}(\mathbf{r}) = R_{n\ell}(r) Y_{\ell m}(\vartheta, \varphi) \tag{7.15}$$

are orthonormal in all three quantum numbers n, ℓ, m,

$$\int \varphi^*_{n'\ell'm'}(\mathbf{r})\varphi_{n\ell m}(\mathbf{r})\,\mathrm{d}V = \delta_{n'n}\,\delta_{\ell'\ell}\,\delta_{m'm} \quad , \tag{7.16}$$

because of the orthonormality of the radial wave functions (7.14) and of the spherical harmonics (11.18).

The probability density

$$\varrho_{n\ell m}(\mathbf{r}) = |\varphi_{n\ell m}(\mathbf{r})|^2 = |R_{n\ell}(r)|^2|Y_{\ell m}(\vartheta,\varphi)|^2 \tag{7.17}$$

of a bound state described by the wave function $\varphi_{n\ell m}(\mathbf{r})$ is a function of r and ϑ only. Because of (11.16) and (11.17) the φ dependence vanishes upon taking the absolute square of the spherical harmonics

$$|Y_{\ell m}(\vartheta,\varphi)|^2 = \frac{2\ell+1}{4\pi}\frac{(\ell-|m|)!}{(\ell+|m|)!}\left(P_\ell^{|m|}(\cos\vartheta)\right)^2 \quad . \tag{7.18}$$

Moreover, the probability density is a function of the modulus $|m|$ of the magnetic quantum number. Altogether we have

$$\varrho_{n\ell m}(r,\vartheta) = |R_{n\ell}(r)|^2\frac{2\ell+1}{4\pi}\frac{(\ell-|m|)!}{(\ell+|m|)!}\left(P_\ell^{|m|}(\cos\vartheta)\right)^2 \quad , \tag{7.19}$$

r: radial variable,
ϑ: polar angle,
$R_{n\ell}(r)$: radial wave function,
$P_\ell^{|m|}(\cos\vartheta)$: associated Legendre function,
n: principal quantum number,
ℓ: quantum number of angular momentum,
m: magnetic quantum number.

The zeros of the radial wave function, $R_{n\ell}(r_i) = 0$, appear in a plot of $\varrho_{n\ell m}(r,\vartheta)$ over the coordinate plane of r and ϑ as *circular node lines* of radius r_i. The zeros of the spherical harmonics, $Y_{\ell m}(\vartheta_j,\varphi) = 0$, appear as *node rays* originating at the origin under the polar angles ϑ_j.

7.1.3 The Infinitely Deep Square-Well Potential

The *infinitely deep square-well potential*

$$V(r) = \begin{cases} 0 & , r \le a \\ \infty & , r > a \end{cases}$$

confines the particle to a sphere of radius a. The solutions $R_{n\ell}(r)$ of the radial Schrödinger equation have to vanish for values $r > a$, otherwise they would give rise to infinite contributions of the potential energy for $r > a$. Thus, we are looking for solutions of the radial Schrödinger equation

$$\left(-\frac{\hbar^2}{2M}\frac{1}{r}\frac{\mathrm{d}^2}{\mathrm{d}r^2}r + V_\ell^{\mathrm{eff}}(r) \right) R_{n\ell}(r) = E_{n\ell}(r) \quad , \quad 0 < r \le a \quad , \tag{7.20}$$

in the range $0 \le r \le a$ only. The solution has to be free of singularities at $r = 0$ and has to vanish at $r = a$. This allows only for spherical Bessel functions of the first kind $j_\ell(kr)$. The wave number k has to be determined in accordance with the boundary condition at $r = a$. Thus, $k = k_{n\ell}$ has to be chosen so that it is a zero of the Bessel function, i.e., $j_\ell(k_{n\ell}a) = 0$. With the normalization to one the solution is given by

$$R_{n\ell}(r) = \begin{cases} \left[\frac{a^3}{2}(j_{\ell+1}(k_{n\ell}a))^2 \right]^{-1/2} j_\ell(k_{n\ell}r) \,, & 0 \le r \le a \\ 0 & , r > a \end{cases} \quad . \tag{7.21}$$

The energy eigenvalue is determined by the wave number,

$$E_{n\ell} = \hbar^2 k_{n\ell}^2 / 2M \quad . \tag{7.22}$$

The *principal quantum number* $n = 0, 1, 2, \ldots$ is equal to the number of nodes (zeros) of the spherical Bessel function $j_\ell(k_{n\ell}r)$ in the domain $0 < r < a$. The wave numbers $k_{n\ell}$ increase monotonically with n, as does the energy eigenvalue $E_{n\ell}$. A simple heuristic argument behind this is that the radial kinetic-energy contribution increases with increasing curvature of the wave function. The curvature itself grows monotonically with the number of nodes.

7.1.4 The Spherical Step Potential

For many applications a potential with stepwise constant values with N regions is a sufficient approximation:

$$V(r) = \begin{cases} V_1 & , 0 \le r < r_1 & \text{region } 1 \\ V_2 & , r_1 \le r < r_2 & \text{region } 2 \\ \vdots & & \vdots \\ V_{N-1} & , r_{N-2} \le r < r_{N-1} & \text{region } N-1 \\ V_N & , r_{N-1} \le r & \text{region } N \end{cases} \quad . \tag{7.23}$$

The eigenfunction

$$\varphi_{n\ell m}(\mathbf{r}) = R_{n\ell}(r) Y_{\ell m}(\vartheta, \varphi)$$

belonging to the energy eigenvalue $E_{n\ell}$ is determined by the radial wave function $R_{n\ell}(r)$, which is a solution of the radial Schrödinger equation (7.12).

Because the potential (7.23) is a stepwise constant function with N regions, $R_{n\ell}(r)$ consists of N pieces $R_{n\ell q}(r), q = 1, \ldots, N$,

$$R_{n\ell}(r) = \begin{cases} R_{n\ell 1}(r) & , 0 \le r < r_1 & \text{region } 1 \\ R_{n\ell 2}(r) & , r_1 \le r < r_2 & \text{region } 2 \\ \vdots & & \vdots \\ R_{n\ell N-1}(r) & , r_{N-2} \le r < r_{N-1} & \text{region } N-1 \\ R_{n\ell N}(r) & , r_{N-1} \le r & \text{region } N \end{cases} \quad . \tag{7.24}$$

The piece $R_{n\ell q}(r)$ fulfills the free radial Schrödinger equation

$$\left(-\frac{\hbar^2}{2M}\frac{1}{r}\frac{\mathrm{d}^2}{\mathrm{d}r^2}r + \frac{\ell(\ell+1)\hbar^2}{2Mr^2} \right) R_{n\ell q}(r) = (E_{n\ell} - V_q) R_{n\ell q}(r) \quad . \tag{7.25}$$

For the solution $R_{n\ell q}$ three cases have to be distinguished:

i) In regions with $E_{n\ell} > V_q$ we obtain a real wave number

$$k_{n\ell q} = \left| \sqrt{2M(E_{n\ell} - V_q)}/\hbar \right| \quad . \tag{7.26}$$

The solution is a linear superposition of the two linearly independent spherical Bessel functions j_ℓ and n_ℓ, (11.31), (11.32),

$$R_{n\ell q}(r) = A_{n\ell q} j_\ell(k_{n\ell q} r) + B_{n\ell q} n_\ell(k_{n\ell q} r) \quad . \tag{7.27}$$

ii) In regions with $E_{n\ell} < V_q$ the wave number is purely imaginary:

$$k_{n\ell q} = \mathrm{i}\kappa_{n\ell q} \quad , \quad \kappa_{n\ell q} = \left| \sqrt{2M(V_q - E_{n\ell})}/\hbar \right| \quad . \tag{7.28}$$

The solution is a linear combination of the two linearly independent Hankel functions $h_\ell^{(\pm)}$, (11.33), (11.34),

$$R_{n\ell q}(r) = A_{n\ell q} h_\ell^{(+)}(\mathrm{i}\kappa_{n\ell q} r) + B_{n\ell q} h_\ell^{(-)}(\mathrm{i}\kappa_{n\ell q} r) \quad . \tag{7.29}$$

For the imaginary argument the Hankel functions are products of functions $C_\ell^\pm(\mathrm{i}\kappa_{n\ell q} r)$, which are complex factors multiplied by real polynomials of r^{-1}, and of decreasing or increasing exponential functions,

$$h_\ell^{(\pm)}(\mathrm{i}\kappa_{n\ell q} r) = C_\ell^\pm \frac{\exp(\mp\kappa_{n\ell q} r)}{\kappa_{n\ell q} r} \quad , \tag{7.30}$$

see (11.35), (11.36), (11.41).

iii) The special case $E_{n\ell} = V_q$ of vanishing wave number leads to the solution

$$R_{n\ell q}(r) = A_{n\ell q} r^{\ell} + B_{n\ell q} r^{-(\ell+1)} \quad . \tag{7.31}$$

The solution (7.24) has to satisfy two boundary conditions:

i) At the origin $r = 0$: there is an absence of singularities, as already discussed in this section. Because the spherical Bessel function n_ℓ is singular at $r = 0$, see (11.40), this requires

$$B_{n\ell 1} = 0 \quad , \quad \text{i.e.,} \quad R_{n\ell 1}(r) = A_{n\ell 1} j_\ell(k_{n\ell 1} r) \quad . \tag{7.32}$$

ii) In region N, $r_{N-1} \le r$: for bound states the energy eigenvalue $E_{n\ell}$ is smaller than V_∞, i.e., $E_{n\ell} < V_N$, so that the wave function (7.29) in this region is given by (7.30). The radial wave function has to vanish sufficiently fast for $r \to \infty$ to allow for the normalization (7.13). This is taken care of by putting

$$B_{n\ell N} = 0 \quad , \quad \text{i.e.,} \quad R_{n\ell N}(r) = A_{n\ell N} h^{(+)}(\mathrm{i}\kappa_{n\ell q} r) \quad . \tag{7.33}$$

The remaining discussion runs very much parallel to Sect. 3.1. The continuity of the function $R_{n\ell}(r)$ and its derivative at the positions $r_1, \ldots, r_{N-1}$ poses $2(N-1)$ conditions analogous to (3.22),

$$R_{n\ell q}(r_q) = R_{n\ell q+1}(r_q) \tag{7.34}$$

and

$$\frac{\mathrm{d}R_{n\ell q}}{\mathrm{d}r}(r_q) = \frac{\mathrm{d}R_{n\ell q+1}}{\mathrm{d}r}(r_q) \tag{7.35}$$

for the $2N-2$ unknown coefficients $A_{n\ell 1}, A_{n\ell 2}, B_{n\ell 2}, \ldots, A_{n\ell N-1}, B_{n\ell N-1}, A_{n\ell N}$. For every value ℓ of angular momentum this is a homogeneous system of $2N-2$ linear equations for an equal number of unknowns. It has a nontrivial solution only if its determinant

$$D_\ell = D_\ell(E) \tag{7.36}$$

vanishes. This leads to a transcendental equation for the energy eigenvalues $E_{n\ell}$:

$$D_\ell(E_{n\ell}) = 0 \quad . \tag{7.37}$$

In general, its solutions can only be found numerically; they are calculated by the computer. Once the eigenvalue $E_{n\ell}$ is determined as a single zero of (7.37) the system of linear equations (7.34), (7.35) can be solved yielding the coefficients $A_{n\ell q}, B_{n\ell q}$ as a function of one of them. This last undetermined coefficient is then fixed by the normalization condition (7.13).

The set of eigenfunctions of step potentials with $V_N < \infty$ is finite, thus they do not form a complete set. In Chap. 8 we present the continuum eigenfunctions supplementing the discrete ones to a complete set of functions.

7.1.5 The Harmonic Oscillator

The *three-dimensional harmonic oscillator* is described by the spherically harmonic potential

$$V(r) = \frac{1}{2} M\omega^2 r^2 \quad , \tag{7.38}$$

r: radius,
ω: angular frequency,
M: mass of particle.

The radial eigenfunctions of the harmonic oscillator are, see (11.52),

$$R_{n_r\ell} = N_{n_r\ell} \left(\frac{r^2}{\sigma_0^2} \right)^{\ell/2} \exp\left(-\frac{r^2}{2\sigma_0^2} \right) L_{n_r}^{\ell+1/2} \left(\frac{r^2}{\sigma_0^2} \right) \quad , \tag{7.39}$$

$\sigma_0^2 = \hbar/(M\omega)$: ground-state width of oscillator,
$n_r = (n - \ell)/2$: radial quantum number,
n: principal quantum number,
ℓ: angular-momentum quantum number,
$L_{n_r}^{\ell+1/2}$: Laguerre polynomial, see (11.49),

$$N_{n_r\ell} = \sqrt{\frac{n_r! 2^{n_r+\ell+2}}{\sqrt{\pi}(2(n_r + \ell) + 1)!!\sigma_0^3}} \quad : \text{ normalization constant.} \tag{7.40}$$

The *principal quantum number*

$$n = 2n_r + \ell \quad , \quad n = 0, 1, 2, \ldots \quad , \tag{7.41}$$

determines the energy eigenvalues of the bound states,

$$E_{n\ell} = \left(n + \tfrac{3}{2}\right)\hbar\omega \quad . \tag{7.42}$$

The eigenfunctions (7.39) form a complete set of functions of the radial variable r. The full three-dimensional wave function is again obtained as a product of the radial wave function $R_{n_r\ell}$ and the spherical harmonic $Y_{\ell m}$:

$$\varphi_{n\ell m}(\mathbf{r}) = R_{n_r\ell}(r) Y_{\ell m}(\vartheta, \varphi) \quad , \quad n = 2n_r + \ell \quad . \tag{7.43}$$

They form a complete set of functions of the radius vector $\mathbf{r}$.

Of course, the Hamiltonian of the three-dimensional harmonic oscillator can be treated as a sum of three Hamiltonians of one-dimensional oscillators (3.13) in the Cartesian coordinates x, y, and z. This leads to eigenfunctions

$$\varphi'_{n_1 n_2 n_3}(\mathbf{r}) = \varphi_{n_1}(x)\varphi_{n_2}(y)\varphi_{n_3}(z) \quad , \tag{7.44}$$

which are products of one-dimensional oscillator wave functions φ_n (3.14). Clearly, they belong to energy eigenvalues

$$E_n = \left(n + \tfrac{3}{2}\right)\hbar\omega \quad , \quad n = n_1 + n_2 + n_3 \quad , \tag{7.45}$$

determined by the principal quantum number n, which is now simply the sum of the three principal quantum numbers n_1, n_2, n_3 of the three one-dimensional oscillators. Also the set (7.44) of eigenfunctions is complete. In fact, the eigenfunctions $\varphi_{n\ell m}(\mathbf{r})$, (7.43), to a given eigenvalue $E_{n\ell}$ can be superimposed as a linear combination of eigenfunctions $\varphi'_{n_1 n_2 n_3}(\mathbf{r})$ belonging to the same energy eigenvalue. Thus, their indices n_1, n_2, n_3 have to satisfy the relation $n_1 + n_2 + n_3 = n$.

7.1.6 The Coulomb Potential. The Hydrogen Atom

In the hydrogen atom an electron of mass M_e carrying elementary charge $(-e)$ moves under the attractive Coulomb force of a proton of a mass M_p about 2000 times as heavy as the electron. The *Coulomb potential*

$$U(r) = \frac{e}{4\pi\varepsilon_0}\frac{1}{r} \tag{7.46}$$

yields the potential energy of the electron upon multiplication with the charge of the electron,

$$V(r) = -eU(r) = -\frac{e^2}{4\pi\varepsilon_0}\frac{1}{r} \quad , \tag{7.47}$$

r: radial variable,
e: elementary charge,
ε_0: electric-field constant.

The constant $e^2/(4\pi\varepsilon_0)$, having the dimension of action times velocity, can be expressed in units $\hbar c$ having the same dimension:

$$\frac{e^2}{4\pi\varepsilon_0} = \alpha\hbar c \quad , \tag{7.48}$$

$\hbar = h/2\pi$: Planck's constant,
c: velocity of light.

The proportionality factor

$$\alpha = 1/137 \tag{7.49}$$

is Sommerfeld's *fine-structure constant.* The Coulomb potential energy now reads

$$V(r) = -\hbar c\frac{\alpha}{r} \quad . \tag{7.50}$$

The bound-state solutions $R_{n\ell}(r)$ of the radial Schrödinger equation (7.12), fulfilling the boundary condition without singularities at $r = 0$ and sufficient decrease for $r \to \infty$, are given by

$$R_{n\ell}(r) = \frac{2}{n^2 a^{3/2}} \sqrt{\frac{(n-\ell-1)!}{(n+\ell)!}} \left(\frac{2r}{na}\right)^{\ell} \exp\left(-\frac{r}{na}\right) L_{n-\ell-1}^{2\ell+1}\left(\frac{2r}{na}\right) , \tag{7.51}$$

r: radial variable,
n: principal quantum number $n = 1, 2, 3, \ldots$,
ℓ: angular-momentum quantum number $\ell = 0, 1, \ldots, n-1$,
$a = \hbar/(\alpha M c) = 0.5292 \times 10^{-10}$ m: Bohr radius,
$M = M_{\mathrm{p}} M_{\mathrm{e}}/(M_{\mathrm{p}} + M_{\mathrm{e}}) \approx M_{\mathrm{e}}$: reduced mass of electron,
M_{p}: proton mass,
M_{e}: electron mass,
$\alpha = e^2/(4\pi\varepsilon_0\hbar c) = 1/137$: Sommerfeld's fine-structure constant,
e: elementary charge,
$\hbar$: Planck's constant,
ε_0: electric-field constant,
L_p^k: Laguerre polynomial, see (11.49), (11.54).

The energy eigenvalues depend solely on the *principal quantum number* n:

$$E_n = -\frac{1}{2} M c^2 \frac{\alpha^2}{n^2} \quad . \tag{7.52}$$

The factor in front of n^{-2} has the numerical value $Mc^2\alpha/2 = 13.65$ eV.

7.1.7 Harmonic Particle Motion

In Sect. 6.1 we introduced the three-dimensional Gaussian wave packet of momentum expectation value p_0. Its probability distribution can be characterized by the probability ellipsoid, as discussed in Chap. 6. We calculate the motion of a Gaussian wave packet with uncorrelated initial widths $\sigma_{x0}, \sigma_{y0}, \sigma_{z0}$ of initial momentum expectation value $\mathbf{p}_0$ under the action of a harmonic force (7.38). The center of the probability ellipsoid, which is initially at rest, moves like

$$\begin{aligned} x(t) &= x_0 \cos\omega t + \frac{p_{x0}}{M\omega} \sin\omega t \quad , \\ y(t) &= y_0 \cos\omega t + \frac{p_{y0}}{M\omega} \sin\omega t \quad , \\ z(t) &= z_0 \cos\omega t + \frac{p_{z0}}{M\omega} \sin\omega t \quad , \end{aligned} \tag{7.53}$$

which represents the motion on an ellipse about the origin in the plane containing the point $\mathbf{r} = 0$ and the initial position $\mathbf{r}_0 = (x_0, y_0, z_0)$ and being tangential to the initial momentum $\mathbf{p}_0 = (p_{x0}, p_{y0}, p_{z0})$. As we have learned in Sect. 3.1, (3.35), the width of a one-dimensional harmonic oscillator oscillates with 2ω, twice the oscillator frequency.

Further Reading

Alonso, Finn: Vol. 3, Chap. 3
Berkeley Physics Course: Vols. 4, 8
Brandt, Dahmen: Chaps. 11, 13
Feynman, Leighton, Sands: Vol. 3, Chap. 19
Flügge: Vol. 1, Chap. 1D
Gasiorowicz: Chaps. 12, 17
Merzbacher: Chap. 10
Messiah: Vol. 1, Chaps. 9, 11
Schiff: Chaps. 3, 4

7.2 Radial Wave Functions in Simple Potentials

Aim of this section: Illustration of the radial wave function $R_{n\ell}(r)$, the energy eigenvalue spectrum $E_{n\ell}$, the potential $V(r)$ and the effective potential $V_\ell^{\mathrm{eff}}(r)$ for four types of potential: the infinitely deep square well, (7.21), (7.22), the square well of finite depth, (7.27) for $N = 2$, the harmonic oscillator, (7.39), (7.42), and the Coulomb potential, (7.51), (7.52).

A plot similar to Figs. 7.1–7.4 is produced, which may contain the following items in a plane spanned by the radial coordinate r and the energy E:

- the *potential* $V(r)$ and the *effective potential* $V_\ell^{\mathrm{eff}}(r)$ as dashed lines of different dash lengths,
- the *eigenvalues* $E_{n\ell}$ shown as short-stroke horizontal dashed lines,
- the *radial wave functions* $R_{n\ell}(r)$ (or simple functions of these) as 2D function graphs for which the graphical representations of the eigenvalues serve as zero lines,
- the *term scheme* shown as a series of short lines at the positions $E_{n\ell}$ on the right-hand side of the scale in E.

On the subpanel **Physics—Comp. Coord.** you can select the **Type of Potential** (deep square well, square well of finite depth, harmonic oscillator, Coulomb) and the way the **Eigenfunction is Shown** ($R_{n\ell}^2$, $R_{n\ell}$, $r^2 R_{n\ell}^2$, $r R_{n\ell}$).

On the subpanel **Physics—Variables** there are four items:

- **Graphical Item** – The parameter ℓ_{DASH} determines the length of the dashes in the graphical representation of the potentials and the eigenvalues.

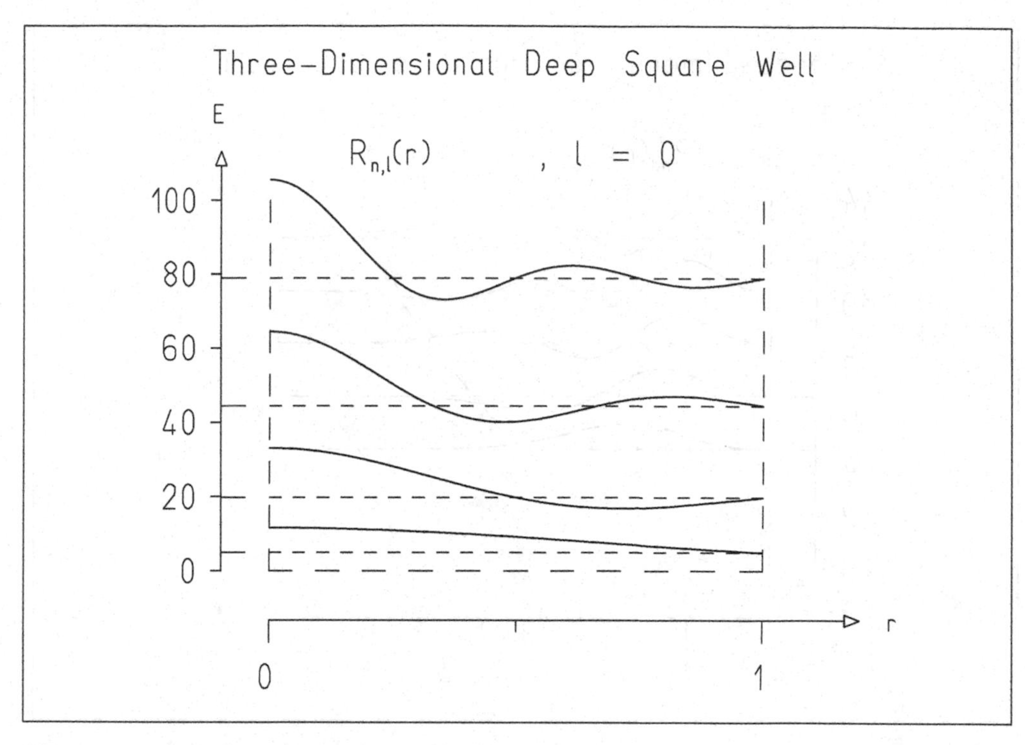

Fig. 7.1. Plot produced with descriptor `Radial wave functions: deep square-well potential` on file `3D_Bound_States.des`

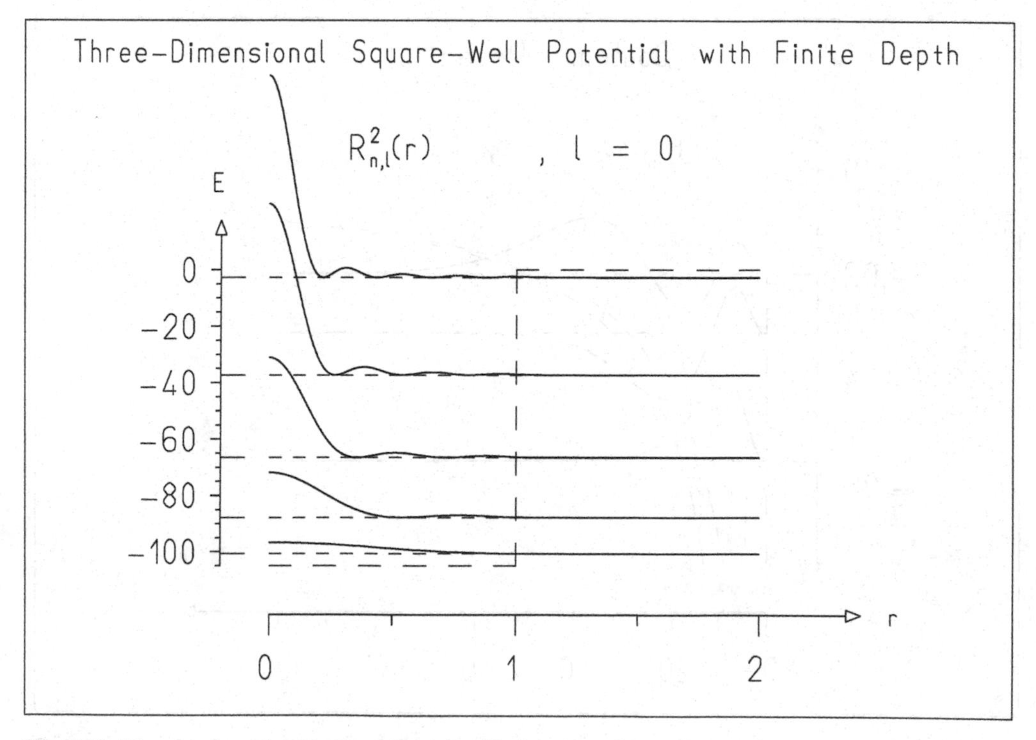

Fig. 7.2. Plot produced with descriptor `Radial wave functions: square well of finite depth` on file `3D_Bound_States.des`

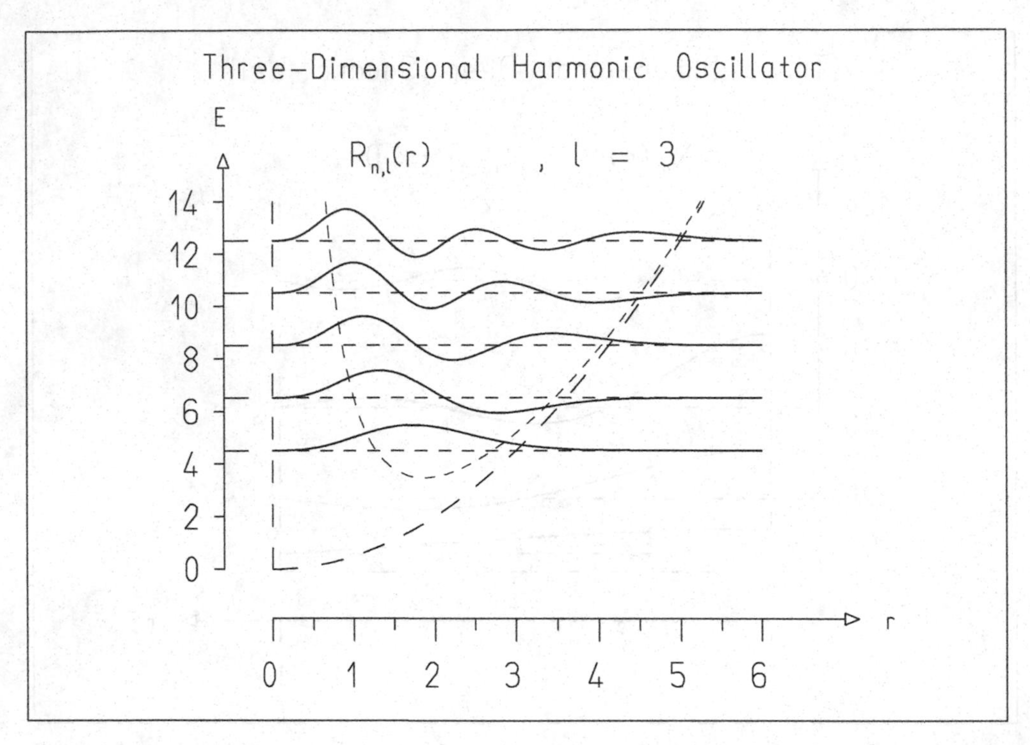

Fig. 7.3. Plot produced with descriptor `Radial wave functions: harmonic oscillator` on file `3D_Bound_States.des`

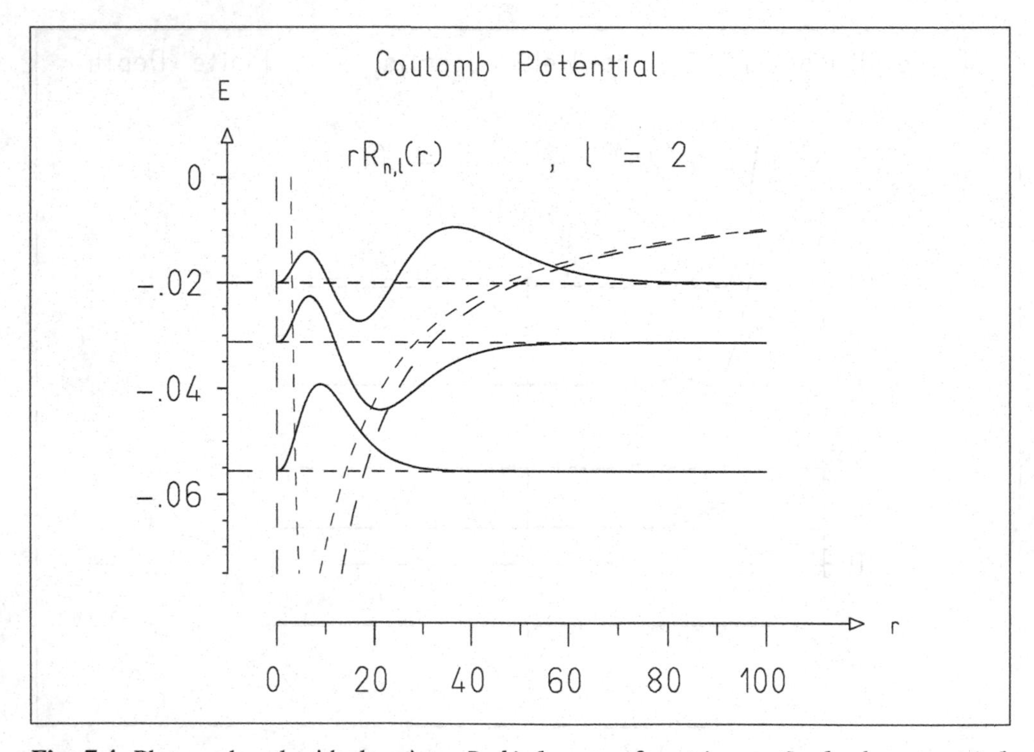

Fig. 7.4. Plot produced with descriptor `Radial wave functions: Coulomb potential` on file `3D_Bound_States.des`

- **Scale Factor** – The factor s determines the scale for the graphical representation of the wave functions. [Because these are plotted in the r, E plane, technically speaking $E = E_{n\ell} + sR_{n\ell}(r)$ is plotted.]
- **Clipping Rectangle for Potential** – The lines showing the potential and the effective potential are normally drawn only inside the 'box' given by the ranges of the world coordinates X and Y. You can extend that box to the left by the fraction **f_X-** of its original length in X. Similarly, **f_X+**, **f_Y-**, **f_Y+** extend the box to the right, bottom, and top, respectively.
- **Items to be Plotted** – Here you find four *check boxes* allowing you to select some or all of the items listed at the beginning of this section.

The subpanel **Physics—Potential** contains three items:

- **Angular-Momentum Quantum Number** – For a multiple plot the value ℓ is taken for the first plot and incremented by one successively from plot to plot.
- **Potential** – Here you find the parameters for the chosen potential (the radius a for the deep square well, the radius a and the depth V_0 for the square well of finite depth, the circular frequency ω for the harmonic oscillator, and the fine-structure constant α for the Coulomb potential).
- **Search Parameters** – There are three parameters determining the display and the accuracy in the computation of eigenvalues and eigenfunctions:
  - A maximum number of **N_max** eigenvalues is displayed. (Counting begins with the lowest eigenvalue for the given angular momentum ℓ.)
  - Eigenvalues are computed and shown only for energies smaller than **E_max**.
  - For the square-well potential of finite depth there is also the parameter **N_Search** used for the numerical computation of the eigenvalues. The range between the minimum of the potential and E_{max} is divided into N_{Search} intervals. The algorithm can find at most one eigenvalue in an interval. Thus, if eigenvalues are dense N_{Search} has to be large.

Example Descriptors on File `3D_Bound_States.des`

- `Radial wave functions: deep square-well potential` (see Fig. 7.1)
- `Radial wave functions: square well of finite depth` (see Fig. 7.2)
- `Radial wave functions: harmonic oscillator` (see Fig. 7.3)
- `Radial wave functions: Coulomb potential` (see Fig. 7.4)

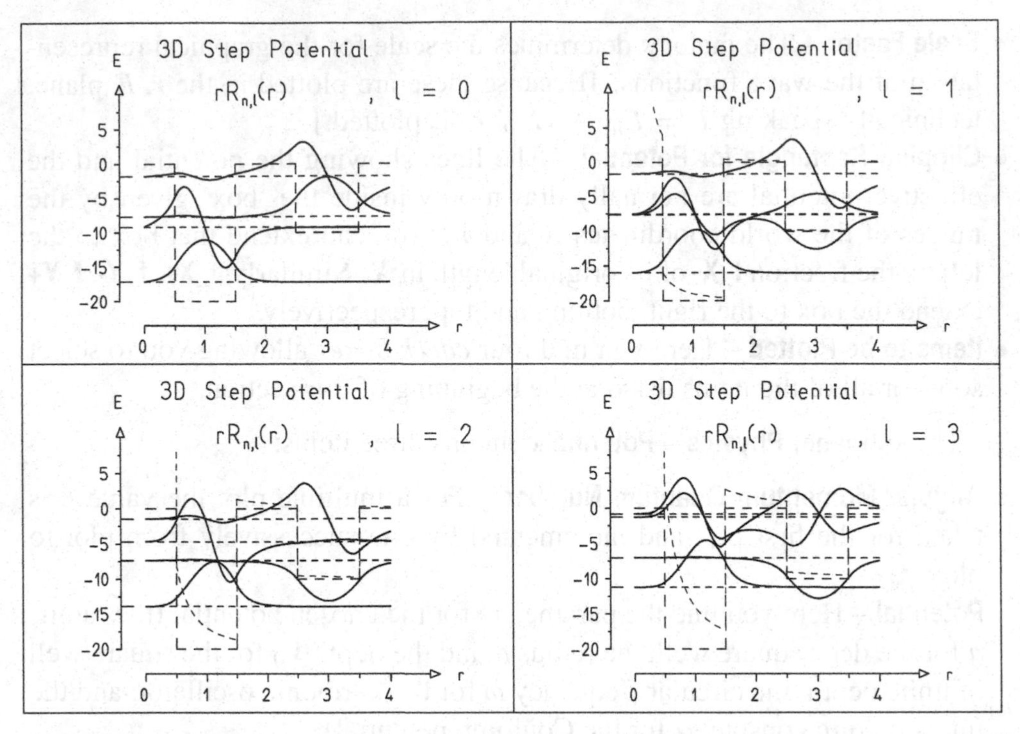

Fig. 7.5. Plot produced with descriptor `Radial wave functions in step potential` on file `3D_Bound_States.des`

7.3 Radial Wave Functions in the Step Potential

Aim of this section: Computation and presentation of the radial wave function $R_{n\ell}$, (7.27), the energy-eigenvalue spectrum $E_{n\ell}$, the potential $V(r)$ and the effective potential $V_\ell^{\text{eff}}(r)$ for the spherical step potential (7.23).

A plot similar to Fig. 7.5 is produced. As in Sect. 7.2 it may contain in an r, E plane:

- the *potential* $V(r)$ and the *effective potential* $V_\ell^{\text{eff}}(r)$,
- the *eigenvalues* $E_{n\ell}$,
- the *radial wave functions* $R_{n\ell}$ (or functions thereof),
- the *term scheme* next to the E axis.

On the subpanel **Physics—Comp. Coord.** you can select the way the **Eigenfunction is Shown** ($R_{n\ell}^2$, $R_{n\ell}$, $r^2R_{n\ell}^2$, $rR_{n\ell}$).

The subpanel **Physics—Variables** is as described in Sect. 7.2.

The subpanel **Physics—Potential** contains five items:

- **Angular-Momentum Quantum Number** – Contains the value ℓ. In a multiple plot it is taken for first plot and incremented by one for every additional plot.

- **Number of Regions** – Contains the number N of different regions in the spherical step potential (7.23), $2 \leq N \leq 5$.
- **Regions** – Here you find the boundaries r_i between regions and the potential values V_i for the different regions of the potential (7.23).
- **Search Parameters** – As described in Sect. 7.2 **N_max** is the maximum number of eigenvalues displayed. Eigenvalues are computed only for $E <$ **E_max**. The range between the minimum of the potential and $E_{\max}$ is divided into **N_Search** intervals. The algorithm searching for eigenvalues can find at most one eigenvalue in an interval.
- **Function D=D(E)** – The eigenvalues $E_{n\ell}$ are found numerically as the zeros of the functions $D_\ell(E) = 0$, (7.36). The function $D_\ell(E)$ may be **Shown** or **Not Shown**. Because it is plotted in the r, E plane, technically speaking the function $r = r_D + f_D \sinh^{-1}[s_D D_\ell(E)]$ is presented. Here **r_D** is the position in r corresponding to $D = 0$ and **f_D** and **s_D** are scale factors.

Example Descriptor on File `3D_Bound_States.des`

- `Radial wave functions in step potential` (see Fig. 7.5)

7.4 Probability Densities

Aim of this section: Illustration of the probability density (7.17) describing a particle in an eigenstate of a spherically symmetric potential.

A plot similar to Figs. 7.6 to 7.9 is produced. It shows the probability density $\varrho_{n\ell m}(r, \vartheta)$ of the eigenstate with quantum numbers n, ℓ, m in a simple spherically symmetric potential. The plot is of the type *surface over polar grid in 3D*. In addition to the probability density the plot contains

- the *radial nodes* as dashed lines,
- the *polar nodes* as dashed lines,
- in the case of a square-well potential the *edge of the potential* $r = a$ as a continuous line.

On the subpanel **Physics—Comp. Coord.** you can select the **Type of Potential** (deep square well, square well of finite depth, harmonic oscillator, or Coulomb).

On the subpanel **Physics—Autom. Scale** you can switch on (or off) an **Automatic Scale**. If it is on, the range in the computing coordinate z is set automatically to extend from $z_{\text{beg}} = 0$ to $z_{\text{end}} = s\varrho_{\max}$. Here s is a **Scale Factor** on the bottom of the subpanel and $\varrho_{\max}$ is the maximum value which $\varrho_{n\ell m}(r, \vartheta)$ can assume. This autoscale facility is particularly useful in multiple plots.

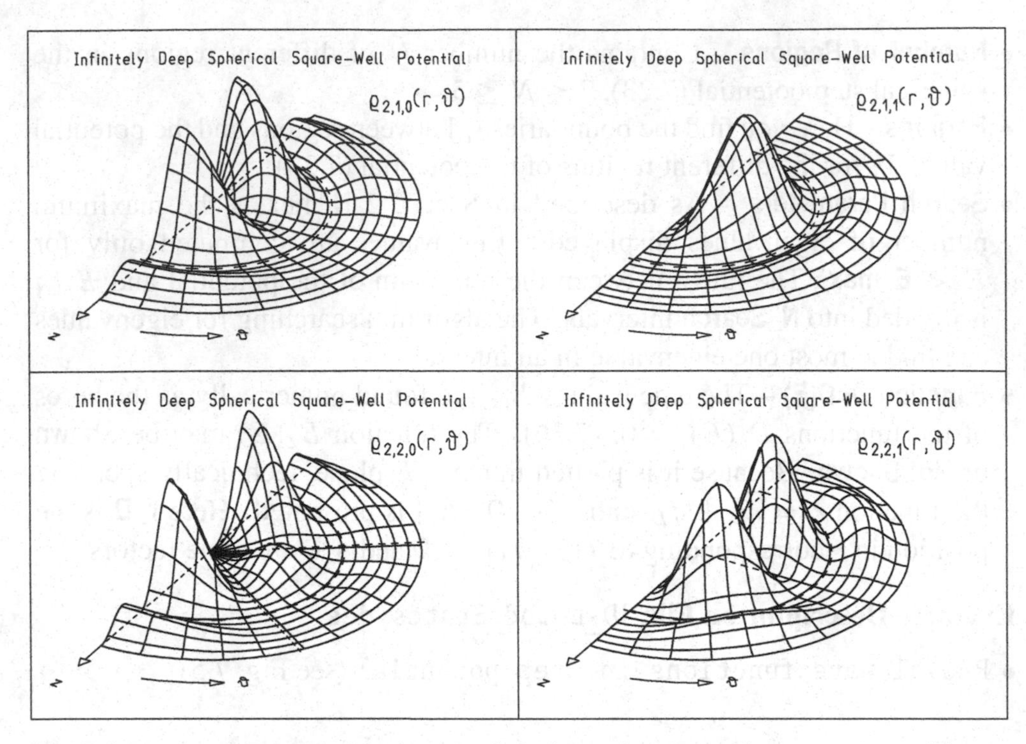

Fig. 7.6. Plot produced with descriptor `Probability density: deep square-well potential` on file `3D_Bound_States.des`

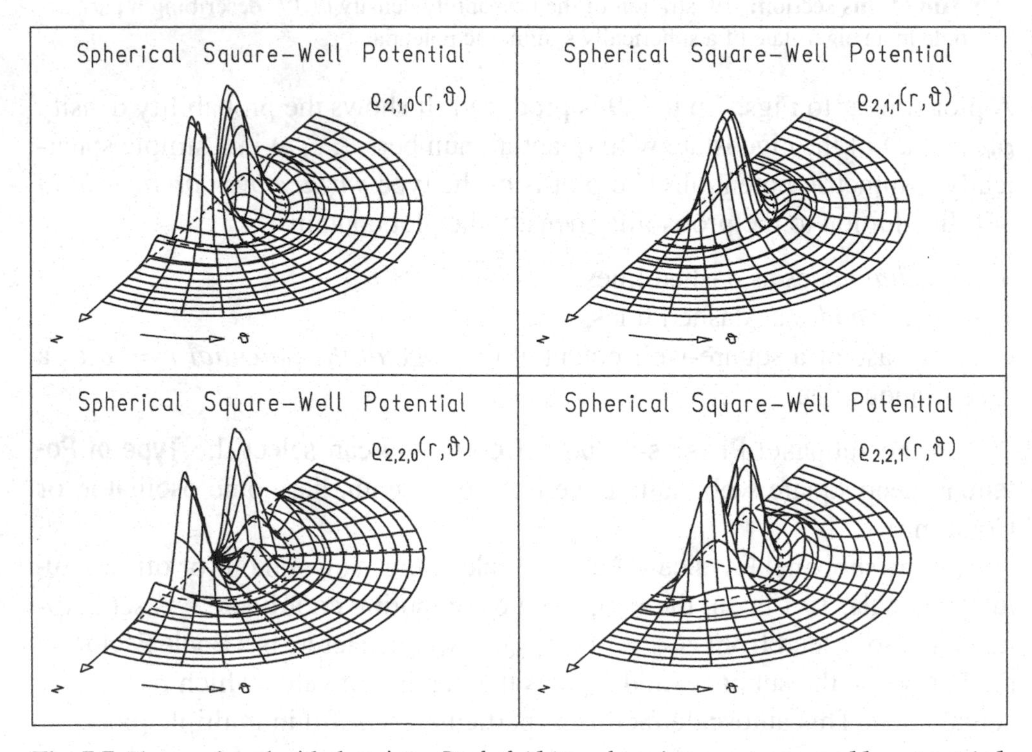

Fig. 7.7. Plot produced with descriptor `Probability density: square-well potential of finite depth` on file `3D_Bound_States.des`

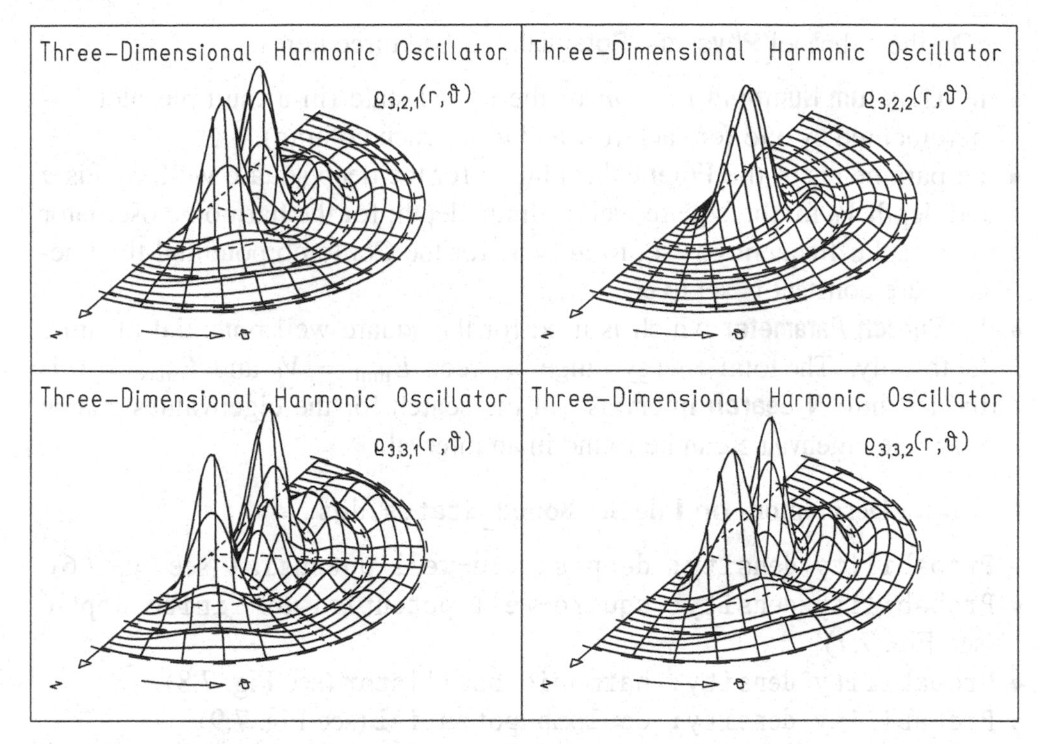

Fig. 7.8. Plot produced with descriptor `Probability density: harmonic oscillator` on file `3D_Bound_States.des`

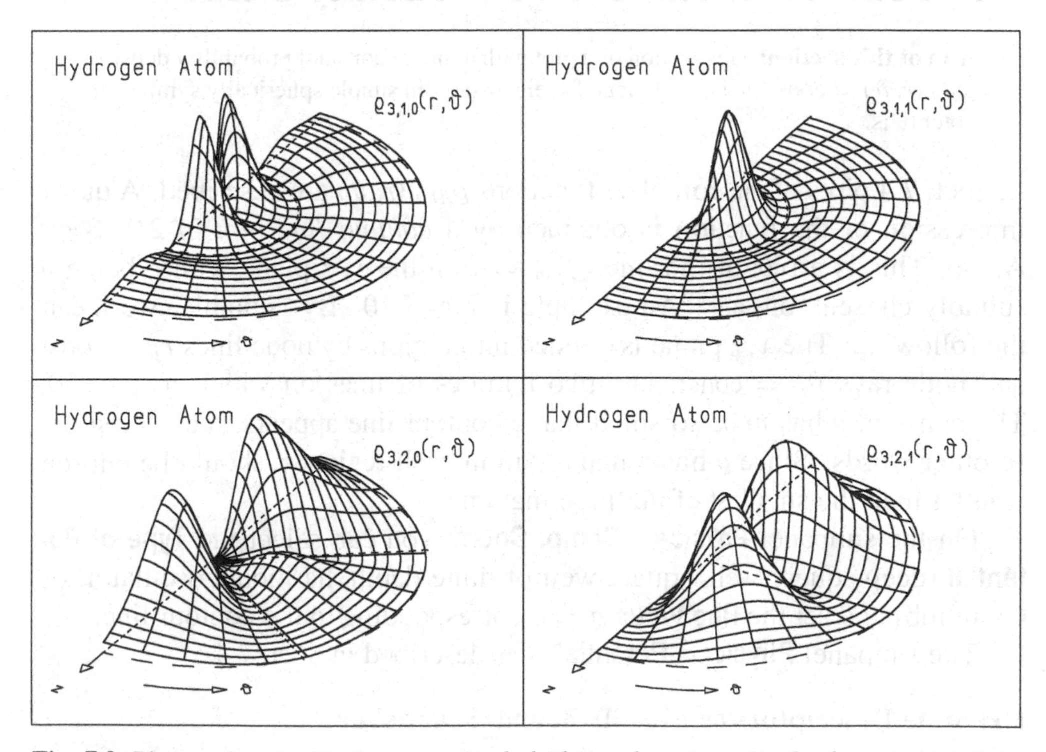

Fig. 7.9. Plot produced with descriptor `Probability density: Coulomb potential` on file `3D_Bound_States.des`

On the subpanel **Physics—Potential** you find three items:

- the **Quantum Numbers** n, ℓ, m of the bound state (in a multiple plot ℓ is incremented by one for each row and m for each column),
- the parameters of the **Potential** (radius a for the deep square well, radius a and depth V_0 for the square well of finite depth; for the harmonic oscillator the circular frequency is set to $\omega = 1$; for the Coulomb potential the fine-structure constant is set to $\alpha = 1$),
- the **Search Parameter** which is used for the square well potential of finite depth only. The total energy range between $E_{\min} = V_0$ and $E_{\max} = 0$ is divided into **N_Search** intervals. In the search for the eigenvalues $E_{n\ell}$ at most one eigenvalue can be found in an interval.

Example Descriptors on File `3D_Bound_States.des`

- `Probability density: deep square-well potential` (see Fig. 7.6)
- `Probability density: square-well potential of finite depth` (see Fig. 7.7)
- `Probability density: harmonic oscillator` (see Fig. 7.8)
- `Probability density: Coulomb potential` (see Fig. 7.9)

7.5 Contour Lines of the Probability Density

Aim of this section: Generation of a plot with a line of constant probability density, $\varrho_{n\ell m}(r, \vartheta) = \text{const}$, in the x, z plane for eigenstates in simple spherically symmetric potentials.

In Sect. 7.4 plots of the complete functions $\varrho_{n\ell m}(r, \vartheta)$ are described. A quick impression of the function is obtained by a *contour-line plot in 2D*, Sect. A.3.4. This is a plot of the line $\varrho_{n\ell m} = c$ in the x, z plane with c being a suitably chosen constant. An example is Fig. 7.10. By 'suitable' we mean the following: The x, z plane is divided into regions by node lines $r_i = \text{const}$ and node rays $\vartheta_i = \text{const}$, i.e., two families of lines on which $\varrho_{n\ell m} = 0$. The constant c has to be so small that a contour line appears in each region. In other words: Since ϱ has a maximum in each region, c should be chosen smaller than the smallest of all these maxima.

On the subpanel **Physics—Comp. Coord.** you can select the **Type of Potential** (deep square well, square well of finite depth, harmonic oscillator, or Coulomb) and set the **Rho Value** $\varrho = c$ corresponding to the contour line.

The subpanel **Physics—Potential** is as described in Sect. 7.4.

Example Descriptors on File `3D_Bound_States.des`

- `Contour lines: deep square well`
- `Contour lines: square well of finite depth`

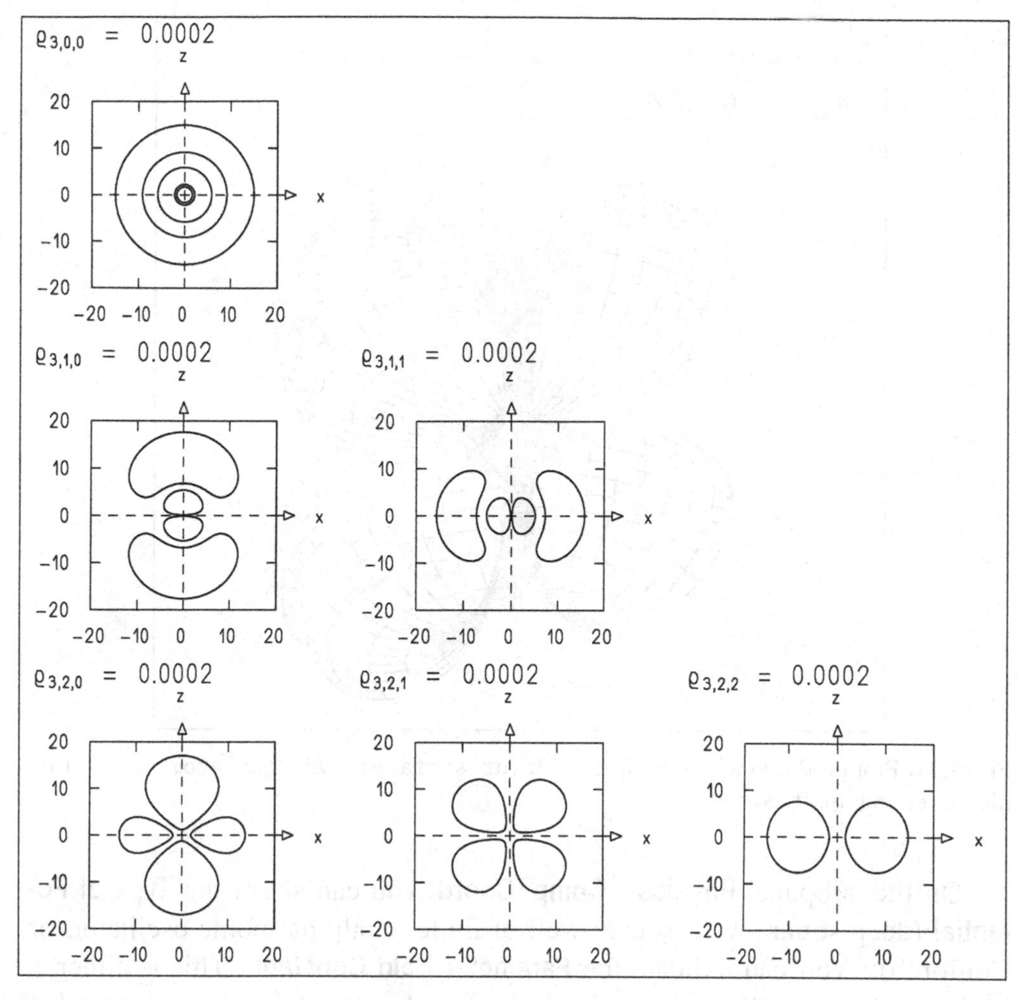

Fig. 7.10. Plot produced with descriptor `Contour lines: hydrogen atom` on file `3D_-Bound_States.des`

- `Contour lines: harmonic oscillator`
- `Contour lines: hydrogen atom` (see Fig. 7.10)

7.6 Contour Surface of the Probability Density

Aim of this section: Generation of a plot with a surface of constant probability density $\varrho_{n\ell m}$ = const, in x, y, z space for eigenstates in simple spherically symmetric potentials.

The subject of Sect. 7.5 which was the construction of a line $\varrho_{n\ell m}(r, \vartheta) = c$ in the x, z plane is extended to the construction of a surface $\varrho_{n\ell m}(x, y, z) = c$ in x, y, z space. For the choice of a suitable value of the constant c we refer to Sect. 7.5. The plot produced is of the type *contour-surface plot in 3D*, Sect. A.3.5. An example is Fig. 7.11.

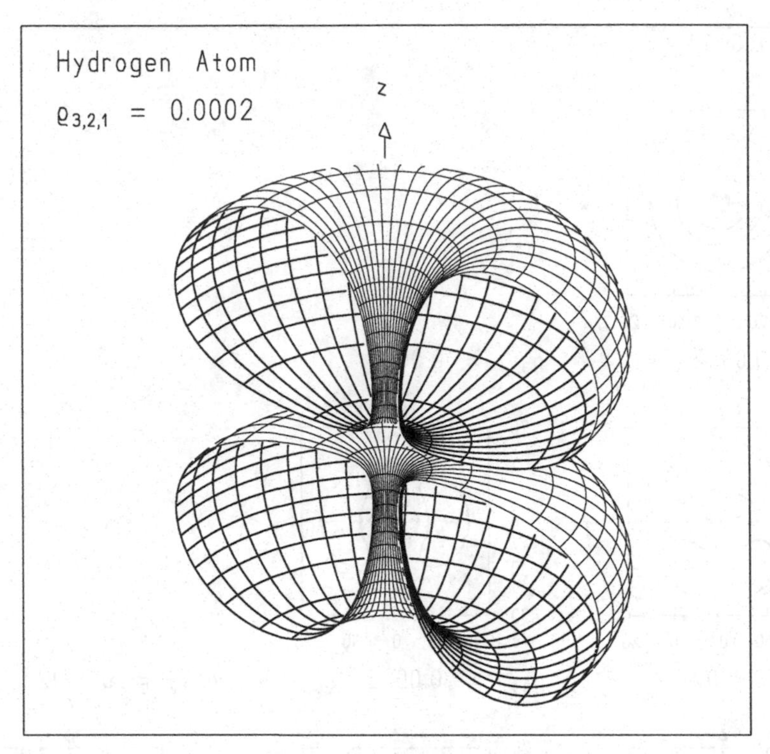

Fig. 7.11. Plot produced with descriptor `Contour surface: hydrogen atom, rho321` on file `3D_Bound_States.des`

On the subpanel **Physics—Comp. Coord.** you can select the **Type of Potential** (deep square well, square well of finite depth, harmonic oscillator, or Coulomb). You can indicate the **Parameter Held Constant**. This is either ϱ itself or, often simpler to determine, $\varrho/\varrho_{\max}$, where $\varrho_{\max}$ is the maximum that the probability density can take. Under **Parameter Value** the numerical value of that parameter is entered.

The subpanel **Physics—Potential** is as described in Sect. 7.4.

Movie Capability: Direct. The value of ϱ (or $\varrho/\varrho_{\max}$) is stepped up by an increment from frame to frame. On the subpanel **Physics—Movie** (see Sect. A.4) you find the increment $\Delta\varrho$ (or $\Delta(\varrho/\varrho_{\max})$). (Attention: The creation of the movie (and also its saving in animated GIF format) will take quite some time. You may want to let the computer work on it while you do something else.)

Example Descriptors on File `3D_Bound_States.des`

- `Contour surface: deep square well, rho321`
- `Contour surface: square well, rho321`
- `Contour surface: harmonic oscillator, rho321`
- `Contour surface: hydrogen atom, rho321` (see Fig. 7.11)

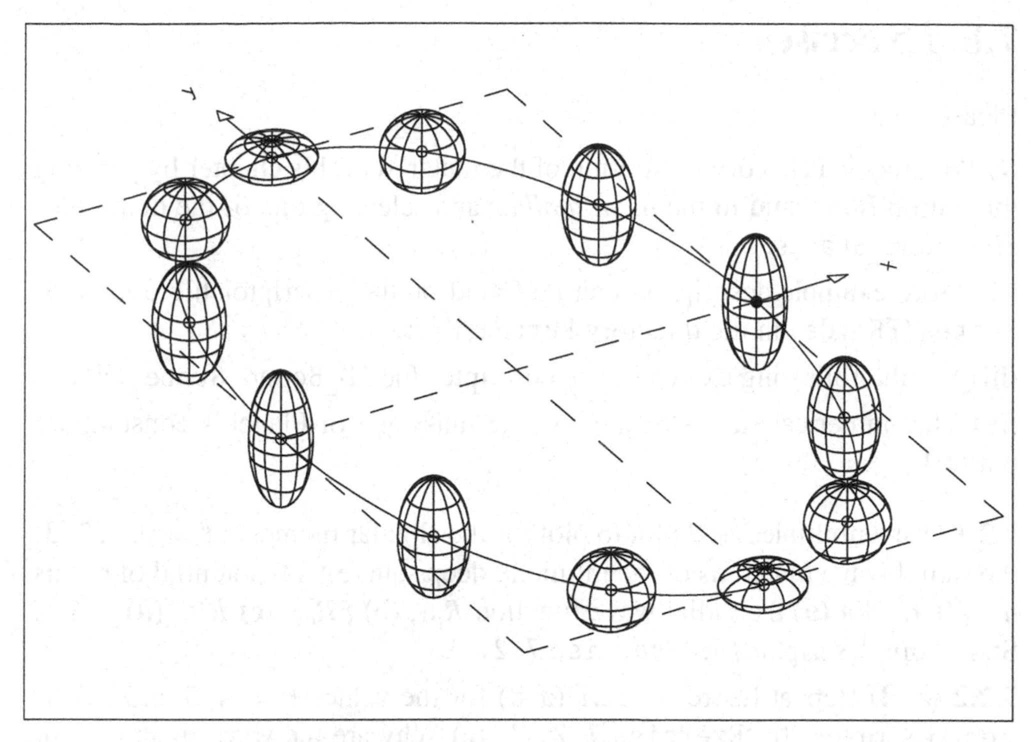

Fig. 7.12. Plot produced with descriptor `Harmonic particle motion (ellipsoids)` on file `3D_Bound_States.des`

7.7 Harmonic Particle Motion

Aim of this section: Illustration of the motion (7.53) of a Gaussian wave packet in a harmonic-oscillator potential by plotting the probability ellipsoid at times t_0, $t_0 + \Delta t, \ldots$. (The illustration is restricted to wave packets with uncorrelated widths $\sigma_x, \sigma_y, \sigma_z$. The probability density is then simply a product of three one-dimensional densities as treated in Sect. 3.7.)

A plot similar to Fig. 7.12 is produced. It is of the type *probability-ellipsoid plot*, Sects. A.3.7 and 6.5.

On the bottom of the subpanel **Physics—Comp. Coord.** you find the **Period of the Harmonic Oscillator**.

The subpanel **Physics—Wave Packet** is as in Sect. 6.5.

Movie Capability: Direct. On the subpanel **Movie** (see Sect. A.4) of the parameters panel you may choose to show or not to show the initial and intermediate positions of the ellipsoid.

Example Descriptor on File `3D_Bound_States.des`

- `Harmonic particle motion (ellipsoids)` (see Fig. 7.12)

7.8 Exercises

Please note:

(i) You may watch a demonstration of the material of this chapter by pressing the button **Run Demo** in the *main toolbar* and selecting one of the demo files `3D_Bound_States`.

(ii) More example descriptors can be found on the descriptor file `3DBound-States(FE).des` in the directory `FurtherExamples`.

(iii) For the following exercises use descriptor file `3D_Bound_States.des`.

(iv) The numerical values of the particle mass and of Planck's constant are put to 1.

7.2.1 Use a multiple 2×2 plot to plot for the angular momenta $\ell = 0, 1, 2, 3$, the radial wave functions of the infinitely deep square-well potential of radius $a = 0.9$. Plot **(a)** the radial wave function $R_{n\ell}$, **(b)** $rR_{n\ell}$, **(c)** $R^2_{n\ell}$, **(d)** $r^2R^2_{n\ell}$. Start from descriptor 19: '`Exercise 7.2.1`'.

7.2.2 (a–d) Repeat Exercise 7.2.1 (a–d) for the values $\ell = 4, 5, 6, 7$. Start from descriptor 20: '`Exercise 7.2.2`'. **(e)** Why are the wave functions for small values of r the more suppressed the higher the values ℓ of angular momentum?

7.2.3 (a–d) Repeat Exercise 7.2.1 (a–d) with a width $a = 0.5$ of the infinitely deep potential for an energy range $0 \leq E \leq 200$. Start from descriptor 19: '`Exercise 7.2.1`'. **(e)** Why are the energy eigenvalues bigger for a smaller radius of the well?

7.2.4 (a–d) Repeat Exercise 7.2.3 (a–d) for the angular momenta $\ell = 4, 5, 6, 7$ in the energy range $0 \leq E \leq 1000$. Start from descriptor 20: '`Exercise 7.2.2`'.

7.2.5 (a–d) Repeat Exercise 7.2.1 (a–d) for a potential depth $V_0 = -100$. Start from descriptor 19: '`Exercise 7.2.1`'. **(e)** Why are the differences $\Delta_i = E_i - V_0$ of the energy eigenvalues E_i and the potential depth V_0 in this potential in the range $0 \leq E \leq 100$ smaller than the eigenvalues in the infinitely deep square well of Exercise 7.2.1?

7.2.6 (a–d) Repeat Exercise 7.2.2 (a–d) for a potential of finite depth $V_0 = -100$. Start from descriptor 19: '`Exercise 7.2.1`'. **(e)** Why do the wave functions exhibit for small r values the same behavior as those of the infinitely deep square well?

7.2.7 Plot for a harmonic-oscillator potential $V = M\omega^2r^2/2$ with the oscillator frequency $\omega = 1$ in the energy range $0 \leq E \leq 15$ for $\ell = 0, 1, 2, 3$ **(a)** the radial wave function $R_{n\ell}$, **(b)** $rR_{n\ell}$, **(c)** $R^2_{n\ell}$, **(d)** $r^2R^2_{n\ell}$. Start from descriptor

3: `'Radial wave functions: harmonic oscillator'`. **(e)** Explain in classical terms why the probability density per radial shell, $r^2 R_{n\ell}^2$, for large ℓ is largest close to the maximal elongation of the radial oscillator.

7.2.8 (a–d) Repeat Exercise 7.2.7 (a–d) for the angular momenta $\ell = 4, 5, 6, 7$. **(e)** Why do the lowest energy eigenvalues no longer occur for larger ℓ?

7.2.9 (a–d) Repeat Exercise 7.2.7 (a–d) for the angular momenta $\ell = 20, 21, 22, 23$ for the lowest energy eigenvalues.

7.2.10 Repeat Exercise 7.2.7 for the angular momenta $\ell = 40, 41, 42, 43$. Plot for the few lowest energy eigenvalues **(a)** the radial wave functions $R_{n\ell}$, **(b)** $r R_{n\ell}$, **(c)** $R_{n\ell}^2$, **(d)** $r^2 R_{n\ell}^2$. **(e)** Approximate the effective potential for higher values ℓ of the angular momentum by an oscillator potential having its minimum at the minimum of $V_{\text{eff}}(r)$. **(f)** Explain why the plotted wave functions look very similar to one-dimensional oscillator wave functions. **(g)** What is the effective oscillator frequency ω_{eff} for the shifted approximate oscillator? **(h)** How are the ℓ independence of ω_{eff} and the equal spacing of the energy levels related?

7.2.11 Plot for a Coulomb potential with $\alpha = 1$ in the range $0 \le r \le 30$ for the lowest three energy levels **(a)** the radial wave function $R_{n\ell}$, **(b)** $r R_{n\ell}$, **(c)** $R_{n\ell}^2$, **(d)** $r^2 R_{n\ell}^2$. Start from descriptor 4: `'Radial wave functions: Coulomb potential'`. Use double plots for the presentation of the angular momenta $\ell = 0, 1$. **(e)** Calculate the energies of the lowest three levels for the two angular momenta.

7.2.12 (a–e) Repeat Exercise 7.2.11 (a–e) for $\ell = 2, 3$.

7.2.13 (a–d) Repeat Exercise 7.2.11 (a–d) for $\ell = 20$. **(e)** Calculate the equivalent oscillator potential by an expansion of the effective potential about its minimum. **(f)** Calculate the energy levels in the approximating oscillator potential for $n_r \ll \ell$ and compare them with the exact Balmer formula. **(g)** Compare the position of the minimum with the value according to the formula for the Bohr radii.

7.2.14 (a–d) Repeat Exercise 7.2.11 (a–d) for $\ell = 40$.

7.3.1 The potential has the form

$$V(r) = \begin{cases} -20\,, & 0 \le r < 1.5 \\ 0\,, & 1.5 \le r < 2.5 \\ -10\,, & 2.5 \le r < 3.5 \\ 0\,, & 3.5 \le r \end{cases} .$$

Plot for $\ell = 0$ **(a)** the wave function $R_{n\ell}$, **(b)** $r R_{n\ell}$, **(c)** $R_{n\ell}^2$, **(d)** $r^2 R_{n\ell}^2$. Start from descriptor 5: `'Radial wave functions in step potential'`. **(e)** Why is the wave function of the third state essentially different from zero only in the second well?

7.3.2 (a–d) Repeat Exercise 7.3.1 (a–d) for $\ell = 1$.

7.3.3 (a–d) Repeat Exercise 7.3.1 (a–d) for $\ell = 2$.

7.3.4 For the potential

$$V(r) = \begin{cases} 100\,, & 0 \le r < 0.5 \\ -20\,, & 0.5 \le r < 1.5 \\ 0\,, & 1.5 \le r \end{cases}$$

plot in a 2×2 plot for the angular momenta $\ell = 0, 1, 2, 3$ **(a)** the wave function $R_{n\ell}$, **(b)** $R^2_{n\ell}$. Start from descriptor 5: ‘`Radial wave functions in step potential`’. **(c)** Why does the energy of the second state vary so much less with increasing angular momentum than in a single potential well without the repulsive hard core below $r = 0.5$?

7.4.1 (a) Plot the probability densities $\varrho_{n\ell m}$ in an infinitely deep square well of radius $a = 0.9$ with a multiple plot for $n = 1$, $\ell = 0$, $m = 0$ and $n = 1$, $\ell = 1$, $m = 0, 1$. Start from descriptor 21: ‘`Exercise 7.4.1`’. **(b)** Explain the significance of the node lines in the polar angle ϑ.

7.4.2 Repeat Exercise 7.4.1 (a) for $n = 2$.

7.4.3 (a) Repeat Exercise 7.4.1 (a) for $n = 3$. **(b)** What are the node half-circles correlated to?

7.4.4 Repeat Exercise 7.4.1 (a) for $n = 1$, $\ell = 2$, $m = 0, 1$ and $n = 1$, $\ell = 3$, $m = 0, 1$. Which is the correlation between ϑ node lines and quantum numbers?

7.4.5 Repeat Exercise 7.4.1 (a) for $n = 1$, $\ell = 2$, $m = 2$ and $n = 1$, $\ell = 3$, $m = 2, 3$.

7.4.6 Repeat Exercise 7.4.1 (a) for $n = 1$, $\ell = 15$, $m = 15$ in a single plot.

7.4.7 (a) Repeat Exercise 7.4.1 (a) for $n = 1$, $\ell = 15$, $m = 0$ in a single plot. **(b)** Explain the difference in this plot and the one of Exercise 7.4.6 (a) in terms of classical angular momenta.

7.4.8 (a) Plot the probability densities $\varrho_{n\ell m}$ in a three-dimensional harmonic oscillator with a multiple plot for $n_r = 1$, $\ell = 0$, $m = 0$ and $n_r = 1$, $\ell = 1$, $m = 0, 1$. Start from descriptor 8: ‘`Probability density: harmonic oscillator`’. **(b)** Why is the decrease of the wave functions with growing r slower than for Exercise 7.4.1?

7.4.9 Repeat Exercise 7.4.8 (a) for $n_r = 2$.

7.4.10 Repeat Exercise 7.4.8 (a) for $n_r = 0$, $\ell = 15$, $m = 15$.

7.4.11 Repeat Exercise 7.4.8 (a) for $n_r = 0$, $\ell = 15$, $m = 0$.

7.4.12 (a) Plot the probability densities $\varrho_{n\ell m}$ in a Coulomb potential for the quantum numbers $n = 1$, $\ell = 0$, $m = 0$. Start from descriptor 9: ‘`Probability density: Coulomb potential`’. **(b)** Why do no states exist for $n = 1$, $\ell = 1$, $m = 1$?

7.4.13 (a) Repeat Exercise 7.4.12 using a multiple plot for the probability densities for the quantum numbers $n = 2$, $\ell = 0$, $m = 0$ and $n = 2$, $\ell = 1$,

$m = 0, 1$. **(b)** How do the nodes and the quantum numbers correspond to each other?

7.4.14 **(a)** Repeat Exercise 7.4.12 (a) for $n = 15$, $\ell = 14$, $m = 14$. Make sure you extend your plot to large values of r by changing the C3 window in x and y. **(b)** Why do you need a very large scale factor in z (unless you use the autoscale facility) to see the peak in the probability density? **(c)** Why does the region of large values of $\rho_{n\ell m}$ occur at large r?

7.4.15 Repeat Exercise 7.4.12 (a) for $n = 15$, $\ell = 14$, $m = 0$.

7.5.1 **(a)** Plot the contour lines for the deep square-well eigenstate with principal quantum number $n = 3$ using descriptor 10: '`Contour lines: deep square well`'. **(b)** Keep that plot on the screen and produce the corresponding plot for $n = 2$. **(c)** Now add the plot for $n = 1$. **(d)** Compare the plots in terms of radial and polar nodes.

7.5.2 Repeat Exercise 7.5.1 for a square well of finite depth using descriptor 11: '`Contour lines: square well of finite depth`'.

7.5.3 Plot descriptor 10: '`Contour lines: deep square well`' (deep square well) and leave the plot on the screen. Now plot descriptor 11: '`Contour lines: square well of finite depth`' (square well of finite depth). **(a)** Compare the radial extension of the contour lines. **(b)** Reduce the depth of the well to $V_0 = -3$. Discuss the slight change in radial extension (best visible for ϱ_{310}). Why are there no plots for $\ell = 2$?

7.5.4 Plot descriptor 12: '`Contour lines: harmonic oscillator`' (harmonic oscillator). You will get a plot for $n_r = 1$. Produce the corresponding plots for $n_r = 0, 2, 3$. Describe the change in radial extension.

7.5.5 Plot descriptor 13: '`Contour lines: hydrogen atom`' (Coulomb potential). Produce similar plots for $n = 1, 2, 3, 4$.

7.6.1 Plot descriptor 17: '`Contour surface: hydrogen atom, rho321`'. After a computing time of up to several minutes (!) you will get a contour-surface plot of the state $n = 3$, $\ell = 2$, $m = 1$ for the electron in the hydrogen atom. **(a)** Switch off the hidden-line technique in the subpanel **Graphics—Hidden Lines** and plot again. Plotting will be much faster. **(b)** Change to $n = 3$, $\ell = 2$, $m = 0$. The surface does not fit into the range of computing coordinates. Therefore, on the subpanel **Physics—Comp. Coord.** increase the magnitudes of x_{beg}, x_{end}, y_{end}, z_{beg}, z_{end} and plot again. **(c)** Change to $n = 4$, $\ell = 2$, $m = 0$ and plot. Next, set $\varrho = 0.0005$. If needed, again adapt the range of computing coordinates. **(d)** Switch the hidden-line technique on again and plot. **(e)** All plots displayed so far show the contour surface in the half-space $y > 0$. To see it in full space set $y_{\mathrm{beg}} = -y_{\mathrm{end}}$, $Y_{\mathrm{beg}} = -1$, $\varphi_{\mathrm{end}} = 360$. (You find these quantities on the subpanels **Physics—Comp. Coord.**, **Graphics—Geometry**, and **Graphics—Accuracy**, respectively.) To check these settings you may first plot with hidden lines off and later, if satisfied, with hidden lines on.

7.6.2 Plot **(a)** descriptor 14: ‘`Contour surface: deep square well, rho321`’, **(b)** descriptor 15: ‘`Contour surface: square well, rho321`’, **(c)** descriptor 16: ‘`Contour surface: harmonic oscillator, rho231`’. With each of them you may perform a set of steps similar to those of Exercise 7.6.1. Note: For (b) begin with ϱ_{321}, proceed with ϱ_{320}, and then study ϱ_{322}; the potential is too shallow to accommodate eigenstates with $n = 4$.

7.7.1 **(a)** Plot the motion of a three-dimensional Gaussian wave packet in a spherically symmetric harmonic-oscillator potential. As initial conditions use $x_0 = 3$, $y_0 = 1$, $z_0 = 2$, $p_{x0} = 1$, $p_{y0} = 2$, $p_{z0} = 3$ for $T = 1$, $t_0 = 0$, $\Delta t = 0.1$, $N = 1$. Start from descriptor 18: ‘`Harmonic particle motion (ellipsoids)`’. **(b)** Calculate the classical angular momentum ($\hbar = 1$) for the initial conditions under (a).

7.7.2 Repeat Exercise 7.7.1 (a) for $T = 0.5$.

7.7.3 **(a)** Repeat Exercise 7.7.1 (a) for the initial conditions $x_0 = 5$, $y_0 = 0$, $z_0 = 0$, $p_{x0} = -2$, $p_{y0} = 0$, $p_{z0} = 0$ for the time intervals $\Delta t = 0.1667$ for $N = 4$ positions. **(b)** Calculate the classical angular momentum.

7.7.4 Repeat Exercise 7.7.1 (a) for the initial conditions $x_0 = 0$, $y_0 = 0$, $z_0 = 0$, $p_{x0} = 5$, $p_{y0} = 0$, $p_{z0} = 0$ and the time intervals $\Delta t = 0.0833$ for $N = 10$ positions.

7.7.5 **(a)** Repeat Exercise 7.7.4 for a Gaussian wave packet at rest, $x_0 = 0$, $y_0 = 0$, $z_0 = 0$, $p_{x0} = 0$, $p_{y0} = 0$, $p_{z0} = 0$, $\sigma_{x0} = 0.2$, $\sigma_{y0} = 0.3$, $\sigma_{z0} = 1.4$, for the time interval $\Delta t = 0.25$ for two positions ($N = 2$). **(b)** Why does the shape of the wave packet change from prolate to oblate?

8. Scattering in Three Dimensions

Contents: Radial scattering wave functions. Boundary and continuity conditions. Solutions for step potentials. Scattering of plane harmonic waves. Scattering-matrix element. Partial scattering amplitude. Scattered wave as sum over partial waves. Scattering amplitude as sum over partial scattering amplitudes. Differential cross section. Total cross section. Partial cross sections. Scattering amplitude and phase. Unitarity and the Argand diagram.

8.1 Physical Concepts

8.1.1 Radial Scattering Wave Functions

Besides the bound states as discussed in Chap. 7 there are continuum states for potentials with $V_\infty < \infty$, (7.11). These states will be studied in this chapter for a spherically symmetric step potential

$$V(r) = \begin{cases} V_1 \ , 0 \le r < r_1 & \text{region 1} \\ V_2 \ , r_1 \le r < r_2 & \text{region 2} \\ \vdots & \vdots \\ V_N \ , r_{N-1} \le r & \text{region } N \end{cases} \quad . \tag{8.1}$$

Scattering states are continuum eigenfunctions of the stationary Schrödinger equation (7.12) for eigenvalues $E \ge V_\infty$, i.e., in the case of the step potential (8.1) for all the values $E \ge V_N$.

The wave number in the N regions is

i) if $E > V_q$:

$$k_q = \left|\sqrt{2M(E - V_q)}/\hbar\right| \quad , \tag{8.2}$$

or

ii) if $E < V_q$:

$$k_q = \mathrm{i}\kappa_q \quad , \quad \kappa_q = \left|\sqrt{2M(V_q - E)}/\hbar\right| \quad . \tag{8.3}$$

S. Brandt et al., *Interactive Quantum Mechanics: Quantum Experiments on the Computer*, DOI 10.1007/978-1-4419-7424-2_8, © Springer Science+Business Media, LLC 2011

The radial wave function of angular momentum ℓ consists again of N pieces

$$R_\ell(k,r) = \begin{cases} R_{\ell 1}(k_1,r) & , 0 \le r < r_1 & \text{region } 1 \\ R_{\ell 2}(k_2,r) & , r_1 \le r < r_2 & \text{region } 2 \\ \vdots & & \vdots \\ R_{\ell N-1}(k_{N-1},r) & , r_{N-2} \le r < r_{N-1} & \text{region } N-1 \\ R_{\ell N}(k_N,r) & , r_{N-1} \le r & \text{region } N \end{cases} \quad . \tag{8.4}$$

Here we have put

$$k = k_N \quad , \quad 0 \le k < \infty \quad ,$$

as the wave number in the region N of incident and reflected wave.

The pieces $R_{\ell q}(k_q, r)$, $q = 1, \ldots, N$, of the wave function in the N regions have the form

$$R_{\ell q}(k_q, r) = A_{\ell q} j_\ell(k_q r) + B_{\ell q} n_\ell(k_q r) \quad . \tag{8.5}$$

For k_q real, i.e., $E > V_q$, the spherical Bessel functions j_ℓ and n_ℓ are real so that $R_{\ell q}(k_q, r)$ is a real function if $A_{\ell q}$ and $B_{\ell q}$ are real coefficients, as will turn out in the following. Alternatively, with the help of (11.38), (11.39) the pieces $R_{\ell q}(k_q, r)$, $q = 1, \ldots, N$, of the wave function can be expressed in terms of spherical Hankel functions, see Sect. 11.1.6,

$$R_{\ell q}(k_q, r) = D_{\ell q} h_\ell^{(-)}(k_q r) + F_{\ell q} h_\ell^{(+)}(k_q r) \tag{8.6}$$

with the coefficients

$$D_{\ell q} = -\frac{1}{2\mathrm{i}}(A_{\ell q} - \mathrm{i}B_{\ell q}) \quad \text{and} \quad F_{\ell q} = \frac{1}{2\mathrm{i}}(A_{\ell q} + \mathrm{i}B_{\ell q}) \quad . \tag{8.7}$$

For k_q real, i.e., $E > V_q$,

$$h_\ell^{(-)}(k_q r) = C_\ell^- \frac{\mathrm{e}^{-\mathrm{i}k_q r}}{k_q r} \tag{8.8}$$

is an *incoming spherical wave*, i.e., a spherical wave propagating from large values of the radial distance r in region q toward the origin $r = 0$. Analogously, for k_q real

$$h_\ell^{(+)}(k_q r) = C_\ell^+ \frac{\mathrm{e}^{\mathrm{i}k_q r}}{k_q r} \tag{8.9}$$

is an *outgoing spherical wave*, i.e., a spherical wave propagating from small values r in region q outward. Thus, in analogy to the one-dimensional case, the wave function in a region q with k_q real can be interpreted as a superposition of an incoming, $h_\ell^{(-)}$, and an outgoing, $h_\ell^{(+)}$, complex spherical wave.

For scattering wave functions we have $E > V_N$, so that in region N the wave number k_N is real and $h_\ell^{(-)}(k_N r)$ is the incident spherical wave, whereas $h_\ell^{(+)}(k_N r)$ is the reflected spherical wave, of angular momentum ℓ. For imaginary wave numbers $k_q = \mathrm{i}\kappa_p$, i.e., $E < V_q$, $q \neq N$, the scattering wave $R_{\ell q}(\mathrm{i}\kappa_q, r)$ in region q is a linear superposition of the real functions, see (11.41),

$$\mathrm{i}^{-(\ell-1)} h_\ell^{(-)}(\mathrm{i}\kappa_q r) = \mathrm{i}^{-(\ell-1)} C_\ell^- \frac{\mathrm{e}^{\kappa_q r}}{r} \quad , \quad \mathrm{i}^{\ell+1} h_\ell^{(+)}(\mathrm{i}\kappa_q r) = \mathrm{i}^{\ell+1} C_\ell^+ \frac{\mathrm{e}^{-\kappa_q r}}{r} \quad . \tag{8.10}$$

Thus, for real coefficients $(\pm\mathrm{i})^{-(\ell\pm1)}\mathrm{i}(A_{\ell q} \pm \mathrm{i}B_{\ell q})/2$ in

$$R_{\ell q}(\mathrm{i}\kappa_q, r) = \frac{\mathrm{i}}{2}\left[(A_{\ell q} - \mathrm{i}B_{\ell q}) h_\ell^{(-)}(\mathrm{i}\kappa_q r) - (A_{\ell q} + \mathrm{i}B_{\ell q}) h_\ell^{(+)}(\mathrm{i}\kappa_q r)\right] \tag{8.11}$$

the radial wave functions $R_{\ell q}(\mathrm{i}\kappa_q, r)$ are real. The physical interpretation of $R_{\ell q}(\mathrm{i}\kappa_q, r)$ in a region q with $E < V_q$ is again the tunnel effect. Even though the total radial energy of the particle is lower than the potential barrier the wave penetrates the wall of height V_q.

8.1.2 Boundary and Continuity Conditions. Solution of the System of Inhomogeneous Linear Equations for the Coefficients

The two *boundary conditions* for scattering solutions are

i) At $r = 0$: $R_{\ell 1}(k_1, r)$ is free of singularities, Sect. 7.1. This requires

$$B_{\ell 1} = 0 \quad , \quad \text{i.e.,} \quad R_{\ell 1}(k_1, r) = A_{\ell 1} j_\ell(k_1 r) \quad , \tag{8.12}$$

since the $n_\ell(k_1 r)$ possess a singularity at $r = 0$.

ii) At $r \to \infty$, i.e., in region N we have to have an incoming and an outgoing spherical wave. This is fulfilled by (8.6) in region N. For given boundary conditions for large r we may assume that the coefficients $A_{\ell N}$ of the Bessel functions $j_\ell(kr)$ in region N are known quantities. Further below in this section we shall discuss the choice of the $A_{\ell N}$ for the boundary condition posed by an incoming plane wave.

As in Sect. 7.1, for the scattering solutions, the radial wave function also has to be *continuous* and *continuously differentiable* at the points $r_1, r_2, \ldots, r_{N-1}$ where the pieces $R_{\ell q}(k_q, r)$ have to be matched. The continuity conditions pose $2(N-1)$ inhomogeneous linear algebraic equations for the $2(N-1)$ coefficients $A_{\ell 1}$, $A_{\ell 2}$, $B_{\ell 2}$, ..., $A_{\ell N-1}$, $B_{\ell N-1}$, $B_{\ell N}$. The coefficient $A_{\ell N}$ given by the boundary condition constitutes the inhomogeneity of the system. Thus the $2(N-1)$ unknown coefficients are uniquely determined by the inhomogeneous linear equations in terms of the coefficient $A_{\ell N}$. The coefficient $A_{\ell N}$ can be chosen to be real. The functions j_ℓ and n_ℓ are real in the regions with real k_q, i.e., $E > V_q$, and the $(\pm\mathrm{i})^{\ell\pm 1} h_\ell^{(\pm)}$ are real in the regions with imaginary $k_q = \mathrm{i}\kappa_q$, i.e., $E < V_q$. Because the coefficients of the linear system of $2(N-1)$ equations are real, the solutions $A_{\ell 1}, A_{\ell 2}, B_{\ell 2}, \ldots, A_{\ell N-1}, B_{\ell N-1}, B_{\ell N}$ are real coefficients and thus real functions of the incoming wave number k. Thus, the radial wave function $R_\ell(k,r)$ is a real function of k and r.

8.1.3 Scattering of a Plane Harmonic Wave

For the usual scattering experiment the particles possess a momentum $\mathbf{p}$ sufficiently sharp to describe the incoming particles by a three-dimensional plane harmonic wave with wave vector $\mathbf{k} = \mathbf{p}/\hbar$. We choose the z direction $\mathbf{e}_z$ of the polar coordinate frame parallel to the momentum, i.e., $\mathbf{p} = \hbar\mathbf{k} = \hbar k \mathbf{e}_z$. The plane wave has the form ($\cos\vartheta = \mathbf{k}\cdot\mathbf{r}/kr$)

$$\varphi(\mathbf{k},\mathbf{r}) = \mathrm{e}^{\mathrm{i}\mathbf{k}\cdot\mathbf{r}} = \mathrm{e}^{\mathrm{i}kz} = \mathrm{e}^{\mathrm{i}kr\cos\vartheta} \quad . \tag{8.13}$$

According to (6.53) it can be decomposed into partial waves,

$$\varphi(\mathbf{k},\mathbf{r}) = \mathrm{e}^{\mathrm{i}kr\cos\vartheta} = \sum_{\ell=0}^{\infty}(2\ell+1)\mathrm{i}^\ell j_\ell(kr) P_\ell(\cos\vartheta) \quad . \tag{8.14}$$

This means that the incoming plane wave in region N is equivalently well described by a set of partial waves of angular momentum ℓ and magnetic quantum number $m = 0$. The radial wave function $j_\ell(kr)$ is a superposition (11.38),

$$j_\ell(kr) = \frac{1}{2\mathrm{i}}\left[h_\ell^{(+)}(kr) - h_\ell^{(-)}(kr)\right] \quad , \tag{8.15}$$

of incoming and outgoing spherical waves $\frac{1}{2\mathrm{i}} h_\ell^{(-)}(kr)$ and $\frac{1}{2\mathrm{i}} h_\ell^{(+)}(kr)$. The incoming radial wave in region N is thus

$$R_{\ell N}^{\mathrm{in}}(k,r) = -\frac{1}{2\mathrm{i}} h_\ell^{(-)}(kr) \quad , \tag{8.16}$$

for the moment leaving aside the weight factor $(2\ell+1)\mathrm{i}^\ell$ in the partial-wave decomposition (8.14). To have the term (8.16) as the incoming spherical wave in the solution of the Schrödinger equation we divide $R_\ell(k,r)$, (8.4), by $(A_{\ell N} - \mathrm{i}B_{\ell N})$ and obtain for the ℓth radial wave function

$$R_\ell^{(+)}(k,r) = \frac{1}{A_{\ell N} - \mathrm{i}B_{\ell N}} R_\ell(k,r) \quad . \tag{8.17}$$

Its piece in region N is then given by

$$R^{(+)}_{\ell N}(k,r) = -\frac{1}{2\mathrm{i}}h^{(-)}_{\ell}(kr) + \frac{1}{2\mathrm{i}}S_{\ell}(k)h^{(+)}_{\ell}(kr) \quad . \tag{8.18}$$

The coefficient

$$S_{\ell}(k) = \frac{A_{\ell N} + \mathrm{i}B_{\ell N}}{A_{\ell N} - \mathrm{i}B_{\ell N}} \ , \tag{8.19}$$

being a function of the incoming wave number $k = p/\hbar$, is called the *scattering-matrix element* S_{ℓ} of the ℓth partial wave. It is the angular-momentum projection of the S matrix. The function $R^{(+)}_{\ell N}(k,r)$ can also be rephrased in terms of $j_{\ell}(kr)$:

$$R^{(+)}_{\ell N}(k,r) = j_{\ell}(kr) + \frac{1}{2\mathrm{i}}(S_{\ell}(k) - 1)h^{(+)}_{\ell}(kr) \quad . \tag{8.20}$$

The coefficient

$$f_{\ell}(k) = \frac{1}{2\mathrm{i}}(S_{\ell}(k) - 1) \tag{8.21}$$

is called the *partial scattering amplitude*. It determines the effect of the potential $V(r)$ on the ℓth partial wave $j_{\ell}(kr)$ in the decomposition (8.14) of the incoming plane wave. The representation (8.20) is the appropriate form for the construction of the full three-dimensional *stationary wave*

$$\varphi^{(+)}(\mathbf{k},\mathbf{r}) = \sum_{\ell=0}^{\infty} \varphi_{\ell}(\mathbf{k},\mathbf{r}) \tag{8.22}$$

with

$$\varphi_{\ell}(k,r) = (2\ell+1)\mathrm{i}^{\ell} R^{(+)}_{\ell}(k,r) P_{\ell}(\cos\vartheta) \tag{8.23}$$

being the *partial stationary wave* of angular momentum ℓ. The stationary wave is the superposition

$$\varphi^{(+)}(\mathbf{k},\mathbf{r}) = \mathrm{e}^{\mathrm{i}\mathbf{k}\cdot\mathbf{r}} + \eta(\mathbf{k},\mathbf{r}) \tag{8.24}$$

of the incoming three-dimensional plane harmonic wave and the *scattered wave*

$$\eta(\mathbf{k},\mathbf{r}) = \sum_{\ell=0}^{\infty} \eta_{\ell}(\mathbf{k},\mathbf{r}) \quad . \tag{8.25}$$

The ℓth *scattered partial wave* $\eta_{\ell}(\mathbf{k},\mathbf{r})$ is given by

$$\eta_{\ell}(\mathbf{k},\mathbf{r}) = (2\ell+1)\mathrm{i}^{\ell}[R^{(+)}_{\ell}(k,r) - j_{\ell}(kr)]P_{\ell}(\cos\vartheta) \quad . \tag{8.26}$$

In region N the piece of the scattered partial wave $\eta_{\ell N}(\mathbf{k},\mathbf{r})$ has the explicit representation

$$\eta_{\ell N}(\mathbf{k},\mathbf{r}) = (2\ell+1)\mathrm{i}^{\ell} f_{\ell}(k) h^{(+)}_{\ell}(kr) P_{\ell}(\cos\vartheta) \quad . \tag{8.27}$$

For far out distances $kr \gg 1$ in region N, the function $\eta_{\ell N}(\mathbf{k}, \mathbf{r})$, and thus $\eta_\ell(\mathbf{k}, \mathbf{r})$, is dominated by the asymptotically leading term of the Hankel function

$$h_\ell^{(+)}(kr) \to \mathrm{i}^{-\ell}\frac{\mathrm{e}^{\mathrm{i}kr}}{kr} \quad , \quad kr \to \infty \quad . \tag{8.28}$$

This leads to the asymptotic representation

$$\eta_\ell(\mathbf{k}, \mathbf{r}) \to (2\ell+1)\frac{1}{k} f_\ell(k)\frac{\mathrm{e}^{\mathrm{i}kr}}{r} P_\ell(\cos\vartheta) \quad , \quad kr \to \infty \quad . \tag{8.29}$$

In region N the total scattered wave is given by (8.25) and (8.27):

$$\begin{aligned} \eta(\mathbf{k}, \mathbf{r}) &= \sum_{\ell=0}^{\infty} \eta_\ell(\mathbf{k}, \mathbf{r}) \\ &= \sum_{\ell=0}^{\infty} (2\ell+1)\mathrm{i}^\ell f_\ell(k) h_\ell^{(+)}(kr) P_\ell(\cos\vartheta) \quad , \quad r_{N-1} \le r \quad . \end{aligned} \tag{8.30}$$

For asymptotic r values in region N, i.e., for $kr \gg 1$, the asymptotic representation is again obtained from (8.28) yielding

$$\eta(\mathbf{k}, \mathbf{r}) \approx f(k, \vartheta)\frac{\mathrm{e}^{\mathrm{i}kr}}{r} \tag{8.31}$$

with the *scattering amplitude* read off (8.29):

$$f(k, \vartheta) = \frac{1}{k}\sum_{\ell=0}^{\infty}(2\ell+1) f_\ell(k) P_\ell(\cos\vartheta) \quad . \tag{8.32}$$

This determines the modulation of the scattered spherical wave $r^{-1}\exp(\mathrm{i}kr)$ in the polar angle ϑ.

Because of the fall-off of (8.31) the infinite sum (8.32) can be approximated by a finite sum

$$\eta(\mathbf{k}, \mathbf{r}) \approx \sum_{\ell=0}^{L} \eta_\ell(\mathbf{k}, \mathbf{r}) \quad , \quad L \ge kr_{N-1} \quad . \tag{8.33}$$

Here L is a dimensionless index; the corresponding angular momentum is $\hbar L$. The stationary wave $\varphi^{(+)}(\mathbf{k}, \mathbf{r})$ is best approximated by inserting (8.33) in (8.24),

$$\varphi^{(+)}(\mathbf{k}, \mathbf{r}) \approx \mathrm{e}^{\mathrm{i}\mathbf{k}\cdot\mathbf{r}} + \sum_{\ell=0}^{L} \eta_\ell(\mathbf{k}, \mathbf{r}) \quad . \tag{8.34}$$

The density of particles driven by the potential out of the original beam into the direction ϑ is given by $|\eta(\mathbf{k}, \mathbf{r})|^2$, which is asymptotically ($kr \gg 1$)

$$|\eta(\mathbf{k},\mathbf{r})|^2 = |f(k,\vartheta)|^2/r^2 \quad . \tag{8.35}$$

The current ΔI of particles passing through a small area Δa vertical to the ray from the scattering potential to the position of Δa at angle ϑ and distance r is

$$\Delta I = |\eta(\mathbf{k},\mathbf{r})|^2 v \Delta a = |f(k,\vartheta)|^2 v \frac{\Delta a}{r^2} = |f(k,\vartheta)|^2 v \Delta\Omega \quad . \tag{8.36}$$

The quantity

$$\Delta\Omega = \frac{\Delta a}{r^2} \tag{8.37}$$

is the solid angle under which Δa appears, seen from the origin.

The incident current density is the incident particle density times its velocity,

$$j = |\mathrm{e}^{\mathrm{i}\mathbf{k}\cdot\mathbf{r}}|^2 v = v \quad . \tag{8.38}$$

Thus, the current ΔI of particles scattered into the solid angle $\Delta\Omega$ can be written as

$$\Delta I = |f(k,\vartheta)|^2 \Delta\Omega\, j \quad . \tag{8.39}$$

The proportionality constant between the initial current density j and the current through the solid-angle element $\Delta\Omega$ is the *differential scattering cross section* $\mathrm{d}\sigma/\mathrm{d}\Omega$:

$$\Delta I = \frac{\mathrm{d}\sigma}{\mathrm{d}\Omega} \Delta\Omega\, j \quad . \tag{8.40}$$

Thus we identify

$$\frac{\mathrm{d}\sigma}{\mathrm{d}\Omega} = |f(k,\vartheta)|^2 \tag{8.41}$$

as the differential cross section for particles of momentum $\mathbf{p} = \hbar\mathbf{k}$ on a scatterer described by the potential $V(r)$. The scattering amplitude $f(k,\vartheta)$ is given by (8.32).

The *total scattering cross section* is the integral of (8.41) over the total solid angle 4π:

$$\sigma_{\mathrm{tot}} = \int \frac{\mathrm{d}\sigma}{\mathrm{d}\Omega}\,\mathrm{d}\Omega = 2\pi \int_{-1}^{+1} |f(k,\vartheta)|^2\,\mathrm{d}\cos\vartheta \quad . \tag{8.42}$$

Using the orthogonality (11.10) of the Legendre polynomials,

$$\int_{-1}^{+1} P_{\ell'}(\cos\vartheta) P_\ell(\cos\vartheta)\,\mathrm{d}\cos\vartheta = \frac{2}{2\ell+1}\delta_{\ell'\ell} \quad , \tag{8.43}$$

we get, using (8.32),

$$\sigma_{\mathrm{tot}} = \frac{4\pi}{k^2} \sum_{\ell=0}^{\infty} (2\ell+1)|f_\ell(k)|^2 = \sum_{\ell=0}^{\infty} \sigma_\ell \tag{8.44}$$

with the *partial cross section* of angular momentum ℓ:

$$\sigma_\ell = \frac{4\pi}{k^2}(2\ell+1)|f_\ell(k)|^2 \quad . \tag{8.45}$$

8.1.4 Scattering Amplitude and Phase. Unitarity. The Argand Diagram

In Chap. 4 we derived current conservation as the basis for the conservation of probability. In three dimensions the same chain of arguments is valid. However, simple physical arguments lead to the same conclusion without calculations. Elastic scattering of particles on a spherically symmetric potential conserves particle number, energy, and angular momentum. Thus the magnitude of the velocity remains unaltered in the scattering process. Because the particle number and angular momentum are conserved, this leads to conservation of the current density of the spherical waves of angular momentum ℓ. Thus the incoming current of a spherical wave is only reflected upon scattering but keeps its magnitude. Therefore, the complex scattering-matrix element S_ℓ determining the relative factor between incoming and outgoing spherical waves in region N must have the absolute value one. This is the *unitarity relation for the scattering-matrix element* S_ℓ:

$$S_\ell^* S_\ell = 1 \quad . \tag{8.46}$$

The scattering-matrix element can only be a complex phase factor, which is conventionally written as

$$S_\ell = \mathrm{e}^{2\mathrm{i}\delta_\ell} \tag{8.47}$$

with the *scattering phase* δ_ℓ of the ℓth partial wave. This is directly verified by (8.19), which also shows that S_ℓ has modulus one:

$$S_\ell = \mathrm{e}^{2\mathrm{i}\delta_\ell} = \frac{A_{\ell N} + \mathrm{i}B_{\ell N}}{A_{\ell N} - \mathrm{i}B_{\ell N}} \quad . \tag{8.48}$$

The phase itself is then given by

$$\cos\delta_\ell = \frac{A_{\ell N}}{\sqrt{A_{\ell N}^2 + B_{\ell N}^2}} \quad , \quad \sin\delta_\ell = \frac{B_{\ell N}}{\sqrt{A_{\ell N}^2 + B_{\ell N}^2}} \quad . \tag{8.49}$$

The partial scattering amplitude $f_\ell(k)$ is given by (8.21) in terms of the scattering-matrix element S_ℓ. With (8.47) one easily expresses $f_\ell(k)$ in terms of the scattering phase δ_ℓ:

$$f_\ell(k) = \mathrm{e}^{\mathrm{i}\delta_\ell}\sin\delta_\ell \quad . \tag{8.50}$$

Starting from (8.46), (8.21) yields the *unitarity relation for the scattering amplitude*

$$\operatorname{Im} f_\ell(k) = |f_\ell(k)|^2 \quad . \tag{8.51}$$

This relation is verified by using (8.50) directly. The unitarity relation is easily interpreted in the complex plane spanned by the real and imaginary parts of $f_\ell(k)$:

$$(\operatorname{Re} f_\ell(k))^2 + \left(\operatorname{Im} f_\ell(k) - \frac{1}{2}\right)^2 = \frac{1}{4} \quad . \tag{8.52}$$

This represents a circle of radius 1/2 about the center (0, 1/2) in that plane. This circle is again referred to as the *Argand diagram* of elastic potential scattering. The analogy to Sect. 4.1 is obvious.

8.1.5 Coulomb Scattering

We consider the scattering of a particle of charge $q = \pm e$ by a Coulomb field originating from a charge $Q = \pm e$ at the origin. The potential is

$$V(r) = \pm \frac{e^2}{4\pi\varepsilon_0}\frac{1}{r} = \pm\alpha\hbar c\frac{1}{r} \quad , \tag{8.53}$$

see Sect. 7.1.6. The potential is *attractive*, i.e., negative if q and Q carry different signs. Otherwise it is *repulsive*. The corresponding effective potential is

$$V_\ell^{\mathrm{eff}}(r) = \frac{\hbar^2}{2M}\frac{\ell(\ell+1)}{r^2} \pm \frac{\alpha\hbar c}{r} \quad . \tag{8.54}$$

The radial Schrödinger equation reads

$$\left[-\frac{\hbar^2}{2M}\frac{1}{r}\frac{\mathrm{d}^2}{\mathrm{d}r^2}r + V_\ell^{\mathrm{eff}}(r)\right] R_\ell(k,r) = E R_\ell(k,r)$$

with $k = \sqrt{2ME}/\hbar$. Solutions exist for all positive energies E,

$$R_\ell(k,r) = \frac{A_\ell}{r}\mathrm{e}^{\mathrm{i}kr}(kr)^{\ell+1} F(\ell+1+\mathrm{i}\eta \,|\, 2(\ell+1) \,|\, z) \quad . \tag{8.55}$$

Here

$$F(a\,|\,b\,|\,z) = 1 + \frac{a}{b}\frac{z}{1!} + \frac{a(a+1)}{b(b+1)}\frac{z^2}{2!} + \cdots$$

is the *confluent hypergeometric function*. The series defining it converges for all complex values of z. The factor A_ℓ is

$$A_\ell = \frac{2^\ell}{(2\ell+1)!}\mathrm{e}^{-\frac{1}{2}\pi\eta}|\Gamma(\ell+1+\mathrm{i}\eta)|$$

with $\Gamma(z)$ being Euler's Gamma function. The dimensionless parameter η is

$$\eta = \pm\frac{1}{ka_{\mathrm{B}}} \quad , \quad a_{\mathrm{B}} = \frac{\hbar}{\alpha Mc} \quad ,$$

with a_{B} the Bohr radius.

In analogy to (8.22) the stationary partial waves are

$$\varphi_\ell(\mathbf{k}, \mathbf{r}) = (2\ell + 1)\mathrm{i}^\ell \mathrm{e}^{\mathrm{i}\delta_\ell} R_\ell(k, r) P_\ell(\cos\vartheta) \quad , \tag{8.56}$$

where the so-called *Coulomb scattering phase* δ_ℓ resulting from the asymptotic behavior of $R_\ell(k, r)$ is

$$\delta_\ell = \frac{1}{2\mathrm{i}} \ln \frac{\Gamma(\ell + 1 + \mathrm{i}\eta)}{\Gamma(\ell + 1 - \mathrm{i}\eta)} \quad .$$

Here, as in Sect. 8.1.3, it is assumed that the incoming wave runs in z direction, $\mathbf{k} = k\mathbf{e}_z$.

The total stationary wave, the *Coulomb wave function*, is

$$\varphi(\mathbf{k}, \mathbf{r}) = \sum_{\ell=0}^{\infty} \varphi_\ell(\mathbf{k}, \mathbf{r}) \quad . \tag{8.57}$$

Its asymptotic form is given by

$$\varphi(\mathbf{k}, \mathbf{r}) \to \mathrm{e}^{\mathrm{i}\{kr\cos\vartheta + \eta\ln[kr(1-\cos\vartheta)]\}} \quad , \quad |r(1 - \cos\vartheta)| \to \infty \quad ,$$

and exhibits the long range of the Coulomb potential $\hbar c\alpha/r$ through the logarithmic term in the exponent. The infinite sum (8.57) can be replaced by a finite sum

$$\varphi(\mathbf{k}, \mathbf{r}) \approx \sum_{\ell=0}^{L} \varphi_\ell(\mathbf{k}, \mathbf{r}) \tag{8.58}$$

for small r, $r \leq L/k$. L, again, is a dimensionless index.

Further Reading

Alonso, Finn: Vol. 3, Chap. 7
Berkeley Physics Course: Vol. 4, Chaps. 8, 9
Brandt, Dahmen: Chaps. 11, 12, 14, 15
Feynman, Leighton, Sands: Vol. 3, Chap. 19
Flügge: Vol. 1, Chap. 1D
Gasiorowicz: Chaps. 11, 19
Merzbacher: Chaps. 11, 19
Messiah: Vol. 1, Chaps. 10, 11
Schiff: Chap. 5

8.2 Radial Wave Functions

Aim of this section: Presentation of the spherical step potential $V(r)$, (8.1), and of the radial wave function $R_\ell(k, r)$, (8.4), in that potential.

A plot similar to Fig. 8.1 or Fig. 8.2 is produced displaying the spherical step potential $V(r)$ as a long-dash broken line, the total energy E as a short-dash horizontal line, and the radial wave function $R_\ell(k, r)$ (or a simple function thereof) as a continuous line.

On the subpanel **Physics—Comp. Coord.** you can select the way the **Radial Wave Function is Shown** (as $R^2_{E\ell}$, $R_{E\ell}$, $r^2 R^2_{E\ell}$, or $r R_{E\ell}$). Moreover, you can choose to **Plot the Result for**

1. fixed energy E, various values of angular momentum ℓ (i.e., over an r, ℓ plane like in Fig. 8.1) or for
2. fixed angular momentum ℓ, various values of energy E (i.e., over an r, E plane like in Fig. 8.2).

Please note: If you choose option 1, make sure that you draw $R_\ell(r)$ for every integer value of ℓ in your range of ℓ: Choose integers for y_{beg}, y_{end} (on the subpanel **Physics—Comp. Coord.**) and set n_y (on the subpanel **Graphics—Accuracy**) equal to $|y_{\text{beg}} - y_{\text{end}}| + 1$.

On the subpanel **Physics—Variables** there are seven items:

- You can choose whether the **Input Below** (in the next item) **is Taken as** wave number k or energy E.
- **Wave Number / Energy** – Corresponding to the choice in the item above the input field is labeled **k** or **E**. Input is enabled only if the option 'fixed energy' was chosen on the subpanel **Physics—Comp. Coord.**.
- **Angular-Momentum Quantum Number** – Input for the quantum number ℓ is enabled only if the option 'fixed angular momentum' was chosen on the subpanel **Physics—Comp. Coord.**. In a multiple plot the value of ℓ is taken for the first plot and incremented by one from plot to plot.
- **Graphical Item** – contains the quantity ℓ_{DASH} which determines the dash length of the broken lines.
- **Scale Factor** – Because the wave function $R_{E\ell}(r)$ is plotted in an r, E plane with the total energy E_{tot} as zero line, technically speaking the function $E = E_{\text{tot}} + s R_{E\ell}(r)$ is shown. The scale factor s can be adjusted here.
- **Clipping Region for Potential** – The lines showing the potential and the effective potential are normally drawn only inside the range of the world coordinate Y. You can extend that upwards by the fraction **f_Y+** of its original range in in Y. Similarly, **f_Y-**, extends the region downwards.
- **Shown as Dashed Line** – Here you can choose to display
  - Potential and Effective Potential,

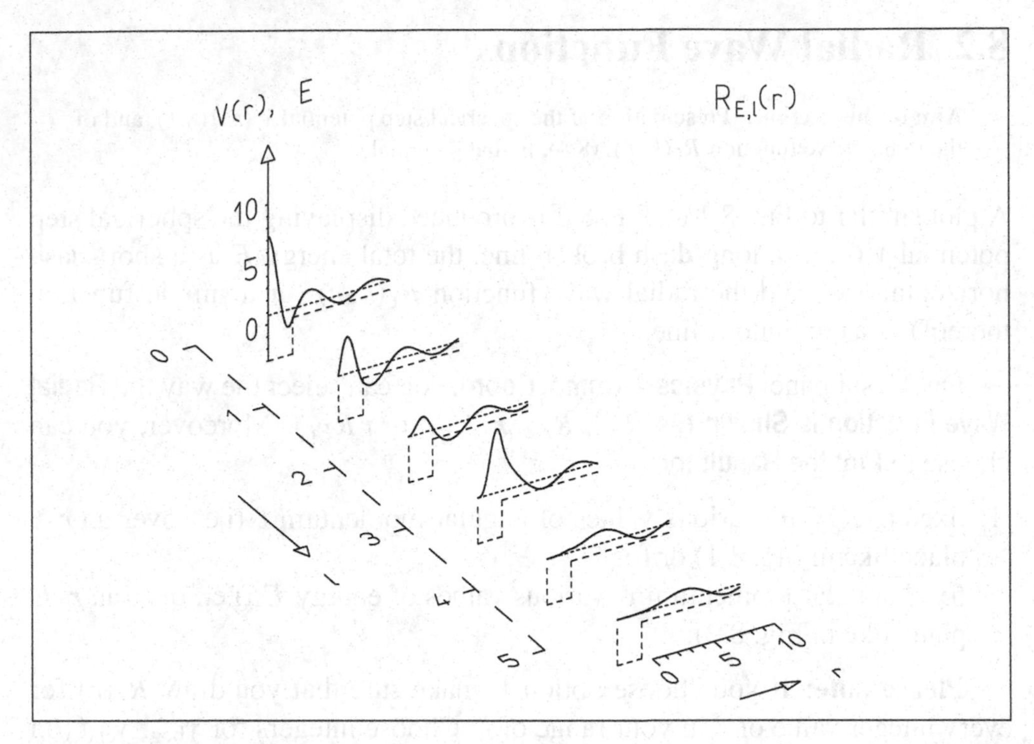

Fig. 8.1. Plot produced with descriptor `3D scattering in step pot., R_El, E fixed, l running` on file `3D_Scattering.des`

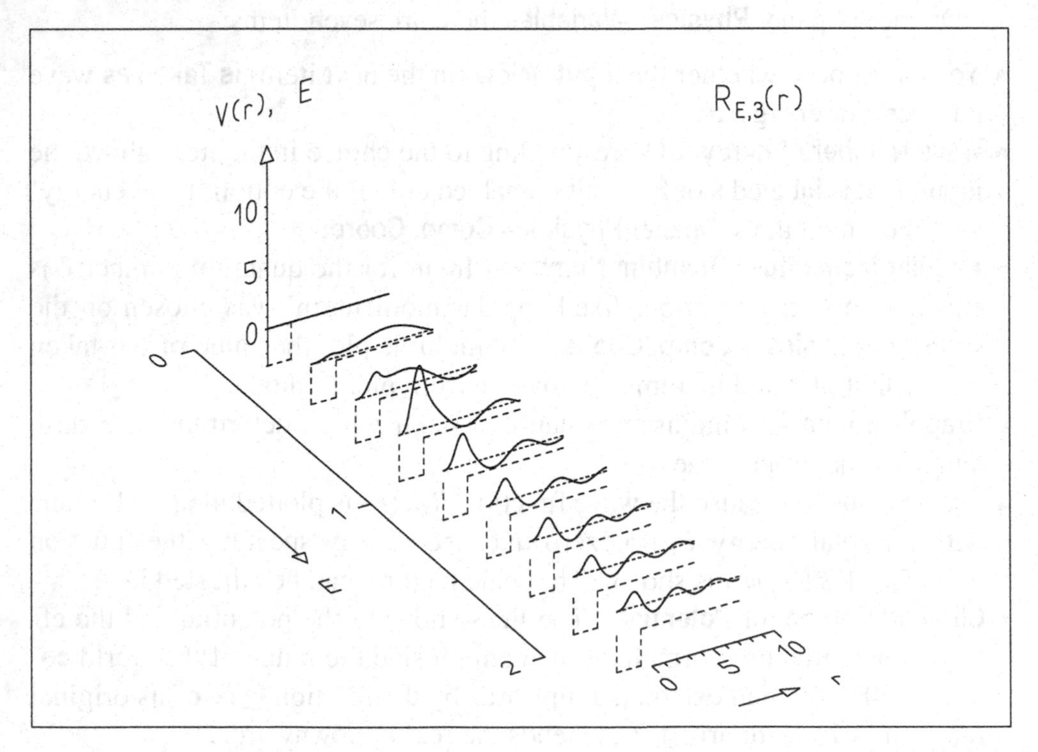

Fig. 8.2. Plot produced with descriptor `3D scattering in step pot., R_El, l fixed, E running` on file `3D_Scattering.des`

– Potential Only,
– None of the Above.

On the subpanel **Physics—Potential** you find the **Number of Regions** N of the spherical step potential (8.1), $2 \leq N \leq 5$. Under the heading **Regions** you find the boundaries r_i between regions and the potential values V_i for the different regions.

Movie Capability (only for fixed ℓ): Indirect. After conversion to direct movie capability the start and end values of energy can be changed in the subpanel **Movie** (see Sect. A.4) of the parameter panel. In the resulting movie the energy of the stationary state changes with time. Such a movie, in particular, is useful for the demonstration of resonances.

Example Descriptors on File `3D_Scattering.des`

- `3D scattering in step pot., R_El, E fixed, l running` (see Fig. 8.1)
- `3D scattering in step pot., R_El, l fixed, E running` (see Fig. 8.2)
- `Movie: 3D scattering in step pot., R_El , l fixed, E running`

8.3 Stationary Wave Functions and Scattered Waves

Aim of this section: Presentation of the stationary wave function $\varphi^{(+)}(r, \vartheta)$, (8.22), approximated as $e^{ikz} + \sum_{\ell=0}^{L} \eta_\ell(r, \vartheta)$, (8.34), the partial waves $\varphi_\ell(r, \vartheta)$, (8.23), the scattered spherical wave $\eta(r, \vartheta)$, (8.30), approximated as $\sum_{\ell=0}^{L} \eta_\ell(r, \vartheta)$, the partial scattered waves $\eta_\ell(r, \vartheta)$, (8.26), (8.29).

A plot is produced showing the function selected and in addition one or several circular arcs corresponding to the step positions r_i of the step potential.

On the subpanel **Physics—Comp. Coord.** you can select as the **Function Computed**

- the partial stationary wave φ_ℓ,
- the stationary wave φ approximated by $e^{ikz} + \sum_{\ell=0}^{L} \eta_\ell$,
- the scattered wave η approximated by $\sum_{\ell=0}^{L} \eta_\ell$.

Moreover, you can choose the way the selected **Function is Shown** (absolute square, real part, imaginary part).

On the subpanel **Physics—Variables** there are three items:

- You can choose whether the **Input Below** (in the next item) **is Taken as** wave number k or energy E.

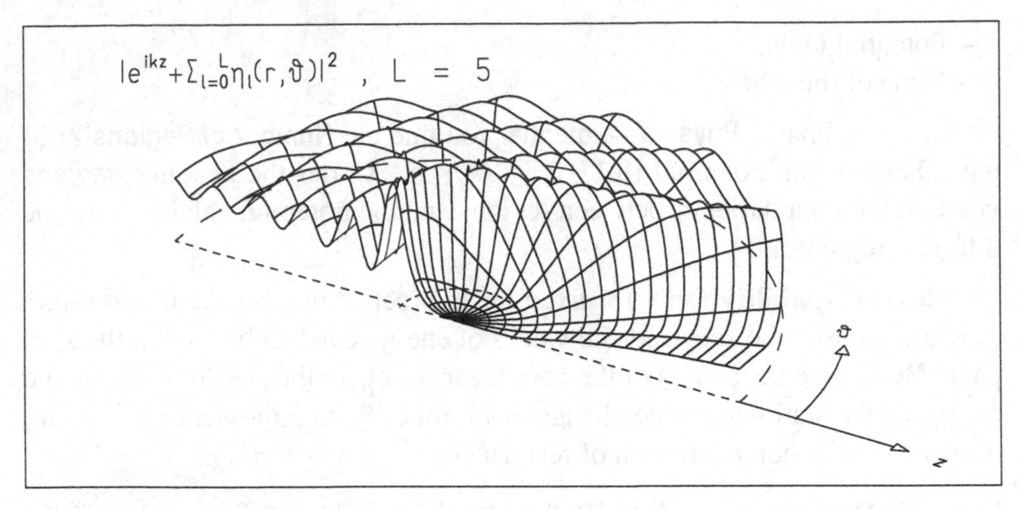

Fig. 8.3. Plot produced with descriptor `3D scattering in step pot., |phi|**2` on file `3D_Scattering.des`

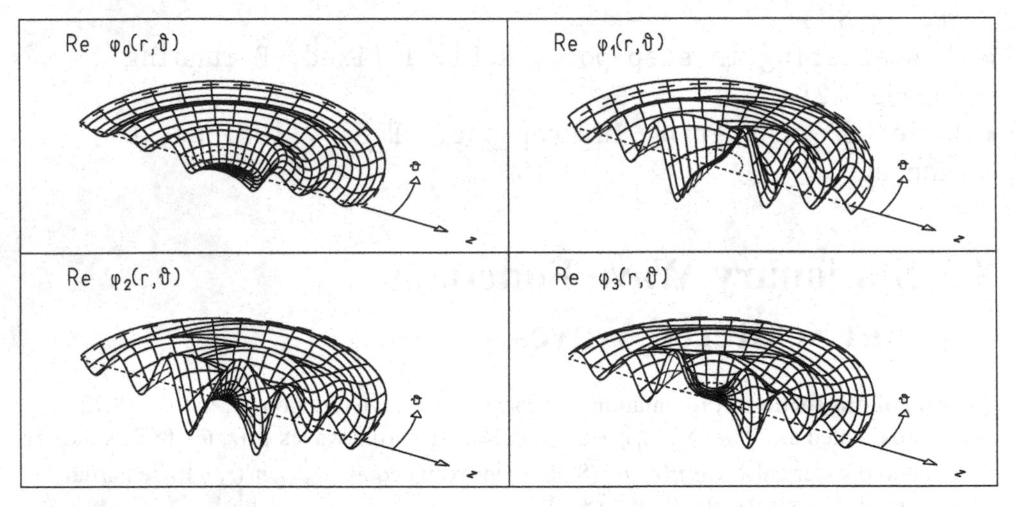

Fig. 8.4. Plot produced with descriptor `3D scattering in step pot., phi_l` on file `3D_Scattering.des`

- **Wave Number / Energy** – Corresponding to the choice in the item above the input field is labeled **k** or **E**.
- **Angular-Momentum Quantum Number** – This is the quantum number ℓ (if you choose to plot a partial wave φ_ℓ or η_ℓ) or the upper index L in the sum approximating φ or η (if you choose to plot one of these two functions). In a multiple plot the input value of ℓ (or L) is taken for the first plot. It is successively incremented by one from plot to plot.

The subpanel **Physics—Potential** is as described in Sect. 8.2.

Example Descriptors on File `3D_Scattering.des`

- `3D scattering in step pot., |phi|**2` (see Fig. 8.3)

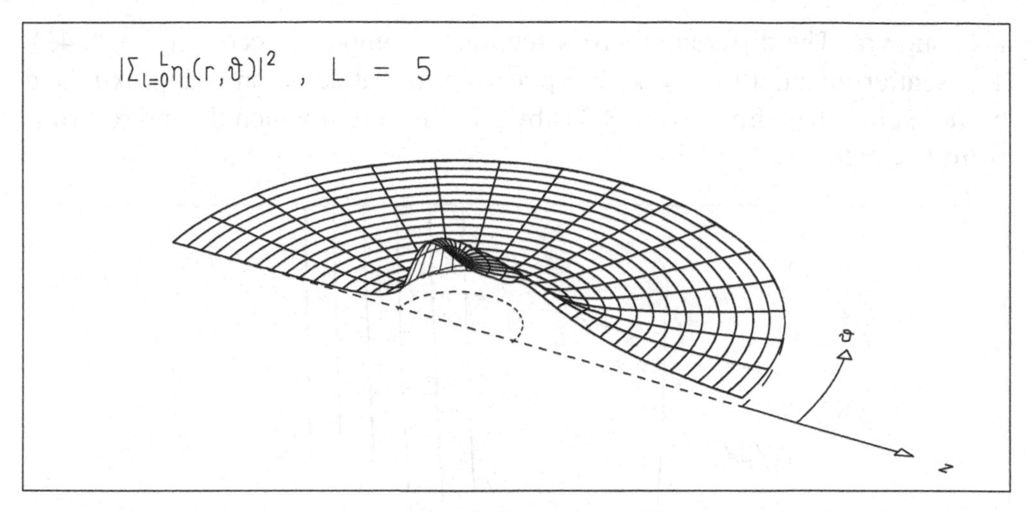

Fig. 8.5. Plot produced with descriptor `3D scattering in step pot., |eta|**2` on file `3D_Scattering.des`

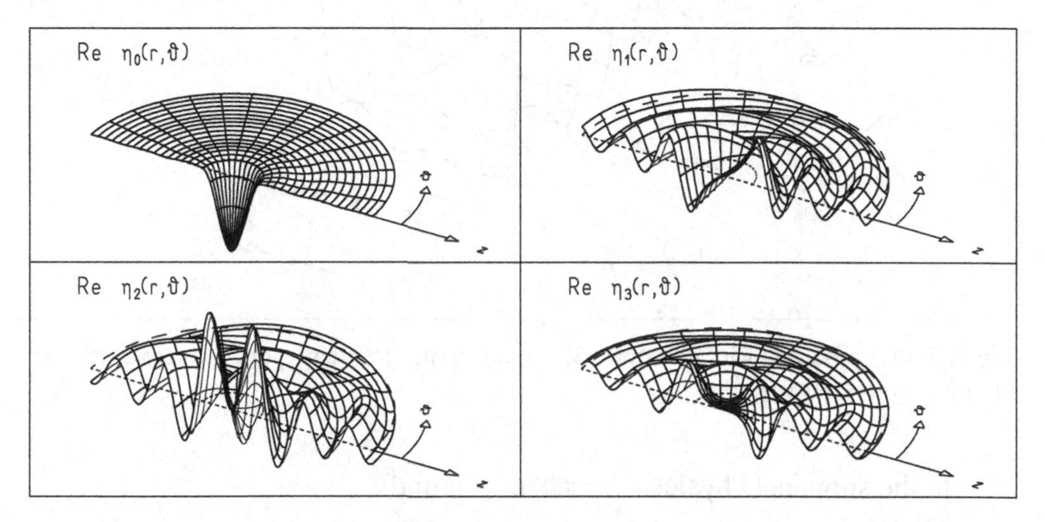

Fig. 8.6. Plot produced with descriptor `3D scattering in step pot., eta_l` on file `3D_Scattering.des`

- `3D scattering in step pot., phi_l` (see Fig. 8.4)
- `3D scattering in step pot., |eta|**2` (see Fig. 8.5)
- `3D scattering in step pot., eta_l` (see Fig. 8.6)

8.4 Differential Cross Sections

Aim of this section: Illustration of the differential cross section $\mathrm{d}\sigma/\mathrm{d}\Omega$, (8.41).

A plot similar to Fig. 8.7 is produced showing, over a plane spanned by the variables $\cos\vartheta$ and E, various curves $\mathrm{d}\sigma(\cos\vartheta)/\mathrm{d}\Omega$ for equidistant fixed values of E. For the same fixed values of E dashed lines $\mathrm{d}\sigma/\mathrm{d}\Omega = 0$ are also shown. The differential cross section is computed according to (8.41). The scattering amplitude $f(k, \vartheta)$ appearing in that equation is approximated by replacing the infinite sum (8.32) by a finite sum in which the index ℓ runs from $\ell = 0$ to $\ell = L$.

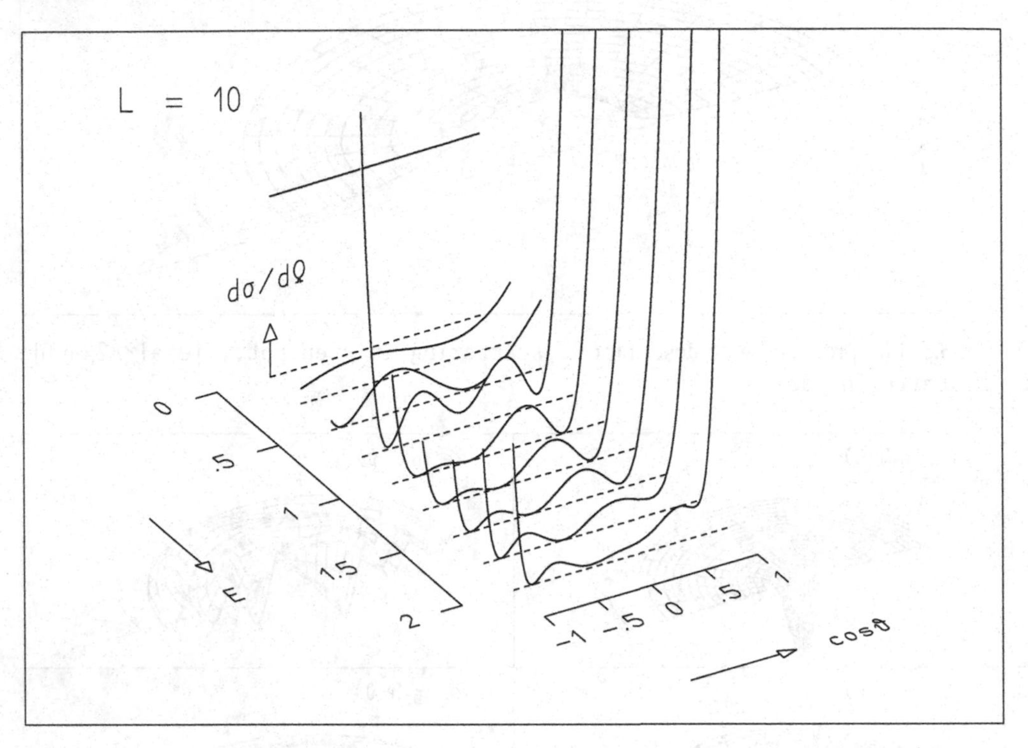

Fig. 8.7. Plot produced with descriptor `3D scattering in step pot., dsigma/dOmega` on file `3D_Scattering.des`

On the subpanel **Physics—Variables** you find

- the **Angular-Momentum Quantum Number** L, i.e., the upper index of the sum used in the approximation of $\mathrm{d}\sigma/\mathrm{d}\Omega$ (In a multiple plot this value of L is used for the first plot and incremented by one for every following plot.),
- as **Graphical Item** the dash length ℓ_{DASH} of the zero lines.

The subpanel **Physics—Potential** is as described in Sect. 8.2.

Movie Capability: Indirect. After conversion to direct movie capability the start and end values of the energy can be changed in the subpanel **Movie** (see Sect. A.4) of the parameter panel. In the resulting movie the energy changes with time.

Example Descriptor on File `3D_Scattering.des`

- `3D scattering in step pot., dsigma/dOmega` (see Fig. 8.7)

8.5 Scattering Amplitude. Phase Shift. Partial and Total Cross Sections

Aim of this section: Illustration of the complex partial scattering amplitude f_ℓ, (8.21), (8.50), in the following graphs: the Argand diagram $\mathrm{Im}\{f_\ell(E)\}$ vs. $\mathrm{Re}\{f_\ell(E)\}$; $\mathrm{Re}\{f_\ell(E)\}$ as function of energy E; $\mathrm{Im}\{f_\ell(E)\}$ as function of energy E; $|f_\ell(E)|^2$ as function of energy E as well as the phase shift $\delta_\ell(E)$, (8.49), as a function of energy E; the partial cross section $\sigma_\ell(E)$, (8.45), as a function of energy E; and the total cross section σ_{tot}, (8.44), approximated by $\sigma_{\mathrm{tot}} \approx \sum_{\ell=0}^{L} \sigma_\ell(E)$ as a function of energy E.

On the bottom of the subpanel **Physics—Comp. Coord.** there are two fields with choices to select from:

- **Curve Shown is of Type**. The type can be
  - Argand diagram: $\mathrm{Im}\, f_\ell(E)$ vs. $\mathrm{Re}\, f_\ell(E)$,
  - $y = y(x)$ – normally chosen if Argand diagram is not selected,
  - $x = x(y)$ – only chosen to produce plot $\mathrm{Re}\, f_\ell(E)$ as plot below Argand diagram (bottom-left plot in Fig. 8.8).
- **Function Computed is** – This choice is enabled only if the Argand diagram was not selected. You then choose to plot one of the six functions $|f_\ell(E)|^2$, $\mathrm{Re}\, f_\ell(E)$, $\mathrm{Im}\, f_\ell(E)$, $\delta_\ell(E)$, $\sigma_\ell(E)$, $\sum_{\ell=0}^{L} \sigma_\ell(E)$.

On the subpanel **Physics—Variables** you find the

- **Angular-Momentum Quantum Number** ℓ [or L if you chose to plot $\sum_{\ell=0}^{L} \sigma_\ell(E)$], which, in case of a multiple plot, is taken for the first plot and incremented by one for every successive plot, and the
- **Range of E in Argand Diagram** – input for the boundaries E_{beg}, E_{end} of that range is enabled only if Argand diagram was selected as curve to be shown.

The subpanel **Physics—Potential** is as described in Sect. 8.2.

Remarks: It is customary to draw an Argand diagram [$\mathrm{Im}\{f_\ell(E)\}$ vs. $\mathrm{Re}\{f_\ell(E)\}$] and graphs $\mathrm{Im}\{f_\ell(E)\}$ and $\mathrm{Re}\{f_\ell(E)\}$ in such a way that the graphs appear to be projections to the right and below the Argand diagram, respectively. You can do that by using a mother descriptor, which in turn quotes several individual descriptors (see Appendix A.10) as in the example plot, Fig. 8.8.

Movie Capability (only for the Argand diagram proper): Direct. The trajectory in the complex plane is seen to develop as the range of the independent variable is increased proportional to time.

Example Descriptors on File `3D_Scattering.des`

- `3D scattering in step pot., Argand, combined plot`
  (see Fig. 8.8, this is a mother descriptor quoting the next four descriptors listed below, each describing one of the four plots in the combined plot)

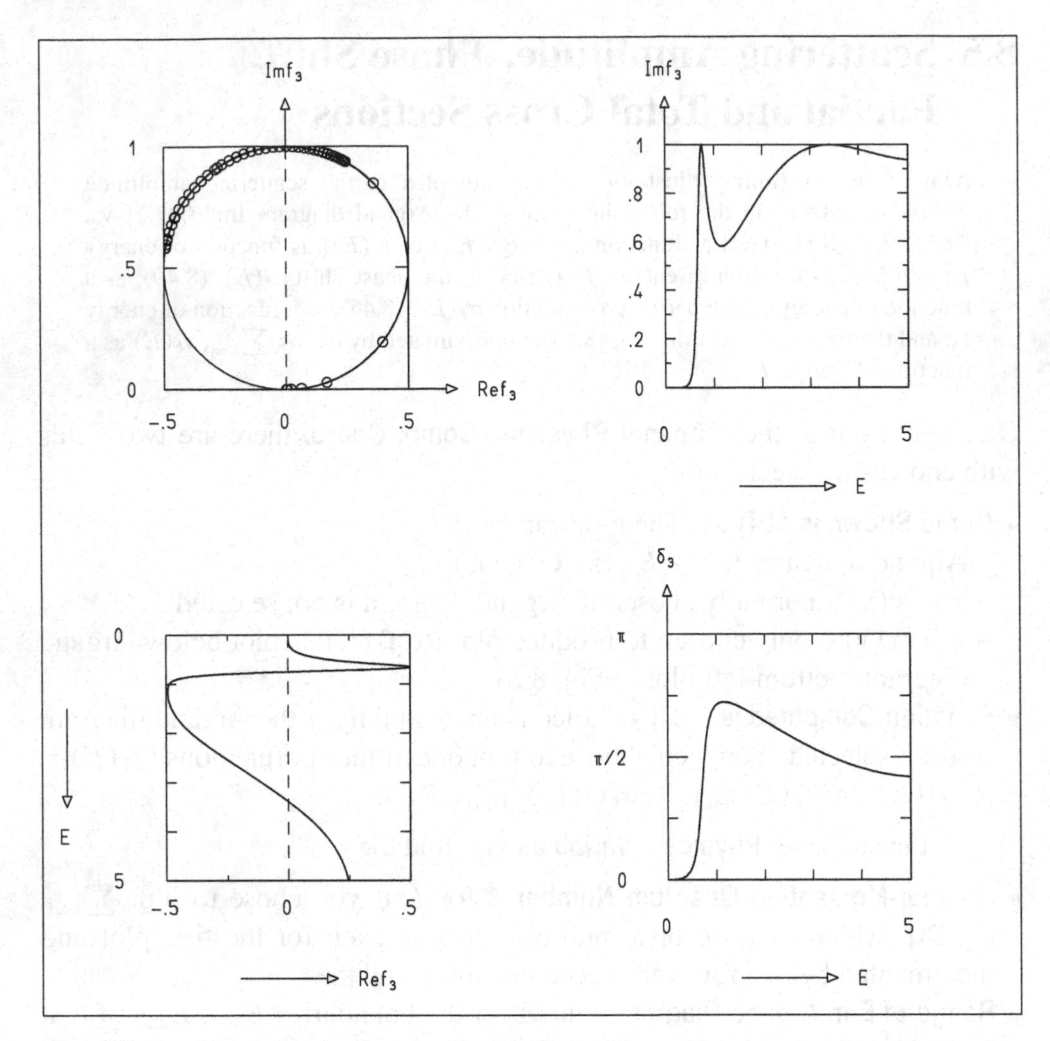

Fig. 8.8. Combined plot produced with descriptor `3D scattering in step pot., Argand, combined plot` on file `3D_Scattering.des`. This descriptor quotes four other descriptors which generate the four individual plots

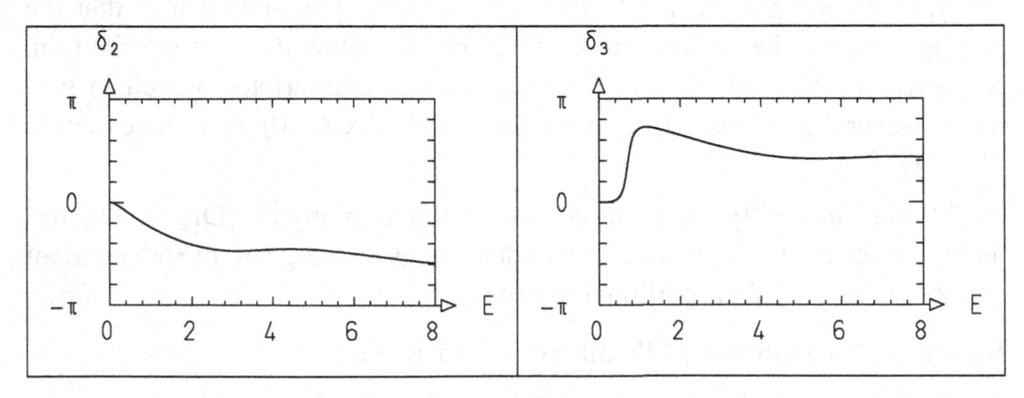

Fig. 8.9. Part of plot produced with descriptor `3D scattering in step pot., delta_l` on file `3D_Scattering.des`

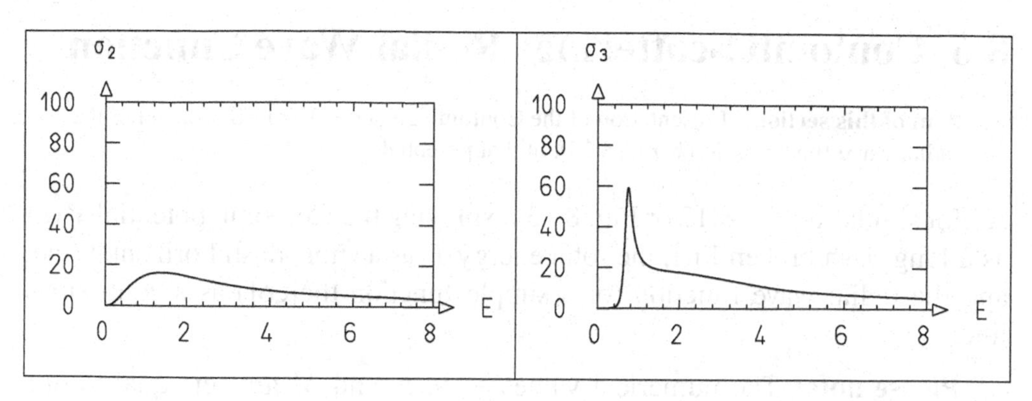

Fig. 8.10. Part of plot produced with descriptor `3D scattering in step pot., sigma_l` on file `3D_Scattering.des`

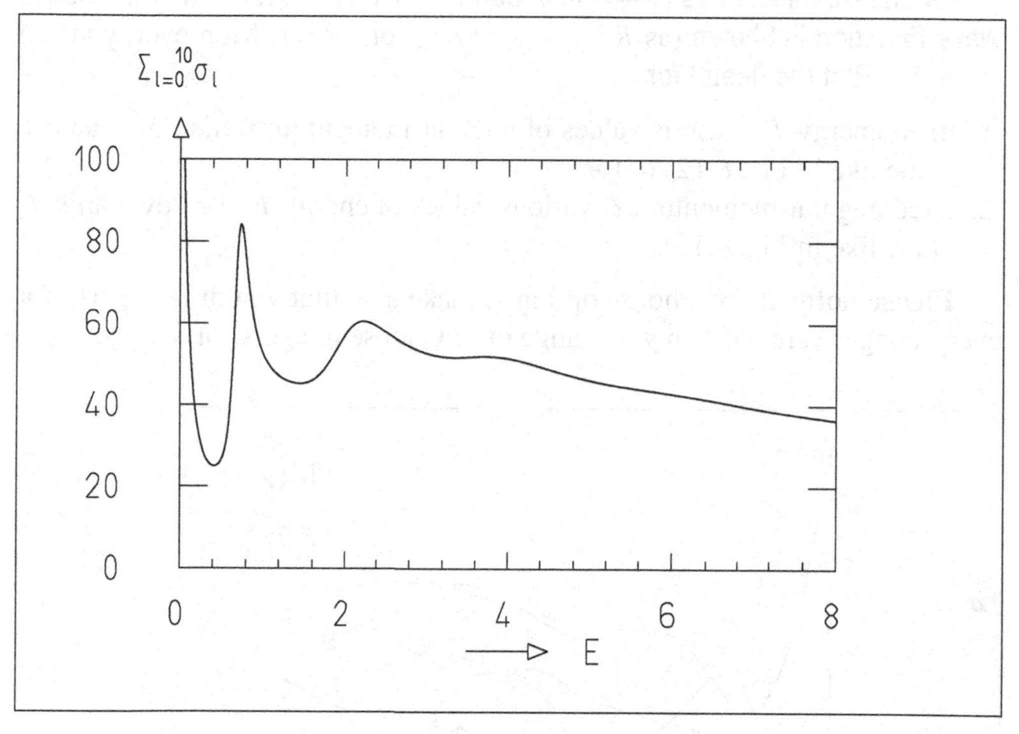

Fig. 8.11. Plot produced with descriptor `3D scattering in step pot., sigma_tot` on file `3D_Scattering.des`

- `3D scattering in step pot., Argand, imaginary part vs. real part`
- `3D scattering in step pot., Argand, imaginary part`
- `3D scattering in step pot., Argand, real part`
- `3D scattering in step pot., Argand, delta_l`
- `3D scattering in step pot., delta_l` (see Fig. 8.9)
- `3D scattering in step pot., sigma_l` (see Fig. 8.10)
- `3D scattering in step pot., sigma_tot` (see Fig. 8.11)

8.6 Coulomb Scattering: Radial Wave Function

Aim of this section: Presentation of the Coulomb potential $V(r)$, (8.53), and of the radial wave functions $R_\ell(k, r)$, (8.55), in that potential.

A plot similar to Fig. 8.12 or Fig. 8.13 displaying the Coulomb potential $V(r)$ as a long-dash broken line, the total energy E as a short-dash horizontal line, and the radial wave function (or a simple function thereof) as a continuous line.

Please note: The numerical values of α, $\hbar$, and M are set equal to one and cannot be changed.

On the subpanel **Physics—Comp. Coord.** you can select the way the **Radial Wave Function is Shown** (as $R_{E\ell}^2$, $R_{E\ell}$, $r^2 R_{E\ell}^2$, or $r R_{E\ell}$). Moreover, you can choose to **Plot the Result for**

1. fixed energy E, various values of angular momentum ℓ (i.e., over an r, ℓ plane like in Fig. 8.12) or for
2. fixed angular momentum ℓ, various values of energy E (i.e., over an r, E plane like in Fig. 8.13).

Please note: If you choose option 1, make sure that you draw $R_{E\ell}(r)$ for every integer value of ℓ in your range of ℓ: Choose integers for $y_{\rm beg}$, $y_{\rm end}$ (on

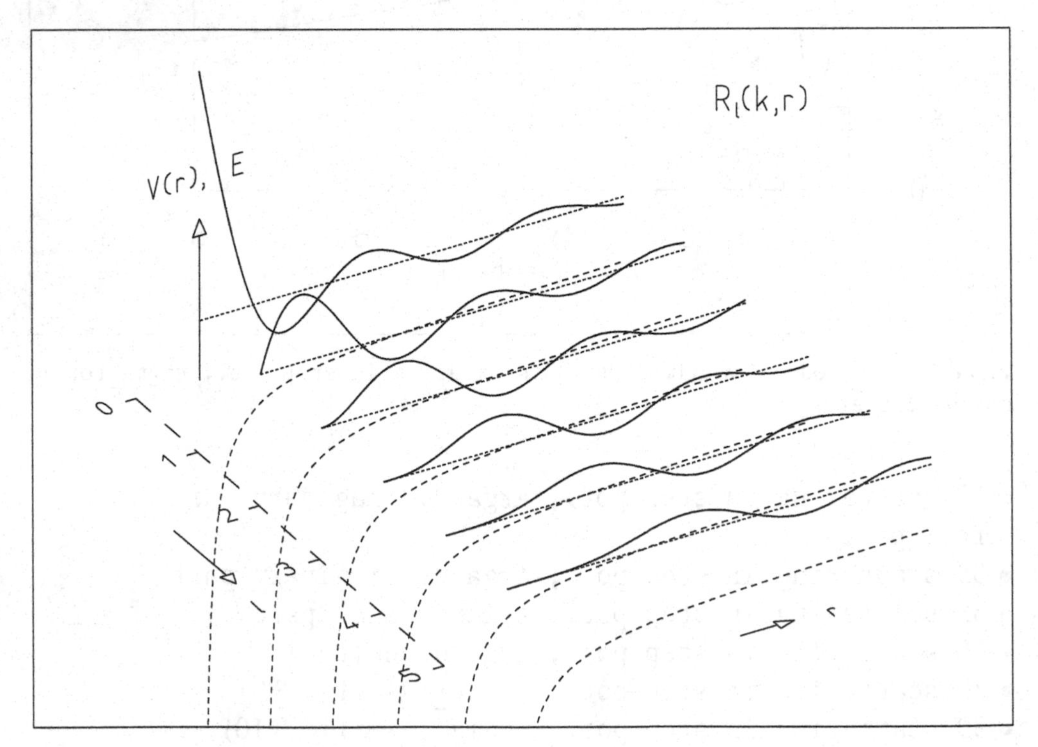

Fig. 8.12. Plot produced with descriptor `Coulomb scattering, R_El, E fixed, l running` on file `3D_Scattering.des`

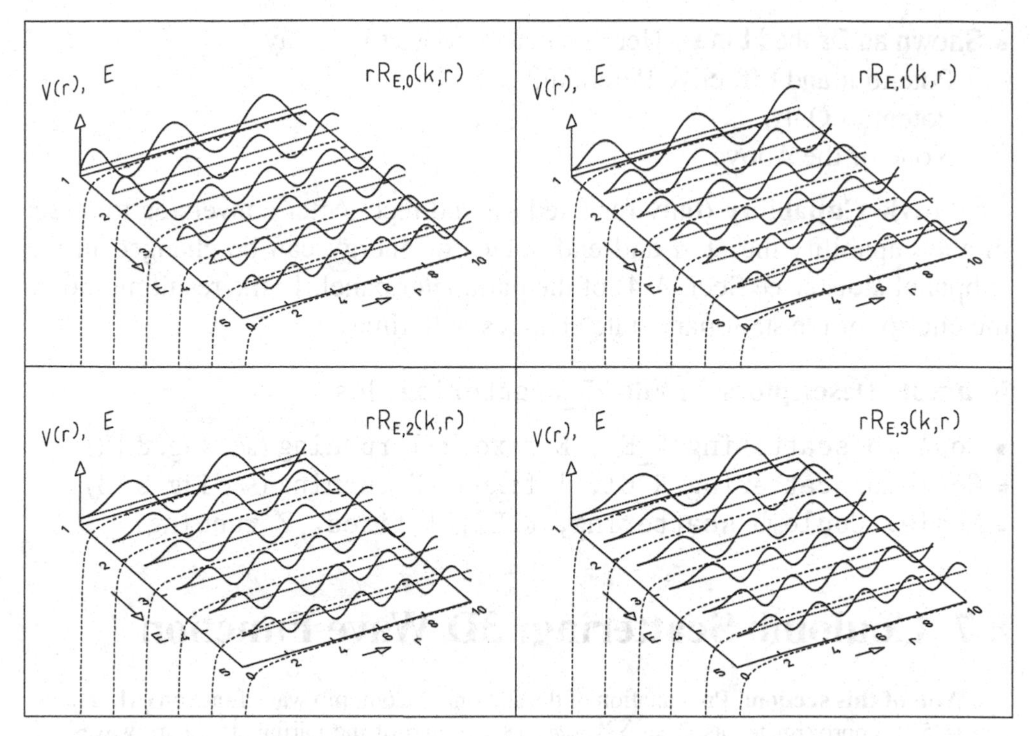

Fig. 8.13. Plot produced with descriptor `Coulomb scattering, R_El, l fixed, E running` on file `3D_Scattering.des`

the subpanel **Physics—Comp. Coord.**) and set n_y (on the subpanel **Graphics—Accuracy**) equal to $|y_{\mathrm{beg}} - y_{\mathrm{end}}| + 1$.

On the subpanel **Physics—Variables** there are seven items:

- Under the heading **Coulomb Potential is** you can choose between **Attractive** and **Repulsive**.
- You can choose whether the **Input Below** (in the next item) **is Taken as** wave number k or energy E.
- **Wave Number / Energy** – Corresponding to the choice in the item above the input field is labeled **k** or **E**. Input is enabled only if the option 'fixed energy' was chosen on the subpanel **Physics—Comp. Coord.**.
- **Angular-Momentum Quantum Number** – Input for the quantum number ℓ is enabled only if the option 'fixed angular momentum' was chosen on the subpanel **Physics—Comp. Coord.**. In a multiple plot the value of ℓ is taken for the first plot and incremented by one from plot to plot.
- **Graphical Item** – contains the quantity ℓ_{DASH} which determines the dash length of the broken lines.
- **Scale Factor** – Because the wave function $R_{E\ell}(r)$ is plotted in an r, E plane with the total energy E_{tot} as zero line, technically speaking the function $E = E_{\mathrm{tot}} + sR_{E\ell}(r)$ is shown. The scale factor s can be adjusted here.

- **Shown as Dashed Line** – Here you can choose to display
  - Potential and Effective Potential,
  - Potential Only,
  - None of the Above.

Movie Capability (only for fixed ℓ): Indirect. After conversion to direct movie capability the start and end values of energy can be changed in the subpanel **Movie** (see Sect. A.4) of the parameter panel. In the resulting movie the energy of the stationary state changes with time.

Example Descriptors on File `3D_Scattering.des`

- `Coulomb scattering R_El, E fixed, l running` (see Fig. 8.12)
- `Coulomb scattering R_El, l fixed, E running` (see Fig. 8.13)
- `Movie: Coulomb scattering, R_El, l fixed, E running`

8.7 Coulomb Scattering: 3D Wave Function

Aim of this section: Presentation of the stationary Coulomb wave function $\varphi(\mathbf{k}, \mathbf{r})$, (8.57), approximated as $\varphi = \sum_{\ell=0}^{L} \varphi_\ell$, (8.58), and of the partial stationary waves $\varphi_\ell(\mathbf{k}, \mathbf{r})$, (8.56).

On the subpanel **Physics—Comp. Coord.** you can select as the **Function Computed:**

- the partial stationary wave φ_ℓ or
- the Coulomb wave φ approximated as $\sum_{\ell=0}^{L} \varphi_\ell$.

Moreover, you can choose the way the selected **Function is Shown** (absolute square, real part, imaginary part).

On the subpanel **Physics—Variables** there are four items:

- Under the heading **Coulomb Potential is** you can choose between **Attractive** and **Repulsive**.
- You can choose whether the **Input Below** (in the next item) **is Taken as** wave number k or energy E.
- **Wave Number / Energy** – Corresponding to the choice in the item above the input field is labeled **k** or **E**.
- **Angular-Momentum Quantum Number** – This is the quantum number ℓ (if you chose to plot a partial wave φ_ℓ) or the upper index L in the sum approximating φ (if you chose to plot φ). In a multiple plot the input value of ℓ (or L) is taken for the first plot. It is successively incremented by one from plot to plot.

Please note: The numerical values of α, $\hbar$, and M are set equal to one and cannot be changed.

Example Descriptor on File 3D_Scattering.des

- Coulomb scattering |phi|**2 (see Fig. 8.14)

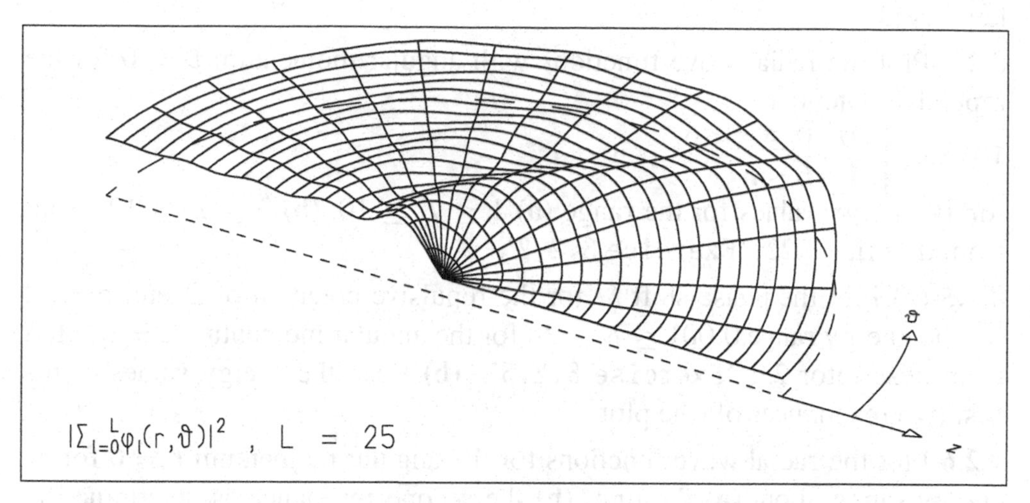

Fig. 8.14. Plot produced with descriptor Coulomb scattering, |phi|**2 on file 3D_Scattering.des

8.8 Exercises

Please note:

(i) You may watch a demonstration of the material of this chapter by pressing the button **Run Demo** in the *main toolbar* and selecting one of the demo files 3D_Scattering.

(ii) More example descriptors can be found on the descriptor files 3DScatt-(FE).des, 3DResonanceScatt(FE).des, and 3DCoulombScatt(FE).des in the directory FurtherExamples.

(iii) For the following exercises use descriptor file 3D_Scattering.des.

(iv) The numerical values of the particle mass and of Planck's constant are put to 1.

8.2.1 Plot the free radial wave functions for the momentum $p = 0.75$ and the angular momenta **(a)** $\ell = 0, 1, 2, \ldots, 10$, **(b)** $\ell = 11, 12, \ldots, 20$, **(c)** $\ell = 21, 22, \ldots, 30$. Start from descriptor 21: 'Exercise 8.2.1'. **(d)** Why is the wave function close to zero in a range adjacent to $r = 0$ and why does this range grow with increasing ℓ?

8.2.2 Plot the free radial wave functions for the energy range $0.01 \le E \le 10$ for **(a)** $\ell = 0$, **(b)** $\ell = 1$, **(c)** $\ell = 2$, **(d)** $\ell = 3$, **(e)** $\ell = 10$. Start from descriptor 22: 'Exercise 8.2.2'.

8.2.3 Repeat Exercise 8.2.2 for the energy range $0.01 \leq E \leq 1$ and for the angular momenta **(a)** $\ell = 6$, **(b)** $\ell = 8$, **(c)** $\ell = 10$. **(d)** Why does the range of very small values of the wave function decrease with increasing energy if ℓ is kept fixed?

8.2.4 Plot the radial wave functions with angular momentum $\ell = 0$ for the repulsive potential

$$V(r) = \begin{cases} 10\,,\ 0 \leq r < 6 \\ 0\,,\ 6 \leq r \end{cases}$$

for 10 energy values for the range **(a)** $1 \leq E \leq 10$, **(b)** $9 \leq E \leq 12$. Start from descriptor 22: '`Exercise 8.2.2`'.

8.2.5 **(a)** Plot the phase shift δ_ℓ for the repulsive potential of Exercise 8.2.4 for the energy range $0.001 \leq E \leq 30$ for the angular momentum $\ell = 0$. Start from descriptor 23: '`Exercise 8.2.5`'. **(b)** Read the energy values of the first two resonances off the plot.

8.2.6 Plot the radial wave functions for the angular momentum $\ell = 0$ for an energy range about **(a)** the first, **(b)** the second resonance as determined in Exercise 8.2.5 (b) for the potential of Exercise 8.2.4. Start from descriptor 22: '`Exercise 8.2.2`'.

8.2.7 Repeat Exercise 8.2.5 for angular momentum $\ell = 1$.

8.2.8 Repeat Exercise 8.2.6 for angular momentum $\ell = 1$.

8.2.9 **(a,b)** Repeat Exercise 8.2.5 (a) and (b) for the lowest resonance in angular momentum $\ell = 2$. **(c)** Why do the resonance energies for resonance wave functions with the same number of radial nodes increase with increasing angular momentum?

8.2.10 Repeat Exercise 8.2.6 for the lowest resonance in angular momentum $\ell = 2$.

8.2.11 Plot the radial wave functions for the potential

$$V(r) = \begin{cases} 0\,,\ 0 \leq r < 1.5 \\ 10\,,\ 1.5 \leq r < 2 \\ 0\,,\ 2 \leq r \end{cases}$$

in the energy range $0.01 \leq E \leq 20$ for **(a)** $\ell = 0$, **(b)** $\ell = 1$, **(c)** $\ell = 2$, **(d)** $\ell = 10$. Start from descriptor 22: '`Exercise 8.2.2`'.

8.2.12 Plot the radial wave functions for the potential of Exercise 8.2.11 for the values $0 \leq \ell \leq 10$ for the energies **(a)** $E = 0.1$, **(b)** $E = 1$, **(c)** $E = 9.9$, **(d)** $E = 20$. Start from descriptor 21: '`Exercise 8.2.1`'.

8.2.13 Plot the phase shift δ_ℓ for the potential of Exercise 8.2.11 for the energy range $0.01 \leq E \leq 40$ for the angular momenta **(a)** $\ell = 0$, **(b)** $\ell = 1$, **(c)** $\ell = 2$, **(d)** $\ell = 3$, **(e)** $\ell = 6$, **(f)** $\ell = 8$ and for the energy range $0.01 \leq E \leq 60$ for the angular momenta **(g)** $\ell = 10$, **(h)** $\ell = 13$. Start from descriptor 23: '`Exercise 8.2.5`'.

8.3.1 Plot the absolute square of the wave function, $|\varphi(r, \vartheta)|^2$, as a function of the radial variable r and the polar angle ϑ for the repulsive potential

$$V(r) = \begin{cases} 6\,, 0 \le r < 2 \\ 0\,, 2 \le r \end{cases}$$

for the energy values **(a)** $E = 3$, **(b)** $E = 6.5$, **(c)** $E = 10$. Start from descriptor 24: '`Exercise 8.3.1`'.

8.3.2 Plot the absolute square of the wave function, $|\varphi(r, \vartheta)|^2$, for the potential of Exercise 8.3.1 for the energy of the lowest resonance in **(a)** $\ell = 0$ ($E_{\text{res}} = 7$), **(b)** $\ell = 1$ ($E_{\text{res}} = 8.5$), **(c)** $\ell = 2$ ($E_{\text{res}} = 10$), **(d)** $\ell = 3$ ($E_{\text{res}} = 11.9$). Start from descriptor 24: '`Exercise 8.3.1`'.

8.3.3 Plot the partial wave functions $\varphi_\ell(r, \vartheta)$ for the potential of Exercise 8.3.1 for the energy $E = 3$ and the angular momenta **(a)** $\ell = 0$, **(b)** $\ell = 1$, **(c)** $\ell = 2$, **(d)** $\ell = 3$, **(e)** $\ell = 4$, **(f)** $\ell = 5$, **(g)** $\ell = 6$, **(h)** $\ell = 7$, **(i)** $\ell = 8$. Start from descriptor 24: '`Exercise 8.3.1`'.

8.3.4 Repeat Exercise 8.3.3 for a free particle [i.e., $V(r) = 0$ everywhere] and compare the results with the ones of 8.3.3.

8.3.5 Repeat Exercise 8.3.3 for the energy $E = 9$ and compare the results with the ones of 8.3.4.

8.3.6 Repeat Exercise 8.3.1 for the absolute square of the scattering wave $|\eta(r, \vartheta)|^2$.

8.3.7 Repeat Exercise 8.3.2 for the absolute square of the scattering wave $|\eta(r, \vartheta)|^2$.

8.3.8 Repeat Exercise 8.3.3 for the partial scattering waves $\eta_\ell(r, \vartheta)$.

8.3.9 Repeat Exercise 8.3.5 for the partial scattering waves $\eta_\ell(r, \vartheta)$.

8.3.10 Plot the partial scattering waves $\eta_\ell(r, \vartheta)$ for the potential of Exercise 8.3.1 for **(a)** $\ell = 0, E = 7$, **(b)** $\ell = 1, E = 7$, **(c)** $\ell = 2, E = 7$, **(d)** $\ell = 3, E = 7$, **(e)** $\ell = 0, E = 8.5$, **(f)** $\ell = 1, E = 8.5$, **(g)** $\ell = 2, E = 8.5$, **(h)** $\ell = 3, E = 8.5$, **(i)** $\ell = 0, E = 10$, **(j)** $\ell = 1, E = 10$, **(k)** $\ell = 2, E = 10$, **(l)** $\ell = 3, E = 10$, **(m)** $\ell = 0, E = 11.9$, **(n)** $\ell = 1, E = 11.9$, **(o)** $\ell = 2, E = 11.9$, **(p)** $\ell = 3, E = 11.9$. Start from descriptor 24: '`Exercise 8.3.1`'.

8.4.1 For the repulsive spherically symmetric potential

$$V(r) = \begin{cases} 3\,, 0 \le r < 2 \\ 0\,, 2 \le r \end{cases}$$

plot the differential cross section $\mathrm{d}\sigma/\mathrm{d}\Omega$ for the energy range $0.001 \le E \le 6$ divided into 10 intervals for the summation of angular momenta up to **(a)** $L = 0$, **(b)** $L = 1$, **(c)** $L = 2$, ..., **(i)** $L = 8$. Start from descriptor 8: '`3D scattering in step pot., dsigma/dOmega`'. **(j)** Why is the differential cross section obtained in (a) independent of $\cos\vartheta$?

8.4.2 Repeat Exercise 8.4.1 for a summation up to $L = 30$ for the energy ranges **(a)** $40 \le E \le 50$ and **(b)** $4000 \le E \le 5000$. Start from descriptor 8:

`3D scattering in step pot., dsigma/dOmega`. **(c)** Calculate the wavelengths ($M = 1$, $\hbar = 1$) for $E = 50$ and $E = 5000$. **(d)** How does the cross section in the forward direction decrease for increasing energy?

8.4.3 For the attractive spherically symmetric potential

$$V(r) = \begin{cases} -3\,, 0 \leq r < 2 \\ 0\,, 2 \leq r \end{cases}$$

plot the differential cross section $\mathrm{d}\sigma/\mathrm{d}\Omega$ for the energy range $0.001 \leq E \leq 6$ divided into 10 intervals for the summation of angular momenta up to **(a)** $L = 0$, **(b)** $L = 1$, **(c)** $L = 2$, ..., **(i)** $L = 8$. Start from descriptor 8: `3D scattering in step pot., dsigma/dOmega`.

8.4.4 Repeat Exercise 8.4.3 for a summation up to $L = 30$ for the energy ranges **(a)** $40 \leq E \leq 50$ and **(b)** $4000 \leq E \leq 5000$. Start from descriptor 8: `3D scattering in step pot., dsigma/dOmega`. **(c)** Compare the results with those in Exercise 8.4.2.

8.4.5 Compare the differential cross sections for the repulsive and attractive potentials of Exercises 8.4.1 and 8.4.3 for low energies $0.001 \leq E \leq 0.1$. Plot for the repulsive potential **(a)** $L = 0, \ldots,$ **(d)** $L = 3$ and for the attractive potential **(e)** $L = 0, \ldots,$ **(h)** $L = 3$. Start from descriptor 8: `3D scattering in step pot., dsigma/dOmega`. **(i)** Why are the cross sections for the attractive potential at low energies larger than those for the repulsive potential?

8.5.1 For the potential

$$V(r) = \begin{cases} 3\,, 0 \leq r < 2 \\ 0\,, 2 \leq r \end{cases}$$

plot for $\ell = 0$ and $0.001 \leq E \leq 6$ **(a)** the Argand plot $f_0(E)$, **(b)** Im $f_0(E)$, **(c)** Re $f_0(E)$, **(d)** $\delta_0(E)$, **(e)** the combined plot of (a–d). Start from the descriptors (a) 10: `3D scattering in step pot., Argand, imaginary part vs. real part`, (b) 11: `3D scattering in step pot., Argand, imaginary part`, (c) 12: `3D scattering in step pot., Argand, real part`, (d) 13: `3D scattering in step pot., Argand, delta_l`, (e) 9: `3D scattering in step pot., Argand, combined plot`. **(f)** Read the lowest resonance energy off the plot of the phase δ_0.

8.5.2 (a–e) Repeat Exercise 8.5.1 (a–e) for the energy range $0.001 \leq E \leq 20$. **(f)** Read the two lowest resonance energies E_1, E_2 and the corresponding phase values δ_{01}, δ_{02} off the plot for δ_0. **(g)** Calculate the values $f_0(E_1)$, $f_0(E_2)$ from the phases δ_{01}, δ_{02}.

8.5.3 (a–e) Repeat Exercise 8.5.2 (a–e) for $\ell = 1$. **(f)** Read the lowest resonance energy off the plot of the phase δ_1.

8.5.4 (a–e) Repeat Exercise 8.5.2 (a–e) for $\ell = 2$. **(f)** Read the lowest resonance energy off the plot of the phase δ_2. **(g)** Explain the increase of the lowest resonance energies as ℓ increases from 0 to 2.

8.5.5 For the potential

$$V(r) = \begin{cases} 0\,, 0 \le r < 2 \\ 4\,, 2 \le r < 2.5 \\ 0\,, 2.5 \le r \end{cases}$$

plot for $\ell = 0$ and $0.001 \le E \le 20$ **(a)** the Argand plot $f_0(E)$, **(b)** Im $f_0(E)$, **(c)** Re $f_0(E)$, **(d)** $\delta_0(E)$, **(e)** the combined plot of (a–d). Start from the descriptors (a) 10: '`3D scattering in step pot., Argand, imaginary part vs. real part`', (b) 11: '`3D scattering in step pot., Argand, imaginary part`', (c) 12: '`3D scattering in step pot., Argand, real part`', (d) 13: '`3D scattering in step pot., Argand, delta_l`', and (e) 9: '`3D scattering in step pot., Argand, combined plot`'. **(f)** Read the lowest resonance energies off the plot of the phase δ_0. **(g)** Give a rough estimate of the resonance energies by calculating the bound-state energies of an infinitely deep square well of width 2.

8.5.6 (a–e) Repeat Exercise 8.5.5 (a–e) for $\ell = 1$.

8.5.7 (a–e) Repeat Exercise 8.5.5 (a–e) for $\ell = 2$.

8.5.8 Plot for the potential of Exercise 8.5.5 in the energy range $0.001 \le E \le 8$ the partial cross sections σ_ℓ for **(a)** $\ell = 0, 1, 2, 3$, **(b)** $\ell = 4, 5, 6, 7$. Start from descriptor 15: '`3D scattering in step pot., sigma_l`'.

8.5.9 Plot for the potential

$$V(r) = \begin{cases} 0\,, 0 \le r < 2 \\ 10\,, 2 \le r < 2.5 \\ 0\,, 2.5 \le r \end{cases}$$

in the energy range $0.001 \le E \le 20$ the partial cross sections σ_ℓ for **(a)** $\ell = 0, 1, 2, 3$, **(b)** $\ell = 4, 5, 6, 7$. Start from descriptor 15: '`3D scattering in step pot., sigma_l`'. **(c)** What is the significance of the small peaks in the plot? **(d)** For zero angular momentum calculate the first few energy eigenvalues of an infinitely deep square well of equal width. **(e)** Why are the energy values of the peaks for $\ell = 0$ in the plot (a) smaller than the energies calculated under (d)?

8.5.10 (a) Plot for the potential of Exercise 8.5.5 in the energy range $0.001 \le E \le 8$ the total cross section obtained by summation of angular momenta up to $L = 15$. Start from descriptor 15: '`3D scattering in step pot., sigma_l`'. **(b)** Compare the value for $E = 0.001$ with the value of the partial cross section σ_0 at this energy. **(c)** Calculate the total classical cross section for the scattering on a hard sphere of radius 2.5.

8.5.11 Plot for the repulsive square-well potential

$$V(r) = \begin{cases} V_0\,, 0 \le r < 2 \\ 0\;\;\,, 2 \le r \end{cases}$$

in the energy range $10^{-5} \le E \le 0.01$ the total cross section σ_{tot} for the summation up to $L = 3$ for **(a)** $V_0 = 10$, **(b)** $V_0 = 100$, **(c)** $V_0 = 1000$,

(d) $V_0 = 10\,000$, **(e)** $V_0 = 100\,000$. Start from descriptor 15: ‘`3D scattering in step pot., sigma_l`’. Read the values of $\sigma_{\rm tot}$ at $E = 10^{-5}$ off the plots. **(f)** Compare the values with the limiting formula $\sigma_{\rm tot}(E = 0) = 4\pi d^2$ for an infinitely high repulsive square-well potential of width d.

8.6.1 Produce a plot with descriptor 17: ‘`Coulomb scattering, R_El, E fixed, l running`’. Leave it on the screen. Produce a second plot and change the potential from attractive to repulsive. The two plots on the screen look rather similar but there are subtle differences. Explain them.

8.6.2 Produce a plot with descriptor 18: ‘`Coulomb scattering, R_El, l fixed, E running`’. Set $\ell = 10$. Explain the behavior of the radial position $r_0(E)$ where the function $rR_{10}(k, r)$ starts to be significantly different from zero.

8.7.1 Produce a plot with descriptor 20: ‘`Coulomb scattering, |phi|**2`’. Study the contributions of the different partial waves φ_ℓ by setting the angular momentum to zero and producing a multiple plot with 5 rows and 5 columns **(a)** for φ_ℓ, **(b)** for $\sum_{\ell=0}^{L} \varphi_\ell$.

8.7.2 Repeat Exercise 8.7.1 for an attractive potential.

9. Spin and Magnetic Resonance

Contents: Brief introduction of spin formalism. Spin and magnetic moment. Pauli equation, describing behavior of spin in magnetic field. Motion in constant field $\mathbf{B}_0$. Principle of magnetic resonance: combination of large constant field $\mathbf{B}_0$ with small field $\mathbf{B}_1(t)$, rotating in the plane perpendicular to $\mathbf{B}_0$. Motion of spin in a coordinate system which rotates with $\mathbf{B}_1$ (and in which the total field is constant) and in a coordinate system which is at rest in the laboratory. At resonance the direction of the spin expectation value is completely inverted. Rabi formula describing the resonance form.

9.1 Physical Concepts

9.1.1 Spin Operators. Eigenvectors and Eigenvalues

In Sect. 6.1.3 we introduced the operator of orbital angular momentum, $\hat{\mathbf{L}} = \hat{\mathbf{r}} \times \hat{\mathbf{p}}$, as vector product of the position operator $\hat{\mathbf{r}} = \mathbf{r}$ and the momentum operator $\hat{\mathbf{p}} = (\hbar/\mathrm{i})\nabla$. We found the components $\hat{L}_x, \hat{L}_y, \hat{L}_z$ of $\hat{\mathbf{L}}$ to fulfill the commutation relations (6.34). Moreover, we found the eigenvalue equations (6.40), (6.41) for the two operators $\hat{\mathbf{L}}^2$ and $\hat{L}_z$. The eigenfunctions of both equations are the spherical harmonics $Y_{\ell m}$.

Besides the operators $\hat{L}_x, \hat{L}_y, \hat{L}_z$ there are also matrices S_x, S_y, S_z which satisfy commutation relations of the type (6.34), i.e.,

$$[S_x, S_y] = \mathrm{i}\hbar S_z \quad , \qquad [S_y, S_z] = \mathrm{i}\hbar S_x \quad , \qquad [S_z, S_x] = \mathrm{i}\hbar S_y \quad . \tag{9.1}$$

In the simplest case, to which we confine ourselves, they are 2×2 matrices,

$$S_x = \frac{\hbar}{2}\sigma_1 \quad , \qquad S_y = \frac{\hbar}{2}\sigma_2 \quad , \qquad S_z = \frac{\hbar}{2}\sigma_3 \quad , \tag{9.2}$$

which are the *Pauli matrices*,

$$\sigma_1 = \begin{pmatrix} 0 & 1 \\ 1 & 0 \end{pmatrix} \quad , \qquad \sigma_2 = \begin{pmatrix} 0 & -\mathrm{i} \\ \mathrm{i} & 0 \end{pmatrix} \quad , \qquad \sigma_3 = \begin{pmatrix} 1 & 0 \\ 0 & -1 \end{pmatrix} \quad , \tag{9.3}$$

multiplied by $\hbar/2$.

S. Brandt et al., *Interactive Quantum Mechanics: Quantum Experiments on the Computer*,
DOI 10.1007/978-1-4419-7424-2_9, © Springer Science+Business Media, LLC 2011

The S_x, S_y, S_z are the operators of the *intrinsic angular-momentum* or *spin* components of a particle and can be regarded as the components of the vector

$$\mathbf{S} = S_x \mathbf{e}_x + S_y \mathbf{e}_y + S_z \mathbf{e}_y \quad . \tag{9.4}$$

Its square is

$$\mathbf{S}^2 = \frac{3}{4}\hbar\sigma_0 \quad , \tag{9.5}$$

where

$$\sigma_0 = \begin{pmatrix} 1 & 0 \\ 0 & 1 \end{pmatrix} \tag{9.6}$$

is the 2×2 unit matrix. The matrix operators $\mathbf{S}^2$ and S_z fulfill eigenvalue equations analogous to (6.40), (6.41),

$$\mathbf{S}^2 \eta_r = \frac{3}{4}\hbar^2 \eta_r \quad , \qquad S_z \eta_r = \frac{1}{2} r \hbar \eta_r \quad , \qquad r = 1, -1 \quad . \tag{9.7}$$

Here the eigenvectors – which correspond to the eigenfunctions $Y_{\ell m}$ in (6.40), (6.41) – are

$$\eta_1 = \begin{pmatrix} 1 \\ 0 \end{pmatrix} \quad , \qquad \eta_{-1} = \begin{pmatrix} 0 \\ 1 \end{pmatrix} \quad . \tag{9.8}$$

They are orthogonal to each other and normalized to one. For later use we already introduce here the notation

$$\eta_1^+ = (1, 0) \quad , \qquad \eta_{-1}^+ = (0, 1) \quad .$$

The eigenvalues in (9.7) can be written as $s(s+1)\hbar^2$ and $s_z\hbar$, respectively, where $s = 1/2$ and $s_z = \pm 1/2$ are the quantum number of spin and of its z component, respectively.

The general spin state is the complex linear combination

$$\chi = \chi_1 \eta_1 + \chi_{-1} \eta_{-1} \quad , \qquad \chi^+ = \chi_1^* \eta_1^+ + \chi_{-1}^* \eta_{-1}^+ \tag{9.9}$$

of the two basic states η_1, η_{-1}. The normalization of χ requires

$$\chi^+ \cdot \chi = |\chi_1|^2 + |\chi_{-1}|^2 = 1 \quad , \tag{9.10}$$

so that the moduli $|\chi_1|$, $|\chi_{-1}|$ can be parametrized as $|\chi_1| = \cos\theta/2$, $|\chi_{-1}| = \sin\theta/2$. Since the phase can be chosen arbitrarily we may write

$$\chi_1 = \mathrm{e}^{-\mathrm{i}\phi/2} \cos\theta/2, \quad , \qquad \chi_{-1} = \mathrm{e}^{\mathrm{i}\phi/2} \sin\theta/2 \quad .$$

The expectation value of the spin vector,

$$\langle \mathbf{S} \rangle = \langle S_x \rangle \mathbf{e}_x + \langle S_y \rangle \mathbf{e}_y + \langle S_z \rangle \mathbf{e}_z \quad , \tag{9.11}$$

for a general spin state is the vector formed by the expectation values of the three components

$$\langle S_x \rangle = \frac{\hbar}{2} \chi^+ \sigma_1 \chi \quad , \qquad \langle S_y \rangle = \frac{\hbar}{2} \chi^+ \sigma_2 \chi \quad , \qquad \langle S_z \rangle = \frac{\hbar}{2} \chi^+ \sigma_3 \chi \quad .$$

9.1.2 Magnetic Moment and Its Motion in a Magnetic Field. Pauli Equation

The electron (charge $-e$, mass M_e) possesses a *magnetic moment* related to its spin. The magnetic-moment operator $\boldsymbol{\mu}$ is proportional to the spin operator **S**,

$$\boldsymbol{\mu} = -\gamma \mathbf{S} \quad . \tag{9.12}$$

The proportionality factor, also called the *gyromagnetic ratio*, is usually written as

$$\gamma = \frac{g_e \mu_B}{\hbar} \quad . \tag{9.13}$$

Here the constant

$$\mu_B = \frac{e}{2M_e}\hbar \tag{9.14}$$

is the *Bohr magneton.* The *gyromagnetic factor* g_e of the free electron can be computed with high precision in the framework of quantum electrodynamics and is (very nearly) equal to 2.

For the proton (charge $+e$, mass M_p) these relations are replaced by

$$\boldsymbol{\mu} = \gamma \mathbf{S} \quad , \qquad \gamma = \frac{g_p \mu_N}{\hbar} \quad , \qquad \mu_N = \frac{e}{2M_p}\hbar \quad ,$$

μ_N and g_p being the *nuclear magneton* and the gyromagnetic factor of the proton, respectively. The latter, dominated by the internal structure of the proton, has to be determined experimentally.

If electrons (or protons) are not free but bound in matter, their gyromagnetic factors appear to be slightly changed, the change providing valuable information on the structure of that matter. It is obtained by precise measurements of γ through magnetic resonance. The formulae below apply to electrons. The change to protons only requires the replacement $\gamma \to -\gamma$.

The potential energy operator describing a static magnetic moment in a field of magnetic induction **B** is

$$H = -\boldsymbol{\mu} \cdot \mathbf{B} = \gamma \mathbf{B} \cdot \mathbf{S} \quad . \tag{9.15}$$

In analogy to the time-dependent Schrödinger equation (3.3) we write down the *Pauli equation*

$$\mathrm{i}\hbar \frac{\mathrm{d}}{\mathrm{d}t}\chi(t) = H\chi(t) = \gamma \mathbf{B} \cdot \mathbf{S}\chi(t) \quad . \tag{9.16}$$

From this we get the equation

$$\frac{\mathrm{d}}{\mathrm{d}t}\left(\chi^+(t)\mathbf{S}\chi(t)\right) = \frac{\mathrm{i}}{\hbar}\chi^+(t)(H\mathbf{S} - \mathbf{S}H)\chi(t) \quad .$$

With (9.15) and (9.1) we obtain the equation of motion for the spin expectation vector (9.11),

$$\frac{\mathrm{d}}{\mathrm{d}t}\langle\mathbf{S}\rangle = \frac{\mathrm{d}}{\mathrm{d}t}\left(\chi^+(t)\mathbf{S}\chi(t)\right) = \gamma\mathbf{B}\times\left(\chi^+(t)\mathbf{S}\chi(t)\right) = \gamma\mathbf{B}\times\langle\mathbf{S}\rangle \quad . \tag{9.17}$$

The simplest case, described by this equation, is that of a constant field, which we take to be oriented in z direction,

$$\mathbf{B} = \mathbf{B}_0 = B_0\mathbf{e}_z \quad , \qquad \boldsymbol{\Omega} = \gamma\mathbf{B}_0 = \Omega_0\mathbf{e}_z \quad .$$

Initially, i.e., at $t = 0$, the vector $\langle\mathbf{S}\rangle$ can be written as

$$\langle\mathbf{S}\rangle_0 = \langle S_x\rangle_0\,\mathbf{e}_x + \langle S_z\rangle_0\,\mathbf{e}_z$$

by orienting $\mathbf{e}_x$ and $\mathbf{e}_z$ appropriately. The time dependence of $\langle\mathbf{S}\rangle$ reads

$$\langle\mathbf{S}\rangle = \langle S_x\rangle_0\,(\mathbf{e}_x\cos\Omega_0 t + \mathbf{e}_y\sin\Omega_0 t) + \langle S_z\rangle_0\,\mathbf{e}_z \quad , \tag{9.18}$$

as is verified by inserting (9.18) into (9.17). Thus, in a constant field, the vector $\langle\mathbf{S}\rangle$ performs a *Larmor precession* around the direction of $\mathbf{B}_0$ with constant angular frequency $\Omega_0 = \gamma B_0$. Both the absolute value of $\langle\mathbf{S}\rangle$ and the angle between $\langle\mathbf{S}\rangle$ and $\mathbf{B}_0$ stay unchanged. This behavior can be pictured as follows. If we construct a sphere of radius $|\langle\mathbf{S}\rangle| = \hbar/2$ around the origin and consider its north pole, i.e., the point where it is pierced by the z axis (which is the direction of the field), then the tip of the vector $\langle\mathbf{S}\rangle$ rotates with angular frequency Ω_0 around the pole. Its trajectory is a circle of constant polar angle, i.e., a parallel in geographic terms.

9.1.3 Magnetic Resonance

Magnetic-resonance experiments are performed in a magnetic field which consists of a relatively large constant field $\mathbf{B}_0$ and a relatively small time-dependent field $\mathbf{B}_1(t)$,

$$\mathbf{B} = \mathbf{B}_0 + \mathbf{B}_1(t) \quad .$$

The field $\mathbf{B}_0$ points in z direction,

$$\mathbf{B}_0 = B_0\mathbf{e}_z \quad ,$$

the direction of the field $\mathbf{B}_1(t)$ rotates with angular velocity ω in the x, y plane,

$$\mathbf{B}_1 = B_1(\mathbf{e}_x\cos\omega t + \mathbf{e}_y\sin\omega t) \quad .$$

Besides the fixed coordinate system spanned by the unit vectors $\mathbf{e}_x$, $\mathbf{e}_y$, $\mathbf{e}_z$ we introduce a coordinate system, rotating with $\mathbf{B}_1$ and spanned by $\mathbf{e}'_x$, $\mathbf{e}'_y$, $\mathbf{e}'_z$. The two are related by

$$
\begin{aligned}
\mathbf{e}'_x &= \mathbf{e}_x \cos\omega t + \mathbf{e}_y \sin\omega t \quad , & \mathbf{e}_x &= \mathbf{e}'_x \cos\omega t - \mathbf{e}'_y \sin\omega t \quad , \\
\mathbf{e}'_y &= -\mathbf{e}_x \sin\omega t + \mathbf{e}_y \cos\omega t \quad , & \mathbf{e}_y &= \mathbf{e}'_x \sin\omega t + \mathbf{e}'_y \cos\omega t \quad , \\
\mathbf{e}'_z &= \mathbf{e}_z \quad , & \mathbf{e}_z &= \mathbf{e}'_z \quad .
\end{aligned}
\tag{9.19}
$$

From classical mechanics we know that the time derivatives of an arbitrary vector $\langle\mathbf{S}\rangle$ differ if the differentiation of its coordinates are taken in the two different frames. The two derivatives are related by

$$\frac{\mathrm{d}\langle\mathbf{S}\rangle}{\mathrm{d}t} = \frac{\mathrm{d}'\langle\mathbf{S}\rangle}{\mathrm{d}t} + \boldsymbol{\omega} \times \langle\mathbf{S}\rangle \quad , \tag{9.20}$$

with

$$\boldsymbol{\omega} = \omega \mathbf{e}_z \quad .$$

By introducing (9.20) into (9.17) we get the equation of motion in the rotating frame

$$\frac{\mathrm{d}'\langle\mathbf{S}\rangle}{\mathrm{d}t} = (\gamma\mathbf{B} - \boldsymbol{\omega}) \times \langle\mathbf{S}\rangle = \gamma\mathbf{B}_{\mathrm{eff}} \times \langle\mathbf{S}\rangle \quad . \tag{9.21}$$

The *effective field*

$$\mathbf{B}_{\mathrm{eff}} = \mathbf{B} - \frac{\boldsymbol{\omega}}{\gamma} = \left(B_0 - \frac{\omega}{\gamma}\right)\mathbf{e}'_z + B_1\mathbf{e}'_x$$

is, of course, constant in the rotating coordinate system.

From the discussion in Sect. 9.1.2 it is now clear, that, in the rotating frame, $\langle\mathbf{S}\rangle$ precesses around the direction of $\mathbf{B}_{\mathrm{eff}}$ with the angular frequency $\Omega' = \gamma B_{\mathrm{eff}}$. In experiments the vector $\langle\mathbf{S}\rangle$ is initially aligned along the field $\mathbf{B}_0$, i.e., the z axis,

$$\langle\mathbf{S}\rangle_0 = \langle\mathbf{S}\rangle(t=0) = \frac{\hbar}{2}\mathbf{e}_z = \frac{\hbar}{2}\mathbf{e}'_z \quad . \tag{9.22}$$

Thus, on the sphere of radius $|\langle\mathbf{S}\rangle| = \hbar/2$ around the origin which we constructed in Sect. 9.1.2 the tip of $\langle\mathbf{S}\rangle$ traces a circle through the north pole, its initial position, around the direction of $\mathbf{B}_{\mathrm{eff}}$, see Fig. 9.1. The circle has a small radius as long as $\mathbf{B}_{\mathrm{eff}}$ is essentially parallel to the z direction (for small ω) or antiparallel (to it for large ω), i.e., for $|B_0 - \omega/\gamma| \gg B_1$. However, at the *resonance frequency* of the rotating field,

$$\omega = \omega_{\mathrm{R}} = \gamma B_0 \quad ,$$

the effective field points in the direction of $\mathbf{e}'_x$: The circle passes through both north and south pole and reaches its maximum radius, that of the sphere itself. This complete change of direction of $\langle\mathbf{S}\rangle$ can be detected by the energy which has to be provided to bring it about, i.e., the energy taken out of the rotating field $\mathbf{B}_1$. The resonance frequency is measured with high precision and thus is the gyromagnetic ratio γ.

Before presenting explicitely the time dependence of

$$\langle \mathbf{S} \rangle = \langle S'_x \rangle \mathbf{e}'_x + \langle S'_y \rangle \mathbf{e}'_y + \langle S'_z \rangle \mathbf{e}'_z = \langle S_x \rangle \mathbf{e}_x + \langle S_y \rangle \mathbf{e}_y + \langle S_z \rangle \mathbf{e}_z \tag{9.23}$$

we introduce the definitions

$$\begin{aligned} &\boldsymbol{\Omega}_0 = \gamma \mathbf{B}_0 \quad , \qquad \boldsymbol{\Omega}_1 = \gamma \mathbf{B}_1 \quad , \qquad \boldsymbol{\Omega} = \gamma \mathbf{B} = \gamma(\mathbf{B} + \mathbf{B}_1) \quad , \\ &\boldsymbol{\Omega}' = \boldsymbol{\Omega} - \boldsymbol{\omega} = (\Omega_0 - \omega)\mathbf{e}'_z + \Omega_1 \mathbf{e}'_x \quad , \qquad \Delta = \omega - \Omega_0 \quad , \\ &\Omega' = |\boldsymbol{\Omega}'| = \sqrt{(\omega - \Omega_0)^2 + \Omega_1^2} = \sqrt{\Delta^2 + \Omega_1^2} \quad . \end{aligned} \tag{9.24}$$

Using these we compute the vector product in the right-hand side of (9.21) in the rotating frame and write down (9.21) in components,

$$\begin{aligned} \frac{\mathrm{d}\langle S'_x \rangle}{\mathrm{d}t} &= \Delta \langle S'_y \rangle \quad , \\ \frac{\mathrm{d}\langle S'_y \rangle}{\mathrm{d}t} &= -\Delta \langle S'_x \rangle - \Omega_1 \langle S'_z \rangle \quad , \\ \frac{\mathrm{d}\langle S'_z \rangle}{\mathrm{d}t} &= \Omega_1 \langle S'_y \rangle \quad . \end{aligned} \tag{9.25}$$

Taking into account the initial conditions (9.22), we find the solutions

$$\begin{aligned} \langle S'_x \rangle &= \frac{\hbar}{2} \frac{\Delta}{\Omega'} \frac{\Omega_1}{\Omega'} (\cos \Omega' t - 1) \quad , \\ \langle S'_y \rangle &= -\frac{\hbar}{2} \frac{\Omega_1}{\Omega'} \sin \Omega' t \quad , \\ \langle S'_z \rangle &= \frac{\hbar}{2} \left(\frac{\Delta^2}{\Omega'^2} + \frac{\Omega_1^2}{\Omega'^2} \cos \Omega' t \right) \quad . \end{aligned} \tag{9.26}$$

Finally, using (9.23) and (9.19), we obtain the components of $\langle \mathbf{S} \rangle$ in the original coordinate frame,

$$\begin{aligned} \langle S_x \rangle &= \frac{\hbar}{2} \frac{\Omega_1}{\Omega'} \left(\sin \omega t \sin \Omega' t + \frac{\Delta}{\Omega'} \cos \omega t \cos \Omega' t - \frac{\Delta}{\Omega'} \cos \omega t \right) \quad , \\ \langle S_y \rangle &= -\frac{\hbar}{2} \frac{\Omega_1}{\Omega'} \left(\cos \omega t \sin \Omega' t - \frac{\Delta}{\Omega'} \sin \omega t \cos \Omega' t + \frac{\Delta}{\Omega'} \sin \omega t \right) \quad , \\ \langle S_z \rangle &= \frac{\hbar}{2} \left(\frac{\Delta^2}{\Omega'^2} + \frac{\Omega_1^2}{\Omega'^2} \cos \Omega' t \right) \quad . \end{aligned} \tag{9.27}$$

For illustrations see Fig. 9.1.

9.1.4 Rabi Formula

In 1937 Rabi published the theory of magnetic resonance, a field which he also pioneered experimentally. The *Rabi formula* gives the probability $P_{-1/2}(t)$ of finding at time t the state η_{-1} if the initial state at time $t = 0$ was η_1. With the formalism of Sect. 9.1.1 we find

$$P_{-\frac{1}{2}}(t) = |\eta_{-1}^{+}\chi(t)|^2 = |\chi_{-1}(t)|^2 \quad . \tag{9.28}$$

Using

$$\chi^{+}(t)\sigma_0\chi(t) = |\chi_1(t)|^2 + |\chi_{-1}(t)|^2 = 1$$

and

$$\chi^{+}(t)\frac{2}{\hbar}S_z\chi(t) = \chi^{+}(t)\sigma_3\chi(t) = |\chi_1(t)|^2 - |\chi_{-1}(t)|^2$$

we obtain

$$\frac{1}{2}\left(1 - \frac{2}{\hbar}\langle S_z\rangle\right) = \frac{1}{2}\chi^{+}(t)\left(\sigma_0 - \frac{2}{\hbar}S_z\right)\chi(t) = |\chi_{-1}(t)|^2 \quad ,$$

so that with (9.27)

$$P_{-\frac{1}{2}}(t) = \frac{1}{2}\left(1 - \frac{2}{\hbar}\langle S_z\rangle\right) = \frac{\Omega_1^2}{\Omega'^2}\sin^2\frac{\Omega'}{2}t \quad . \tag{9.29}$$

We introduce the *Rabi period*

$$T = \frac{2\pi}{\Omega'} = \frac{2\pi}{\sqrt{\Delta^2 + \Omega_1^2}} \quad , \qquad \Delta = \omega - \Omega_0 \quad . \tag{9.30}$$

For $t = T/2, 3T/2, \ldots$ the probability (9.29) assumes its maximum value

$$A = P_{-\frac{1}{2}}\left(\frac{T}{2}\right) = \frac{\Omega_1^2}{\Omega'^2} = \frac{\Omega_1^2}{\Omega_1^2 + (\omega - \Omega_0)^2} = \frac{(\Omega_1/\Omega_0)^2}{(\Omega_1/\Omega_0)^2 + (1 - \omega/\Omega_0)^2} \quad . \tag{9.31}$$

As a function of the angular frequency ω the *Rabi amplitude* A is of the typical Breit–Wigner resonance form. At the resonance frequency $\omega_R = \Omega_0$ it reaches its maximum value $A = 1$, see Fig. 9.4

Further Reading

Brandt, Dahmen: Chap. 16

9.2 The Spin-Expectation Vector near and at Resonance

Aim of this section: Illustration of the time dependences of the spin-expectation vector in the rotating coordinate frame, (9.26), and in the fixed frame, (9.27). To this end the trajectory of the tip of $\langle \mathbf{S}' \rangle$ (or $\langle \mathbf{S} \rangle$) is traced on a sphere of radius $|\langle \mathbf{S} \rangle|$.

On the subpanel **Physics—Comp. Coord.**, under the heading **Physics parameters** you enter 4 quantities:

- ω/Ω_0, the frequency of the rotating field (in units of Larmor frequency Ω_0 implied by the fixed field),
- Ω_1/Ω_0, the Larmor frequency of the rotating field (in units of Ω_0),
- f, the fraction of the Rabi period T, defined in (9.30), for which the trajectory is shown,
- an **Increment** δ, see below.

If you produce a **Multiple Plot**, then you can choose the individual plots to show the spin trajectory on the unit sphere

- **with one parameter incremented** or
- **in rotating and in fixed frame.**

For the first case you can determine the **Coordinate frame** as either **fixed** or **rotating** and, under the heading, **Incremented in multiple plot** you can select one of the three parameters ω/Ω_0, Ω_1/Ω_0, f to be incremented by δ from plot to plot.

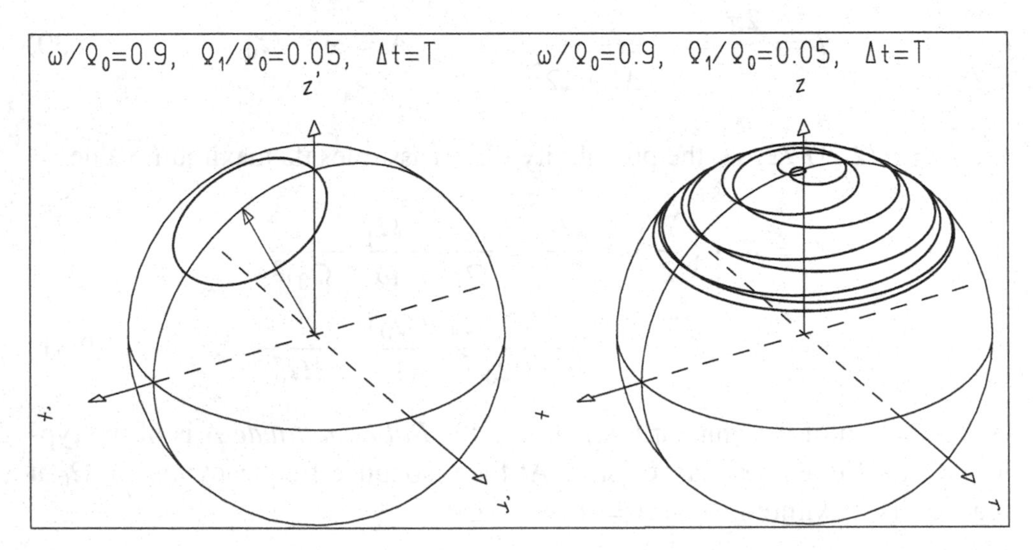

Fig. 9.1. Plot produced with descriptor `Spin in rotating and in fixed frame, off resonance` on file `Magnetic_Resonance.des`. Left: Trajectory of the tip of the spin-expectation vector in the rotating coordinate frame. The arrow, pointing away from the coordinate origin, has the direction of the effective field $\mathbf{B}_{\mathrm{eff}}$. Right: Same trajectory in the fixed frame.

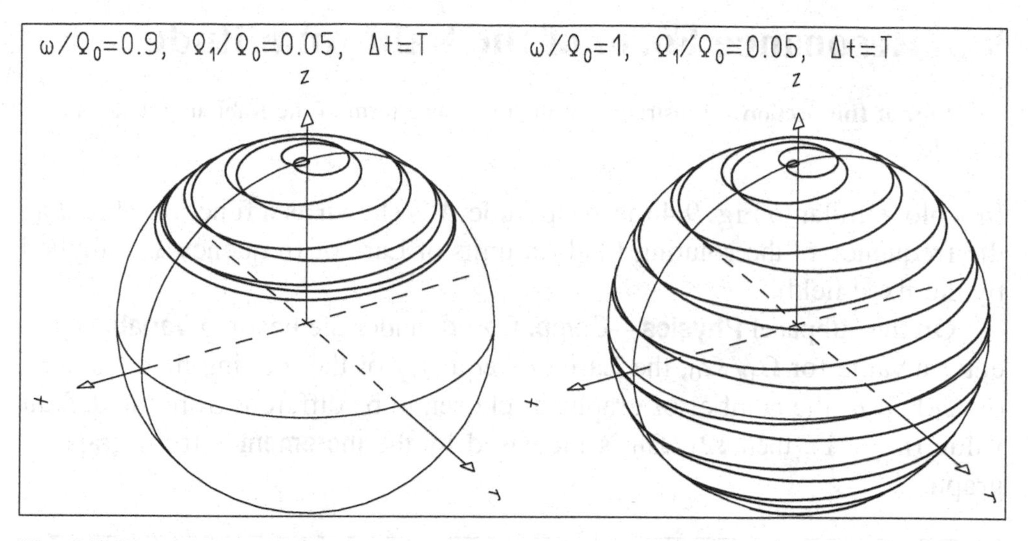

Fig. 9.2. Plot produced with descriptor `Spin in fixed frame, omega incremented` on file `Magnetic_Resonance.des`.

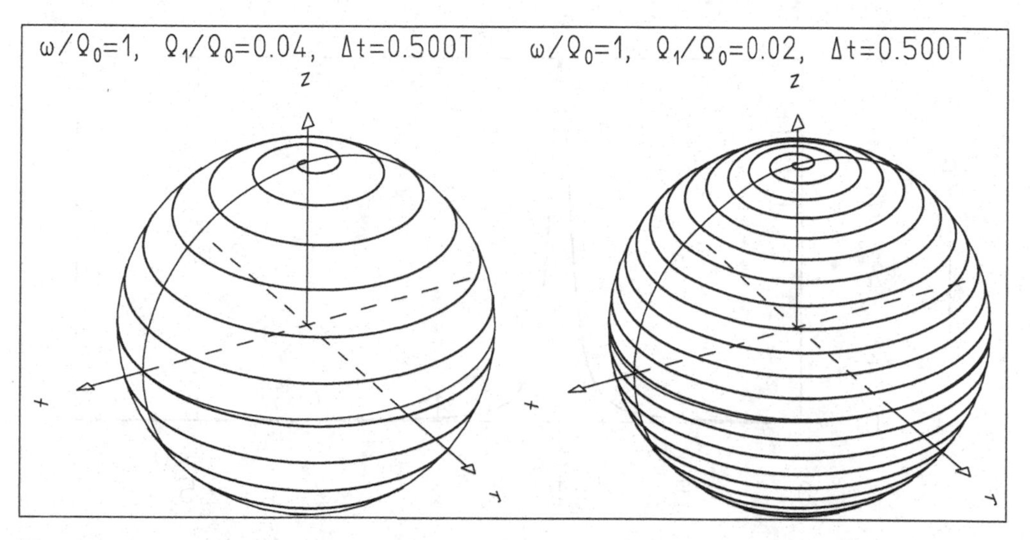

Fig. 9.3. Plot produced with descriptor `Spin in fixed frame, Omega_1 decremented` on file `Magnetic_Resonance.des`.

Movie Capability: Direct. The trajectory on the unit sphere is seen to develop from $t = 0$ to $t = fT$.

Example Descriptors on File `Magnetic_Resonance.des`

- `Spin in rotating and in fixed frame, off resonance` (see Fig. 9.1)
- `Spin in fixed frame, omega incremented` (see Fig. 9.2)
- `Spin in fixed frame, Omega_1 decremented` (see Fig. 9.3)

9.3 Resonance Form of the Rabi Amplitude

Aim of this section: Illustration of the resonance form of the Rabi amplitude A, (9.4).

In a plot similar to Fig. 9.4 the Amplitude A is shown as a function of ω/Ω_0, the frequency of the rotating field (in units of Larmor frequency Ω_0 implied by the fixed field).

On the subpanel **Physics—Comp. Coord.** under the heading **Variables** you enter a value for Ω_1/Ω_0, the Larmor frequency of the rotating field (in units of Ω_0). If n, the number of graphs, is chosen to be different from the default value ($n = 1$), then Ω_1/Ω_0 is increased by the **Increment** δ from graph to graph.

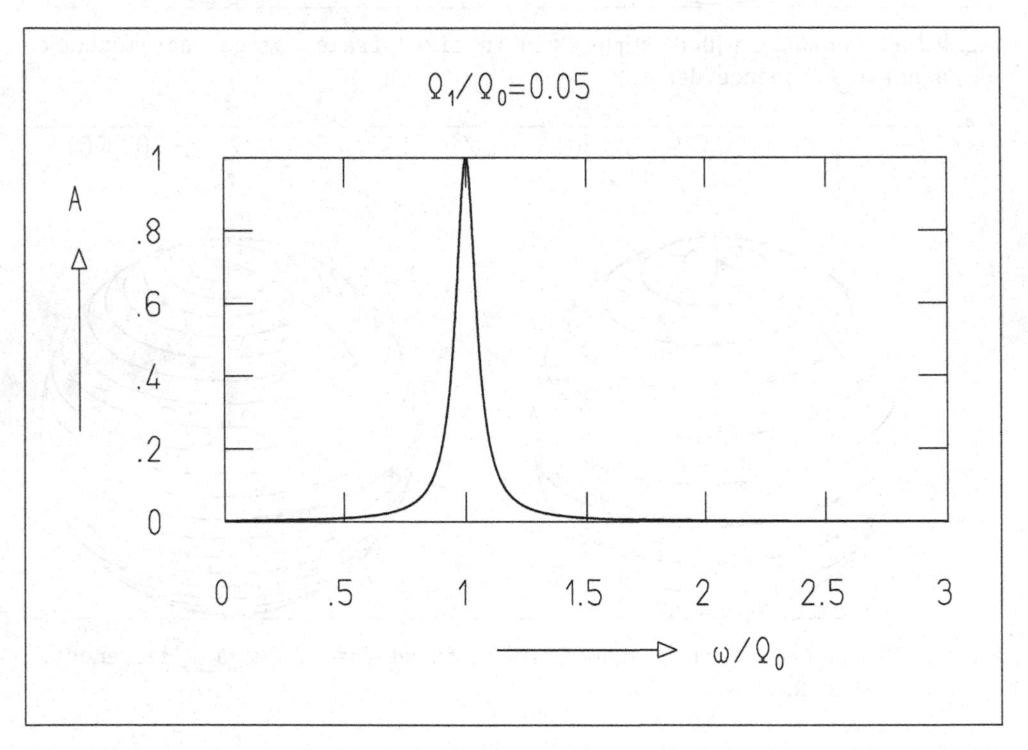

Fig. 9.4. Plot produced with descriptor `Resonance form of the Rabi amplitude` on file `Magnetic_Resonance.des`

Example Descriptor on File `Magnetic_Resonance.des`

- `Resonance form of the Rabi amplitude` (see Fig. 9.4)

9.4 Exercises

Please note:

(i) You may watch a demonstration of the material of this chapter by pressing the button **Run Demo** in the *main toolbar* and selecting one of the demo files `Magnetic_Resonance`.

(ii) More example descriptors can be found on the descriptor file `MagRes-(FE).des` in the directory `FurtherExamples`.

(iii) For the following exercises use descriptor file `Magnetic_Resonance.des`.

9.2.1 Plot the time dependence of the spin-expectation vector in the rotating frame for values of ω/Ω_0 near and at resonance. Start from descriptor 5: '`Exercise 9.2.1`'.

9.2.2 Plot the time dependence of the spin-expectation vector in the fixed frame for values of ω/Ω_0 near and at resonance. Start from descriptor 6: '`Exercise 9.2.2`'.

9.2.3 Plot the time dependence of the spin-expectation vector in the fixed frame at resonance the for values $\Omega_1/\Omega_0 = 0.01, 0.02, 0.03$. Use $f = t/T = 0.5$. Start from descriptor 7: '`Exercise 9.2.3`'

9.3.1 Adapt the descriptor 4: '`Resonance form of the Rabi amplitude`' so that you can produce a plot like Fig. 9.4 with several curves corresponding to several values of Ω_1/Ω_0. Discuss the result.

10. Hybridization

Contents: Construction of hybrid wave functions as normalized superpositions of stationary wave functions of the electron in the hydrogen atom with angular-momentum quantum numbers $\ell = 0, m = 0$ and $\ell = 1, m = 0$. Qualitative discussion of the hybridization model of chemical bonds.

10.1 Physical Concepts

10.1.1 Hybrid States in the Coulomb Potential

The stationary wave functions or eigenstates of an electron in the Coulomb potential of the hydrogen nucleus (Sect. 7.1.6) can be written as

$$\psi_{n,\ell,m} = R_{n,\ell} Y_{\ell,m} \quad . \tag{10.1}$$

(Denoting a stationary state by ψ cannot be confusing, since no time dependence is discussed in this chapter; we need the letter φ for angular variables.) Here n is the principal quantum number, ℓ the quantum number of angular momentum, and m that of its z component; $R_{n,\ell} = R_{n,\ell}(r)$ is the radial wave function (7.51), which depends on the radial position r, and $Y_{\ell,m} = Y_{\ell,m}(\vartheta, \varphi)$ the spherical harmonic (Sect. 11.1.4), which depends on the angles ϑ and φ. The energy eigenvalues E_n, (7.52), depend on n only. Accordingly, in general, there are several eigenstates of the same eigenvalue (which are called *degenerate*) and therefore linear combinations of these are again eigenstates.

In the following we consider exclusively the superposition of a state $\ell = 0, m = 0$ (called an *s state*) with a state $\ell = 1, m = 0$ (a *p state*) and introduce the notation

$$s_n = \psi_{n,0,0} = R_{n,0} Y_{0,0} = \frac{1}{\sqrt{4\pi}} R_{n,0} \quad , \tag{10.2}$$

$$p_n = \psi_{n,1,0} = R_{n,1} Y_{1,0} = \sqrt{\frac{3}{4\pi}} \cos\vartheta \, R_{n,1} \quad . \tag{10.3}$$

Both s_n and p_n are real functions; s_n has no angular dependence whereas p_n is proportional to $\cos\vartheta$, i.e., it is antisymmetric with respect to the equato-

S. Brandt et al., *Interactive Quantum Mechanics: Quantum Experiments on the Computer*, DOI 10.1007/978-1-4419-7424-2_10, © Springer Science+Business Media, LLC 2011

rial plane $\vartheta = \pi/2$. There is no φ dependence. Because of the orthonormality of the $R_{n,\ell}$ and the $Y_{\ell,m}$ the states s_n and p_n are orthonormal, i.e.,

$$\int |s_n|^2 \, dV = 1 \quad , \qquad \int |p_n|^2 \, dV = 1 \quad , \qquad \int s_n p_n \, dV = 0 \quad ,$$

where integration is performed over all space.

The normalized linear combination of s_n and p_n,

$$h_n = \frac{1}{\sqrt{1+\lambda^2}}(s_n + \lambda p_n) \quad , \tag{10.4}$$

is called a *hybrid state* with the *hybridization parameter* λ. Figure 10.3 shows, in a plane containing the z axis, the functions s_2, p_2, h_2, and $|h_2|^2$ for a particular value of λ. Due to the different symmetry properties of s_2 and p_2 the hybrid h_2 is neither symmetric nor antisymmetric with respect to $\vartheta = \pi/2$. The square $|h_2|^2$, which is a probability density, extends much farther along the negative z axis than along the positive one. This feature of hybrid states, namely, to possess a single preferred direction, is used in some theories of chemical bonds: An electron in a hybrid state in one atom reaches out to its partner in another atom along that preferred direction and vice versa. The hybrid state is rotationally symmetric around the z axis, which we call the *orientation axis*. We denote it by the unit vector $\hat{\mathbf{a}}$; here $\hat{\mathbf{a}} = \mathbf{e}_z$. Since the choice of the z axis was arbitrary, the orientation axis can be given any direction. The direction $\hat{\mathbf{a}}$ is then characterized by a polar angle ϑ_a and an azimuthal angle φ_a with respect to the x, y, z coordinate frame.

We write the general form of a hybrid state as

$$h_n(\lambda; \vartheta_a, \varphi_a) = \frac{1}{\sqrt{1+\lambda^2}}(s_n + \lambda p_n(\vartheta_a, \varphi_a)) \quad , \tag{10.5}$$

where $p_n(\vartheta_a, \varphi_a)$ is a p state of principal quantum number n with the symmetry axis

$$\hat{\mathbf{a}} = \sin\vartheta_a \cos\varphi_a \, \mathbf{e}_x + \sin\vartheta_a \sin\varphi_a \, \mathbf{e}_y + \cos\vartheta_a \, \mathbf{e}_z \quad . \tag{10.6}$$

In (10.3), where the symmetry axis is the z axis, the angular dependence shows up in the form of $\cos\vartheta$ or, written as scalar product, as $\hat{\mathbf{r}} \cdot \mathbf{e}_z$. Its generalization reads $\hat{\mathbf{r}} \cdot \hat{\mathbf{a}}$, from which the general state $p_n(\vartheta_a, \varphi_a)$ in (10.5) can be written as

$$p_n(\vartheta_a, \varphi_a) = \sqrt{\frac{3}{4\pi}}(\hat{\mathbf{r}} \cdot \hat{\mathbf{a}}) R_{n,1} \quad . \tag{10.7}$$

All atoms other than hydrogen have more than one electron. Quantitative discussion and computation of their properties is outside the scope of this book. Nevertheless, the study of the simple hybrid states (10.5) provides insight into hybrid bonds. In the following we summarize the assumptions made in the simplest model. In Sect. 10.1.2 we try to justify them by qualitative arguments. In Sect. 10.1.3 we shall compute the hybridization parameter and the orientations for some hybrid states in situations of particularly high symmetry.

10.1.2 Some Qualitative Details of Hybridization

An atom with atomic number Z consists of a nucleus with electric charge Ze and Z electrons, each of charge $-e$. The structure of the *Periodic Table* is explained by the *Pauli principle*, which states that there cannot be two electrons in an atom with exactly the same set of quantum numbers n, ℓ, m, s_z. For a given principal quantum number n there are $2n - 1$ different values of the angular-momentum quantum number ℓ ($\ell = -n+1, -n+2, \ldots, n-1$); for each value of ℓ there are $2\ell + 1$ different values of the quantum number m of the z component of angular momentum ($m = -\ell, -\ell + 1, \ldots, \ell$); there are two possible values, $s_z = \pm 1/2$, of the quantum number of the z component of spin for every electron. In all, for a given value of n there are $2n^2$ different sets of quantum numbers ℓ, m, s_z.

For $n = 1$ there are 2 states. One is occupied in the hydrogen atom ($Z = 1$), both are filled in the helium atom ($Z = 2$). In the case of the lithium atom ($Z = 3$) there is no free place for an electron in the inner *shell* of lowest energy. The third electron necessarily has the principal quantum number $n = 2$. Its probability density is shifted further away from the nucleus. This electron "sees" a potential similar to that of the hydrogen nucleus, because the nuclear charge $3e$ is "shielded" by the charge $-2e$ of the two electrons in the inner shell. If the shielding were perfect, all states with $n = 2$ were degenerate. Since it is not, the states with $n = 2, \ell = 0$, called $2s$ states, are of lowest energy. (For an illustration, in which we use a square-well rather than a Coulomb potential, see Fig. 10.1.) For the lithium atom, one of these $2s$ states is filled; in the beryllium atom ($Z = 4$) both are occupied. Boron ($Z = 5$), in addition to the two $2s$ electrons, has one $2p$ electron ($n = 2, \ell = 1$); carbon ($Z = 6$) has two; nitrogen ($Z = 7$) has three. In all, there are 6 different $2p$ states. They are all filled in neon ($Z = 10$), which concludes the second ($n = 2$) period of the Periodic Table. In the third period the states with $n = 3$ are filled. In particular, we note that silicon ($Z = 14$), situated directly below carbon in the Periodic Table, has in its outer shell two $3s$ and two $3p$ electrons. Its situation is that of carbon but with $n = 3$ instead of $n = 2$.

As a first qualitative example we consider a molecule of lithium hydride LiH, formed of an atom of lithium Li and one of hydrogen H. The single electron of hydrogen is in the $1s$ state. Since there is no $1p$ state, there can

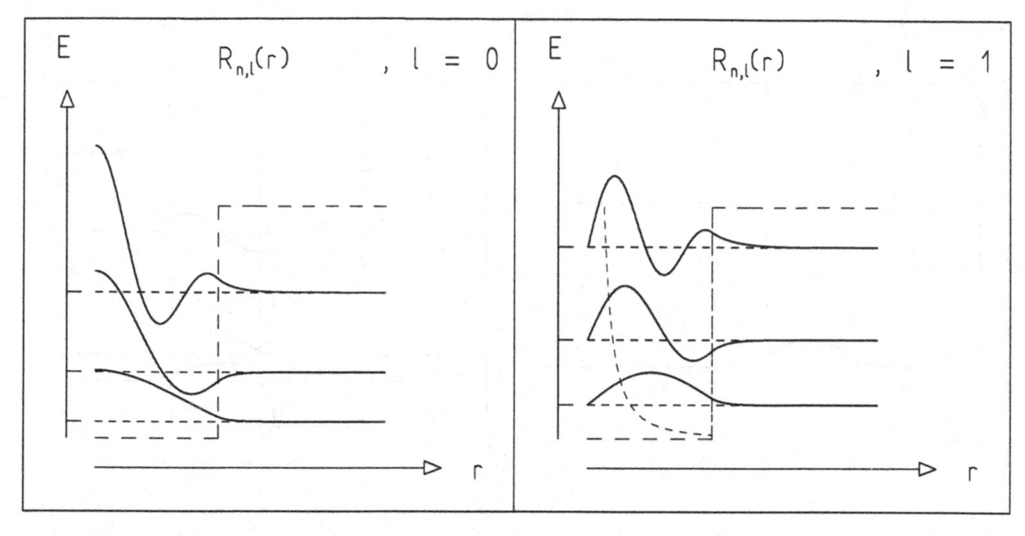

Fig. 10.1. Radial wave functions and their eigenvalues in a 3D square-well potential. The eigenvalues are systematically higher for states with $\ell = 1$ (right) compared to the corresponding states with $\ell = 0$ (left).

be no question of hybridization in that atom. The outer electron of lithium is in the $2s$ state, which is nearly degenerate with the $2p$ state, since shielding in the atom with only one outer electron is nearly perfect. The electron can be in a hybrid state formed by a superposition of the states $2s$ and $2p$. There are models in which the binding energy of the LiH molecule is computed from the overlap of this hybrid wave function with the $1s$ wave function in the hydrogen atom. The binding energy outweighs the slight extra energy of the lithium hybrid state compared to its original $2s$ state. The hybrid state is of the form (10.5); the orientation axis $\hat{\mathbf{a}}$ points along the line connecting the hydrogen nucleus with that of the lithium atom. The hybridization parameter λ depends on the details of the model used.

In our computations of hybrid wave functions we do assume perfect shielding: We simply compute the hydrogen wave functions for given quantum numbers and then superimpose them to obtain the desired hybrid wave function.

10.1.3 Hybridization Parameters and Orientations of Highly Symmetric Hybrid States

In this section we concentrate on the atoms carbon C and silicon Si. Both have four electrons in their outer shell with principal quantum number $n = 2$ for C and $n = 3$ for Si. In a solitary C or Si atom there are two s electrons and two p electrons in the outer shell. A diamond crystal is a symmetric arrangement of carbon atoms. Every atom is in the center of a tetrahedron with carbon atoms at its four corners. Because of the presence of the four neighbor atoms

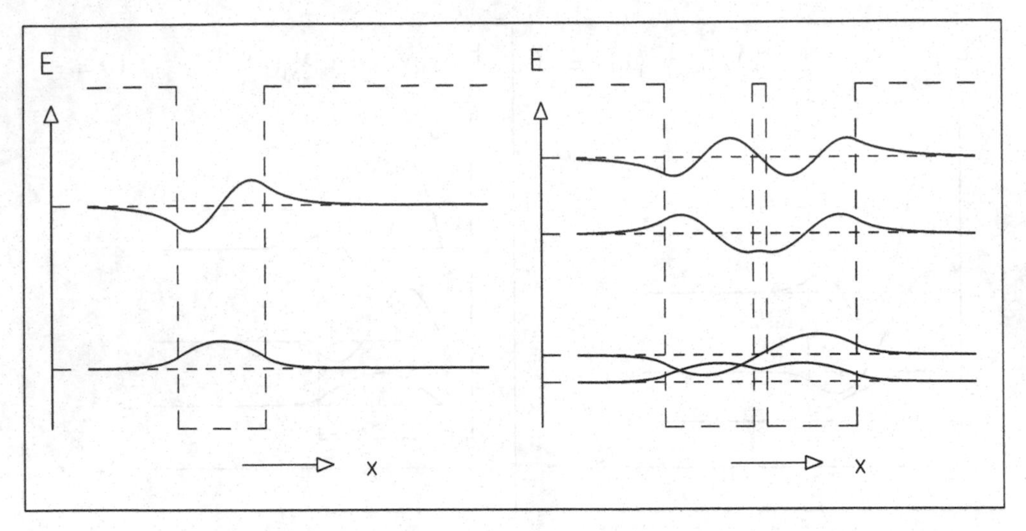

Fig. 10.2. Eigenvalues and eigenfunctions in a 1D potential. For every state in a single well (left), there are two states in the double well: one with lower, the other with higher energy than in the single well. The state of lower energy is *binding*, that of higher energy is *anti-binding*.

the electrons can assume "binding" states of lower energy, see Fig. 10.2. Part of this energy is used to promote one of the two s electrons from the s state to the p state; that is possible as long as the binding energy per atom exceeds the promotion energy. With this *promotion* there are one s electron and three p electrons in the outer shell. Other arrangements of atoms, leading to promotion, are possible. In graphite, a hexagonal planar structure, each C atom has three nearest neighbors to which it is tightly bound. In comparison, the binding between planes is rather loose. In a linear molecule like carbon dioxide CO_2 with the structure O–C–O the carbon atom is bound to two oxygen atoms which are situated in exactly opposite positions.

The states s_n and p_n are normalized and all are orthogonal to each other. We align the p_n states along the coordinate directions z, x, y, i.e., we write them in the form

$$\begin{aligned} p_{nz} &= p_n(\vartheta_a = 0) = \sqrt{\frac{3}{4\pi}} R_{n,1} \cos\vartheta \quad , \\ p_{nx} &= p_n(\vartheta_a = 90°, \varphi_a = 0) = \sqrt{\frac{3}{4\pi}} R_{n,1} \sin\vartheta \cos\varphi \quad , \\ p_{ny} &= p_n(\vartheta_a = 90°, \varphi_a = 90°) = \sqrt{\frac{3}{4\pi}} R_{n,1} \sin\vartheta \sin\varphi \quad , \end{aligned} \qquad (10.8)$$

see Fig. 10.7.

We now consider superpositions of the three p_n states (10.8) along the Cartesian coordinate axes with coefficients a_z, a_x, a_y:

$$a_z p_{nz} + a_x p_{nx} + a_y p_{ny} =$$

$$
\begin{aligned}
&= \sum_{f=x,y,z} a_f p_{nf} = \sum_{f=x,y,z} a_f \sqrt{\frac{3}{4\pi}} R_{n,1}(\hat{\mathbf{r}} \cdot \mathbf{e}_f) \\
&= \sqrt{\frac{3}{4\pi}} R_{n,1} \left(\hat{\mathbf{r}} \cdot \sum_{f=x,y,z} a_f \mathbf{e}_f \right) = \sqrt{\frac{3}{4\pi}} R_{n,1} \left(\hat{\mathbf{r}} \cdot \mathbf{a}\right) \\
&= |\mathbf{a}| \sqrt{\frac{3}{4\pi}} R_{n,1} \left(\hat{\mathbf{r}} \cdot \hat{\mathbf{a}}\right) = |\mathbf{a}| p_n(\vartheta_a, \varphi_a) \quad , \qquad (10.9)
\end{aligned}
$$

where the coefficients a_x, a_y, a_z have been interpreted as those of an (unnormalized) vector $\mathbf{a}$ defining an orientation axis (10.6) of a general p_n state (10.7). With $|\mathbf{a}| = \sqrt{a_x^2 + a_y^2 + a_z^2}$ and $|\mathbf{a}_\perp| = \sqrt{a_x^2 + a_y^2}$ the relations of these coefficients to the spherical angles ϑ_a, φ_a defining the orientation axis in (10.6) read

$$
\cos \vartheta_a = \frac{a_z}{|\mathbf{a}|} \quad , \quad \sin \vartheta_a = \frac{|\mathbf{a}_\perp|}{|\mathbf{a}|} \quad , \quad \cos \varphi_a = \frac{a_x}{|\mathbf{a}_\perp|} \quad , \quad \sin \varphi_a = \frac{a_y}{|\mathbf{a}_\perp|} \quad . \qquad (10.10)
$$

States resulting from hybridization again have to be normalized and to be orthogonal to each other (and to those states which do not hybridize). With these assumptions the following, particularly symmetric, situations are possible.

***sp* Hybridization** The s_n state and *one* of the p_n states form two hybrid states. We assume that p_n state to be oriented in z. The other p_n states stay unchanged. We consider the most symmetric case, in which the two hybrids are identical in form but oriented back to back. This is a picture for the bonds (as far as they are due to hybrids of the carbon atom) in the carbon dioxide molecule CO_2.

We first construct two p_n states oriented in z and in $-z$ direction, i.e., with $\vartheta_a = 0$ and $\vartheta_a = 180°$, respectively. (We assign $\varphi_a = 0$ to these states even though at the given values of ϑ_a the angle φ_a need not be specified.) From (10.9) we get

$$
\begin{aligned}
p_{n1} &= p_n(\vartheta_a = 0, \varphi_a = 0) = p_{nz} \quad , \\
p_{n2} &= p_n(\vartheta_a = 180°, \varphi_a = 0) = -p_{nz} \quad .
\end{aligned}
$$

Using these and the s_n state we form the two orthonormal superpositions

$$
h_1 = \frac{1}{\sqrt{2}}(s_n + p_{nz}) \quad , \qquad h_2 = \frac{1}{\sqrt{2}}(s_n - p_{nz}) \quad .
$$

Written in the general form (10.5), these hybrids are determined by the following hybridization parameters and orientation axes:

$$\lambda_1 = 1 \quad , \quad \vartheta_{a1} = 0 \quad , \quad \varphi_{a1} = 0 \quad ,$$
$$\lambda_2 = 1 \quad , \quad \vartheta_{a2} = 180° \quad , \quad \varphi_{a2} = 0 \quad .$$

The hybrids were constructed such that the term $p_n(\vartheta_a, \varphi_a)$, in turn, is oriented towards each of the neighboring atoms and that the hybridization parameter λ ensures orthonormalization. In the present case and the two cases discussed below, because of the high symmetry, λ has the same value for all hybrids.

sp^2 Hybridization The s_n state and *two* of the p_n states form three hybrid states. We assume these p_n states to be oriented in x and in y. Again, we consider a particularly symmetric case: The hybrid orientation axes lie in the x, y plane, each forming angles of 120° with the other two. We begin by constructing three p_n states with this orientation, again using (10.9),

$$p_{n1} = p_n(\vartheta_a = 90°, \varphi_a = 0) = p_{nx} \quad ,$$
$$p_{n2} = p_n(\vartheta_a = 90°, \varphi_a = 120°) = -\frac{1}{2}p_{nx} + \frac{\sqrt{3}}{2}p_{ny} \quad ,$$
$$p_{n3} = p_n(\vartheta_a = 90°, \varphi_a = 240°) = -\frac{1}{2}p_{nx} - \frac{\sqrt{3}}{2}p_{ny} \quad .$$

Superposition with s_n and proper normalization yield the three hybrids

$$h_1 = \frac{1}{\sqrt{3}}\left(s_n + \sqrt{2}p_{nx}\right) \quad ,$$
$$h_2 = \frac{1}{\sqrt{3}}\left(s_n - \frac{1}{2}\sqrt{2}p_{nx} + \frac{1}{2}\sqrt{6}p_{ny}\right) \quad ,$$
$$h_2 = \frac{1}{\sqrt{3}}\left(s_n - \frac{1}{2}\sqrt{2}p_{nx} - \frac{1}{2}\sqrt{6}p_{ny}\right) \quad .$$

They are determined by the parameters

$$\lambda_1 = \sqrt{2} \quad , \quad \vartheta_{a1} = 90° \quad , \quad \varphi_{a1} = 0 \quad ,$$
$$\lambda_2 = \sqrt{2} \quad , \quad \vartheta_{a2} = 90° \quad , \quad \varphi_{a2} = 120° \quad ,$$
$$\lambda_3 = \sqrt{2} \quad , \quad \vartheta_{a3} = 90° \quad , \quad \varphi_{a3} = 240° \quad .$$

An example for sp^2 hybridization is graphite, which we mentioned above. The rather strong bonds within a layer are explained by these hybrids. Responsible for the weaker binding between layers is the single electron in each atom, remaining in the p_{nz} state.

sp^3 Hybridization The s_n state and *three* p_n states form four hybrid states. We assume the p_n states to be oriented in x, in y, and in z. Yet again, we consider only the most symmetric case: The hybrid orientation axes point from the center to the corners of a tetrahedron. An equivalent formulation is that they point from the coordinate origin to four out of the eight corners of a cube surrounding it. Such hybrid states serve, for instance, to explain the binding of carbon atoms in a diamond crystal and of silicon atoms in a crystal with diamond structure. Again, with the help of (10.9), we construct p_n states with the desired orientations,

$$p_{n1} = p_n(\vartheta_a = 54.74°, \varphi_a = 45°) = \frac{1}{\sqrt{3}}(p_{nx} + p_{ny} + p_{nz}) \quad ,$$

$$p_{n2} = p_n(\vartheta_a = 125.26°, \varphi_a = 135°) = \frac{1}{\sqrt{3}}(p_{nx} - p_{ny} - p_{nz}) \quad ,$$

$$p_{n3} = p_n(\vartheta_a = 54.74°, \varphi_a = 225°) = \frac{1}{\sqrt{3}}(-p_{nx} + p_{ny} - p_{nz}) \quad ,$$

$$p_{n4} = p_n(\vartheta_a = 125.26°, \varphi_a = 315°) = \frac{1}{\sqrt{3}}(-p_{nx} - p_{ny} + p_{nz}) \quad .$$

Superposition with the s_n state yields the hybrids

$$h_1 = \frac{1}{2}(s_n + p_{nx} + p_{ny} + p_{nz}) \quad , \qquad h_2 = \frac{1}{2}(s_n + p_{nx} - p_{ny} - p_{nz}) \quad ,$$
$$h_3 = \frac{1}{2}(s_n - p_{nx} + p_{ny} - p_{nz}) \quad , \qquad h_4 = \frac{1}{2}(s_n - p_{nx} - p_{ny} + p_{nz})$$

with the parameters

$$\begin{array}{lll} \lambda_1 = \sqrt{3} \quad , & \vartheta_{a1} = 54.74° \quad , & \varphi_{a1} = 45° \quad , \\ \lambda_2 = \sqrt{3} \quad , & \vartheta_{a2} = 125.26° \quad , & \varphi_{a2} = 135° \quad , \\ \lambda_3 = \sqrt{3} \quad , & \vartheta_{a3} = 54.74° \quad , & \varphi_{a3} = 225° \quad , \\ \lambda_4 = \sqrt{3} \quad , & \vartheta_{a3} = 125.26° \quad , & \varphi_{a3} = 315° \quad . \end{array}$$

10.2 Hybrid Wave Functions and Probability Densities

Aim of this section: Illustration, as surface over a polar grid, of the wave functions s_n (10.2), p_n (10.3), and h_n (10.4) as well as their absolute squares.

A plot similar to those shown in Fig. 10.3 is produced. On the subpanel **Physics—Comp. Coord.** you can select the **Function Shown** as either the wave function ψ or the probability density $\varrho = |\psi|^2$. You can switch on (or off) an **Automatic Scale**. If it is on, the range in the computing coordinate z is set automatically to extend from $z_{\mathrm{beg}} = 0$ to $z_{\mathrm{end}} = |f|_{\mathrm{max}}$. Here $|f|_{\mathrm{max}}$ is the maximum value which the absolute value of the function to be plotted ($f = \psi$ or $f = |\psi|^2$) can assume.

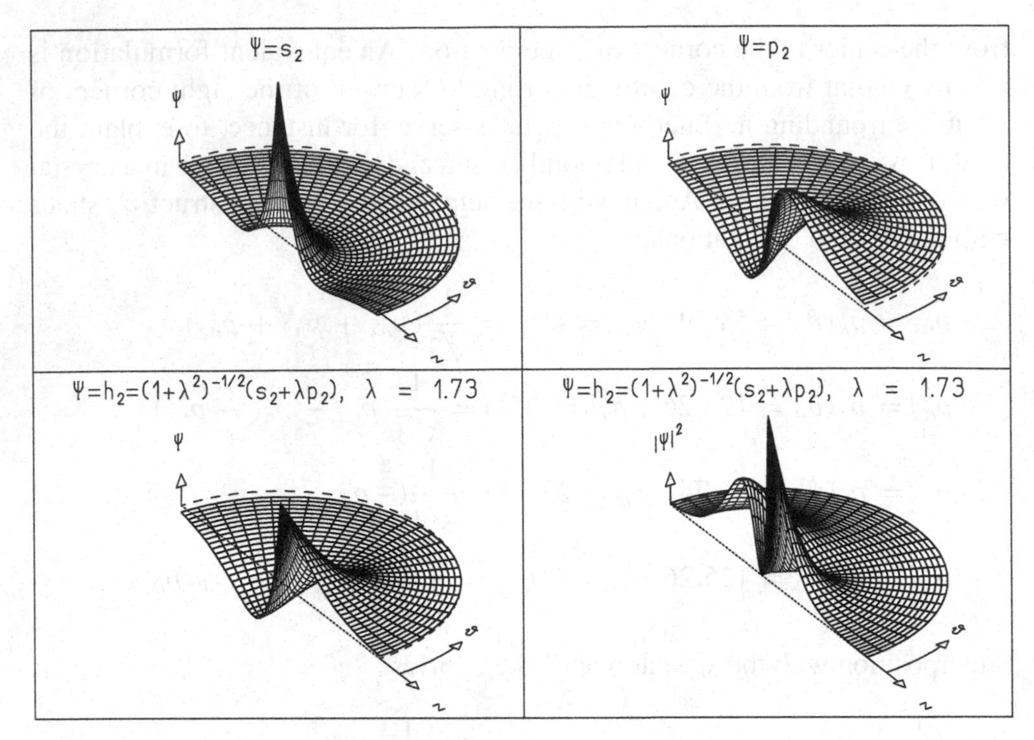

Fig. 10.3. Plot produced with descriptor `Mother descriptor showing s, p, and hybrid wave functions` on file `Hybrid_States.des`

The subpanel **Physics—Hybrid** allows you

- to enter the **Principal Quantum Number** n,
- to indicate the **State Chosen** as either s, or p, or hybrid,
- to enter the **Hybridization Parameter** λ.

Example Descriptors on File `Hybrid_States.des`

- `Mother descriptor showing s, p, and hybrid wave functions` (see Fig. 10.3)
- `Wave function s2` (see Fig. 10.3, top left)
- `Wave function p2` (see Fig. 10.3, top right)
- `sp3 hybrid wave function h2` (see Fig. 10.3, bottom left)
- `Square of sp3 hybrid wave function h2` (see Fig. 10.3, bottom right)

10.3 Contour Lines of Hybrid Wave Functions and Probability Densities

Aim of this section: Illustration, as contour lines in the x, y plane, of the wave functions s_n (10.2), p_n (10.3), and h_n (10.4) as well as their absolute squares.

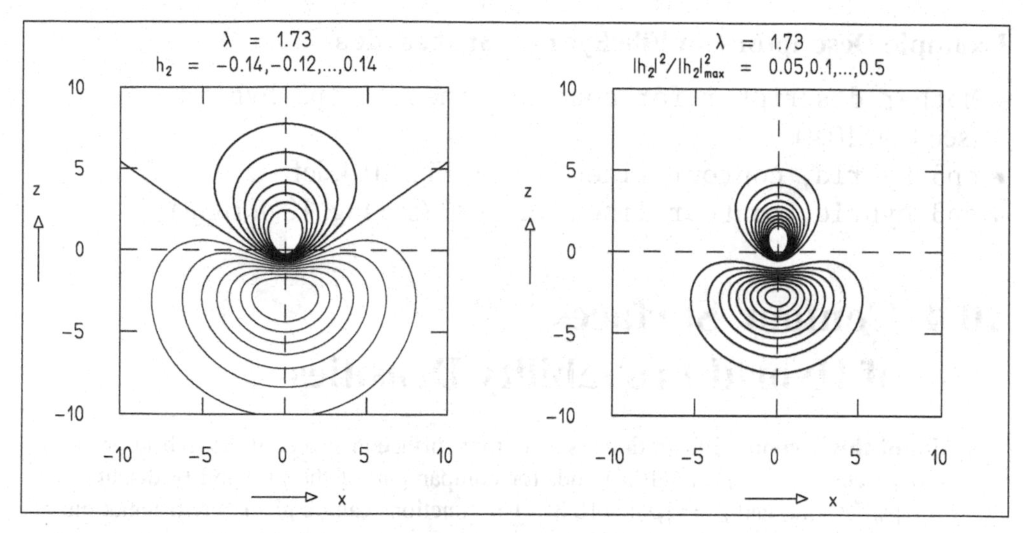

Fig. 10.4. Plot produced with descriptor `Mother descriptor for contour lines of sp3 hybrid` on file `Hybrid_States.des`

A plot similar to either of those shown in Fig. 10.4 is produced. Contour lines are lines of constant function value. On the subpanel **Physics—Comp. Coord.** you can enter a **Function value** f and an **Increment** Δf. You may then either choose to plot n lines for

$$f, f + \Delta f, \ldots, f + (n-1)\Delta f$$

or $2n+1$ lines for

$$f - n\Delta f, \ldots, f, \ldots, f + n\Delta f \quad .$$

(The number of lines is specified in the subpanel **Graphics—Accuracy**.) If the latter possibility is chosen, lines $f < 0$, $f = 0$, and $f > 0$ are shown in different colors. Finally, you can choose the **Contour Lines to Correspond to Constant Values of**

- ψ (the wave function),
- $|\psi|^2$ (the probability density),
- $\psi/|\psi|_{\mathrm{max}}$,
- $|\psi|^2/|\psi|^2_{\mathrm{max}}$.

Here $|\psi|_{\mathrm{max}}$ and $|\psi|^2_{\mathrm{max}}$ are the maximum values the functions $|\psi|$ and $|\psi|^2$ can take, respectively.

The subpanel **Physics—Hybrid** allows you

- to enter the **Principal Quantum Number** n,
- to indicate the **State Chosen** as either s, or p, or hybrid,
- to enter the **Hybridization Parameter** λ.

Example Descriptors on File `Hybrid_States.des`

- `Mother descriptor for contour lines of sp3 hybrid` (see Fig. 10.4)
- `sp3 hybrid, contour lines h2` (see Fig. 10.4, left)
- `sp3 hybrid, contour lines |h2|**2` (see Fig. 10.4, right)

10.4 Contour Surfaces of Hybrid Probability Densities

Aim of this section: Illustration, as a contour surface in space, of the hybrid probability density $\varrho = |h_n|^2$ (10.4) and, for comparison, of the probability densities $\varrho = |s_n|^2$ (10.2) and $\varrho = |p_n|^2$ (10.3). The functions can be given any orientation in space.

A plot similar to those shown in Figs. 10.5 and 10.6 is produced. Shown are surfaces of constant probability density ϱ. On the subpanel **Physics—Comp. Coord.** you indicate the **Parameter Held Constant**. This is either ϱ itself or, often simpler to determine, $\varrho/\varrho_{\max}$, where $\varrho_{\max}$ is the maximum which the probability density can take. Under **Parameter Value** the numerical value of that parameter is entered. You can also determine the **Orientation of the Symmetry Axis** of a p state or a hybrid state by entering values (in degrees) for the **Polar Angle** ϑ_a and the **Azimuthal Angle** φ_a of that axis.

The subpanel **Physics—Hybrid** allows you

- to enter the **Principal Quantum Number** n,
- to indicate the **State Chosen** as either s, or p, or hybrid,

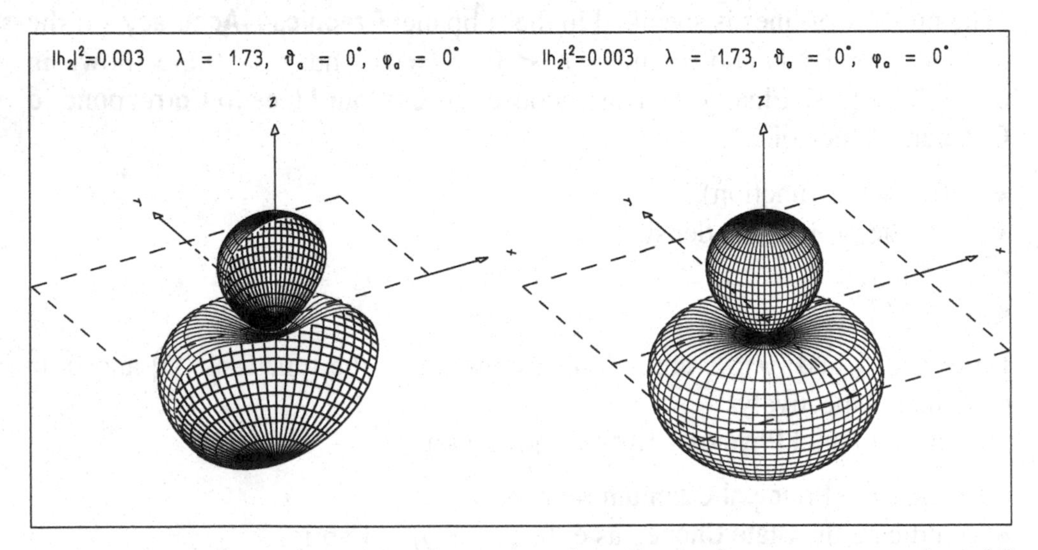

Fig. 10.5. Plot produced with descriptor `Mother descriptor for contour surface h2 (open and closed)` on file `Hybrid_States.des`

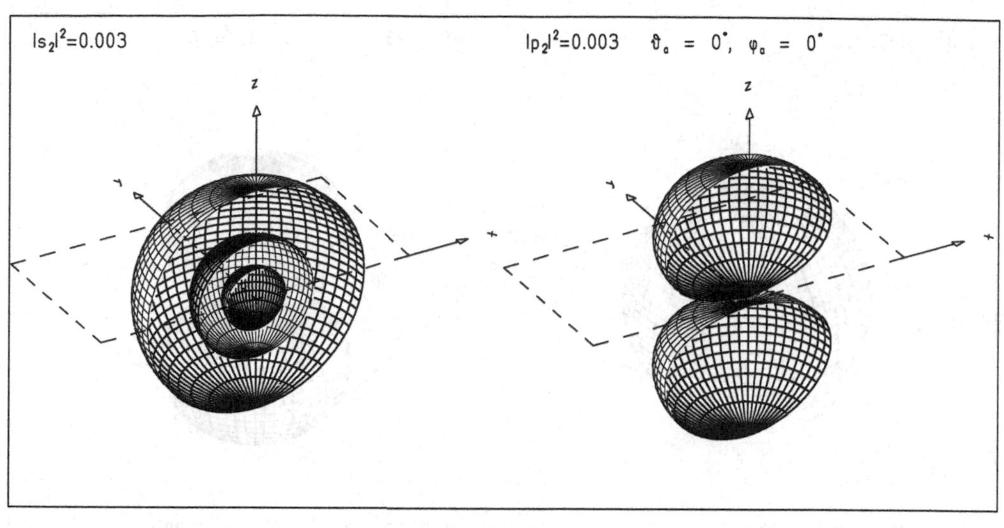

Fig. 10.6. Plot produced with descriptor `Mother descriptor for contour surfaces s2 and p2, cut open` on file `Hybrid_States.des`

- to enter the **Hybridization Parameter** λ.

Movie Capability: Direct. The value of ϱ (or $\varrho/\varrho_{\max}$) is stepped up by an increment from frame to frame. On the subpanel **Physics—Movie** (see Sect. A.4) you find the increment $\Delta\varrho$ (or $\Delta(\varrho/\varrho_{\max})$). (Attention: The creation of the movie (and also its saving in animated GIF format) will take quite some time. You may want to let the computer work on it while you do something else.)

Example Descriptors on File `Hybrid_States.des`

- `Mother descriptor for contour surface h2 (open and closed)` (see Fig. 10.5)
- `Contour surface for sp3 hybrid, theta=0, phi=0, cut open` (see Fig. 10.5, left)
- `Contour surface for sp3 hybrid, theta=0` (see Fig. 10.5, right)
- `Mother descriptor for contour surfaces s2 and p2, cut open` (see Fig. 10.6)
- `Contour surface of s2, cut open` (see Fig. 10.6, left)
- `Contour surface of p2, cut open` (see Fig. 10.6, right)
- `Mother descriptor for contour surfaces s2, p2(z), p2(x), p2(y)` (see Fig. 10.7)
- `Contour surface of s2` (see Fig. 10.7, top left)
- `Contour surface of p2` (see Fig. 10.7, top right)
- `Contour surface of p2, theta=90, phi=0` (see Fig. 10.7, bottom left)
- `Contour surface of p2, theta=90, phi=90` (see Fig. 10.7, bottom right)

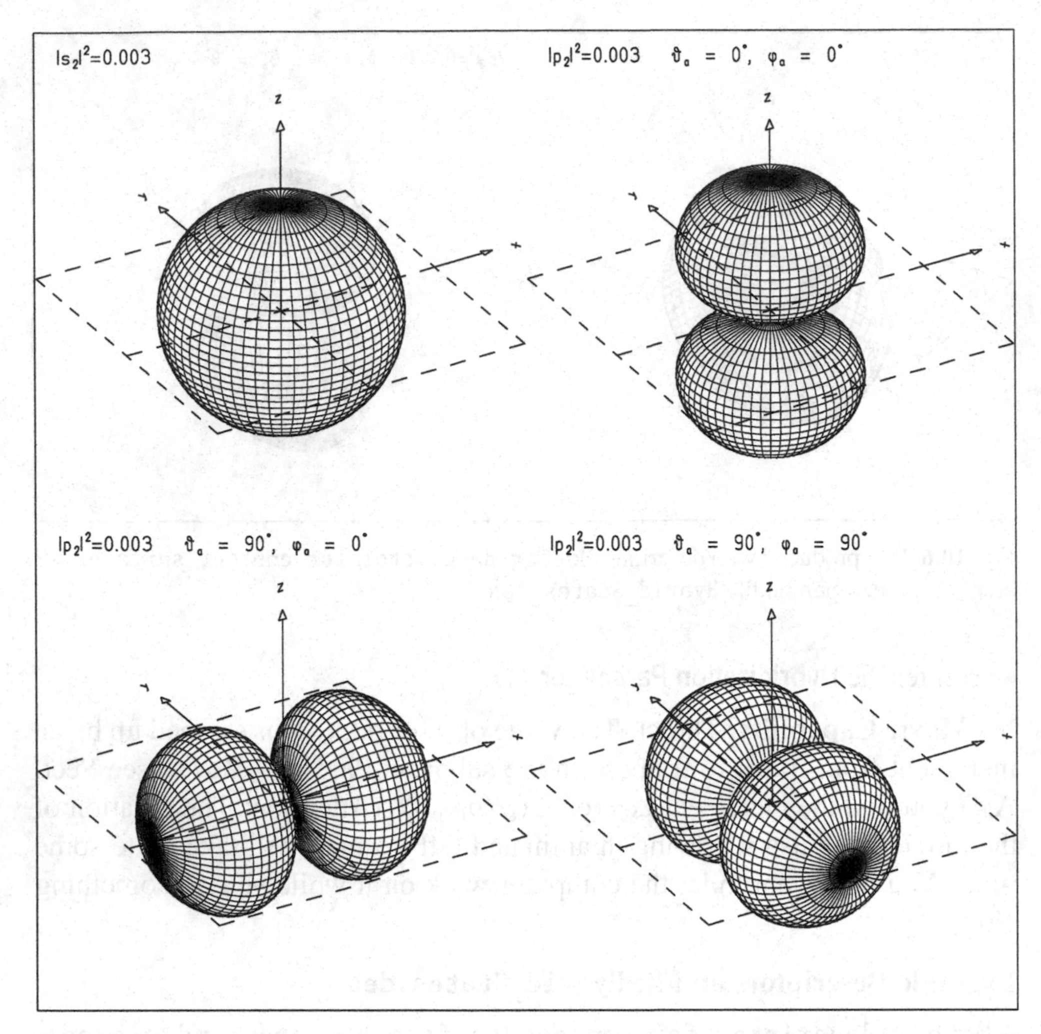

Fig. 10.7. Plot produced with descriptor `Mother descriptor for contour surfaces s2, p2(z), p2(x), p2(y)` on file `Hybrid_States.des`

10.5 Exercises

Please note:

(i) You may watch a demonstration of the material of this chapter by pressing the button **Run Demo** in the *main toolbar* and selecting one of the demo files `Hybrids`.

(ii) More example descriptors can be found on the descriptor file `Hybrids-(FE).des` in the directory `FurtherExamples`.

(iii) For the following exercises use descriptor file `Hybrid_States.des`.

10.2.1 Figure 10.3 shows the functions h_2 and $|h_2|^2$ for an sp^3 hybrid, i.e., for the value $\lambda = \sqrt{3} \approx 1.73$ of the hybridization parameter. Produce such plots for sp and sp^2 hybrids.

10.2.2 Plot the functions as shown in Fig. 10.3 for the principal quantum number $n = 3$. Start out from the descriptors defining the constituent plots of Fig. 10.3 and remember to adapt the computing coordinates.

10.3.1 Figure 10.4 shows contour plots of the functions h_2 and $|h_2|^2$ for an sp^3 hybrid, i.e., for the value $\lambda = \sqrt{3} \approx 1.73$ of the hybridization parameter. Produce such plots for sp and sp^2 hybrids. Which of the three has the largest extension in the direction $-z$?

10.3.2 Produce contour-line plots as shown in Fig. 10.4 for the principal quantum number $n = 3$. Start out from the descriptors defining the constituent plots of Fig. 10.4 and remember to adapt the computing coordinates.

10.4.1 The right-hand part of Fig. 10.5 shows the contour surface of an sp^3 hybrid for $n = 2$, but oriented along the z direction, i.e., for $\vartheta_a = 0$, $\varphi_a = 0$. Produce such plots for the properly oriented sp, sp^2, and sp^3 hybrids as listed in Sect. 10.1.3.

10.4.2 Produce contour-surface plots as shown in Fig. 10.5 for the principal quantum number $n = 3$. Start out from the descriptors defining the constituent plots of Fig. 10.5 and remember to adapt the computing coordinates.

11. Special Functions of Mathematical Physics

Contents: Discussion of the most important formulae and construction of plots for some functions of mathematical physics relevant to quantum mechanics. These functions are Hermite polynomials, Legendre polynomials and Legendre functions, spherical harmonics, Bessel functions and spherical Bessel functions, Airy functions, and Laguerre polynomials. Directly related to some of these and also discussed are the eigenfunctions of the one-dimensional harmonic oscillator and the radial eigenfunctions of the harmonic oscillator in three dimensions and of the hydrogen atom. From statistics the Gaussian distribution of one and two variables and the error function as well as the binomial and Poisson distributions are described.

11.1 Basic Formulae

11.1.1 Hermite Polynomials

The *Hermite polynomials* are solutions of the differential equation

$$\frac{\mathrm{d}^2 H_n}{\mathrm{d}x^2} - 2x\frac{\mathrm{d}H_n}{\mathrm{d}x} + 2nH_n = 0 \quad , \quad n = 0, 1, \ldots \quad . \tag{11.1}$$

They can be computed from the recurrence relation

$$H_0(x) = 1 \quad , \quad H_1(x) = 2x \quad ,$$

$$H_n(x) = 2xH_{n-1}(x) - 2(n-1)H_{n-2}(x) \quad , \quad n = 2, 3, \ldots \quad , \tag{11.2}$$

or from *Rodrigues' formula*

$$H_n(x) = (-1)^n \mathrm{e}^{x^2} \frac{\mathrm{d}^n}{\mathrm{d}x^n} \mathrm{e}^{-x^2} \tag{11.3}$$

and satisfy the orthogonality relation

$$\int_{-\infty}^{\infty} H_n(x) H_m(x) \mathrm{e}^{-x^2}\,\mathrm{d}x = 0 \quad , \quad n \neq m \quad . \tag{11.4}$$

S. Brandt et al., *Interactive Quantum Mechanics: Quantum Experiments on the Computer*,
DOI 10.1007/978-1-4419-7424-2_11, © Springer Science+Business Media, LLC 2011

11.1.2 Harmonic-Oscillator Eigenfunctions

The *eigenfunctions of the one-dimensional harmonic oscillator* also known as the *Hermite–Weber functions* are

$$\varphi_n(x) = (\sqrt{\pi}2^n n!\sigma_0)^{-1/2} H_n\left(\frac{x}{\sigma_0}\right) \exp\left(-\frac{x^2}{2\sigma_0^2}\right) \quad , \quad n = 0, 1, 2, \ldots \quad , \tag{11.5}$$

with

$$\sigma_0 = \sqrt{\hbar/m\omega} \quad ,$$

where m and ω are the mass and angular frequency, respectively. The eigenfunctions are orthonormal,

$$\int_{-\infty}^{\infty} \varphi_n(x)\varphi_m(x)\,\mathrm{d}x = \delta_{nm} \quad . \tag{11.6}$$

11.1.3 Legendre Polynomials and Legendre Functions

The *Legendre polynomials* solve the differential equation

$$(1-x^2)\frac{\mathrm{d}^2 P_\ell(x)}{\mathrm{d}x^2} - 2x\frac{\mathrm{d}P_\ell(x)}{\mathrm{d}x} + \ell(\ell+1)P_\ell(x) = 0 \quad , \quad \ell = 0, 1, 2, \ldots \quad . \tag{11.7}$$

They follow from the recurrence relation

$$P_0(x) = 1 \quad , \quad P_1(x) = x \quad ,$$

$$(\ell+1)P_{\ell+1}(x) = (2\ell+1)xP_\ell(x) - \ell P_{\ell-1}(x) \quad , \quad \ell = 1, 2, \ldots \quad , \tag{11.8}$$

or from Rodrigues' formula

$$P_\ell(x) = \frac{1}{2^\ell \ell!}\frac{\mathrm{d}^\ell}{\mathrm{d}x^\ell}\left[(x^2-1)^\ell\right] \tag{11.9}$$

and satisfy the orthogonality relation

$$\int_{-1}^{1} P_\ell(x)P_n(x)\,\mathrm{d}x = 0 \quad , \quad \ell \neq n \quad . \tag{11.10}$$

The *associated Legendre functions* are solutions of the differential equation

$$(1-x^2)\frac{\mathrm{d}^2 P_\ell^m}{\mathrm{d}x^2} - 2x\frac{\mathrm{d}P_\ell^m}{\mathrm{d}x} + \left[\ell(\ell+1) - \frac{m^2}{1-x^2}\right]P_\ell^m = 0 \quad ,$$

$$\ell = 0, 1, 2, \ldots \quad , \quad m = 0, 1, \ldots, \ell \quad . \tag{11.11}$$

With $P_\ell^0(x) = P_\ell(x)$ they can be obtained from the recurrence relation

$$P_m^m(x) = (-1)^m \frac{(2m)!}{2^m m!}(1-x^2)^{m/2} \quad ,$$

$$(2\ell+1)xP_\ell^m(x) = (\ell-m+1)P_{\ell+1}^m(x) + (\ell+m)P_{\ell-1}^m(x) \tag{11.12}$$

or from

$$P_\ell^m(x) = (1-x^2)^{m/2}\frac{\mathrm{d}^m}{\mathrm{d}x^m}P_\ell(x) \quad . \tag{11.13}$$

They are orthogonal:

$$\int_{-1}^{1} P_n^m(x)P_\ell^m(x)\,\mathrm{d}x = 0 \quad , \quad \ell \neq n \quad . \tag{11.14}$$

11.1.4 Spherical Harmonics

The *spherical harmonics* solve the differential equation

$$\frac{1}{\sin\vartheta}\frac{\partial}{\partial\vartheta}\sin\vartheta\frac{\partial Y_{\ell m}}{\partial\vartheta} + \frac{1}{\sin^2\vartheta}\frac{\partial^2 Y_{\ell m}}{\partial\varphi^2} + \ell(\ell+1)Y_{\ell m} = 0 \tag{11.15}$$

and can be expressed by the associated Legendre functions,

$$Y_{\ell m}(\vartheta,\varphi) = (-1)^m\sqrt{\frac{2\ell+1}{4\pi}\frac{(\ell-m)!}{(\ell+m)!}}P_\ell^m(\cos\vartheta)\mathrm{e}^{\mathrm{i}m\varphi} \quad , \quad 0 \le m \le \ell \quad . \tag{11.16}$$

For negative m values one defines

$$Y_{\ell m} = (-1)^m Y_{\ell,|m|}^* \quad , \quad m < 0 \quad . \tag{11.17}$$

The first few spherical harmonics are

$$Y_{0,0} = \frac{1}{\sqrt{4\pi}} \quad , \qquad Y_{1,0} = \sqrt{\frac{3}{4\pi}}\cos\vartheta \quad ,$$
$$Y_{1,1} = -\sqrt{\frac{3}{8\pi}}\sin\vartheta\,\mathrm{e}^{\mathrm{i}\varphi} \quad , \qquad Y_{2,0} = \sqrt{\frac{5}{16\pi}}(3\cos^2\vartheta - 1) \quad ,$$
$$Y_{2,1} = -\sqrt{\frac{15}{8\pi}}\sin\vartheta\cos\vartheta\,\mathrm{e}^{\mathrm{i}\varphi} \quad , \qquad Y_{2,2} = \sqrt{\frac{15}{32\pi}}\sin^2\vartheta\,\mathrm{e}^{2\mathrm{i}\varphi} \quad .$$

The spherical harmonics are orthonormal,

$$\int_{-1}^{+1}\int_0^{2\pi} Y_{\ell' m'}^*(\vartheta,\varphi)Y_{\ell m}(\vartheta,\varphi)\,\mathrm{d}\cos\vartheta\,\mathrm{d}\varphi = \delta_{\ell'\ell}\delta_{m'm} \quad . \tag{11.18}$$

The absolute square of a spherical harmonic is directly proportional to the square of the associated Legendre function with the same indices, (11.16):

$$|Y_{\ell m}(\vartheta,\varphi)|^2 = \frac{2\ell+1}{4\pi}\frac{(\ell-m)!}{(\ell+m)!}[P_\ell^m(\cos\vartheta)]^2 \quad . \tag{11.19}$$

11.1.5 Bessel Functions

Bessel's differential equation

$$x^2\frac{d^2Z_\nu(x)}{dx^2}+x\frac{dZ_\nu(x)}{dx}+(x^2-\nu^2)Z_\nu(x)=0 \tag{11.20}$$

is solved by the *Bessel functions* of the first kind $J_\nu(x)$, of the second kind (also called *Neumann functions*) $N_\nu(x)$, and of the third kind (also called *Hankel functions*) $H_\nu^{(1)}(x)$ and $H_\nu^{(2)}(x)$, which are complex linear combinations of the former two. The Bessel functions of the first kind are

$$J_\nu(x)=\left(\frac{x}{2}\right)^\nu\sum_{k=0}^{\infty}\frac{(-1)^k x^{2k}}{4^k k!\Gamma(\nu+k+1)} \quad , \tag{11.21}$$

where $\Gamma(z)$ is Euler's *Gamma function*

$$\Gamma(z+1)=\int_0^\infty t^z e^{-t}\, dt \quad . \tag{11.22}$$

It fulfills the relations

$$\Gamma(1)=1 \quad , \quad \Gamma(\tfrac{1}{2})=\sqrt{\pi} \quad , \quad \Gamma(z+1)=z\Gamma(z) \quad , \tag{11.23}$$

which lead for integer arguments $n \geq 1$ to the identification

$$\Gamma(n+1)=n! \quad . \tag{11.24}$$

The Bessel functions of the second kind are

$$N_\nu(x)=\frac{1}{\sin\nu\pi}[J_\nu(x)\cos\nu\pi-J_{-\nu}(x)] \quad . \tag{11.25}$$

For integer $\nu = n$

$$J_{-n}(x)=(-1)^n J_n(x) \tag{11.26}$$

and $N_n(x)$ has to be determined from (11.25) by the limit $\nu \to n$. The *modified Bessel functions* are defined as

$$I_\nu(x)=e^{-i\pi\nu/2}J_\nu(e^{i\pi/2}x) \quad , \quad -\pi < \arg x \leq \pi/2 \quad . \tag{11.27}$$

The Hankel functions are defined by

$$H_\nu^{(1)}(x)=J_\nu(x)+iN_\nu(x) \quad , \tag{11.28}$$

$$H_\nu^{(2)}(x)=J_\nu(x)-iN_\nu(x) \quad . \tag{11.29}$$

11.1.6 Spherical Bessel Functions

The differential equation

$$x^2 \frac{\mathrm{d}^2 z_\ell(x)}{\mathrm{d}x^2} + 2x \frac{\mathrm{d}z_\ell(x)}{\mathrm{d}x} + \left[x^2 - \ell(\ell+1)\right] z_\ell(x) = 0 \tag{11.30}$$

with integer ℓ is solved by the *spherical Bessel functions* of the first kind

$$j_\ell(x) = \sqrt{\frac{\pi}{2x}} J_{\ell+1/2}(x) \quad , \tag{11.31}$$

the spherical Bessel functions of the second kind (also called *spherical Neumann functions*)

$$n_\ell(x) = -\sqrt{\frac{\pi}{2x}} N_{\ell+1/2}(x) = (-1)^\ell j_{-\ell-1}(x) \quad , \tag{11.32}$$

and the spherical Bessel functions of the third kind (also called *spherical Hankel functions* of the first and second kind)

$$h_\ell^{(+)}(x) = n_\ell(x) + \mathrm{i}\, j_\ell(x) = \mathrm{i}[j_\ell(x) - \mathrm{i}n_\ell(x)] = \mathrm{i}\sqrt{\frac{\pi}{2x}} H_{\ell+1/2}^{(1)}(x) \, , \tag{11.33}$$

$$h_\ell^{(-)}(x) = n_\ell(x) - \mathrm{i}\, j_\ell(x) = -\mathrm{i}[j_\ell(x) + \mathrm{i}n_\ell(x)] = -\mathrm{i}\sqrt{\frac{\pi}{2x}} H_{\ell+1/2}^{(2)}(x) \quad . \tag{11.34}$$

The spherical Hankel functions can be written in the form

$$h_\ell^{(\pm)}(x) = C_\ell^{\pm} \frac{\mathrm{e}^{\pm \mathrm{i}x}}{x} \quad , \tag{11.35}$$

where

$$C_\ell^{\pm} = (\mp \mathrm{i})^\ell \sum_{s=0}^{\ell} \frac{1}{2^s s!} \frac{(\ell+s)!}{(\ell-s)!} (\mp \mathrm{i}x)^{-s} \tag{11.36}$$

is a polynomial in $1/x$. Explicitly, the first few Hankel functions are

$$h_0^{(\pm)}(x) = \frac{\mathrm{e}^{\pm \mathrm{i}x}}{x} \quad , \quad h_1^{(\pm)} = \left(\mp \mathrm{i} + \frac{1}{x}\right) \frac{\mathrm{e}^{\pm \mathrm{i}x}}{x} \quad . \tag{11.37}$$

By inversion of (11.33) and (11.34) we obtain

$$j_\ell(x) = \frac{1}{2\mathrm{i}} \left[h_\ell^{(+)}(x) - h_\ell^{(-)}(x)\right] \tag{11.38}$$

and

$$n_\ell(x) = \frac{1}{2}\left[h_\ell^{(+)}(x) + h_\ell^{(-)}(x)\right] \quad , \tag{11.39}$$

of which the first few are simply

$$j_0(x) = \frac{\sin x}{x} \quad , \quad j_1(x) = \frac{\sin x}{x^2} - \frac{\cos x}{x} \quad ,$$

$$n_0(x) = \frac{\cos x}{x} \quad , \quad n_1(x) = \frac{\cos x}{x^2} + \frac{\sin x}{x} \quad .$$

The behavior of the spherical Bessel and Neumann functions for small x is

$$j_\ell(x) \to \frac{x^\ell}{(2\ell+1)!!} \quad , \quad n_\ell(x) \to (2\ell-1)!!\, x^{-(\ell+1)} \quad , \quad x \to 0 \quad , \tag{11.40}$$

with

$$(2\ell+1)!! = 1 \times 3 \times 5 \times \cdots \times (2\ell+1) \quad .$$

The $h_\ell^{(\pm)}$ for purely imaginary argument $x = \mathrm{i}\eta$ can be expressed as

$$h_\ell^{(\pm)}(\mathrm{i}\eta) = (\mp\mathrm{i})^{\ell\pm1} \sum_{s=0}^{\ell} \frac{1}{2^s s!} \frac{(\ell+s)!}{(\ell+s)!} (\pm\eta)^{-s} \frac{\mathrm{e}^{\mp\eta}}{\eta} \quad . \tag{11.41}$$

Thus $\mathrm{i}^{\ell+1} h_\ell^{(+)}(\mathrm{i}\eta)$ is a real function of η. Its asymptotic behavior for large η is

$$\mathrm{i}^{\ell+1} h_\ell^{(+)}(\mathrm{i}\eta) \to \frac{\mathrm{e}^{-\eta}}{\eta} \quad , \quad \eta \to \infty \quad . \tag{11.42}$$

Introducing the result (11.41) into (11.38) and (11.39) we get for the spherical Bessel and Neumann functions expressions which can be made explicitly real by appropriate powers of the imaginary unit:

$$(-\mathrm{i})^\ell j_\ell(\mathrm{i}\eta) \quad , \quad \mathrm{i}^{\ell+1} n_\ell(\mathrm{i}\eta) \quad .$$

11.1.7 Airy Functions

Closely related to the Bessel functions are the *Airy functions* $\mathrm{Ai}(x)$ and $\mathrm{Bi}(x)$. They are solutions of the differential equation

$$\left(\frac{\mathrm{d}^2}{\mathrm{d}x^2} - x\right) f(x) = 0$$

and are given by

$$\mathrm{Ai}(x) = \begin{cases} \dfrac{\sqrt{x}}{3}\left\{I_{-1/3}\left(\dfrac{2}{3}x^{3/2}\right) - I_{1/3}\left(\dfrac{2}{3}x^{3/2}\right)\right\} \quad , & x > 0 \\ \dfrac{\sqrt{|x|}}{3}\left\{J_{-1/3}\left(\dfrac{2}{3}|x|^{3/2}\right) + J_{1/3}\left(\dfrac{2}{3}|x|^{3/2}\right)\right\} \quad , & x < 0 \end{cases} \tag{11.43}$$

and

$$\mathrm{Bi}(x) = \begin{cases} \sqrt{\frac{x}{3}} \left\{ I_{-1/3}\left(\frac{2}{3}x^{3/2}\right) + I_{1/3}\left(\frac{2}{3}x^{3/2}\right) \right\} \quad , & x > 0 \\ \sqrt{\frac{|x|}{3}} \left\{ J_{-1/3}\left(\frac{2}{3}|x|^{3/2}\right) - J_{1/3}\left(\frac{2}{3}|x|^{3/2}\right) \right\} \quad , & x < 0 \end{cases} . \tag{11.44}$$

Their derivatives, denoted by $\mathrm{Ai}'(x)$ and $\mathrm{Bi}'(x)$, read

$$\mathrm{Ai}'(x) = \begin{cases} -\frac{x}{3} \left\{ I_{-2/3}\left(\frac{2}{3}x^{3/2}\right) - I_{2/3}\left(\frac{2}{3}x^{3/2}\right) \right\} \quad , & x > 0 \\ -\frac{|x|}{3} \left\{ J_{-2/3}\left(\frac{2}{3}|x|^{3/2}\right) - J_{2/3}\left(\frac{2}{3}|x|^{3/2}\right) \right\} \quad , & x < 0 \end{cases} \tag{11.45}$$

and

$$\mathrm{Bi}'(x) = \begin{cases} \frac{x}{\sqrt{3}} \left\{ I_{-2/3}\left(\frac{2}{3}x^{3/2}\right) + I_{2/3}\left(\frac{2}{3}x^{3/2}\right) \right\} \quad , & x > 0 \\ \frac{|x|}{\sqrt{3}} \left\{ J_{-2/3}\left(\frac{2}{3}|x|^{3/2}\right) + J_{2/3}\left(\frac{2}{3}|x|^{3/2}\right) \right\} \quad , & x < 0 \end{cases} . \tag{11.46}$$

11.1.8 Laguerre Polynomials

The *Laguerre polynomials* solve the differential equation

$$x\frac{\mathrm{d}^2 L_n^\alpha(x)}{\mathrm{d}x^2} + (\alpha + 1 - x)\frac{\mathrm{d}L_n^\alpha(x)}{\mathrm{d}x} + nL_n^\alpha(x) = 0 \quad . \tag{11.47}$$

They are given by the recurrence relation

$$L_0^\alpha(x) = 1 \quad , \quad L_1^\alpha = \alpha + 1 - x \quad ,$$

$$(n+1)L_{n+1}^\alpha(x) = (2n + \alpha + 1 - x)L_n^\alpha(x) - (n+\alpha)L_{n-1}^\alpha(x) \tag{11.48}$$

or by the explicit formula

$$L_n^\alpha(x) = \sum_{j=0}^{n} (-1)^j \frac{\Gamma(\alpha + n + 1)x^j}{\Gamma(n - j + 1)\Gamma(\alpha + j + 1)j!} \tag{11.49}$$

or by Rodrigues' formula

$$L_n^\alpha(x) = \frac{1}{n!}\frac{\mathrm{e}^x}{x^\alpha}\frac{\mathrm{d}^n}{\mathrm{d}x^n}(x^{n+\alpha}\mathrm{e}^{-x}) \tag{11.50}$$

and satisfy the orthogonality relation

$$\int_0^\infty L_n^\alpha(x)L_m^\alpha(x)x^\alpha \mathrm{e}^{-x}\,\mathrm{d}x = 0 \quad , \quad n \neq m \quad . \tag{11.51}$$

11.1.9 Radial Eigenfunctions of the Harmonic Oscillator

The *radial eigenfunctions of the three-dimensional harmonic oscillator* are

$$R_{n\ell}(\varrho) = N_{n\ell}\varrho^{\ell} \exp\left(-\varrho^2/2\right) L_{n_\mathrm{r}}^{\ell+1/2}\left(\varrho^2\right) \tag{11.52}$$

with

$N_{n\ell} = \sqrt{(n_\mathrm{r}!2^{n_\mathrm{r}+\ell+2})/\{[2(\ell+n_\mathrm{r})+1]!!\sqrt{\pi}\sigma_0^3\}}$,
$(2m+1)!! = 1 \times 3 \times 5 \times \cdots \times (2m+1)$,
$\varrho = r/\sigma_0$,
r: radial distance from origin,
$\sigma_0 = \sqrt{\hbar/M\omega}$: ground-state width,
$n_\mathrm{r} = (n-\ell)/2$,
n: principal quantum number,
ℓ: angular-momentum quantum number.

They are orthonormal:

$$\int_0^\infty R_{n\ell}\left(\frac{r}{\sigma_0}\right) R_{m\ell}\left(\frac{r}{\sigma_0}\right) r^2\,\mathrm{d}r = \delta_{nm} \quad . \tag{11.53}$$

11.1.10 Radial Eigenfunctions of the Hydrogen Atom

The *radial eigenfunctions of the electron in the hydrogen atom* are

$$R_{n\ell}(r) = N_{n\ell}\left(\frac{2r}{na}\right)^{\ell} \exp\left(-\frac{r}{na}\right) L_{n-\ell-1}^{2\ell+1}\left(\frac{2r}{na}\right) \tag{11.54}$$

with

$N_{n\ell} = 2\sqrt{(n-\ell-1)!/(n+\ell)!}/(a^{3/2}n^2)$,
n: principal quantum number,
ℓ: angular-momentum quantum number,
a: Bohr's radius,
r: radial distance from origin.

They are orthonormal:

$$\int_0^\infty R_{n\ell}(r) R_{m\ell}(r) r^2\,\mathrm{d}r = \delta_{nm} \quad . \tag{11.55}$$

11.1.11 Gaussian Distribution and Error Function

The *probability density* of a *Gaussian distribution* with *mean* x_0 and *width* σ is

$$f(x) = \frac{1}{\sigma\sqrt{2\pi}} \exp\left\{-\frac{(x-x_0)^2}{2\sigma^2}\right\} \quad . \tag{11.56}$$

Its *distribution function* or *cumulative probability distribution* is the monotonically increasing function

$$F(x) = \int_{-\infty}^{x} f(x')\,\mathrm{d}x' \quad ; \quad F(-\infty) = 0 \quad , \quad F(\infty) = 1 \quad .$$

For the special values $x_0 = 0$, $\sigma = 1/\sqrt{2}$ the probability density takes the simple form

$$f(x) = \frac{1}{\sqrt{\pi}} \mathrm{e}^{-x^2} \quad .$$

Two integrals, closely related to the distribution function $F(x)$ of this particular probability density, are the *error function*

$$\operatorname{erf}(x) = \frac{2}{\sqrt{\pi}} \int_0^x \mathrm{e}^{-x'^2}\,\mathrm{d}x' \tag{11.57}$$

and the *complementary error function*

$$\operatorname{erfc}(x) = \frac{2}{\sqrt{\pi}} \int_x^{\infty} \mathrm{e}^{-x'^2}\,\mathrm{d}x' \quad . \tag{11.58}$$

The joint Gaussian probability density of two variables x_1 and x_2, also called the *bivariate Gaussian probability density*, is defined by

$$\rho(x_1, x_2) = A \exp\left\{-\frac{1}{2(1-c^2)}\left[\frac{(x_1 - \langle x_1\rangle)^2}{\sigma_1^2} - 2c\frac{(x_1 - \langle x_1\rangle)}{\sigma_1}\frac{(x_2 - \langle x_2\rangle)}{\sigma_2} + \frac{(x_2 - \langle x_2\rangle)^2}{\sigma_2^2}\right]\right\} \quad . \tag{11.59}$$

The normalization constant

$$A = \frac{1}{2\pi\sigma_1\sigma_2\sqrt{1-c^2}}$$

ensures that the probability density is properly normalized:

$$\int_{-\infty}^{+\infty}\int_{-\infty}^{+\infty} \rho(x_1, x_2)\,\mathrm{d}x_1\,\mathrm{d}x_2 = 1 \quad .$$

The bivariate Gaussian is completely described by five parameters. They are the *expectation values* $\langle x_1 \rangle$ and $\langle x_2 \rangle$, the *widths* σ_1 and σ_2, and the *correlation coefficient* c. The *marginal distributions* defined by

$$\rho_1(x_1) = \int_{-\infty}^{+\infty} \rho(x_1, x_2)\, \mathrm{d}x_2 \quad ,$$
$$\rho_2(x_2) = \int_{-\infty}^{+\infty} \rho(x_1, x_2)\, \mathrm{d}x_1 \tag{11.60}$$

are for the bivariate Gaussian distribution simply Gaussians of a single variable,

$$\rho_1(x_1) = \frac{1}{\sqrt{2\pi}\sigma_1} \exp\left[-\frac{(x_1 - \langle x_1 \rangle)^2}{2\sigma_1^2} \right] ,$$
$$\rho_2(x_2) = \frac{1}{\sqrt{2\pi}\sigma_2} \exp\left[-\frac{(x_2 - \langle x_2 \rangle)^2}{2\sigma_2^2} \right] . \tag{11.61}$$

Each marginal distribution depends on two parameters only, the expectation value and the width of its variable.

Lines of constant probability density in x_1, x_2 are the lines of intersection between the surface $\rho(x_1, x_2)$ and a plane $\rho = a = \text{const.}$

One particular ellipse, for which

$$\rho(x_1, x_2) = A \exp\left\{ -\frac{1}{2} \right\} \quad ,$$

i.e., the one for which the exponent in the bivariate Gaussian is simply equal to $-1/2$, is called the *covariance ellipse*. Points x_1, x_2 on the covariance ellipse fulfill the equation

$$\frac{1}{1-c^2} \left\{ \frac{(x_1 - \langle x_1 \rangle)^2}{\sigma_1^2} - 2c \frac{(x_1 - \langle x_1 \rangle)}{\sigma_1} \frac{(x_2 - \langle x_2 \rangle)}{\sigma_2} + \frac{(x_2 - \langle x_2 \rangle)^2}{\sigma_2^2} \right\} = 1 \quad . \tag{11.62}$$

Projected on the x_1 axis and the x_2 axis, it yields lines of lengths $2\sigma_1$ and $2\sigma_2$, respectively.

11.1.12 Binomial Distribution and Poisson Distribution

We consider a simple experiment, e.g., the throwing of a coin, which yields one of two possible results labeled

$$\kappa = 0 \quad , \quad \kappa = 1 \quad .$$

The probabilities to obtain these results are

$$P(\kappa = 1) = p \quad , \quad P(\kappa = 0) = 1 - p \quad .$$

If one performs n independent experiments, then the probability to obtain k times the result $\kappa = 1$ is

$$P(k) = W_k^n = \binom{n}{k} p^k (1-p)^{n-k} \quad , \quad k = 0, 1, \ldots, n \quad , \qquad (11.63)$$

with

$$\binom{n}{k} = \frac{n!}{k!(n-k)!} \quad , \quad n! = 1 \times 2 \times 3 \times \cdots \times n \quad , \quad 0! = 1! = 1 \quad .$$

The set of probabilities (11.63) is the *binomial distribution.* In the limit

$$n \to \infty \quad , \quad \lambda = np = \text{const}$$

it turns into the *Poisson distribution*

$$P(k, \lambda) = \frac{\lambda^k}{k!} \mathrm{e}^{-\lambda} \quad . \qquad (11.64)$$

Further Reading

Messiah: Vol. 1, Appendix B
Abramowitz, Stegun, Chaps. 6, 7, 8, 9, 10, 14, 22, 26

11.2 Hermite Polynomials and Related Functions

Aim of this section: Illustration of the Hermite polynomials $H_n(x)$, (11.2), and of the eigenfunctions $\varphi_n(x)$ of the one-dimensional harmonic oscillator, (11.5).

At the bottom of the subpanel **Physics—Comp. Coord.** you can select between

- **Hermite Polynomials** $H_n(x)$ and
- **Eigenfunctions of 1D Harmonic Oscillator** $\varphi_n(x)$.

On the subpanel **Physics—Variables** you find the **Index** n. If you ask for a *multiple plot*, this value of n is taken only for the first plot. It is increased successively by 1 for every further plot. If you choose the harmonic-oscillator functions, you can decide to plot either $\varphi_n(x)$ or $(\varphi_n(x))^2$.

Example Descriptors on File `Math_Functions.des`

- `Hermite polynomials` (Fig. 11.1)
- `Eigenfunctions of 1D harmonic oscillator`

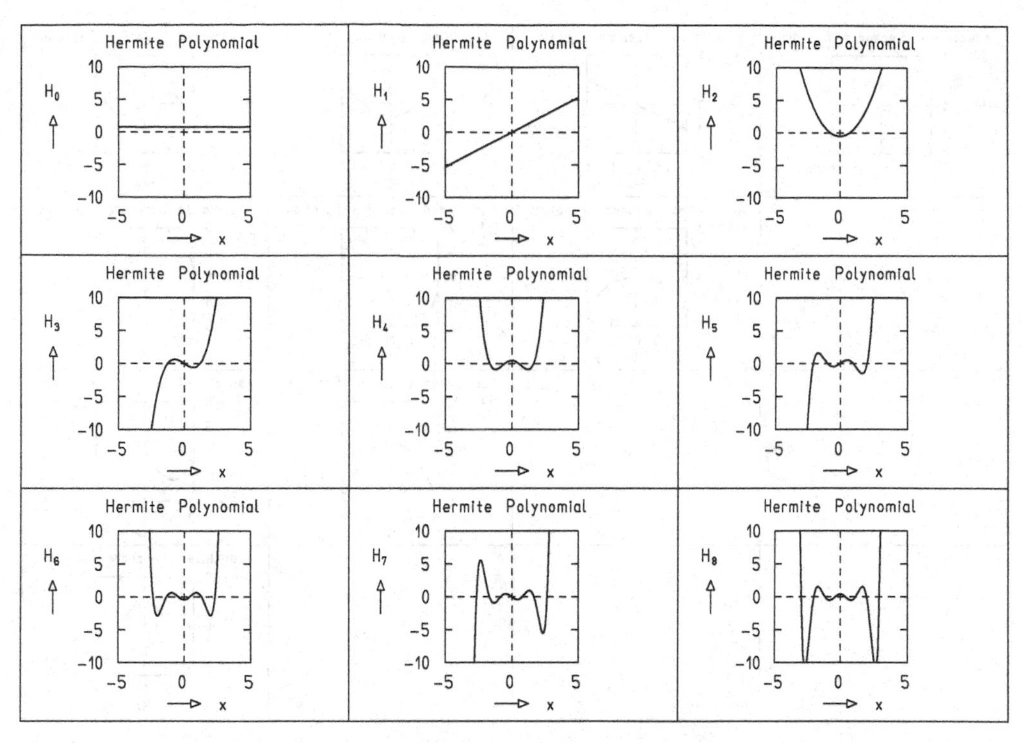

Fig. 11.1. Plot produced with descriptor `Hermite polynomials` on file `Math_Functions-.des`

11.3 Legendre Polynomials and Related Functions

Aim of this section: Illustration of the Legendre polynomials $P_\ell(x)$, (11.8), the associated Legendre functions $P_\ell^m(x)$, (11.12), and the absolute squares $|Y_{\ell m}|^2$, (11.19), of the spherical harmonics.

At the bottom of the subpanel **Physics—Comp. Coord.** you can select between

- **Legendre Polynomials** $P_\ell(x)$,
- **Associated Legendre Functions** $P_\ell^m(x)$,
- **Absolute Squares** $|Y_{\ell m}|^2$ **of the Spherical Harmonics.**

On the subpanel **Physics—Variables** you find – for the choice P_ℓ – the **Index** ℓ. For the choices P_ℓ^m and $|Y_{\ell m}|^2$ you find the **Indices** ℓ and m. In a multiple plot these values are taken for the first plot. For the choice P_ℓ the index ℓ is incremented by 1 for each further plot. For the choices P_ℓ^m and $|Y_{\ell m}|^2$ the index ℓ is increased horizontally (from column to column) and the index m vertically (from row to row).

Example Descriptors on File `Math_Functions.des`

- `Legendre polynomials`

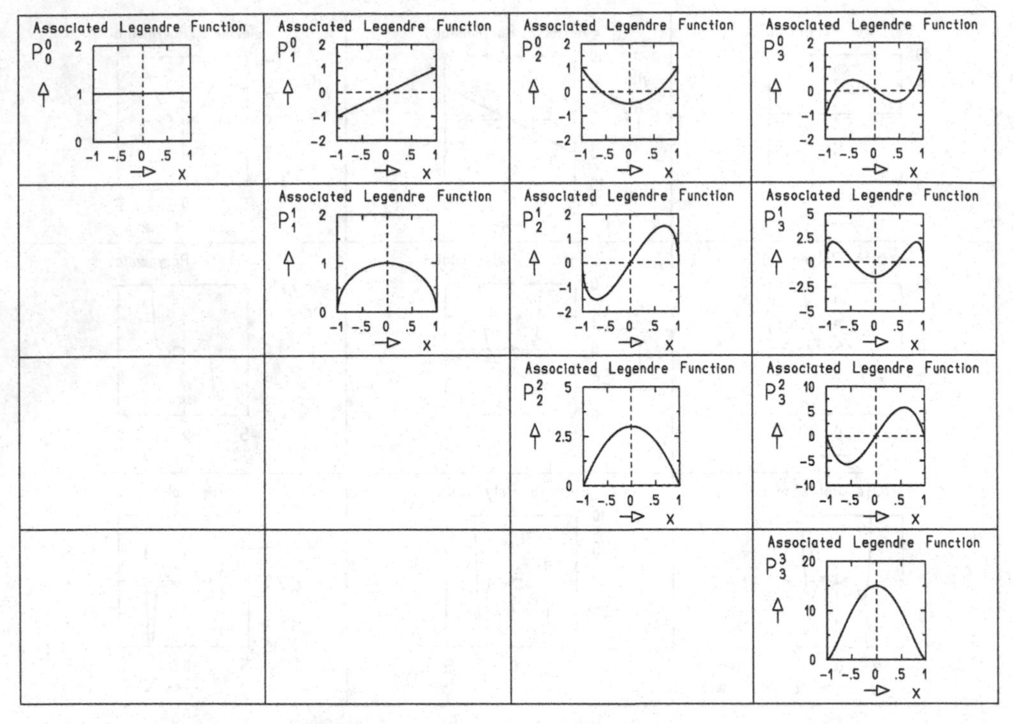

Fig. 11.2. Plot produced with descriptor `Associated Legendre functions` on file `Math_-Functions.des`

- `Associated Legendre functions` (see Fig. 11.2)
- `Absolute squares of spherical harmonics`

11.4 Spherical Harmonics: Surface over Cartesian Grid

Aim of this section: Illustration of the spherical harmonics $Y_{\ell m}(\vartheta, \varphi)$, (11.16), as surfaces over a Cartesian grid spanned by ϑ and φ.

At the bottom of the subpanel **Physics—Comp. Coord.** you can select to display one of three aspects of the function:

- the **Absolute Square** $|Y_{\ell m}(\vartheta, \varphi)|^2$,
- the **Real Part** $\mathrm{Re}\, Y_{\ell m}(\vartheta, \varphi)$, or
- the **Imaginary Part** $\mathrm{Im}\, Y_{\ell m}(\vartheta, \varphi)$.

Under the heading **Indices** you find the two indices ℓ and m.

In a *multiple plot* the index m is incremented by 1 from column to column. In the first row the selected aspect of the function is shown, in the second row the next aspect (i.e., the one following in the list above – if the list is exhausted, the first in the list is taken), and so on.

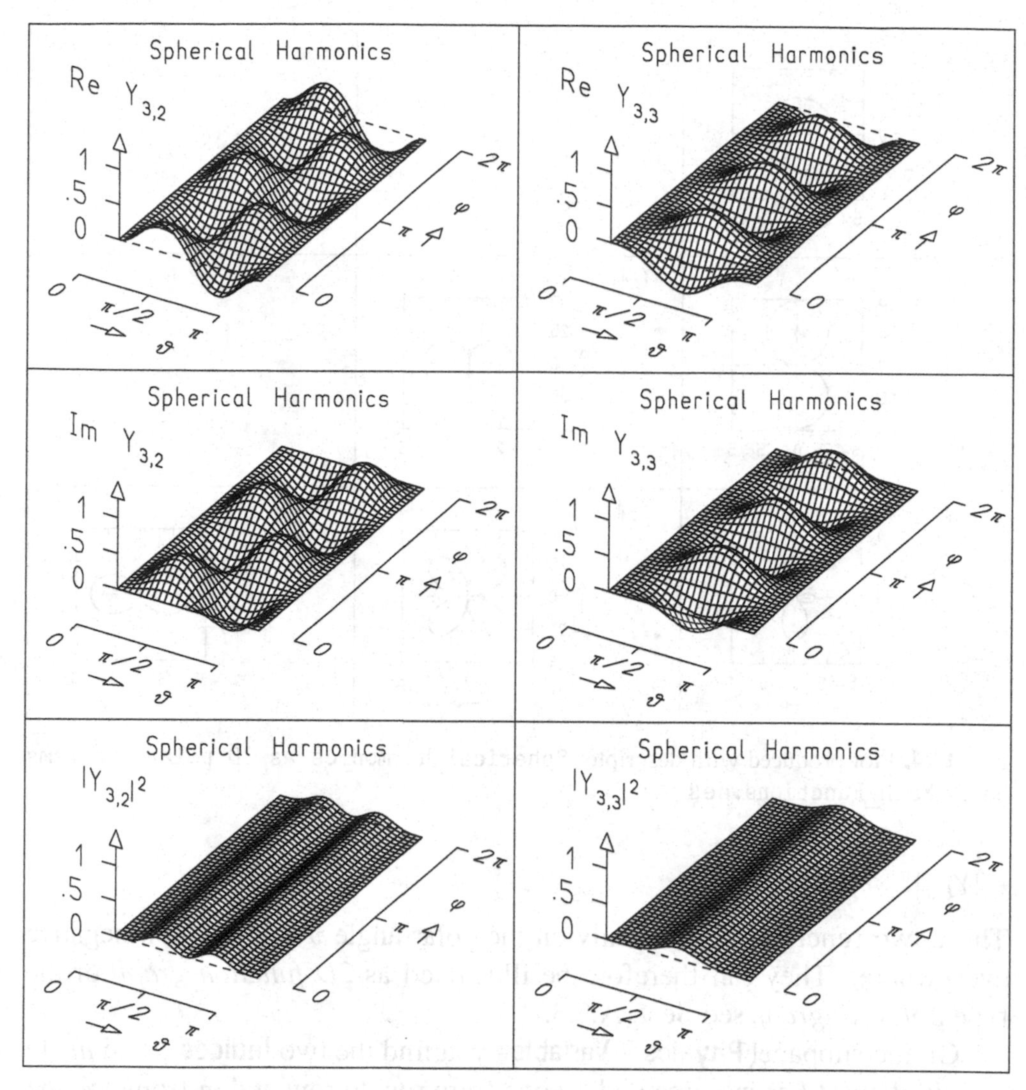

Fig. 11.3. Plot produced with descriptor `Spherical harmonics as surfaces over Cartesian grid` on file `Math_Functions.des`

Example Descriptor on File `Math_Functions.des`

- `Spherical harmonics as surfaces over Cartesian grid` (see Fig. 11.3)

11.5 Spherical Harmonics: 2D Polar Diagram

Aim of this section: Illustration of the absolute value $|Y_{\ell m}|$ or the absolute square $|Y_{\ell m}|^2$, (11.19), of the spherical harmonics as 2D polar diagram.

On the bottom of the subpanel **Physics—Comp. Coord.** you can choose to plot

- $|Y_{\ell m}|^2$ or

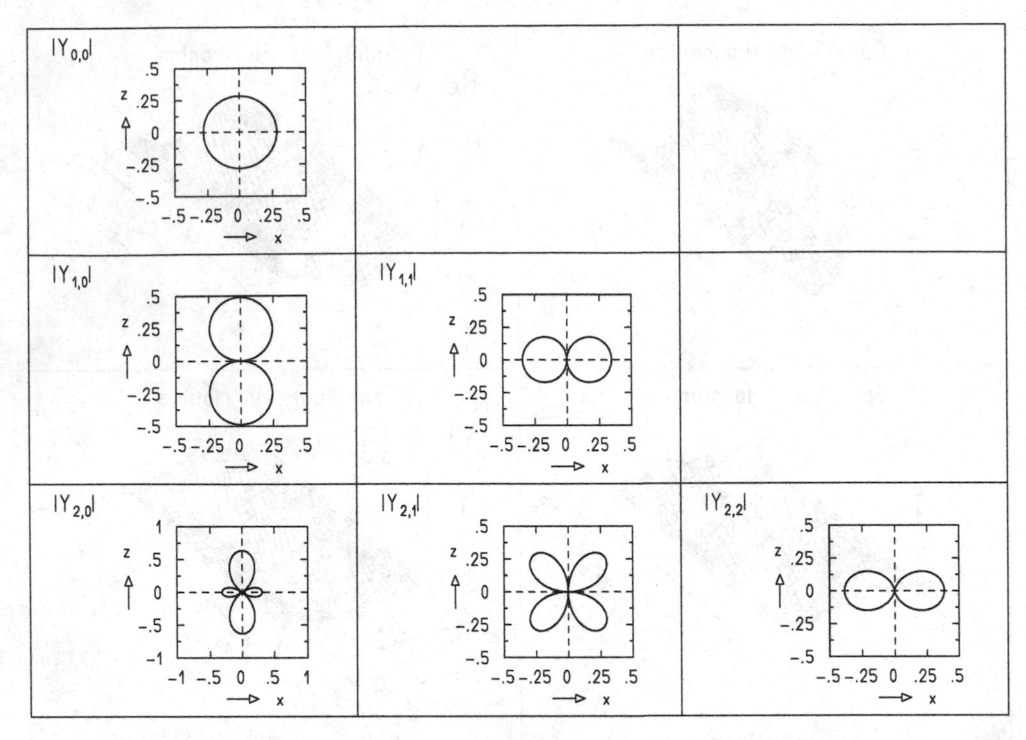

Fig. 11.4. Plot produced with descriptor `Spherical harmonics as 2D polar diagrams` on file `Math_Functions.des`

- $|Y_{\ell m}|$.

These two functions depend only on the polar angle ϑ and are non-negative everywhere. They can therefore be illustrated as *2D function graph* of the type *polar diagram*, see Sect. A.3.3.

On the subpanel **Physics—Variables** you find the two **Indices** ℓ and m. In a *multiple plot* ℓ is incremented by one from row to row and m from column to column.

Example Descriptor on File `Math_Functions.des`

- `Spherical harmonics as 2D polar diagrams` (see Fig. 11.4).

11.6 Spherical Harmonics: Polar Diagram in 3D

Aim of this section: Illustration of the absolute value $|Y_{\ell m}|$ or the absolute square $|Y_{\ell m}|^2$, (11.19), of the spherical harmonics as polar diagram in 3D.

On the bottom of the subpanel **Physics—Comp. Coord.** you can choose to plot

- $|Y_{\ell m}|^2$ or
- $|Y_{\ell m}|$.

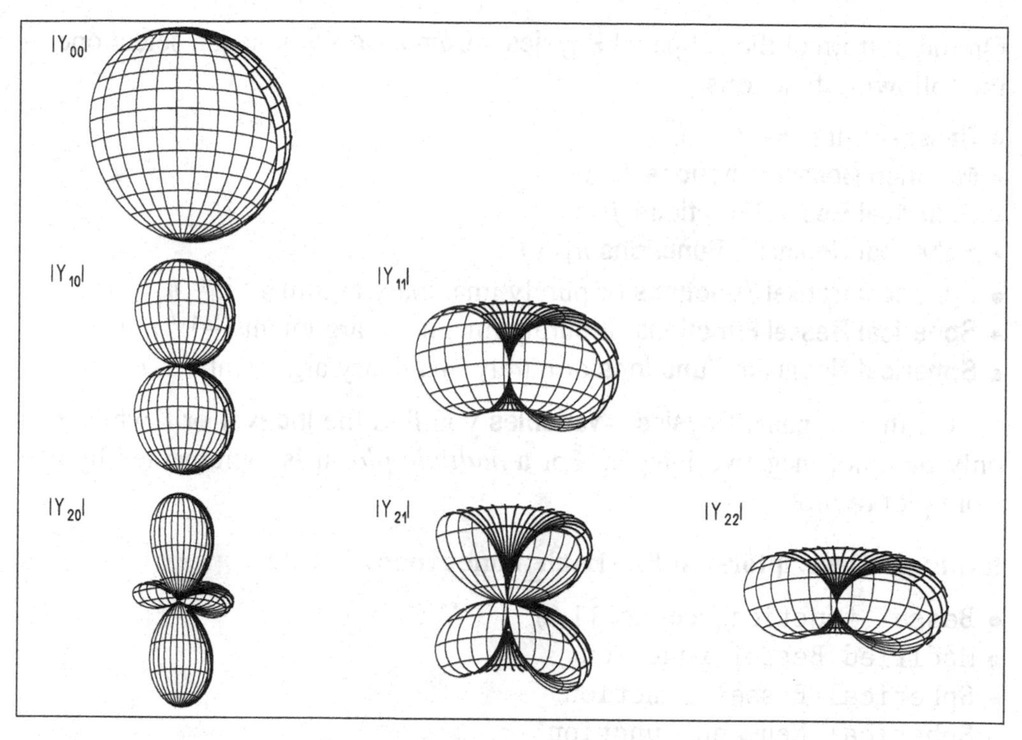

Fig. 11.5. Plot produced with descriptor `Spherical harmonics: polar diagram in 3D, cut open` on file `Math_Functions.des`

There you also find the values of the **Indices** ℓ and m. In a *multiple plot* ℓ is incremented by one from row to row and m from column to column.

There is an **Automatic Scale** facility. If it is switched on, the scale factor between world and computing coordinates is set automatically to a reasonable value. (More precisely, the surface shown in world coordinates will just touch a cube with a half-edge length which is the maximum of the absolute values of X_{beg}, X_{end}, Y_{beg}, Y_{end}, Z_{beg}, Z_{end} which define the ranges of world coordinates.)

Example Descriptors on File `Math_Functions.des`

- `Spherical harmonics: polar diagram in 3D`
- `Spherical harmonics: polar diagram in 3D, cut open` (see Fig. 11.5)

11.7 Bessel Functions and Related Functions

Aim of this section: Illustration of the Bessel function $J_n(x)$, (11.21), and of the modified Bessel function $I_n(x)$, (11.27), for integer index n. Illustration of the spherical Bessel functions $j_\ell(x)$, (11.31), and spherical Neumann functions $n_\ell(x)$, (11.32), for purely real and purely imaginary arguments, and of the spherical Hankel functions of the first kind $h_\ell^{(+)}$, (11.41), for purely imaginary arguments.

On the bottom of the subpanel **Physics—Comp. Coord.** you can select one of the following functions:

- **Bessel Functions** $J_n(x)$,
- **Modified Bessel Functions** $I_n(x)$,
- **Spherical Bessel Functions** $j_\ell(x)$,
- **Spherical Neumann Functions** $n_\ell(x)$,
- **Spherical Hankel Functions** of purely imaginary argument $\mathrm{i}^{\ell+1}h_\ell^{(+)}(\mathrm{i}x)$,
- **Spherical Bessel Functions** of purely imaginary argument $(-\mathrm{i})^\ell j_\ell(\mathrm{i}x)$,
- **Spherical Neumann Functions** of purely imaginary argument $\mathrm{i}^{\ell+1}n_\ell(\mathrm{i}x)$.

On the subpanel **Physics—Variables** you find the **Index** n or ℓ which can only be a non-negative integer. For a *multiple plot* it is incremented by one from plot to plot.

Example Descriptors on File `Math_Functions.des`

- `Bessel function` (see Fig. 11.6)
- `Modified Bessel function`
- `Spherical Bessel function`
- `Spherical Neumann function`
- `Spherical Hankel function of purely imaginary argument`
- `Spherical Bessel function of purely imaginary argument`
- `Spherical Neumann function of purely imaginary argument`

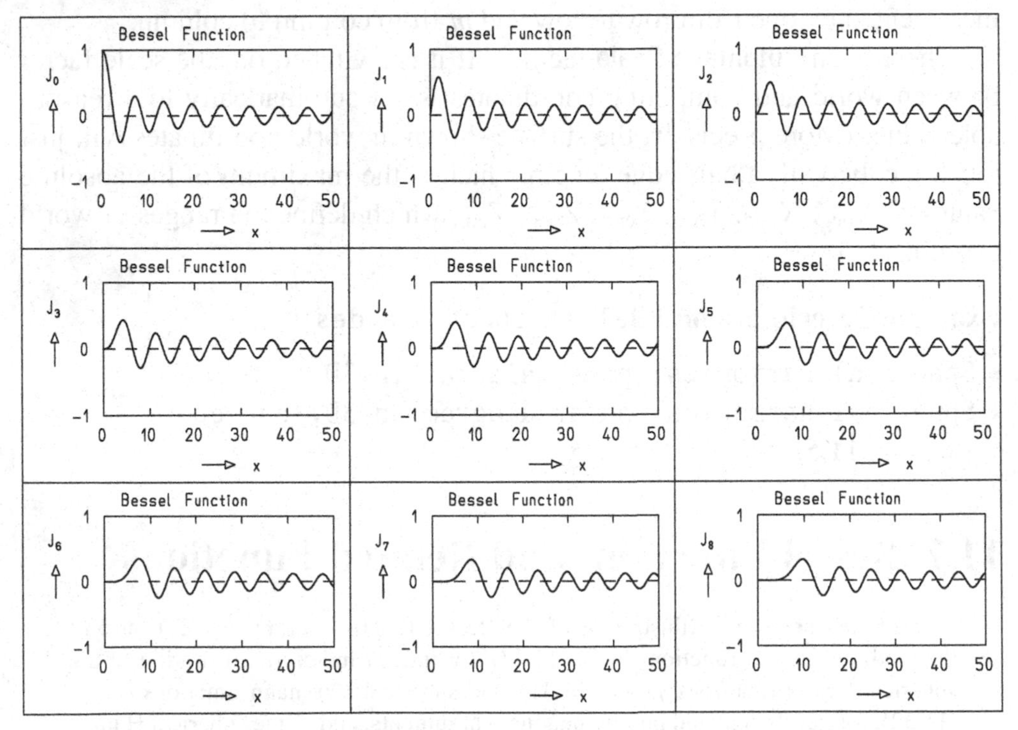

Fig. 11.6. Plot produced with descriptor `Bessel function` on file `Math_Functions.des`

11.8 Bessel Function and Modified Bessel Function with Real Index

Aim of this section: Illustration of the Bessel function $J_\nu(x)$, (11.21), and of the modified Bessel function $I_\nu(x)$, (11.27), as surfaces over a Cartesian grid spanned by the (real) argument x and the (real) index ν.

At the bottom of the subpanel **Physics—Comp. Coord.** you can choose one of the two functions

- **Bessel function** $J_\nu(x)$,
- **Modified Bessel function** $I_\nu(x)$.

There is also a **Cut-Off Facility**, which, if **On**, limits the surface shown to the **Range of Computing Coordinates** in z.

Example Descriptors on File `Math_Functions.des`

- `Bessel function J(x,nu)` (see Fig. 11.7)
- `Modified Bessel function I(x,nu)`

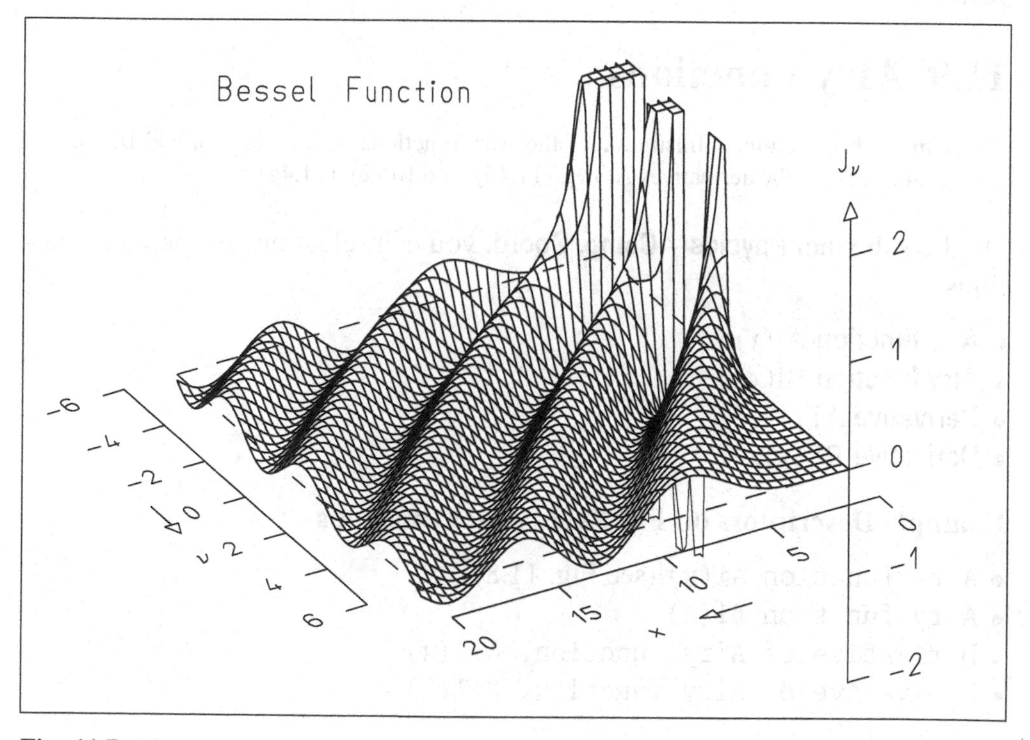

Fig. 11.7. Plot produced with descriptor `Bessel function J(x,nu)` on file `Math_Functions.des`

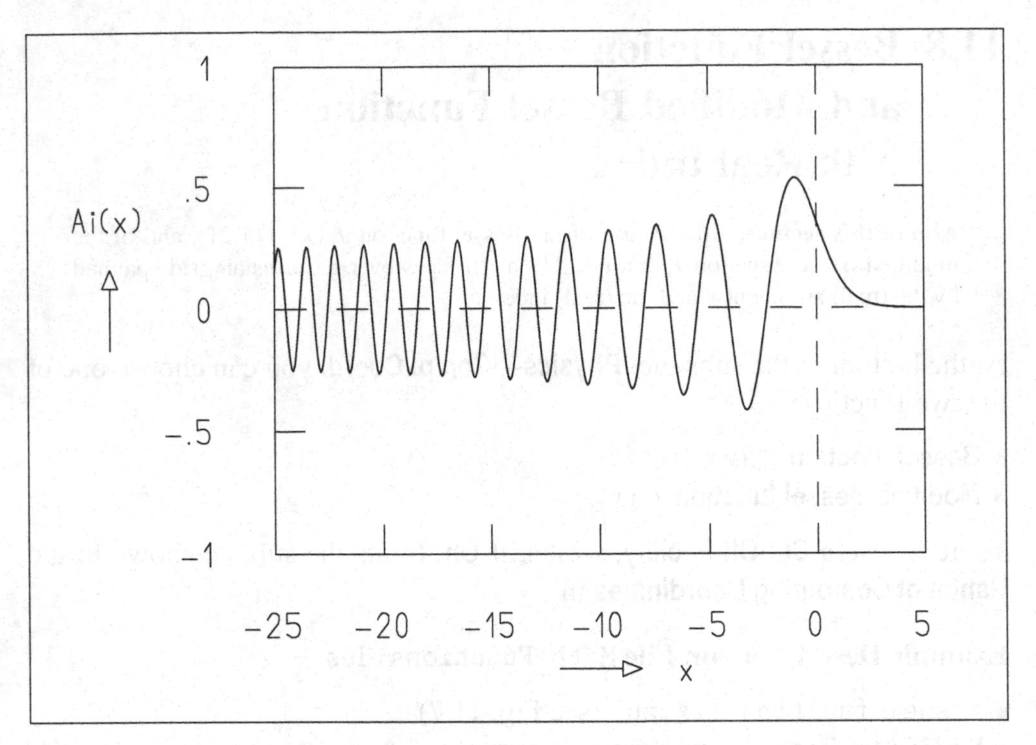

Fig. 11.8. Plot produced with descriptor `Airy function Ai(x)` on file `Math_Functions-.des`

11.9 Airy Functions

Aim of this section: Illustration of the Airy functions Ai(x), (11.43), and Bi(x), (11.44), and of the derivatives Ai′(x), (11.45), and Bi′(x), (11.46).

On the subpanel **Physics—Comp. Coord.** you can select one of the four functions

- **Airy function** Ai(x),
- **Airy function** Bi(x),
- **Derivative** Ai′(x),
- **Derivative** Bi′(x).

Example Descriptors on File `Math_Functions.des`

- `Airy function Ai(x)` (see Fig. 11.8)
- `Airy function Bi(x)`
- `Derivative of Airy function, Ai'(x)`
- `Derivative of Airy function, Bi'(x)`

11.10 Laguerre Polynomials

Aim of this section: Illustration of the Laguerre polynomials $L_n^\alpha(x)$, (11.48). Illustration of the radial eigenfunctions $R_{n_r\ell}(\varrho)$, (11.52), of the three-dimensional harmonic oscillator ($\sigma_0 = 1$). Illustration of the radial eigenfunctions $R_{n\ell}(\varrho)$, (11.54), of the electron in the hydrogen atom, where the Bohr radius is set equal to 1.

On the bottom of the subpanel **Physics—Comp. Coord.** you can select one of the three functions:

- **Laguerre Polynomials** L_n^α,
- **Radial Eigenfunctions of 3D Harmonic Oscillator** $R_{n_r\ell}$,
- **Radial Eigenfunctions of Hydrogen Atom** $R_{n\ell}$.

On the subpanel **Physics—Variables** you find the two **Indices** for the selected function. In a *multiple plot* the first index is incremented by one from column to column and the second index from row to row.

Example Descriptors on File `Math_Functions.des`

- `Laguerre polynomials` (see Fig. 11.9)
- `Radial eigenfunctions of 3D harmonic oscillator`
- `Radial eigenfunctions of hydrogen atom`

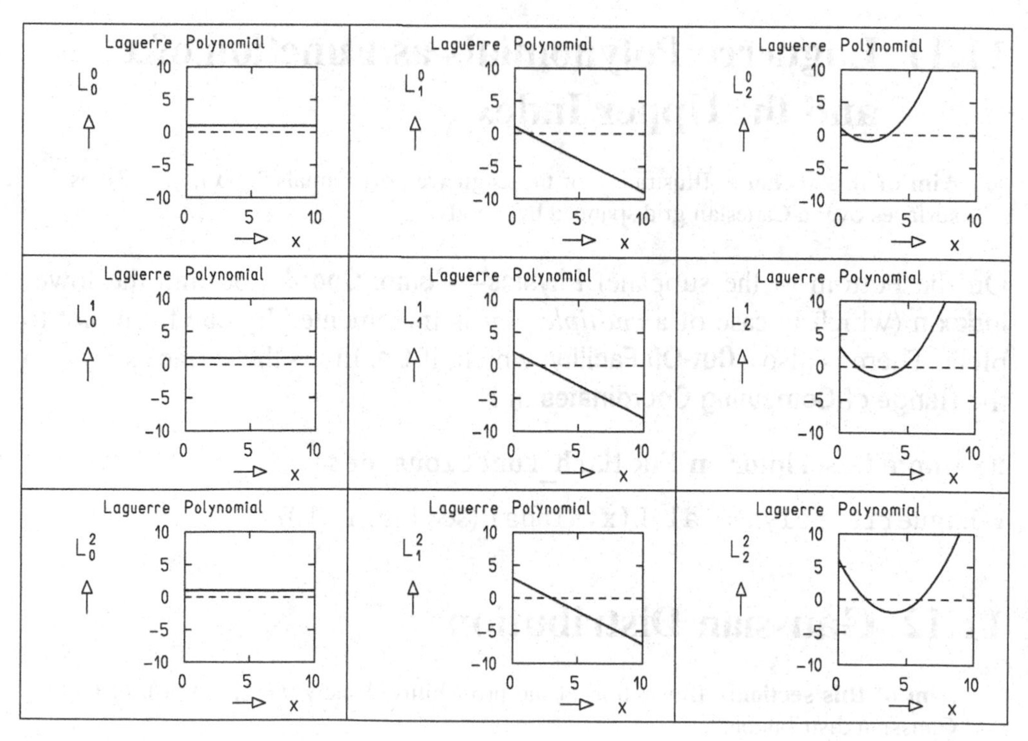

Fig. 11.9. Plot produced with descriptor `Laguerre polynomials` on file `Math_Functions.des`

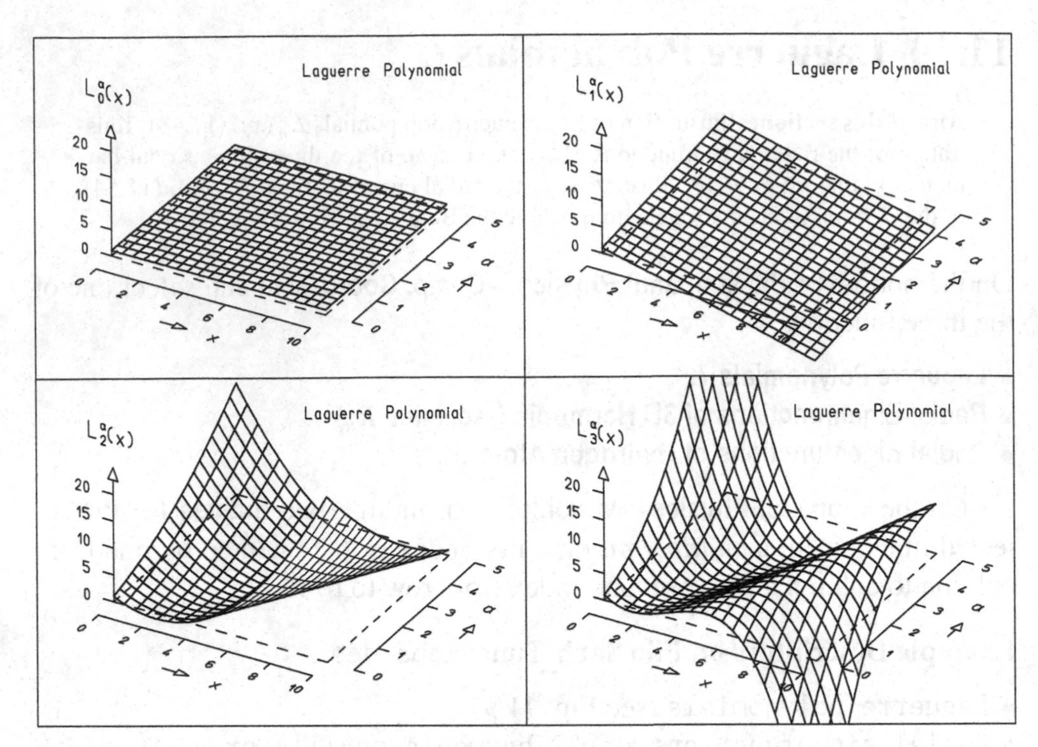

Fig. 11.10. Plot produced with descriptor `Laguerre polynomial L(x,alpha)` on file `Math_Functions.des`

11.11 Laguerre Polynomials as Function of x and the Upper Index α

Aim of this section: Illustration of the Laguerre polynomials $L_n^\alpha(x)$, (11.48), as surfaces over a Cartesian grid spanned by x and α.

On the bottom of the subpanel **Physics—Comp. Coord.** you find the lower **Index n** (which in case of a *multiple plot* is incremented by one from plot to plot). There is also a **Cut-Off Facility**, which, if **On**, limits the surface shown to the **Range of Computing Coordinates** in z.

Example Descriptor on File `Math_Functions.des`

- `Laguerre polynomial L(x,alpha)` (see Fig. 11.10)

11.12 Gaussian Distribution

Aim of this section: Illustration of the probability density $f(x)$, (11.56), of the Gaussian distribution.

On the subpanel **Physics—Comp. Coord.** under the heading **Variables** you find the mean **x_0** and the width **sigma_x** of the distribution.

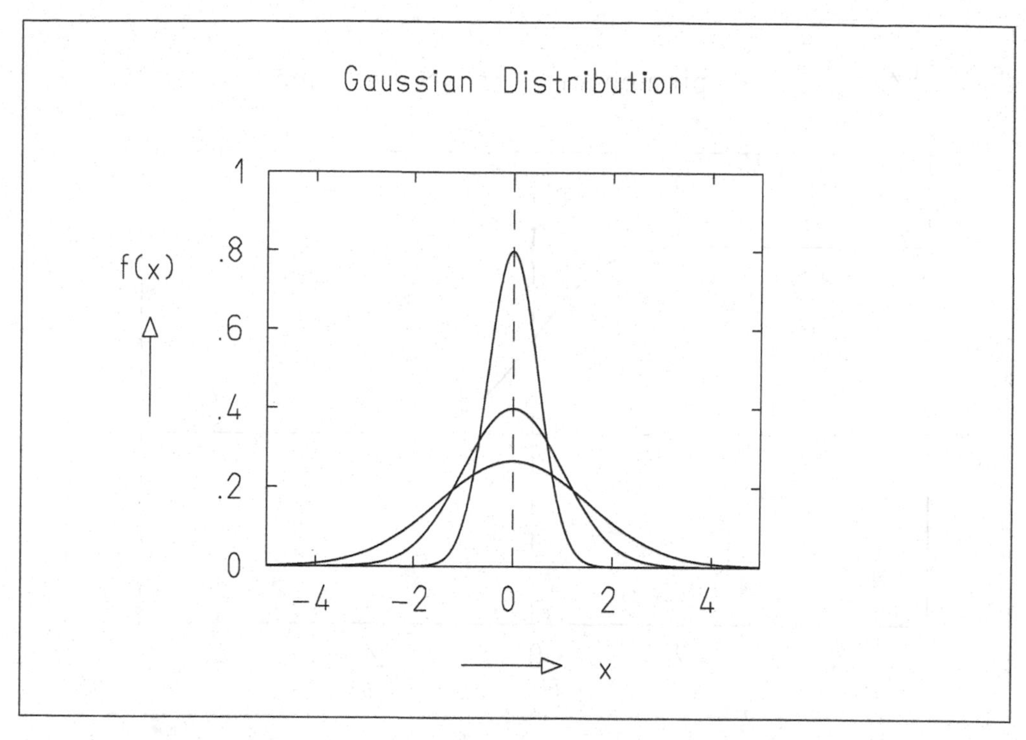

Fig. 11.11. Plot produced with descriptor `Gaussian distribution` on file `Math_Functions.des`

You may show a total of **n** graphs. The first has the width **sigma_x**. It is incremented by **Delta_sigma_x** from graph to graph.

Example Descriptor on File `Math_Functions.des`

- `Gaussian distribution` (see Fig. 11.11)

11.13 Error Function and Complementary Error Function

Aim of this section: Illustration of the error function erf(x), (11.57), and of the complementary error function erfc(x), (11.58).

On the subpanel **Physics—Comp. Coord.** you can select one of the two functions

- **Error Function** erf(x),
- **Complementary Error Function** erfc(x).

Example Descriptors on File `Math_Functions.des`

- `Error function erf(x)`
- `Complementary error function erfc(x)` (see Fig. 11.12)

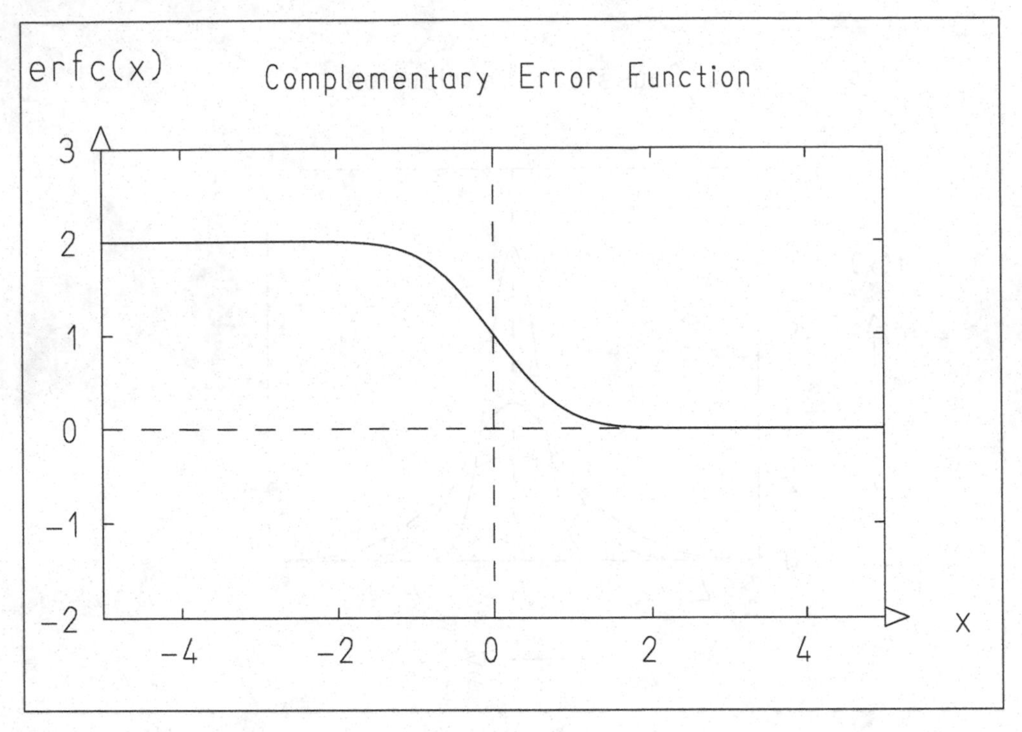

Fig. 11.12. Plot produced with descriptor `Complementary error function erfc(x)` on file `Math_Functions.des`

11.14 Bivariate Gaussian Distribution

Aim of this section: Illustration of the bivariate Gaussian probability density $\rho(x_1, x_2)$, (11.59).

The function $\rho(x_1, x_2)$ is shown as surface over the x_1, x_2 plane. The marginal distributions $\rho_1(x_1)$ and $\rho_2(x_2)$, (11.61), are displayed over the high-x_1 and high-x_2 margin of the x_1, x_2 plane, respectively.

On the subpanel **Physics—2D Gaussian** you can enter the variables $x_{10} = \langle x_1 \rangle$, $x_{20} = \langle x_2 \rangle$, σ_1, σ_2, and c. Moreover, you can choose (or not choose) to show

- the expectation values x_{10}, x_{20} as a filled circle in the x_1, x_2 plane,
- the covariance ellipse as a line $\rho(x_1, x_2) = \text{const}$,
- a "frame", i.e., a rectangle enclosing the ellipse (not normally wanted).

Example Descriptor on File `Math_Functions.des`

- `2D Gaussian` (see Fig. 11.13)

11.15 Bivariate Gaussian: Covariance Ellipse

Aim of this section: Illustration of the covariance ellipse (11.62).

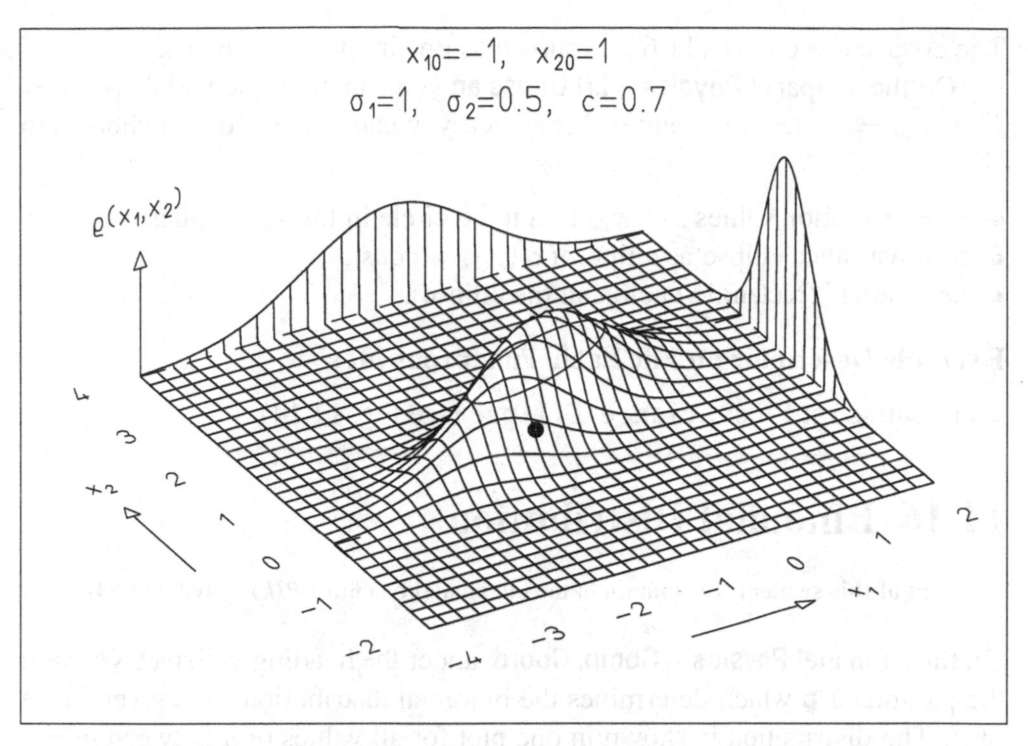

Fig. 11.13. Plot produced with descriptor `2D Gaussian` on file `Math_Functions.des`

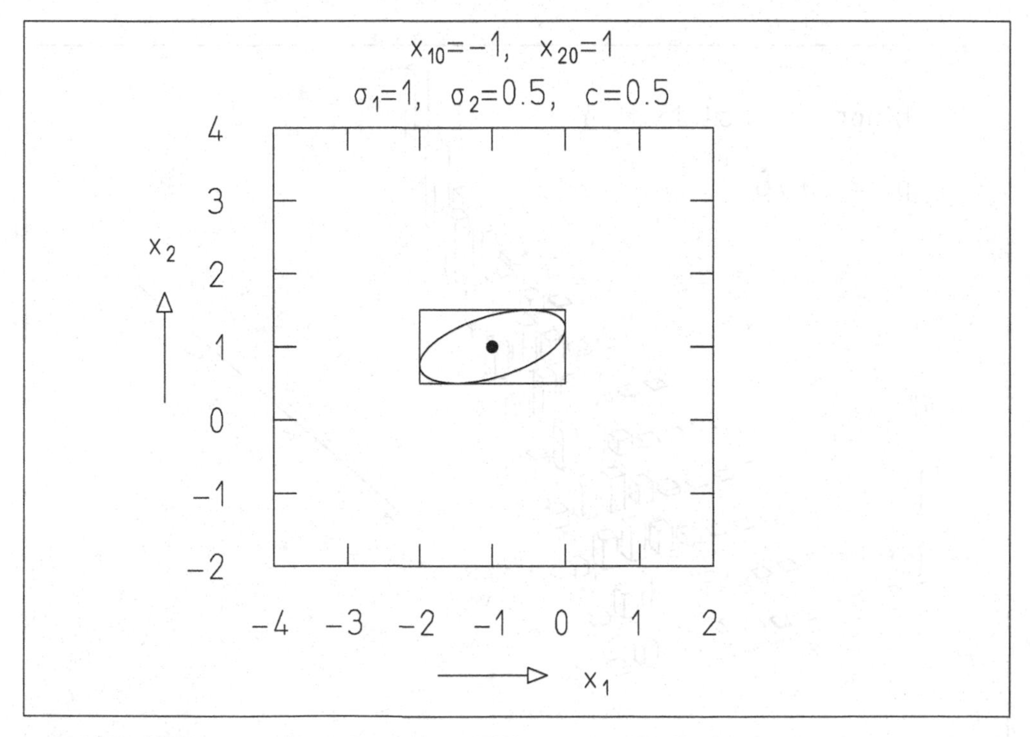

Fig. 11.14. Plot produced with descriptor `2D Gaussian: Covariance Ellipse` on file `Math_Functions.des`

The covariance ellipse (11.62) is shown as line in the x_1, x_2 plane.

On the subpanel **Physics—2D Gaussian** you can enter the variables $x_{10} = \langle x_1 \rangle$, $x_{20} = \langle x_2 \rangle$, σ_1, σ_2, and c. Moreover, you can choose (or not choose) to show

- the expectation values x_{10}, x_{20} as a filled circle in the x_1, x_2 plane,
- the covariance ellipse as a line $\rho(x_1, x_2) = \text{const}$,
- the "frame" (rectangle enclosing the ellipse).

Example Descriptor on File `Math_Functions.des`

- `2D Gaussian: Covariance Ellipse` (see Fig. 11.14)

11.16 Binomial Distribution

Aim of this section: Illustration of the binomial probability $P(k) = W_k^n$, (11.63).

On the subpanel **Physics—Comp. Coord.** under the heading **Variables** you find the parameter **p** which determines the binomial distribution for a given value of n. The distribution is shown in one plot for all values of n between $n = 0$ and $n =$ **n_max**. In a *multiple plot* the parameter p is increased by **Delta p** from plot to plot.

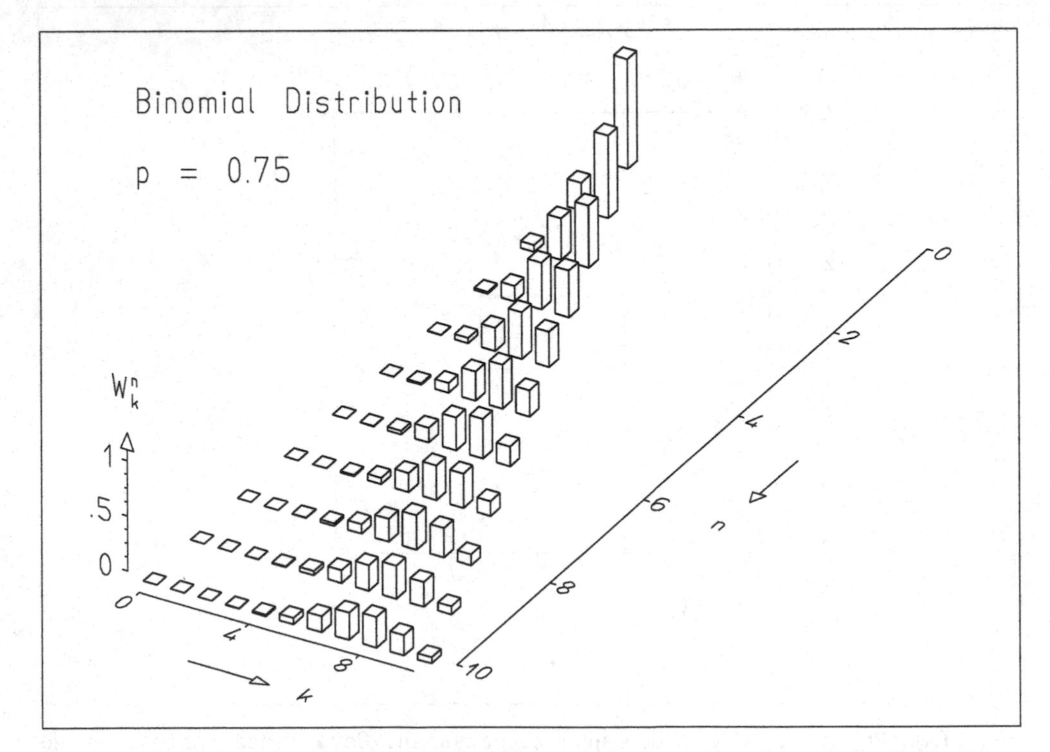

Fig. 11.15. Plot produced with descriptor `Binomial distribution` on file `Math_Functions.des`

Example Descriptor on File `Math_Functions.des`

- `Binomial distribution` (see Fig. 11.15)

11.17 Poisson Distribution

Aim of this section: Illustration of the Poisson probability distribution $P(k, \lambda)$, (11.64).

For a fixed value of the parameter λ, the distribution $P(k, \lambda)$ is represented by a *histogram*, which is a set of columns of the height

$$P(0, \lambda),\ P(1, \lambda),\ \ldots,\ P(k_{\max}, \lambda) \quad .$$

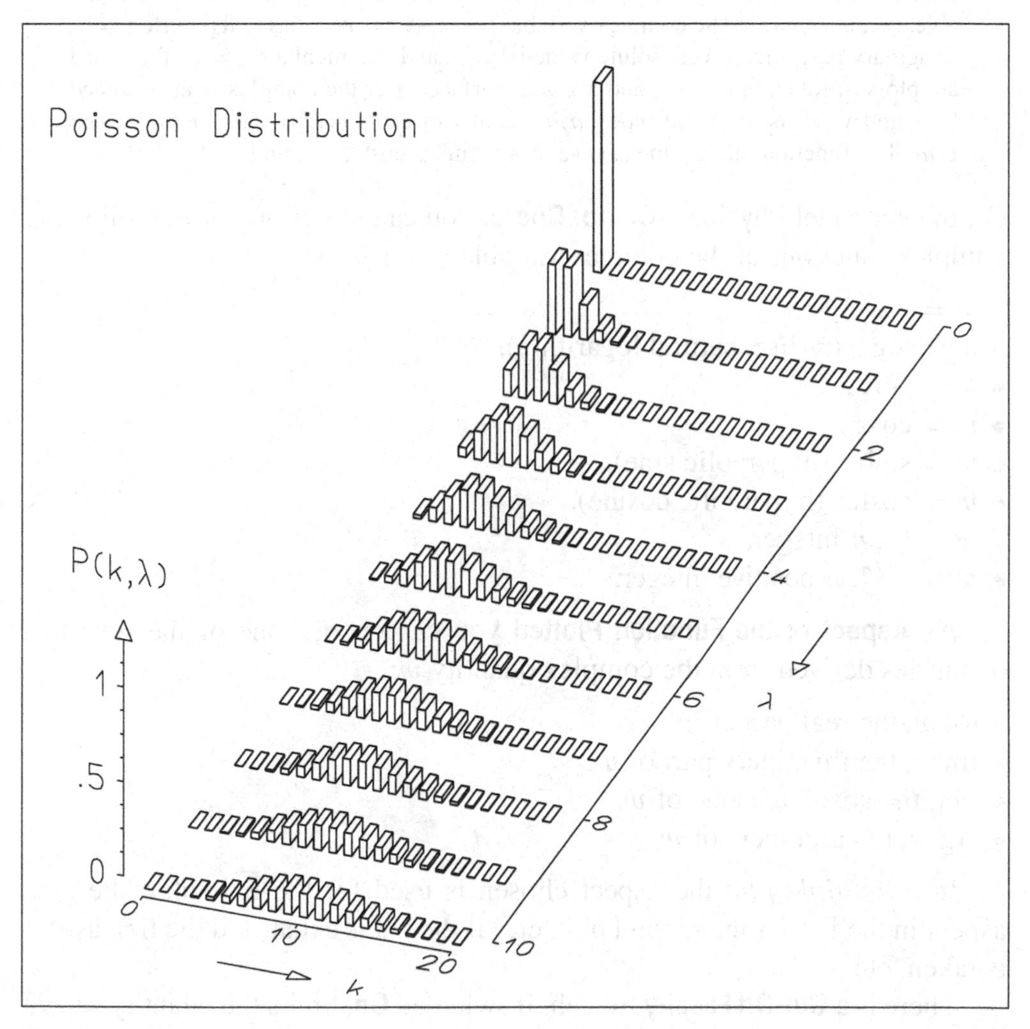

Fig. 11.16. Plot produced with descriptor `Poisson distribution` on file `Math_Functions-.des`

We show several such histograms for different values of λ by placing these sets of columns in a plane spanned by k and λ, see Fig. 11.16.

In the subpanel **Physics—Comp. Coord.** you find the **Range of the Parameter λ** defined by the values **λ_beg** and **λ_end** and the number of histograms **n_HIST** shown within that range. The **Range in k** extends between $k = 0$ and $k =$ **k_max**.

Example Descriptor on File `Math_Functions.des`

• `Poisson distribution` (see Fig. 11.16)

11.18 Simple Functions of a Complex Variable

Aim of this section: Illustration of simple complex functions of one complex variable, i.e., $w = w(z)$. The complex variable $z = x + \mathrm{i}y = r\mathrm{e}^{\mathrm{i}\varphi}$ has real part $\mathrm{Re}\, z = x$, imaginary part $\mathrm{Im}\, z = y$, absolute value $|z| = r$, and argument $\arg z = \varphi$. Presented are plots of $\mathrm{Re}\, w$, $\mathrm{Im}\, w$, $|w|$, and $\arg w$ as surfaces over the complex plane spanned by x and y. Plots can be of type *surface over Cartesian grid* or *surface over polar grid*. The functions are e^z, $\log z$, $\sin z$, $\cos z$, $\sinh z$, $\cosh z$, z^n, and $z^{1/n}$.

On the subpanel **Physics—Comp. Coord.** you can select one of the following complex functions of the complex variable $z = x + \mathrm{i}y$:

• $w = \mathrm{e}^z$,
• $w = \log z$ ($\equiv \ln z$, natural logarithm),
• $w = \sin z$,
• $w = \cos z$,
• $w = \sinh z$ (hyperbolic sine),
• $w = \cosh z$ (hyperbolic cosine),
• $w = z^n$, n integer,
• $w = z^{1/n}$, n positive integer.

As **Aspect of the Function Plotted** you can choose one of the four real quantities derived from the complex quantity w:

• $\mathrm{Re}\, w$, the real part of w,
• $\mathrm{Im}\, w$, the imaginary part of w,
• $|w|$, the absolute value of w,
• $\arg w$, the argument of w.

In a *multiple plot* the aspect chosen is used for the first plot, the next aspect in the list for the second plot, etc. If the list is exhausted the first aspect is taken, etc.

There is a **Cut-Off Facility**, which, if switched **On**, limits the quantity shown to the range of computing coordinates in z.

The subpanel **Physics—Variables** is needed only for the two functions $w = z^n$ and $w = z^{1/n}$. It contains the integer **n** in the exponent of these

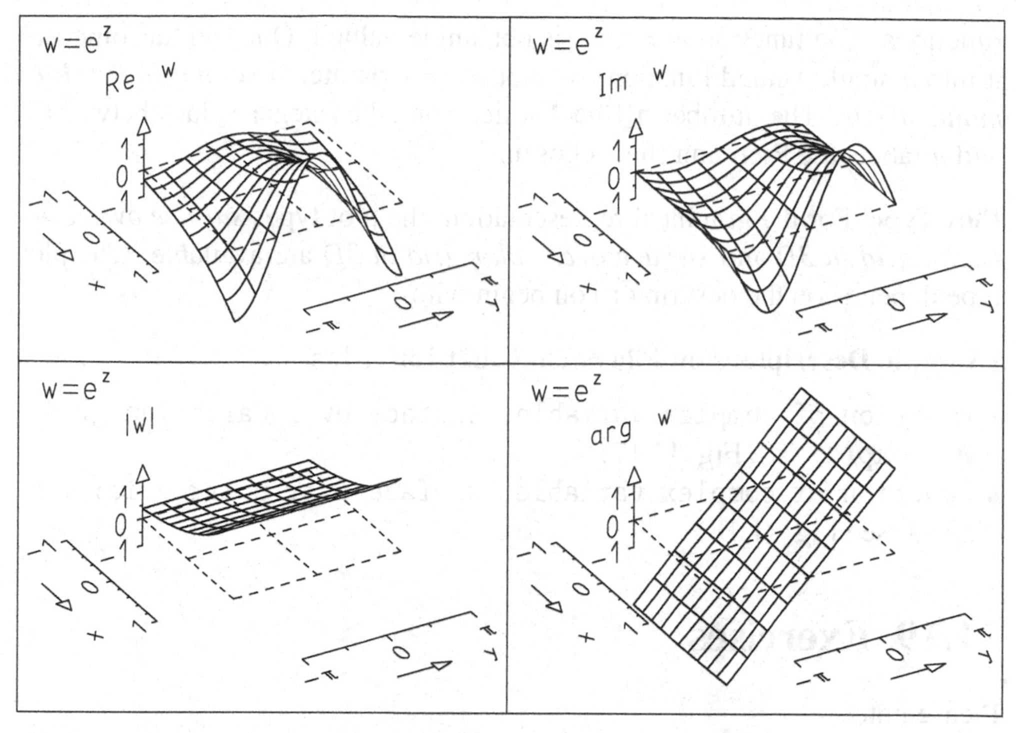

Fig. 11.17. Plot produced with descriptor `Function of complex variable, surface over Cartesian grid: w = exp(z)` on file `Math_Functions.des`

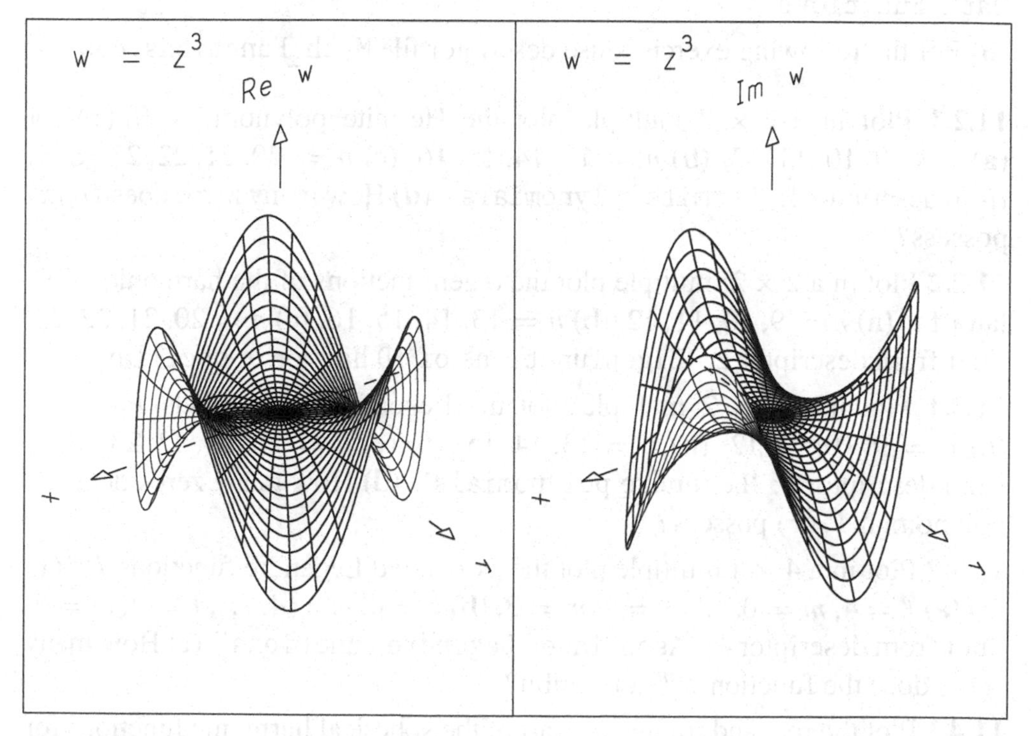

Fig. 11.18. Plot produced with descriptor `Function of complex variable, surface over polar grid: w = z**3` on file `Math_Functions.des`

functions. The function $w = z^{1/n}$ is not single-valued. One can 'decompose' it into n single-valued functions by plotting it separately over n different *Riemann sheets*. The number **n_Sheet** which can take integer values between 1 and n labels the Riemann sheet chosen.

Plot Type For the graphical representation, the plot types *surface over Cartesian grid in 3D* and *surface over polar grid in 3D* are available. The plot type depends on the descriptor you begin with.

Example Descriptors on File `Math_Functions.des`

- `Function of complex variable, surface over Cartesian grid: w = exp(z)` (see Fig. 11.17)
- `Function of complex variable, surface over polar grid: w = z**3` (see Fig. 11.18)

11.19 Exercises

Please note:

(i) You may watch a demonstration of the material of this chapter by pressing the button **Run Demo** in the *main toolbar* and selecting one of the demo files `Math_Functions`.

(ii) For the following exercises use descriptor file `Math_Functions.des`.

11.2.1 Plot in a 2×2 multiple plot the Hermite polynomials $H_n(x)$ for **(a)** $n = 9, 10, 11, 12$, **(b)** $n = 13, 14, 15, 16$, **(c)** $n = 20, 21, 22, 23$. Start from descriptor 1: '`Hermite polynomials`'. **(d)** How many zeros does $H_n(x)$ possess?

11.2.2 Plot in a 2×2 multiple plot the eigenfunctions of the harmonic oscillator for **(a)** $n = 9, 10, 11, 12$, **(b)** $n = 13, 14, 15, 16$, **(c)** $n = 20, 21, 22, 23$. Start from descriptor 2: '`Eigenfunctions of 1D harmonic oscillator`'.

11.3.1 Plot in a 2×2 multiple plot the Legendre polynomials $P_\ell(x)$ for **(a)** $\ell = 9, 10, 11, 12$, **(b)** $\ell = 13, 14, 15, 16$, **(c)** $\ell = 20, 21, 22, 23$. Start from descriptor 3: '`Legendre polynomials`'. **(d)** How many zeros does the polynomial $P_\ell(x)$ possess?

11.3.2 Plot in a 4×4 multiple plot the associated Legendre functions $P_\ell^m(x)$ for **(a)** $\ell = 4, m = 0, \ldots, \ell = 7, m = 3$, **(b)** $\ell = 4, m = 4, \ldots, \ell = 7, m = 7$. Start from descriptor 4: '`Associated Legendre functions`'. **(c)** How many zeros does the function $P_\ell^m(x)$ exhibit?

11.4.1 Plot the real and imaginary part of the spherical harmonic functions for $\ell = 2$ and its absolute square in a 3×3 multiple plot. Start from descriptor 6: '`Spherical harmonics as surfaces over Cartesian grid`'.

11.5.1 Plot as polar diagrams in a 3×3 multiple plot the absolute square $|Y_{\ell m}|^2$ of the spherical harmonic function for **(a)** $\ell = 4$, $m = 0$ as plot in the upper-left field; **(b)** $\ell = 4$, $m = 4$ as plot in the upper-left field. Start from descriptor 7: '`Spherical harmonics as 2D polar diagrams`'.

11.6.1 Repeat Exercise 11.5.1 but produce polar plots in 3D starting from descriptor 8: '`Spherical harmonics, polar diagram in 3D`'.

11.6.2 Repeat Exercise 11.6.1, but start from descriptor 9: '`Spherical harmonics, polar diagram in 3D, cut open`'.

11.7.1 Plot the Bessel functions $J_n(x)$ in a 5×5 multiple plot for $n = 0, 1, \ldots, 24$. Start from descriptor 10: '`Bessel function`'.

11.7.2 Repeat Exercise 11.7.1 for the modified Bessel functions $I_n(x)$ starting from descriptor 11: '`Modified Bessel function`'. (Adjust the scale in x by setting x_{end} to a suitable value.)

11.7.3 Plot in a 2×2 multiple plot the spherical Bessel functions $j_\ell(x)$ in the range $0 \leq x \leq 50$ for **(a)** $\ell = 9, 10, 11, 12$, **(b)** $\ell = 13, 14, 15, 16$. Start from descriptor 12: '`Spherical Bessel function`'.

11.7.4 Plot in a 2×2 multiple plot the spherical Neumann functions $n_\ell(x)$ in the range $0 \leq x \leq 50$ for **(a)** $\ell = 9, 10, 11, 12$, **(b)** $\ell = 13, 14, 15, 16$. Start from descriptor 13: '`Spherical Neumann function`'.

11.7.5 Plot in a 2×2 multiple plot the spherical modified Hankel functions $\mathrm{i}^{\ell+1} h^{(+)}(\mathrm{i}\varrho)$ in the range $0 \leq x \leq 20$ for **(a)** $\ell = 9, 10, 11, 12$, **(b)** $\ell = 13, 14, 15, 16$. Start from descriptor 14: '`Spherical Hankel function of purely imaginary argument`'.

11.7.6 Plot in a 2×2 multiple plot the spherical modified Bessel functions $(-\mathrm{i})^\ell j_\ell(\mathrm{i}\varrho)$ in the range $0 \leq x \leq 20$ for **(a)** $\ell = 9, 10, 11, 12$, **(b)** $\ell = 13, 14, 15, 16$. Start from descriptor 15: '`Spherical Bessel function of purely imaginary argument`'.

11.7.7 Plot in a 2×2 multiple plot the spherical modified Neumann functions $\mathrm{i}^{\ell+1} n_\ell(\mathrm{i}\varrho)$ in the range $0 \leq x \leq 20$ for **(a)** $\ell = 9, 10, 11, 12$, **(b)** $\ell = 13, 14, 15, 16$. Start from descriptor 16: '`Spherical Neumann function of purely imaginary argument`'.

11.9.1 Plot the Airy functions using descriptors 19: '`Airy function Ai(x)`' and 20: '`Airy function Bi(x)`'.

11.10.1 Plot the Laguerre polynomials $L_n^\alpha(x)$ in a 2×2 multiple plot for the index values **(a)** $n = 3$, $\alpha = 0, 1$; $n = 4$, $\alpha = 0, 1$ in the range $0 \leq x \leq 11$, **(b)** $n = 3$, $\alpha = 2, 3$; $n = 4$, $\alpha = 2, 3$ in the range $0 \leq x \leq 13$. Start from descriptor 23: '`Laguerre polynomials`'.

11.10.2 Plot in a 3×3 multiple plot the radial eigenfunctions of a spherically symmetric harmonic oscillator in the range $0 \leq x \leq 8$ for the angular momenta $\ell = 1, 2, 3$ for the principal quantum numbers **(a)** $n = 3, 4, 5$,

(b) $n = 6, 7, 8$. Start from descriptor 24: ‘`Radial eigenfunctions of 3D harmonic oscillator`’.

11.10.3 Plot in a 3×3 multiple plot the radial eigenfunctions of the hydrogen atom in the range $0 \leq x \leq 60$ for the principal quantum numbers $n = 3, 4, 5$ and **(a)** $\ell = 0, 1, 2$, **(b)** $\ell = 3, 4, 5$. Start from descriptor 25: ‘`Radial eigenfunctions of hydrogen atom`’.

11.11.1 Plot the Laguerre polynomials $L_n^\alpha(x)$ in a 2×2 multiple plot for the range $0 \leq x \leq 10$ with α as a running index in the range $0 \leq \alpha \leq 5$ and for $n = 4, 5, 6, 7$. Start from descriptor 26: ‘`Laguerre polynomial L(x,alpha)`’.

11.12.1 Plot a Gaussian distribution using descriptor 27: ‘`Gaussian distribution`’. **(a)** Produce graphs for more values of σ by setting n equal to 5. **(b)** Change x_0 by setting it to $2, -3$.

11.13.1 Plot the two error functions using descriptors 28: ‘`Error function erf(x)`’, 29: ‘`Complementary error function erfc(x)`’.

11.14.1 Plot a bivariate Gaussian distribution using descriptor 30: ‘`2D Gaussian`’. **(a)** Produce plots for $c = -0.5, 0, 0.9$. **(b)** Repeat **(a)** for $\sigma_1 = 0.5, \sigma_2 = 1$ and for $\sigma_1 = 1, \sigma_2 = 1$.

11.15.1 Plot the covariance ellipse of a bivariate Gaussian distribution using descriptor 31: ‘`2D Gaussian, Covariance Ellipse`’. Produce additional plots for the numerical values given in the preceding exercise.

11.16.1 Plot a binomial distribution using descriptor 32: ‘`Binomial distribution`’. Produce a 3×3 multiple plot for $p = 0.1, 0.2, \ldots, 0.9$.

11.17.1 Produce a set of Poisson distributions for several values of λ using descriptor 33: ‘`Poisson distribution`’. Change the range of λ to **(a)** $0 \leq \lambda \leq 1$, **(b)** $9 \leq \lambda \leq 10$. Observe the characteristic difference in the shape of the distribution in these two domains of λ.

11.18.1 **(a)** Plot $w = \mathrm{e}^z$ using descriptor 34: ‘`Function of complex variable, surface over Cartesian grid: w = exp(z)`’. **(b)** Leave the plot on the screen and produce additional plots for $w = \log z$, $\sin z$, $\cos z$, $\sinh z$, $\cosh z$. Compare them.

11.18.2 Plot $w = z^3$ using descriptor 35: ‘`Function of complex variable, surface over polar grid: w = z**3`’. Leave the plot on the screen and produce additional plots for $w = z^0, z^1, z^2, z^4, z^5$. Compare them.

12. Additional Material and Hints for the Solution of Exercises

12.1 Units and Orders of Magnitude

12.1.1 Definitions

Every physical quantity q can be expressed as a product of a dimensionless *numerical value* $q_{\mathrm{N}u}$ and its *unit* q_u,

$$q = q_{\mathrm{N}u}\mathrm{q}_u \quad . \tag{12.1}$$

The index u specifies the particular system of units used. We may factorize the numerical value

$$q_{\mathrm{N}u} = q_{\mathrm{M}u} \times 10^{q_{\mathrm{E}u}} \tag{12.2}$$

into a *mantissa* $q_{\mathrm{M}u}$ and a power of ten with the integer *exponent* $q_{\mathrm{E}u}$. The factorization (12.2) is by no means unique, but it is understood that the mantissa is not too far from one, say $0.001 \le |q_{\mathrm{M}u}| \le 1000$.

It is important that numerical values used in the computer all have similar exponents because, otherwise, the result of computation may have severe rounding errors. Moreover, numbers with very small or very large exponents, typically $|q_{\mathrm{E}u}| \ge 38$, cannot be represented at all with simple techniques. Numerical values that we use as input in computer programs should therefore have exponents close to zero, say $|q_{\mathrm{E}u}| \le 5$.

In mechanics and quantum mechanics one can choose three physical quantities to be *basic quantities*, define units for them as *basic units*, and derive the units of all other quantities from these basic units.

12.1.2 SI Units

In the international system of weights and measures (SI) the basic quantities are *length*, *mass*, and *time* with the basic units *meter* (m), *kilogram* (kg), and *second* (s), respectively.

The SI units of the more important physical quantities and the numerical values of some constants of nature are given in Table 12.1.

S. Brandt et al., *Interactive Quantum Mechanics: Quantum Experiments on the Computer*, DOI 10.1007/978-1-4419-7424-2_12,

Table 12.1. SI units

SI Units (Based on m, kg, s)	
Quantity	Unit
Energy	$1\,\mathrm{E_{SI}} = 1\,\mathrm{m^2\,kg\,s^{-2}}$
Length	$1\,\ell_{\mathrm{SI}} = 1\,\mathrm{m}$
Time	$1\,\mathrm{t_{SI}} = 1\,\mathrm{s}$
Angular Frequency	$1\,\omega_{\mathrm{SI}} = 1\,\mathrm{t_{SI}^{-1}} = 1\,\mathrm{s^{-1}}$
Action	$1\,\mathrm{A_{SI}} = 1\,\mathrm{E_{SI}\,t_{SI}} = 1\,\mathrm{m^2\,kg\,s^{-1}}$
Velocity	$1\,\mathrm{v_{SI}} = 1\,\ell_{\mathrm{SI}}\,\mathrm{t_{SI}^{-1}} = 1\,\mathrm{m\,s^{-1}}$
Mass	$1\,\mathrm{M_{SI}} = 1\,\mathrm{kg}$
Momentum	$1\,\mathrm{p_{SI}} = 1\,\mathrm{M_{SI}\,v_{SI}} = 1\,\mathrm{m\,kg\,s^{-1}}$
Constants	
Planck's Constant	$\hbar = \hbar_{\mathrm{NSI}}\,\mathrm{A_{SI}}$, $\hbar_{\mathrm{NSI}} = 1.055 \times 10^{-34}$
Speed of Light	$c = c_{\mathrm{NSI}}\,\mathrm{v_{SI}}$, $c_{\mathrm{NSI}} = 2.998 \times 10^{8}$
Electron Mass	$M_{\mathrm{e}} = M_{\mathrm{eNSI}}\,\mathrm{M_{SI}}$, $M_{\mathrm{eNSI}} = 9.110 \times 10^{-31}$
Proton Mass	$M_{\mathrm{p}} = M_{\mathrm{pNSI}}\,\mathrm{M_{SI}}$, $M_{\mathrm{pNSI}} = 1.673 \times 10^{-27}$
Bohr Radius	$a = a_{\mathrm{NSI}}\,\ell_{\mathrm{SI}}$, $a_{\mathrm{NSI}} = 0.5292 \times 10^{-10}$
Energy of Hydrogen Ground State	$E_1 = E_{1\mathrm{NSI}}\,\mathrm{E_{SI}}$, $E_{1\mathrm{NSI}} = 2.180 \times 10^{-18}$

12.1.3 Scaled Units

Unfortunately, the constants typical of quantum phenomena, listed at the bottom of Table 12.1, all have very large or small exponents in SI units. A simple way out of this problem is the use of *scaled units*, i.e., units multiplied by suitable powers of 10. We reformulate the factorization (12.1) and (12.2),

$$\begin{aligned} q = q_{\mathrm{N}u}\mathrm{q}_u = q_{\mathrm{M}u} \times 10^{q_{\mathrm{E}u}}\mathrm{q}_u = q_{\mathrm{M}u} \times 10^{q'_{\mathrm{E}u}} \times 10^{\bar{q}_{\mathrm{E}u}}\mathrm{q}_u \\ = q_{\mathrm{M}u} \times 10^{q'_{\mathrm{E}u}}\mathrm{q}'_u = q'_{\mathrm{N}u}\mathrm{q}'_u \quad , \end{aligned} \tag{12.3}$$

by writing the exponent $q_{\mathrm{E}u}$ as a sum of two integers,

$$q_{\mathrm{E}u} = q'_{\mathrm{E}u} + \bar{q}_{\mathrm{E}u} \quad , \tag{12.4}$$

where $q'_{\mathrm{E}u}$ is chosen close to zero, and thus defining a scaled unit

$$\mathrm{q}'_u = 10^{\bar{q}_{\mathrm{E}u}}\mathrm{q}_u \tag{12.5}$$

and the corresponding numerical value

$$q'_{\mathrm{N}u} = q_{\mathrm{M}u} \times 10^{q'_{\mathrm{E}u}} \quad . \tag{12.6}$$

The decomposition (12.4) of the exponent is not unique. It has to be chosen in such a way that the numerical values (12.6), multiplied by a scaled unit, are not too far away from unity, i.e., that exponents $q'_{\mathrm{E}u}$ are close to zero, say

Table 12.2. Scaled SI units

Examples of Scaled SI Units	
with a choice of scale factors for action, mass, velocity	with a choice of scale factors for action, mass, energy
$\mathrm{E}'_{\mathrm{SI}} = \mathrm{M}'_{\mathrm{SI}}\,\mathrm{v}'^2_{\mathrm{SI}} = 10^{-14}\,\mathrm{m^2\,kg\,s^{-2}}$	$\mathrm{E}'_{\mathrm{SI}} = 10^{-18}\,\mathrm{E}_{\mathrm{SI}} = 10^{-18}\,\mathrm{m^2\,kg\,s^{-2}}$
$\ell'_{\mathrm{SI}} = \mathrm{A}'_{\mathrm{SI}}/\mathrm{M}'_{\mathrm{SI}}\,\mathrm{v}'_{\mathrm{SI}} = 10^{-12}\,\mathrm{m}$	$\ell'_{\mathrm{SI}} = \mathrm{A}'_{\mathrm{SI}}/\sqrt{\mathrm{E}'_{\mathrm{SI}}\,\mathrm{M}'_{\mathrm{SI}}} = 10^{-10}\,\mathrm{m}$
$\mathrm{t}'_{\mathrm{SI}} = \mathrm{A}'_{\mathrm{SI}}/\mathrm{M}'_{\mathrm{SI}}\,\mathrm{v}'^2_{\mathrm{SI}} = 10^{-20}\,\mathrm{s}$	$\mathrm{t}'_{\mathrm{SI}} = \mathrm{A}'_{\mathrm{SI}}/\mathrm{E}'_{\mathrm{SI}} = 10^{-16}\,\mathrm{s}$
$\omega'_{\mathrm{SI}} = 1/\mathrm{t}'_{\mathrm{SI}} = 10^{20}\,\mathrm{s^{-1}}$	$\omega'_{\mathrm{SI}} = 1/\mathrm{t}'_{\mathrm{SI}} = 10^{16}\,\mathrm{s^{-1}}$
$\mathrm{A}'_{\mathrm{SI}} = 10^{-34}\,\mathrm{A}_{\mathrm{SI}} = 10^{-34}\,\mathrm{m^2\,kg\,s^{-1}}$	$\mathrm{A}'_{\mathrm{SI}} = 10^{-34}\,\mathrm{A}_{\mathrm{SI}} = 10^{-34}\,\mathrm{m^2\,kg\,s^{-1}}$
$\mathrm{v}'_{\mathrm{SI}} = 10^{8}\,\mathrm{v}_{\mathrm{SI}} = 10^{8}\,\mathrm{m\,s^{-1}}$	$\mathrm{v}'_{\mathrm{SI}} = \sqrt{\mathrm{E}'_{\mathrm{SI}}/\mathrm{M}'_{\mathrm{SI}}} = 10^{6}\,\mathrm{m\,s^{-1}}$
$\mathrm{M}'_{\mathrm{SI}} = 10^{-30}\,\mathrm{M}_{\mathrm{SI}} = 10^{-30}\,\mathrm{kg}$	$\mathrm{M}'_{\mathrm{SI}} = 10^{-30}\,\mathrm{M}_{\mathrm{SI}} = 10^{-30}\,\mathrm{kg}$
$\mathrm{p}'_{\mathrm{SI}} = \mathrm{M}'_{\mathrm{SI}}\,\mathrm{v}'_{\mathrm{SI}} = 10^{-22}\,\mathrm{m\,kg\,s^{-1}}$	$\mathrm{p}'_{\mathrm{SI}} = \sqrt{\mathrm{E}'_{\mathrm{SI}}\,\mathrm{M}'_{\mathrm{SI}}} = 10^{-24}\,\mathrm{m\,kg\,s^{-1}}$
Constants	
$\hbar = 1.055\,\mathrm{A}'_{\mathrm{SI}}$	$\hbar = 1.055\,\mathrm{A}'_{\mathrm{SI}}$
$c = 2.998\,\mathrm{v}'_{\mathrm{SI}}$	$c = 299.8\,\mathrm{v}'_{\mathrm{SI}}$
$M_{\mathrm{e}} = 0.9110\,\mathrm{M}'_{\mathrm{SI}}$	$M_{\mathrm{e}} = 0.9110\,\mathrm{M}'_{\mathrm{SI}}$
$M_{\mathrm{p}} = 1673\,\mathrm{M}'_{\mathrm{SI}}$	$M_{\mathrm{p}} = 1673\,\mathrm{M}'_{\mathrm{SI}}$
$a = 52.92\,\ell'_{\mathrm{SI}}$	$a = 0.5292\,\ell'_{\mathrm{SI}}$
$E_1 = 0.000\,218\,\mathrm{E}'_{\mathrm{SI}}$	$E_1 = 2.180\,\mathrm{E}'_{\mathrm{SI}}$

$|q'_{\mathrm{E}u}| \le 5$. Since three basic units can be chosen, three scaling exponents $q'_{\mathrm{E}u}$ may also be chosen. All other scaling exponents are fixed by this choice.

As examples in Table 12.2 we give two sets of scaled SI units. The set in the left column is based on the choice of scale factors for action, mass, and velocity,

$$\mathrm{A}'_{\mathrm{SI}} = 10^{-34}\,\mathrm{A}_{\mathrm{SI}} \quad , \quad \mathrm{v}'_{\mathrm{SI}} = 10^{8}\,\mathrm{v}_{\mathrm{SI}} \quad , \quad \mathrm{M}'_{\mathrm{SI}} = 10^{-30}\,\mathrm{M}_{\mathrm{SI}} \quad ,$$

which ensures that $\hbar$, c, and M_{e} have numerical values close to unity. For the set in the right column,

$$\mathrm{A}'_{\mathrm{SI}} = 10^{-34}\,\mathrm{A}_{\mathrm{SI}} \quad , \quad \mathrm{E}'_{\mathrm{SI}} = 10^{-18}\,\mathrm{E}_{\mathrm{SI}} \quad , \quad \mathrm{M}'_{\mathrm{SI}} = 10^{-30}\,\mathrm{M}_{\mathrm{SI}}$$

were chosen. Note that in this case the powers of 10 of the scale factors for E'_{SI} and M'_{SI} have to be chosen either even for both or odd for both to ensure that the square roots appearing will again be integer powers of 10.

12.1.4 Atomic and Subatomic Units

Scaling factors, or at least scaling factors with large absolute powers of 10, are unnecessary if one chooses units that are 'natural' to the system studied. One selects three quantities typical for the system and sets their numerical values equal to one. For questions of atomic physics it is most natural to

Table 12.3. Atomic units

Atomic Units (Based on $\hbar_{\mathrm{Na}} = c_{\mathrm{Na}} = M_{\mathrm{eNa}} = 1$)
$1\,\mathrm{E_a} = 1\,\mathrm{M_a}\,\mathrm{v_a^2} = M_\mathrm{e}\,c^2 = 8.188 \times 10^{-14}\,\mathrm{m^2\,kg\,s^{-2}}$
$1\,\ell_\mathrm{a} = 1\,\mathrm{A_a}/\mathrm{M_a}\,\mathrm{v_a} = \hbar/M_\mathrm{e}\,c = \lambda\!\!\!^{-}_\mathrm{e} = 3.863 \times 10^{-13}\,\mathrm{m}$
$1\,\mathrm{t_a} = 1\,\mathrm{A_a}/\mathrm{M_a}\,\mathrm{v_a^2} = \lambda\!\!\!^{-}_\mathrm{e}/c = 1.288 \times 10^{-21}\,\mathrm{s}$
$1\,\omega_\mathrm{a} = 1\,\mathrm{t_a^{-1}} = 0.7764 \times 10^{21}\,\mathrm{s^{-1}}$
$1\,\mathrm{A_a} = \hbar = 1.055 \times 10^{-34}\,\mathrm{m^2\,kg\,s^{-1}}$
$1\,\mathrm{v_a} = c = 2.998 \times 10^{8}\,\mathrm{m\,s^{-1}}$
$1\,\mathrm{M_a} = M_\mathrm{e} = 9.110 \times 10^{-31}\,\mathrm{kg}$
$1\,\mathrm{p_a} = 1\,\mathrm{M_a}\,\mathrm{v_a} = M_\mathrm{e}\,c = 2.731 \times 10^{-22}\,\mathrm{m\,kg\,s^{-1}}$

Constants	
$\hbar = \hbar_\mathrm{Na}\,\mathrm{A_a} = \hbar_\mathrm{NSI}\,\mathrm{A_{SI}}$,	$\hbar_\mathrm{Na} = 1 \rightarrow \mathrm{A_a} = \hbar_\mathrm{NSI}\,\mathrm{A_{SI}} = 1.055 \times 10^{-34}\,\mathrm{m^2\,kg\,s^{-1}}$
$c = c_\mathrm{Na}\,\mathrm{v_a} = c_\mathrm{NSI}\,\mathrm{v_{SI}}$,	$c_\mathrm{Na} = 1 \rightarrow \mathrm{v_a} = c_\mathrm{NSI}\,\mathrm{A_{SI}} = 2.998 \times 10^{8}\,\mathrm{m\,s^{-1}}$
$M_\mathrm{e} = M_\mathrm{eNa}\,\mathrm{M_a} = M_\mathrm{eNSI}\,\mathrm{M_{SI}}$,	$M_\mathrm{eNa} = 1 \rightarrow \mathrm{M_a} = M_\mathrm{eNSI}\,\mathrm{M_{SI}} = 9.110 \times 10^{-31}\,\mathrm{kg}$
$M_\mathrm{p} = M_\mathrm{pNa}\,\mathrm{M_a} = M_\mathrm{pNSI}\,\mathrm{M_{SI}}$,	$M_\mathrm{pNa} = M_\mathrm{pNSI}\,\mathrm{M_{SI}}/\mathrm{M_a} = 1836$
$a = a_\mathrm{Na}\,\ell_\mathrm{a} = a_\mathrm{NSI}\,\ell_\mathrm{SI}$,	$a_\mathrm{Na} = a_\mathrm{NSI}\,\ell_\mathrm{SI}/\ell_\mathrm{a} = 137.0$
$E_1 = E_{1\mathrm{Na}}\,\mathrm{E_a} = E_{1\mathrm{NSI}}\,\mathrm{E_{SI}}$,	$E_{1\mathrm{Na}} = E_{1\mathrm{NSI}}\,\mathrm{E_{SI}}/\mathrm{E_a} = 2.662 \times 10^{-5}$

choose Planck's constant, the velocity of light, and the electron mass as these quantities. The so-defined *atomic units* are listed in Table 12.3. All actions are expressed in multiples of $\hbar$, all velocities in multiples of c, and all masses in multiples of M_e. The unit of length is

$$1\,\ell_a = \hbar/M_\mathrm{e}\,c = \lambda\!\!\!^{-}_\mathrm{e} = 3.863 \times 10^{-13}\,\mathrm{m} \quad ,$$

which is called the *Compton wavelength* of the electron. The Bohr radius is $137\,\lambda\!\!\!^{-}_\mathrm{e}$. The unit of time is the time it takes for a light pulse to traverse one unit of length.

Many phenomena in nuclear physics are best treated in *subatomic units*, which are obtained by using $\hbar$, c, and the proton mass M_p as basic units, Table 12.4.

12.1.5 Data-Table Units

Often energies of atomic systems are given in electron volts,

$$1\,\mathrm{eV} = 1.602 \times 10^{-19}\,\mathrm{m^2\,kg\,s^{-2}} \quad .$$

A system of units based on the electron volt, the meter, and the second, which we call *data-table units* (which we identify by an index d), is presented in Table 12.5. You need scale factors in order to use this system. Table 12.6 contains two useful sets of scaled data-table units.

Table 12.4. Subatomic units

Subatomic Units (Based on $\hbar_{\mathrm{Ns}} = c_{\mathrm{Ns}} = M_{\mathrm{pNs}} = 1$)
$1\,\mathrm{E_s} = 1\,\mathrm{M_s}\,\mathrm{v_s^2} = M_\mathrm{p}\,c^2 = 1.504 \times 10^{-10}\,\mathrm{m^2\,kg\,s^{-2}}$
$1\,\ell_\mathrm{s} = 1\,\mathrm{A_a/M_a\,v_a} = \hbar/M_\mathrm{p}\,c = \lambda\!\!\!^{-}_\mathrm{p} = 2.103 \times 10^{-16}\,\mathrm{m}$
$1\,\mathrm{t_s} = 1\mathrm{A_a/M_a\,v_a^2} = \lambda\!\!\!^{-}_\mathrm{p}/c = 7.015 \times 10^{-25}\,\mathrm{s}$
$1\,\omega_\mathrm{s} = 1\,\mathrm{t_a^{-1}} = 1.426 \times 10^{24}\,\mathrm{s^{-1}}$
$1\,\mathrm{A_s} = \hbar = 1.055 \times 10^{-34}\,\mathrm{m^2\,kg\,s^{-1}}$
$1\,\mathrm{v_s} = c = 2.998 \times 10^{8}\,\mathrm{m\,s^{-1}}$
$1\,\mathrm{M_s} = M_\mathrm{p} = 1.673 \times 10^{-27}\,\mathrm{kg}$
$1\,\mathrm{p_s} = 1\,\mathrm{M_s\,v_s} = M_\mathrm{p}\,c = 5.016 \times 10^{-19}\,\mathrm{m\,kg\,s^{-1}}$
Constants
$\hbar = \hbar_{\mathrm{Ns}}\,\mathrm{A_s} = \hbar_{\mathrm{NSI}}\,\mathrm{A_{SI}}$, $\hbar_{\mathrm{Ns}} = 1 \rightarrow \mathrm{A_s} = \hbar_{\mathrm{NSI}}\,\mathrm{A_{SI}} = 1.055 \times 10^{-34}\,\mathrm{m^2\,kg\,s^{-1}}$
$c = c_{\mathrm{Ns}}\,\mathrm{v_s} = c_{\mathrm{NSI}}\,\mathrm{v_{SI}}$, $c_{\mathrm{Ns}} = 1 \rightarrow \mathrm{v_s} = c_{\mathrm{NSI}}\,\mathrm{A_{SI}} = 2.998 \times 10^{8}\,\mathrm{m\,s^{-1}}$
$M_\mathrm{e} = M_{\mathrm{eNs}}\,\mathrm{M_s} = M_{\mathrm{eNSI}}\,\mathrm{M_{SI}}$, $M_{\mathrm{eNs}} = M_{\mathrm{eNSI}}\,\mathrm{M_{SI}/M_s} = 5.445 \times 10^{-4}$
$M_\mathrm{p} = M_{\mathrm{pNs}}\,\mathrm{M_s} = M_{\mathrm{pNSI}}\,\mathrm{M_{SI}}$, $M_{\mathrm{pNs}} = 1 \rightarrow \mathrm{M_s} = M_{\mathrm{pNSI}}\,\mathrm{M_{SI}} = 1.673 \times 10^{-27}\,\mathrm{kg}$
$a = a_{\mathrm{Ns}}\,\ell_\mathrm{s} = a_{\mathrm{NSI}}\,\ell_{\mathrm{SI}}$, $a_{\mathrm{Ns}} = a_{\mathrm{NSI}}\,\ell_{\mathrm{SI}}/\ell_\mathrm{s} = 2.516 \times 10^{5}$
$E_1 = E_{1\mathrm{Ns}}\,\mathrm{E_s} = E_{1\mathrm{NSI}}\,\mathrm{E_{SI}}$, $E_{1\mathrm{Ns}} = E_{1\mathrm{NSI}}\,\mathrm{E_{SI}/E_s} = 1.449 \times 10^{-8}$

Table 12.5. Data-table units

Data Table Units (Based on eV, m, s)	
Quantity	Unit
Energy	$1\,\mathrm{E_d} = 1\,\mathrm{eV}$
Length	$1\,\ell_\mathrm{d} = 1\,\mathrm{m}$
Time	$1\,\mathrm{t_d} = 1\,\mathrm{s}$
Angular Frequency	$1\,\omega_\mathrm{d} = 1\,\mathrm{t_d^{-1}} = 1\,\mathrm{s^{-1}}$
Action	$1\,\mathrm{A_d} = 1\,\mathrm{E_d}\,\mathrm{t_d} = 1\,\mathrm{eV\,s}$
Velocity	$1\,\mathrm{v_d} = 1\,\ell_\mathrm{d}\,\mathrm{t_d^{-1}} = 1\,\mathrm{m\,s^{-1}}$
Mass	$1\,\mathrm{M_d} = \mathrm{E_d}\,\mathrm{v_d^{-2}} = 1\,\mathrm{eV\,s^2\,m^{-2}}$
Momentum	$1\,\mathrm{p_d} = 1\,\mathrm{M_d}\,\mathrm{v_d} = 1\,\mathrm{eV\,s\,m^{-1}}$
Constants	
Planck's Constant	$\hbar = \hbar_{\mathrm{Nd}}\,\mathrm{A_d}$, $\hbar_{\mathrm{Nd}} = 0.6582 \times 10^{-15}$
Speed of Light	$c = c_{\mathrm{Nd}}\,\mathrm{v_d}$, $c_{\mathrm{Nd}} = 2.998 \times 10^{8}$
Electron Mass	$M_\mathrm{e} = M_{\mathrm{eNd}}\,\mathrm{M_d}$, $M_{\mathrm{eNd}} = 5.685 \times 10^{-12}$
Proton Mass	$M_\mathrm{p} = M_{\mathrm{pNd}}\,\mathrm{M_d}$, $M_{\mathrm{pNd}} = 1.044 \times 10^{-8}$
Bohr Radius	$a = a_{\mathrm{Nd}}\,\ell_\mathrm{d}$, $a_{\mathrm{Nd}} = 0.5292 \times 10^{-10}$
Energy of Hydrogen Ground State	$E_1 = E_{1\mathrm{Nd}}\,\mathrm{E_d}$, $E_{1\mathrm{Nd}} = -13.61$

A different system of units still (which we identify by an index c) measures masses in units $1\,M_c = 1\,\mathrm{eV}/c^2$. It is obtained through the following set of equations:

Table 12.6. Scaled data-table units

Examples of Scaled Data-Table Units	
with a choice of scale factors for energy, length, and mass suitable for problems of atomic physics	with a choice of scale factors for energy, length, and mass suitable for problems of nuclear physics
$\mathrm{E}'_\mathrm{d} = 1\,\mathrm{eV}$	$\mathrm{E}'_\mathrm{d} = 10^6\,\mathrm{E_d} = 1\,\mathrm{MeV}$
$\ell'_\mathrm{d} = 10^{-9}\,\ell_\mathrm{d} = 10^{-9}\,\mathrm{m}$	$\ell'_\mathrm{d} = 10^{-14}\,\ell_\mathrm{d} = 10^{-14}\,\mathrm{m}$
$\mathrm{t}'_\mathrm{d} = \ell'_\mathrm{d}\sqrt{\mathrm{M}'_\mathrm{d}/\mathrm{E}'_\mathrm{d}} = 10^{-15}\,\mathrm{s}$	$\mathrm{t}'_\mathrm{d} = \ell'_\mathrm{d}\sqrt{\mathrm{M}'_\mathrm{d}/\mathrm{E}'_\mathrm{d}} = 10^{-21}\,\mathrm{s}$
$\omega'_\mathrm{d} = 1/\mathrm{t}'_\mathrm{d} = 10^{15}\,\mathrm{s}^{-1}$	$\omega'_\mathrm{d} = 1/\mathrm{t}'_\mathrm{d} = 10^{21}\,\mathrm{s}^{-1}$
$\mathrm{A}'_\mathrm{d} = \ell'_\mathrm{d}\sqrt{\mathrm{M}'_\mathrm{d}\,\mathrm{E}'_\mathrm{d}} = 10^{-15}\,\mathrm{eV\,s}$	$\mathrm{A}'_\mathrm{d} = \ell'_\mathrm{d}\sqrt{\mathrm{M}'_\mathrm{d}\,\mathrm{E}'_\mathrm{d}} = 10^{-15}\,\mathrm{eV\,s}$
$\mathrm{v}'_\mathrm{d} = \sqrt{\mathrm{E}'_\mathrm{d}/\mathrm{M}'_\mathrm{d}} = 10^{6}\,\mathrm{m\,s}^{-1}$	$\mathrm{v}'_\mathrm{d} = \sqrt{\mathrm{E}'_\mathrm{d}/\mathrm{M}'_\mathrm{d}} = 10^{7}\,\mathrm{m\,s}^{-1}$
$\mathrm{M}'_\mathrm{d} = 10^{-12}\,\mathrm{M_d} = 10^{-12}\,\mathrm{eV\,m^{-2}\,s^2}$	$\mathrm{M}'_\mathrm{d} = 10^{-8}\,\mathrm{M_d} = 10^{-8}\,\mathrm{eV\,m^{-2}\,s^2}$
$\mathrm{p}'_\mathrm{d} = \sqrt{\mathrm{E}'_\mathrm{d}\,\mathrm{M}'_\mathrm{d}} = 10^{-6}\,\mathrm{eV\,m^{-1}\,s}$	$\mathrm{p}'_\mathrm{d} = \sqrt{\mathrm{E}'_\mathrm{d}\,\mathrm{M}'_\mathrm{d}} = 10^{-7}\,\mathrm{eV\,m^{-1}\,s}$
Constants	
$\hbar = 0.6582\,\mathrm{A}'_\mathrm{d}$	$\hbar = 0.6582\,\mathrm{A}'_\mathrm{d}$
$c = 299.8\,\mathrm{v}'_\mathrm{d}$	$c = 29.98\,\mathrm{v}'_\mathrm{d}$
$M_\mathrm{e} = 5.685\,\mathrm{M}'_\mathrm{d}$	$M_\mathrm{e} = 0.000\,5685\,\mathrm{M}'_\mathrm{d}$
$M_\mathrm{p} = 10\,440\,\mathrm{M}'_\mathrm{d}$	$M_\mathrm{p} = 1.044\,\mathrm{M}'_\mathrm{d}$
$a = 0.052\,92\,\ell'_\mathrm{d}$	$a = 0.5292 \times 10^{-4}\,\ell'_\mathrm{d}$
$E_1 = -13.61\,\mathrm{E}'_\mathrm{d}$	$E_1 = -13.61 \times 10^{-6}\,\mathrm{E}'_\mathrm{d}$

$$M = M_\mathrm{Nd}\,\mathrm{M_d} = M_\mathrm{Nd}\,\mathrm{eV\,s^2\,m^{-2}} = M_\mathrm{Nd}\frac{\mathrm{eV}}{c^2}(c^2\,\mathrm{s^{-2}\,m^{-2}})$$

$$= M_\mathrm{Nd}(2.998 \times 10^8)^2\,\mathrm{eV}/c^2 = M_\mathrm{Nc}\,\mathrm{eV}/c^2 \quad ,$$

$$M_\mathrm{Nc} = M_\mathrm{Nd} \times 8.988 \times 10^{16}$$

or

$$1\,\mathrm{eV}/c^2 = 8.988 \times 10^{16}\,\mathrm{eV\,s^2\,m^{-2}} = 1.783 \times 10^{-36}\,\mathrm{kg} \quad .$$

In units eV/c^2 the electron and proton mass are

$$M_\mathrm{e} = 0.5110 \times 10^6\,\mathrm{eV}/c^2 \quad , \quad M_\mathrm{p} = 938.3 \times 10^6\,\mathrm{eV}/c^2 \quad .$$

12.1.6 Special Scales

The Hydrogen Atom The energy spectrum in the hydrogen atom is given by (7.52). In Sect. 7.2 we use atomic units but we allow a change of the input value $\sqrt{\alpha}$. What, then, is the meaning of the energy scale in Fig. 7.4 in the default case $\alpha = 1$? If the right-hand side of (7.52) is written with α^2 replaced by 1, it means that a factor α^2 is missing in the equation. We absorb this factor into the energy unit by defining a new unit

$$\mathrm{E_h} = \alpha^2\,\mathrm{E_a} = 5.328 \times 10^{-5}\,\mathrm{E_a} \quad ,$$

where $\mathrm{E_a}$ is the energy in atomic units. For the lowest eigenvalue in Fig. 7.4 (which has $\ell = 2$ and then $n = 3$) we read off $E_3 = 0.0555\,\mathrm{E_h} = 2.957 \times 10^{-6}\,\mathrm{E_a}$ as expected since $E_3 = \frac{1}{9}E_1$ and $E_1 = 2.662 \times 10^{-5}\,\mathrm{E_a}$, see Table 12.3.

The Harmonic Oscillator Similarly the energy scale of the harmonic-oscillator eigenvalue spectrum described by (7.45) and shown in Fig. 7.3 for the default value of $\omega = 1\omega_\mathrm{a}$ in atomic units can be interpreted as the spectrum of an oscillator with arbitrary angular frequency $\omega = \omega_\mathrm{Na}\omega_\mathrm{a}$ if the numerical values on the energy scale are given the units

$$E_\mathrm{o} = E_\mathrm{a}\,\omega_\mathrm{Na} \quad .$$

12.2 Argand Diagrams and Unitarity for One-Dimensional Problems

12.2.1 Probability Conservation and the Unitarity of the Scattering Matrix

Scattering processes in one spatial dimension offer a simple study of the properties of the S matrix. In stepwise constant potentials of the kind (3.16) the wave function is of complex exponential form in the various regions. We represent the wave functions in the region 1 of vanishing potential far to the left as

$$\varphi_1(x) = A_1' \mathrm{e}^{\mathrm{i}k_1 x} + B_1' \mathrm{e}^{-\mathrm{i}k_1 x} \tag{12.7}$$

and in the region N far to the right as

$$\varphi_N(x) = A_N' \mathrm{e}^{\mathrm{i}k_N x} + B_N' \mathrm{e}^{-\mathrm{i}k_N x} \quad . \tag{12.8}$$

Here, we have also included the term $B_N' \mathrm{e}^{-\mathrm{i}k_N x}$ representing a wave coming in from large values of x propagating to the left. Obviously, there are two physical scattering situations:

i) incoming right-moving wave at negative x values represented by the term $A_1' \mathrm{e}^{\mathrm{i}k_1 x}$: the outgoing waves are the transmitted wave $A_N' \mathrm{e}^{\mathrm{i}k_N x}$ and the reflected wave $B_1' \mathrm{e}^{-\mathrm{i}k_1 x}$;
ii) incoming left-moving wave at large positive x values, represented by the term $B_N' \mathrm{e}^{-\mathrm{i}k_N x}$: the outgoing waves are the transmitted wave $B_1 \mathrm{e}^{-\mathrm{i}k_1 x}$ and the reflected wave $A_N \mathrm{e}^{\mathrm{i}k_N x}$.

For real potentials $V(x)$ every solution of the time-dependent Schrödinger equation keeps the normalization at all times. Thus, the integral of the probability density over the whole x axis does not change with time. This is interpreted as probability conservation. It can also be expressed as the conservation of the *probability-current density*

$$j(x,t) = \frac{\hbar}{2mi}\left(\psi^*\frac{\partial}{\partial x}\psi - \psi\frac{\partial}{\partial x}\psi^*\right) \tag{12.9}$$

through a *continuity equation*

$$\frac{\partial \varrho}{\partial t} + \frac{\partial j}{\partial x} = 0 \quad , \tag{12.10}$$

where

$$\varrho(x,t) = \psi^*(x,t)\psi(x,t) \tag{12.11}$$

is the probability density. For stationary states

$$\psi(x,t) = \mathrm{e}^{-\frac{\mathrm{i}}{\hbar}Et}\varphi(x) \tag{12.12}$$

the probability density ϱ and current density j are time independent and probability conservation is tantamount to

$$\frac{\mathrm{d}}{\mathrm{d}x}j(x) = 0 \quad , \tag{12.13}$$

i.e., the probability-current density is constant in x.

For the wave functions (12.7), (12.8) this means

$$k_1(|A_1'|^2 - |B_1'|^2) = k_N(|A_N'|^2 - |B_N'|^2) \tag{12.14}$$

or

$$\left|\sqrt{\frac{k_N}{k_1}}A_N'\right|^2 + |B_1'|^2 = |A_1'|^2 + \left|\sqrt{\frac{k_N}{k_1}}B_N'\right|^2 \tag{12.15}$$

for arbitrary values of A_1' and B_N'. We associate the quantities on either side of the above equation with the absolute squares of the components of two-dimensional complex vectors

$$\begin{pmatrix} A_N \\ B_1 \end{pmatrix} = \begin{pmatrix} \sqrt{\frac{k_N}{k_1}}A_N' \\ B_1' \end{pmatrix} \quad , \quad \begin{pmatrix} A_1 \\ B_N \end{pmatrix} = \begin{pmatrix} A_1' \\ \sqrt{\frac{k_N}{k_1}}B_N' \end{pmatrix} \quad . \tag{12.16}$$

The equation (12.15) derived from current conservation then states the equality of the length of these two vectors. Thus, they may be related by a complex 2×2 matrix

$$S = \begin{pmatrix} S_{11} & S_{12} \\ S_{21} & S_{22} \end{pmatrix}$$

defined by

$$\begin{pmatrix} A_N \\ B_1 \end{pmatrix} = S\begin{pmatrix} A_1 \\ B_N \end{pmatrix} \quad , \tag{12.17}$$

which is unitary, i.e., S and its adjoint

$$S^\dagger = \begin{pmatrix} S_{11}^* & S_{21}^* \\ S_{12}^* & S_{22}^* \end{pmatrix} \tag{12.18}$$

fulfill the relation

$$S\,S^\dagger = 1 \quad , \tag{12.19}$$

or

$$\begin{pmatrix} S_{11}S_{11}^* + S_{12}S_{12}^* & S_{11}S_{21}^* + S_{12}S_{22}^* \\ S_{21}S_{11}^* + S_{22}S_{12}^* & S_{21}S_{21}^* + S_{22}S_{22}^* \end{pmatrix} = \begin{pmatrix} 1 & 0 \\ 0 & 1 \end{pmatrix} \quad . \tag{12.20}$$

The unitary matrix S is called the *scattering matrix* or *S matrix*. If we consider the amplitudes A_1 and B_N given to be one or zero, we have two cases:

i) wave coming in from the left:

$$A_1 = 1 \quad , \quad B_N = 0 \quad , \quad A_N = S_{11} \quad , \quad B_1 = S_{21} \quad , \tag{12.21}$$

i.e., S_{11} is the transmission coefficient and S_{21} the reflection coefficient for a right-moving incoming wave;

ii) wave coming in from the right:

$$A_1 = 0 \quad , \quad B_N = 1 \quad , \quad A_N = S_{12} \quad , \quad B_1 = S_{22} \quad , \tag{12.22}$$

i.e., S_{12} is the reflection coefficient and S_{22} the transmission coefficient for a left-moving wave.

12.2.2 Time Reversal and the Scattering Matrix

Invariance under *time reversal* implies that $\varphi^*(x)$ is also a solution of the stationary Schrödinger equation with the real potential $V(x)$. Because of the change of the sign of the exponents in the wave functions, incoming and outgoing waves are interchanged and we find that the scattering matrix also relates the vectors

$$\begin{pmatrix} B_N^* \\ A_1^* \end{pmatrix} = S \begin{pmatrix} B_1^* \\ A_N^* \end{pmatrix} = \begin{pmatrix} S_{11} & S_{12} \\ S_{21} & S_{22} \end{pmatrix} \begin{pmatrix} B_1^* \\ A_N^* \end{pmatrix} \quad . \tag{12.23}$$

By complex conjugation we find

$$\begin{pmatrix} B_N \\ A_1 \end{pmatrix} = S^* \begin{pmatrix} B_1 \\ A_N \end{pmatrix} = \begin{pmatrix} S_{11}^* & S_{12}^* \\ S_{21}^* & S_{22}^* \end{pmatrix} \begin{pmatrix} B_1 \\ A_N \end{pmatrix} \tag{12.24}$$

or, by rearranging,

$$\begin{pmatrix} A_1 \\ B_N \end{pmatrix} = \begin{pmatrix} S_{22}^* & S_{21}^* \\ S_{12}^* & S_{11}^* \end{pmatrix} \begin{pmatrix} A_N \\ B_1 \end{pmatrix} \quad . \tag{12.25}$$

Putting this into the form of (12.17) we have

$$\begin{pmatrix} A_N \\ B_1 \end{pmatrix} = \begin{pmatrix} S_{22}^* & S_{21}^* \\ S_{12}^* & S_{11}^* \end{pmatrix}^{-1} \begin{pmatrix} A_1 \\ B_N \end{pmatrix} \quad . \tag{12.26}$$

So, by comparison, we find the relation

$$\begin{pmatrix} S_{11} & S_{12} \\ S_{21} & S_{22} \end{pmatrix}^{-1} = \begin{pmatrix} S_{22}^* & S_{21}^* \\ S_{12}^* & S_{11}^* \end{pmatrix} \quad . \tag{12.27}$$

Because of the unitarity of the S matrix, we have $S^{-1} = S^\dagger$ and thus

$$\begin{pmatrix} S_{11}^* & S_{21}^* \\ S_{12}^* & S_{22}^* \end{pmatrix} = \begin{pmatrix} S_{22}^* & S_{21}^* \\ S_{12}^* & S_{11}^* \end{pmatrix} \quad , \tag{12.28}$$

so that time-reversal invariance is equivalent to

$$S_{11} = S_{22} \quad . \tag{12.29}$$

The time-reversal-invariant S matrix has the particular form

$$S = \begin{pmatrix} S_{11} & S_{12} \\ S_{21} & S_{11} \end{pmatrix} \quad . \tag{12.30}$$

For real potentials, time-reversal invariance holds true. Thus (12.21), (12.22), (12.29) show that the transmission coefficients for right-moving and left-moving incoming waves are equal in this case. Just as a side remark we note that *space-reflection* symmetry reduces S further to

$$S_{12} = S_{21} \tag{12.31}$$

so that the spatial-reflection invariance leads to

$$S = \begin{pmatrix} S_{11} & S_{12} \\ S_{12} & S_{11} \end{pmatrix} \quad . \tag{12.32}$$

12.2.3 Diagonalization of the Scattering Matrix

We return to the time-reversal invariant form (12.30) and investigate the restrictions of unitarity (12.20):

$$\begin{aligned} S_{11}S_{11}^* + S_{12}S_{12}^* &= 1 \quad , & S_{11}S_{21}^* + S_{12}S_{11}^* &= 0 \quad , \\ S_{21}S_{11}^* + S_{11}S_{12}^* &= 0 \quad , & S_{21}S_{21}^* + S_{11}S_{11}^* &= 1 \quad , \end{aligned} \tag{12.33}$$

which represent only two independent relations. The off-diagonal relations yield

$$S_{21} = -\frac{S_{11}}{S_{11}^*} S_{12}^* \quad , \tag{12.34}$$

i.e., the second off-diagonal element is determined by the first. Thus, the only relation remaining besides (12.34) is

$$S_{11}S_{11}^{*} + S_{12}S_{12}^{*} = 1 \quad , \tag{12.35}$$

which can be solved by

$$S_{11} = \mathrm{e}^{2\mathrm{i}s_{11}} \cos\sigma \quad , \quad S_{12} = \mathrm{e}^{2\mathrm{i}s_{12}} \sin\sigma \quad , \tag{12.36}$$

yielding

$$S_{21} = \mathrm{e}^{2\mathrm{i}s_{21}} \sin\sigma \quad , \quad s_{21} = 2s_{11} - s_{12} + \pi/2 \quad , \tag{12.37}$$

with real phases s_{11}, s_{12}, s_{21} and the real angle σ. Because S is a unitary matrix, it can be diagonalized by a unitary transformation

$$S^{\mathrm{D}} = U^{\dagger} S U \tag{12.38}$$

or, equivalently,

$$S = U S^{\mathrm{D}} U^{\dagger} \quad . \tag{12.39}$$

The *diagonalized S matrix* turns out to be

$$S^{\mathrm{D}} = \begin{pmatrix} S_1^{\mathrm{D}} & 0 \\ 0 & S_2^{\mathrm{D}} \end{pmatrix} = \begin{pmatrix} \mathrm{e}^{2\mathrm{i}\delta_1} & 0 \\ 0 & \mathrm{e}^{2\mathrm{i}\delta_2} \end{pmatrix} \tag{12.40}$$

with the scattering phases δ_1 and δ_2 determined by

$$\delta_1 = s_{11} + \frac{1}{2}\sigma \quad , \quad \delta_2 = s_{11} - \frac{1}{2}\sigma \quad . \tag{12.41}$$

It is unitary itself, i.e.,

$$S_1^{\mathrm{D}*} S_1^{\mathrm{D}} = 1 \quad , \quad S_2^{\mathrm{D}*} S_2^{\mathrm{D}} = 1 \quad . \tag{12.42}$$

The unitary matrix U diagonalizing S has the form

$$U = \frac{1}{\sqrt{2}} \begin{pmatrix} -\mathrm{i}\,\mathrm{e}^{-2\mathrm{i}s_{11}} & \mathrm{i}\,\mathrm{e}^{-2\mathrm{i}s_{11}} \\ \mathrm{e}^{-2\mathrm{i}s_{12}} & \mathrm{e}^{-2\mathrm{i}s_{12}} \end{pmatrix} \quad . \tag{12.43}$$

The matrix elements of the S matrix can be expressed in terms of S_1^{D} and S_2^{D} by

$$\begin{gathered} S_{11} = \tfrac{1}{2}(S_1^{\mathrm{D}} + S_2^{\mathrm{D}}) \quad , \quad S_{12} = \tfrac{1}{2\mathrm{i}}\mathrm{e}^{-2\mathrm{i}(s_{11}-s_{12})}(S_1^{\mathrm{D}} - S_2^{\mathrm{D}}) \quad , \\ S_{21} = -\tfrac{1}{2\mathrm{i}}\mathrm{e}^{2\mathrm{i}(s_{11}-s_{12})}(S_1^{\mathrm{D}} - S_2^{\mathrm{D}}) \quad , \quad S_{22} = S_{11} \quad . \end{gathered} \tag{12.44}$$

12.2.4 Argand Diagrams

We decompose the S-matrix elements into real and imaginary parts,

$$S_{ik} = \varrho_{ik} + \mathrm{i}\sigma_{ik} \quad \text{and} \quad S_i^{\mathrm{D}} = \varrho_i^{\mathrm{D}} + \mathrm{i}\sigma_i^{\mathrm{D}} \quad . \tag{12.45}$$

From (12.42), (12.44), i.e.,

$$\frac{1}{2}S_1^{\mathrm{D}} = S_{11} - \frac{1}{2}S_2^{\mathrm{D}} \quad \text{and} \quad S_1^{\mathrm{D}}S_1^{\mathrm{D}*} = 1 \quad , \tag{12.46}$$

we have

$$\left(\varrho_{11} - \frac{1}{2}\varrho_2^{\mathrm{D}}\right)^2 + \left(\sigma_{11} - \frac{1}{2}\sigma_2^{\mathrm{D}}\right)^2 = \frac{1}{4} \quad . \tag{12.47}$$

This represents the equation of the curve of $S_{11}(k) = \varrho_{11}(k) + \mathrm{i}\sigma_{11}(k)$ in an *Argand plot* in the complex plane. Here, the wave number k of the incident wave plays the role of the curve parameter.

Whenever the transmission amplitude S_{11} has the absolute square $|S_{11}|^2 = 1$, because of (12.35) the reflection amplitude S_{12} vanishes so that the wave numbers k_i of the intersections of the curve $S_{12}(k_i)$ with the origin of the Argand plot are the transmission resonances.

If $\varrho_2^{\mathrm{D}}(k)$ and $\sigma_2^{\mathrm{D}}(k)$ are slowly varying with k in a range where $\varrho_{11}(k)$ and $\sigma_{11}(k)$ are quickly varying functions,[1] then the above equation describes a circle of radius $\frac{1}{2}$ about a center with the coordinates $\frac{1}{2}\varrho_2^{\mathrm{D}}, \frac{1}{2}\sigma_2^{\mathrm{D}}$. These coordinates define a point on the circle

$$\frac{1}{4}\left(\varrho_2^{\mathrm{D}}\right)^2 + \frac{1}{4}\left(\sigma_2^{\mathrm{D}}\right)^2 = \frac{1}{4}S_2^{\mathrm{D}*}S_2^{\mathrm{D}} = \frac{1}{4} \tag{12.48}$$

of radius $\frac{1}{2}$ about the origin of the complex plane. By the same argument in a range in which $\varrho_1^{\mathrm{D}}(k)$, $\sigma_1^{\mathrm{D}}(k)$ are slowly changing whereas $\varrho_{11}(k)$, $\sigma_{11}(k)$ are quickly varying, we have instead

$$\left(\varrho_{11} - \frac{1}{2}\varrho_1^{\mathrm{D}}\right)^2 + \left(\sigma_{11} - \frac{1}{2}\sigma_1^{\mathrm{D}}\right)^2 = \frac{1}{4} \quad , \tag{12.49}$$

i.e., a circle of radius $\frac{1}{2}$ about the center $\frac{1}{2}\varrho_1^{\mathrm{D}}, \frac{1}{2}\sigma_1^{\mathrm{D}}$. The center itself moves slowly on a circle of radius $\frac{1}{2}$ about the origin.

The absolute square of the off-diagonal element S_{21} is given by the unitarity relation, see (12.33), (12.36),

$$S_{21}S_{21}^* = 1 - S_{11}S_{11}^* = 1 - \cos^2\sigma = \sin^2\sigma \quad . \tag{12.50}$$

In terms of the real and imaginary parts

[1] In this range $\varrho_1^{\mathrm{D}}(k)$, $\sigma_1^{\mathrm{D}}(k)$ are quickly varying because of $S_{11} = (1/2)(S_1^{\mathrm{D}} + S_2^{\mathrm{D}})$.

$$S_{21} = \varrho_{21} + \mathrm{i}\sigma_{21} \quad ; \qquad (12.51)$$

this is equivalent to

$$\varrho_{21}^2 + \sigma_{21}^2 = \sin^2\sigma \quad . \qquad (12.52)$$

The behavior of the reflection coefficient S_{21} is read off the equations (12.37), (12.44),

$$S_{21} - \frac{1}{2}\mathrm{e}^{2\mathrm{i}(s_{21}-s_{11}+\pi/4)}S_2^{\mathrm{D}} = \frac{1}{2}\mathrm{e}^{2\mathrm{i}(s_{21}-s_{11}-\pi/4)}S_1^{\mathrm{D}} \quad , \qquad (12.53)$$

which can be rewritten using the explicit form of S_1^{D} and S_2^{D}, (12.40),(12.41),

$$S_{21} - \frac{1}{2}\mathrm{e}^{\mathrm{i}(2s_{21}-\sigma+\pi/2)} = \frac{1}{2}\mathrm{e}^{\mathrm{i}(2s_{21}+\sigma-\pi/2)} \qquad (12.54)$$

or in terms of the real and imaginary parts (12.51) as ($r_{21} = 2s_{21} - \sigma + \pi/2$)

$$\left[\varrho_{21} - \frac{1}{2}\left(\varrho_2^{\mathrm{D}}\cos 2r_{21} - \sigma_2^{\mathrm{D}}\sin 2r_{21}\right)\right]^2 + \left[\sigma_{21} - \frac{1}{2}\left(\varrho_2^{\mathrm{D}}\sin 2r_{21} + \sigma_2^{\mathrm{D}}\cos 2r_{21}\right)\right]^2 = \frac{1}{4} \quad . \qquad (12.55)$$

Again, this is the equation of a circle in the complex $\varrho_{21}, \sigma_{21}$ plane about a center at $\exp[\mathrm{i}(2s_{21}-\sigma+\pi/2)]$ in the complex plane if this center only slowly varies with k.

12.2.5 Resonances

The nonvanishing matrix elements (12.40), (12.41)

$$S_1^{\mathrm{D}} = \mathrm{e}^{\mathrm{i}(2s_{11}+\sigma)} \quad \text{and} \quad S_2^{\mathrm{D}} = \mathrm{e}^{\mathrm{i}(2s_{11}-\sigma)} \qquad (12.56)$$

of the diagonalized S matrix lie on the unit circle in the complex plane. For varying wave number k the pointers represented by $S_1^{\mathrm{D}}(k)$ or $S_2^{\mathrm{D}}(k)$ move on the unit circle.

A *resonance* phenomenon occurs whenever one of the two matrix elements $S_1^{\mathrm{D}}(k)$ or $S_2^{\mathrm{D}}(k)$ moves through a large part of the unit circle in a small interval to both sides of the wave number k_{r} at resonance. For the scattering phases δ_1, δ_2 of the diagonalized S matrix, this means a fast increase by an angle close to π.

For definiteness we assume that the element $S_1^{\mathrm{D}}(k)$ exhibits this fast variation in the interval

$$k_{\mathrm{r}} - \kappa < k < k_{\mathrm{r}} + \kappa \qquad (12.57)$$

surrounding the resonance at k_r, whereas the other element $S_2^D(k)$ remains practically unchanged in this region,

$$S_2^D(k) \approx S_2^D(k_r) = e^{2i\delta_2(k_r)} \tag{12.58}$$

with

$$\delta_2(k_r) = s_{11}(k_r) - \frac{1}{2}\sigma(k_r) = \delta_{2r} \quad . \tag{12.59}$$

Here δ_{2r} is the phase at the wave number k_r.

Under these assumptions we find for

$$S_{11}(k) = \frac{1}{2}S_1^D(k) + \frac{1}{2}S_2^D(k_r) \tag{12.60}$$

the behavior

$$S_{11} = \frac{1}{2}e^{2i\delta_1(k)} + \frac{1}{2}e^{2i\delta_{2r}} = e^{2is_{11}}\cos\sigma \quad . \tag{12.61}$$

This represents a circle of radius $\frac{1}{2}$ about the center $\frac{1}{2}e^{2i\delta_{2r}}$, see Fig. 4.8. From the condition of a constant phase δ_{2r} we conclude that

$$\sigma(k_r) = 2[s_{11}(k_r) - \delta_{2r}] \tag{12.62}$$

and find that

$$S_{11} = e^{2is_{11}}\cos 2[s_{11}(k_r) - \delta_{2r}] \quad . \tag{12.63}$$

This shows that for $s_{11}(k_r) = \delta_{2r} = \delta_2(k_r)$ the circle of radius $\frac{1}{2}$ about the center $\exp(2i\delta_{2r})/2$ touches the unitarity circle of radius 1 about the origin of the complex plane.

Thus, the wave number k_r of the resonance of the diagonal element S_1^D also determines the resonance of the element S_{11}.

The fast variation of the phase δ_1 over the range $0 < \delta_1 < \pi$ in the neighborhood of a resonance leads to a change of the phase s_{11} of the matrix element S_{11} passing quickly through the value $\delta_{2r} = \delta_2(k_r)$. The size of the jump at k_r in the phase δ_1 depends on how completely the curve $S_1^D(k)$ overlaps with the unit circle in the neighborhood of k_r. For a full circle of $S_1^D(k)$ the phase $\delta_1(k)$ increases by $\pi/2$ from $\delta_2(k_r) - \frac{\pi}{4}$ to $\delta_2(k_r) + \frac{\pi}{4}$. For a half-circle of $S_1^D(k)$ the phase $s_{11}(k)$ of $S_{11}(k)$ changes by $\frac{\pi}{4}$ from $\delta_{2r} - \frac{\pi}{8}$ to $\delta_{2r} + \frac{\pi}{8}$.

The reflection coefficient exhibits a resonance for minimal transmission. According to (12.36)

$$|S_{21}|^2 = \sin^2\sigma \quad ;$$

this means $\sigma \simeq \pi/2$ and $|S_{21}|^2 \leq 1$. For simple square-well potentials, the approximate size of the phase s_{21} of S_{21} can be determined by coarse arguments. As an example we take the single repulsive square well, already

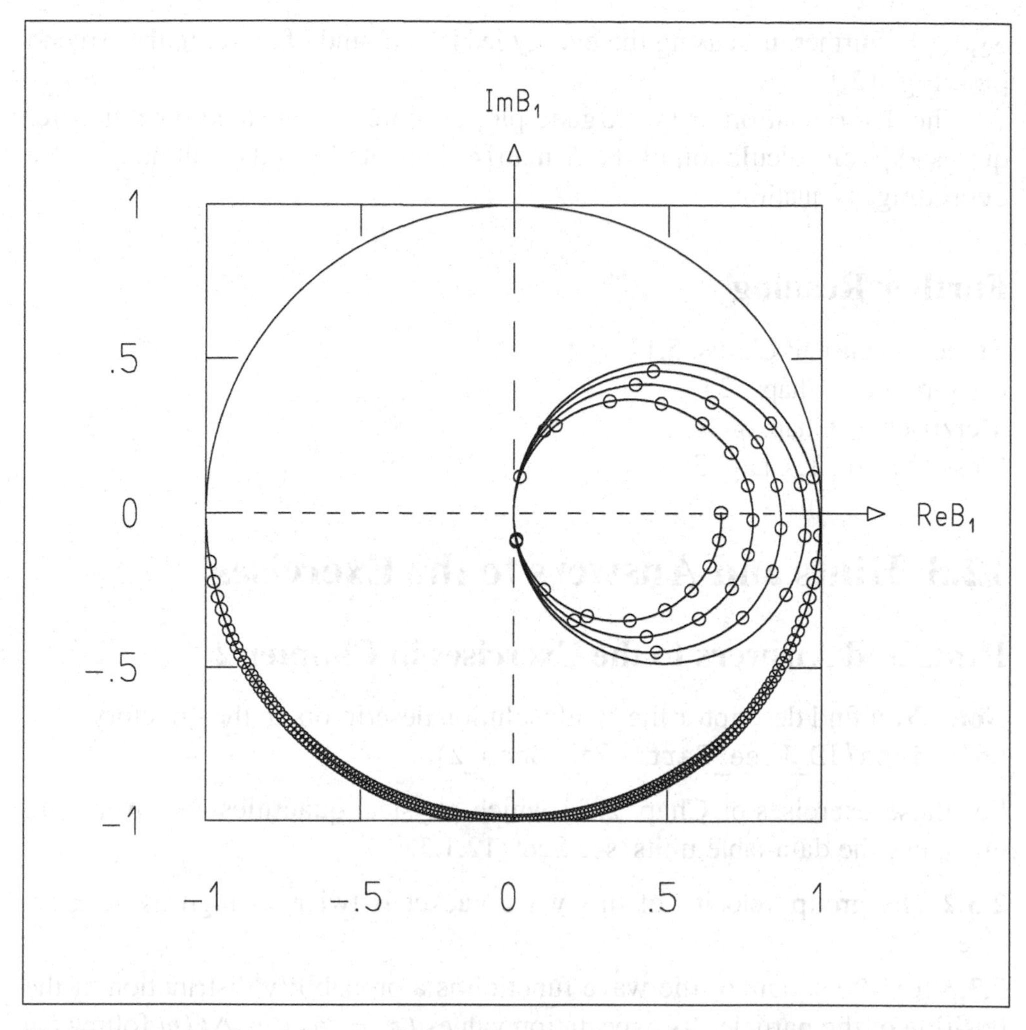

Fig. 12.1. Argand plot of the scattering-matrix element $S_{21} = B_1$ for the physical situation of Fig. 4.8

discussed in (4.51). For a kinetic energy low compared to the height V_0 of the repulsive potential, no transmission ($A_2 = 0$), i.e., only reflection with a phase shift π occurs. Thus, for $A_1 = 1$, this means that the reflection coefficient $B_1 = S_{21}$ starts at $B_1 = -1$ in the Argand plot. As long as transmission remains negligible, i.e., for $E < V_0$, the coefficient B_1 keeps the absolute value $|B_1| = 1$ and thus moves on the unit circle in the Argand plot. For $E = V_0$ reflection at a 'thinner medium' (see Sect. 4.1) occurs causing no phase shift, (4.44), (4.45), so that $S_{21} = B_1 = 1$. For growing energies $E > V_0$, the first transmission resonance at $k_2 = \ell\pi/d$ (4.54) is soon reached, so that there is no reflection, i.e., $S_{21} = B_1 = 0$. As soon as the energy is further increased, transmission goes through a minimum before it reaches the next resonance. This minimum occurs for reflection at a denser medium, so that no phase shift occurs, $s_{21} = 0$ and $S_{21} = B_1 = \sin\sigma$, $\sigma \approx \pi/2$, so that

$S_{21} \approx 1$. Further increasing the energy leads to a kind of spiral in the Argand plot, Fig. 12.1.

The determination of the Argand plot of more complicated potentials requires explicit calculation of the S-matrix elements from the solutions of the Schrödinger equation.

Further Reading

Brandt, Dahmen: Chaps. 5,11
Gasiorowicz: Chaps. 24
Merzbacher: Chap. 6
Messiah: Vol. 1, Chap. 3

12.3 Hints and Answers to the Exercises

Hints and Answers to the Exercises in Chapter 2

Note: You find descriptor files with solution descriptors in the directory `Solutions/1D_Free_Particle_(Chap_2)`.

For those exercises of Chap. 2, in which physical quantities are given with units, use the data-table units, see Sect. 12.1.5.

2.3.2 The group velocity of this wave packet is twice as high as in Exercise 2.3.1.

2.3.3 (c) The square of the wave function is a probability distribution of the position of the particle. Its expectation values $\langle x \rangle = x_0 + p_0 \Delta t/m$ follow the trajectory of free motion with the velocity $v_0 = p_0/m$. If the particle position after the time interval Δt is measured to be larger than its expectation value $\langle x \rangle = x_0 + p_0 \Delta t/m$ at time Δt, its momentum p must have been larger than the momentum expectation value p_0. Thus, the wavelength $\lambda = h/p$ in the spatial region to the right of $\langle x \rangle$ must be smaller than $\lambda_0 = h/p_0$. For the spatial region $x < \langle x \rangle$ the analogous arguments result in $\lambda > \lambda_0$.

2.3.4 The wavelength is halved because of the doubling of the momentum.

2.3.5 The wave packet widens as time passes. The phase velocity of the de Broglie waves is $p/2m$. Thus, the waves superimposed to form the wave packet move with different phase velocities so that the width of the wave packet spreads in time.

2.3.7 The behavior of $\langle x(t) \rangle$ for large t is determined by the asymptotic form of (2.23), which is discussed in Exercise 2.5.4.

2.3.8 (d) $p_0 = 5.685 \times 10^{-12}\,\mathrm{meV\,s\,m^{-1}}$. (e) The momentum width $\sigma_p = 0.5 \times 10^{-12}\,\mathrm{eV\,s\,m^{-1}}$ is relatively small compared to the momentum p_0. Thus, the spreading of the phase velocities is small. Therefore, the dispersion of the wave packet with time is small. (f) The unit at the x axis is mm.

2.3.9 (d) $p_0 = 17.06 \times 10^{-12}\,\mathrm{eV\,s\,m^{-1}}$.

2.3.10 (d) $p_0 = 12\,\mathrm{eV\,s\,m^{-1}}$. (e) See Exercise 2.3.8 (f). (f) The momentum expectation value is $p_0 = 12$, the momentum width of the wave packet is $\sigma_p = 9$. Thus, the Gaussian spectral function extends also significantly into the range of negative momenta. Therefore, the wave packet contains contributions propagating to the left. (g) The small wavelengths to the right belong to high momenta of a particle moving right. The large wavelengths to the left of the center indicate that a particle found at these x values is at rest. The small wavelengths to the left indicate a particle with some momentum pointing to the left.

2.4.1 The quantum-mechanical wave packet widens as time passes, whereas the optical wave packet keeps its initial width.

2.4.2 The group velocity of the optical wave packet remains unchanged. The vacuum speed of light does not depend on the wave number.

2.4.5 As time passes the shape of the wave packet remains unaltered. The phase velocity of electromagnetic waves in the vacuum is independent of their frequencies. Therefore, all waves move with the same speed as light in vacuum and the wave packet does not change its shape.

2.5.3 The average momenta p_0 and the momentum widths σ_p for quantile trajectories with $P = 0.001$ running backward, from slow to fast, read: $p_0 = 3, \sigma_p = 1$; $p_0 = 1, \sigma_p = 0.5$; $p_0 = 1, \sigma_p = 1$.

2.5.4 $x_P(t) \to x_0 + (t/m)\{p_0 + (2\sigma_p^2/\hbar)(x_P(t_0) - x_0)\}$. The sign of the term in curly brackets determines whether $x_P(t)$ runs forward or backward in x.

2.5.5 The quantum potential (2.27) drops off as $1/t^2$ for large times and thus approaches zero. The corresponding force vanishes like $1/t^3$. Therefore, asymptotically there is no force and the quantile trajectories become diverging straight lines.

2.6.4 (b) $p_0 = 5.685 \times 10^{-12}\,\mathrm{eV\,s\,m^{-1}}$ and $p_0 = 17.06 \times 10^{-12}\,\mathrm{eV\,s\,m^{-1}}$. (c) The physical unit at the abscissa is $10^{-12}\,\mathrm{eV\,s\,m^{-1}}$. (d) $E_0 = 2.843 \times 10^{-12}\,\mathrm{eV}$, $E_0 = 25.58 \times 10^{-12}\,\mathrm{eV}$.

2.6.5 (b) $p_0 = 3.372 \times 10^{-6}\,\mathrm{eV\,s\,m^{-1}}$ and $p_0 = 5.84 \times 10^{-6}\,\mathrm{eV\,s\,m^{-1}}$. (c) $10^{-6}\,\mathrm{eV\,s\,m^{-1}}$. (d) $v_0 = 0.5932 \times 10^6\,\mathrm{m\,s^{-1}}$ and $v_0 = 1.027 \times 10^6\,\mathrm{m\,s^{-1}}$. (e) The order of the quotient v^2/c^2 is equal to 10^{-4}.

2.6.6 (b) $p_0 = 10.44 \times 10^{-9}\,\mathrm{eV\,s\,m^{-1}}$ and $p_0 = 31.32 \times 10^{-9}\,\mathrm{eV\,s\,m^{-1}}$. (c) $10^{-9}\,\mathrm{eV\,s\,m^{-1}}$. (d) $E_0 = 5.22 \times 10^{-9}\,\mathrm{eV}$ and $E_0 = 93.96 \times 10^{-9}\,\mathrm{eV}$.

2.6.7 (b) $p_0 = 1.370 \times 10^{-3}\,\mathrm{eV\,s\,m^{-1}}$ and $p_0 = 2.373 \times 10^{-3}\,\mathrm{eV\,s\,m^{-1}}$. (c) $10^{-3}\,\mathrm{eV\,s\,m^{-1}}$. (d) $v_0 = 1.460 \times 10^6\,\mathrm{m\,s^{-1}}$ and $v_0 = 2.529 \times 10^6\,\mathrm{m\,s^{-1}}$. (e) The order of magnitude of the quotient v^2/c^2 is 10^{-4}.

2.7.4 The sum (2.3) over the contributions of the harmonic waves $\psi_{p_n}(x,t)$, (2.1), with the momentum vectors $p_n = p_0 - 3\sigma_p + (n-1)\Delta p$, see Sect. 2.7, is a Fourier series. It is periodic in space. The wavelength of this periodicity is $\Lambda = 2\pi\hbar/\Delta p$. The wave patterns constituting the wave packet reoccur along the x axis at distances Λ. For more densely spaced p_n, i.e., smaller Δp, the reoccurrences are spaced more widely. Only the Fourier integral (2.4) yields the Gaussian wave packet without periodic repetitions.

2.8.1 (a) For small times σ_x decreases as σ_{p0} grows and for larger times it increases. (b) p and x are uncorrelated for $t = 0$, correlated for $t > 0$, and anticorrelated for $t < 0$.

2.9.1 (b) The point of smallest p for $\sigma_{p0} = 2$ has $p = 0$ and therefore stays at rest.

Hints and Answers to the Exercises in Chapter 3

Note: You find descriptor files with solution descriptors in the directory `Solutions/1D_Bound_States_(Chap_3)`.

For those exercises of Chap. 3 in which physical quantities are given in terms of physical units use the data-table units, see Sect. 12.1.5.

3.2.2 For an infinitely deep square-well potential symmetrically placed with respect to the origin of the coordinate frame, the expectation values of position and momentum vanish.

3.2.3 (a) The values of the wave functions $\varphi_n^{(\pm)}(-d/2)$ and $\varphi_n^{\pm}(d/2)$ do not vanish. Thus they violate the boundary conditions to be satisfied by solutions of the Schrödinger equation with an infinitely deep square-well potential. (b) The eigenfunctions $\varphi_n(x)$ are standing waves resulting as even and odd superpositions of the right-moving and left-moving waves $\varphi_n^{(\pm)}(x)$.

3.2.4 (c) Using $\Delta p \Delta x \geq \hbar/2$ and $E = (\Delta p)^2/2m$ for the kinetic energy of the ground state we get $E \geq 1.06 \times 10^{-3}$ eV for (a).

3.2.5 (c) The estimate of E is reduced by the factor $M_e/M_p = 1/1836$ with respect to Exercise 3.2.4.

3.2.6 (a) The energy of the ground state is $E_1 = 14.9\,\text{eV}$. (b) The value of the Coulomb potential at the Bohr radius $a = \hbar c/\alpha c^2$ is $V(a) = -\alpha^2 c^2 = -27.2\,\text{eV}$. The sum of this value and E_1 is $-12.3\,\text{eV}$. The binding energy of the hydrogen ground state is $-13.65\,\text{eV}$. The order of magnitude of the true ground-state energy in hydrogen and the rough estimate is the same.

3.2.7 (g) The unit of the energy scale is eV. (h) The spring constants are for (a) 5.685, (b) 12.97, (c) 22.74 eV nm^{-2}. (i) The unit of the x scale is nanometers (nm). (j) Close to the wall of the harmonic-oscillator potential the speed of the particle is lowest; thus its probability density is largest here.

3.2.8 (c) The angular frequencies for the two spring constants are $\omega = 0.3095 \times 10^4\,\mathrm{s}^{-1}$ and $\omega = 0.3095 \times 10^6\,\mathrm{s}^{-1}$. (f) $\sigma_0 = 0.4513 \times 10^{-5}\,\mathrm{m}$ and $\sigma_0 = 0.4513 \times 10^{-6}\,\mathrm{m}$.

3.3.1 (c) The energy eigenvalues are $E_1 = -5.7$, $E_2 = -5.1$, $E_3 = -3.9$, $E_4 = -2.3$, $E_5 = -0.6$. The differences $\Delta_i = E_i - V_0$ are $\Delta_1 = 0.3$, $\Delta_2 = 0.9$, $\Delta_3 = 2.1$, $\Delta_4 = 3.7$, $\Delta_5 = 5.4$.

3.3.2 (c) The energy eigenvalues for the width $d = 2$ are $E_1 = -5.3$, $E_2 = -3.2$, $E_3 = -0.4$; the differences Δ_i are $\Delta_1 = 0.7$, $\Delta_2 = 2.8$, $\Delta_3 = 5.6$. For the width $d = 6$ we find $E_1 = -5.9$, $E_2 = -5.6$, $E_3 = -5$, $E_4 = -4.2$, $E_5 = -3.2$, $E_6 = -2$, $E_7 = -0.7$; the differences Δ_i are $\Delta_1 = 0.1$, $\Delta_2 = 0.4$, $\Delta_3 = 1$, $\Delta_4 = 1.8$, $\Delta_5 = 2.8$, $\Delta_6 = 4$, $\Delta_7 = 5.3$.

3.3.3 (a) The energy eigenvalues of the stationary wave functions in an infinitely deep square well are given by (3.12). For $d = 2$ the eigenvalues are $E'_1 = 1.23$, $E'_2 = 4.93$, $E'_3 = 11.10$, etc.; for $d = 4$ they are $E'_1 = 0.31$, $E'_2 = 1.23$, $E'_3 = 2.78$, $E'_4 = 4.93$, $E'_5 = 7.71$ etc.; for $d = 6$ they are $E'_1 = 0.14$, $E'_2 = 0.55$, $E'_3 = 1.23$, $E'_4 = 2.19$, $E'_5 = 3.43$, $E'_6 = 4.93$, $E'_7 = 6.72$. (b) The differences Δ_i of the energy eigenvalues to the depth V_0 of the potential are smaller than the eigenvalues of the stationary states in the corresponding potential well of the same width. The separation of the energy eigenvalues are also smaller than for the infinitely deep square well. (c) The eigenfunctions in the infinitely deep square well vanish at the edges of the potential well whereas the eigenfunctions in the well of finite depth extend also into regions outside the well. Their curvature $\mathrm{d}^2\varphi/\mathrm{d}x^2$ is smaller than in the infinitely deep square well, and thus the contribution of the kinetic-energy term $T = -(\hbar^2/2m)\mathrm{d}^2\varphi/\mathrm{d}x^2$ is smaller for the well of finite depth. This difference between the two potentials becomes the more prominent the larger the eigenvalue E_i is.

3.3.4 (c) In the central region of the potential the eigenvalue is equal to the value of the potential. Therefore, the contribution of the kinetic-energy term T in the Schrödinger equation vanishes. Thus, the second derivative of the wave function vanishes in the central region of the potential so that it has to be a straight line. (d) The two lowest eigenfunctions in the double potential well can be closely approximated by a symmetric and an antisymmetric superposition of the two ground-state eigenfunctions of the two single potential wells constituting the double well. The two ground-state wave functions of the single wells have the same energy eigenvalues so that the difference in energy originates from the minor differences between the two wave functions. This explains the narrow spacing of their energy eigenvalues. (e) The double-well potential is symmetric about the point $x = 1.75$. Thus, the Hamiltonian $H = T + V$ is symmetric under a reflection with respect to this point. Therefore, the corresponding parity operator P commutes with H. As a consequence, the eigenfunctions of H to different eigenvalues have to be eigenfunctions of P as well, so that they have to be symmetric or antisymmetric.

3.3.5 (c) The potential of this exercise is no longer symmetric with respect to a point x.

3.3.6 (c) The energy eigenvalue of the ground state is the same as the value of the potential in the second well. The contribution of the kinetic energy T vanishes in the second well so that the second derivative $\mathrm{d}^2\varphi/\mathrm{d}x^2$ vanishes in this region. Therefore, the wave function in this region is a straight line.

3.3.7 (e) The behavior of the eigenstates of a double-well potential with different wells is most easily discussed in terms of the eigenstates of the single wells. The lowest eigenvalues in the two single wells appear in the wide right well. The ground-state eigenvalue in the narrower left well is much higher. Thus the ground state of the double well is given by a wave function closely resembling the ground state in the single right well and is very small in the left well. The same argument is valid for the highest state. However, the ground state in the narrow well to the left has an eigenvalue not very much different from the first excited state with one node of the wide well to the right. Two linear combinations can be formed out of them, one with a number of nodes equal to the sum of the number of nodes of the two wave functions (in this case altogether one) and one with an additional node located in the region between the two wells (in this case altogether two nodes). The combination with the lower number of nodes possesses a lower (average) curvature, i.e., at average a lower second spatial derivative. Thus its kinetic energy is lower and so is the corresponding eigenvalue of the energy. The linear combination with one more node has a higher kinetic energy and thus its energy eigenvalue is slightly higher. In the process of increasing the width of the second well, the distance between the two eigenvalues of the single wells decreases. The eigenvalues of the two eigenstates of the single wells cross each other. For a narrower right well the nodeless eigenstate of the left well has a lower eigenvalue. During the widening of the second well, the eigenvalue of its first excited state with one node decreases so that, eventually, it becomes lower than the eigenvalue of the nodeless ground state in the left well. This process is called *level crossing*. However, because the eigenstate for the double well with the lower number of nodes stays related to the lower energy eigenvalue, there is no level crossing for this energy eigenvalue with the one of the state with one more node. During the widening of the second well, the eigenvalues of the two states of the double well at first approach each other up to a minimum distance. With further broadening of the right well, the eigenvalues of the two states depart again from each other.

3.4.1 (k) The single potential well (a) contains two bound states. The assembly of the nodeless degenerate ground states in the r single wells splits into a band of the r nondegenerate states with no node within the single potential wells. The assembly of the first excited states with one node within the single potential wells analogously splits into a band of states with one node within the single wells. The energy separation of the states within the bands is small compared to the energy separation between the ground state and the first excited state in a single well.

3.4.2 (e) The states in the r single wells constituting the quasiperiodic potential are degenerate. They are linearly independent and can be superimposed into linear combinations that approximate the eigenstates of the quasiperiodic potential. The superpositions follow the node structure of the eigenstates of the wide single-well potential, which remains after all the inner walls have been removed from the quasiperiodic potential. The eigenstates of the wide potential well are narrowly spaced with energy eigenvalues becoming larger as the number of nodes of the eigenstates grows. The nodeless ground states of the single wells are combined to wave functions with $0, 1, \ldots, r-1$ nodes. These states, if they have an even number of nodes, are symmetric with respect to the center of the quasiperiodic potential. The ones with an odd number of nodes are antisymmetric. They form the lowest band of states in the quasiperiodic potential. Their energy separation corresponds to the small energy difference between the states in the wide single-well potential that contains all the narrow wells of the quasiperiodic potential. The eigenstates of the quasiperiodic potential with r, $r+1$, $\ldots$, $2r-1$ nodes are composed of the r one-node first excited states of the single wells and from the second band. The separation in energy of the two bands corresponds to the energy separation of the ground and first excited state of the single well and thus is much larger than the separation of the states within the lowest band.

3.4.3 (e) The lowest band of 10 states splits into a set of eight states narrowly spaced in energy and another of two states closely spaced. The reason for the separation into two sets is that the states in the outermost wells possess a somewhat larger energy since their curvature is larger because of the higher walls at the right and left boundary of the quasiperiodic potential. Because the states in the outer wells have a somewhat larger energy, their superposition with the ground states of the inner wells is less pronounced. Thus, the two states locate the particle mainly close to the right and left boundary of the potential. The two states are narrowly spaced because they are again superpositions of wave functions that are mainly different from zero close to the left and right boundary. The lower one is the symmetric, the higher of the two the antisymmetric superposition of the two wave functions that are mainly different from zero either at the left or at the right boundary. The two eigenstates correspond to the *surface states* in a crystal of finite extension.

3.5.1 The numbers of eigenstates are (a) 17, (b) 9, and (c) 8. (d) Because of the infinitely high wall at the left-hand side in (c) the spectrum just contains the antisymmetric states (i.e., the ones with odd parity) of (a). On the right-hand side these eigenfunctions are identical up to normalization.

3.5.2 (a) The potential possesses 2 eigenstates. (b) Their energies are $E_1 \approx -2.962$ and $E_2 \approx -0.325$. (c) Since two states are bound, the first resonance is expected to possess three bumps. (d) The minimum value for $|V_3|$ turns out to be approximately 1.535.

3.6.1 (a) The average energies of the two separated bands are -7.5 and -5, respectively. (b) The lower band contains five states, the upper one four. The lower band belongs to the five narrow (deep) wells, the upper band to the four wider (less deep) wells. (b) The new potential values are $V_3 = -17.5$, $V_5 = -12.5$. The lower band, now belonging to the four wide wells, has correspondingly four states, and the upper one, now connected to the five narrow wells, has thus five states. (c) Since the outer two of the narrow wells are changed, the quasi-periodicity just holds for the three inner wells, resulting in three states of the lower band for the first case and three states for the upper band in the second case. The centers of the bands are only slightly changed. But the emerging two surface states are, while near to each other, shifted from the remaining bands to higher energies. Their energies are highest for the less deep narrow wells of the second case ($E \approx -3.26$ vs. $E \approx -2.18$).

3.7.1 (d) The period of the wave function is $T = 4$. (e) The period of the absolute square is $T = 2$. (f) $\psi(x, t + T) = \mathrm{e}^{\mathrm{i}\alpha}\psi(x, t)$.

3.7.2 (d) The wave packets plotted are Gaussian so that for the initial state we have $\Delta x \Delta p = \hbar/2$. A wave packet initially wide in x is thus narrow in momentum. The period of a classical particle in a harmonic-oscillator potential is independent of its amplitude. Independent of its actual position in the spatially wide initial state, the particle will reach the point of the potential minimum after the same time interval, because its momentum spread about the initial value $p_0 = 0$ is small. Thus the width of the spatial distribution will shrink and assume its minimal width at the center position in the oscillator. While moving toward the other turning point, the distribution widens in space to the same shape as in the initial state. An initially narrow Gaussian wave packet has a wide momentum distribution. A classical particle the initial position of which is described by a probability distribution will reach the center position of the oscillator after a time interval, in general, dependent on the initial position and momentum. Because the original momentum probability distribution is wide, the spatial probability distribution is wide at the center of the harmonic oscillator. The wave packet shrinks to its original shape at the other turning point. The wave function observed in (c) describes the coherent state with an initial width equal to that of the ground state.

3.7.3 (d) The Gaussian wave packet, which is initially wide in space, possesses a narrow momentum distribution about $p_0 = 0$. If the particle is at an off-center position, the harmonic force drives it back toward the center. Independent of its initial position, the particle reaches the center after the same time interval if its initial momentum is zero. Because there is a small initial momentum spread, the probability distribution does not oscillate between its initial width and zero width but between a maximum and a minimum value.

3.7.4 (d) The initial state at the center position of the oscillator has relative width $f_\sigma = 1$. It is therefore the ground state of the oscillator and is stationary. Its probability distribution is time independent and so is its width.

3.8.1 (a) The motion of the mean position and the motion of the quantile position away from the mean because of the widening of the wave packet approximately compensate each other for nearly half a period.

3.11.1 (d) The phase velocity of a right-moving or left-moving wave in the deep square well is $v_{pi} = \hbar k_i/2m$. It depends on the wave number k_i. A wave packet being a superposition of stationary wave functions of different k_i is thus made up of waves of different phase velocity. Therefore, it widens as time elapses. (e) The wiggly patterns occurring when the wave packet is close to the wall are an interference effect between the incident and the reflected wave. As soon as the wave packet becomes wide enough, the wiggly patterns occur everywhere in the deep well.

3.11.2 (n) The wave number of the lowest state in the infinitely deep square well is $k = \pi/d = 0.31416$ for $d = 10$. The corresponding temporal period is $T_1 = 2\pi/\omega$ with $\omega = E/\hbar$ and $E = \hbar^2 k^2/2m$. Thus we find $T_1 = 4\pi m/\hbar k^2 = 127.32$. (o) At the time $T = T_1/2$ the wave packet looks the same as the initial wave packet. (p) At the beginning of the motion the wave packet is so narrow that it does not touch the walls of the deep square well. Therefore, the classical particle position and the expectation value of the wave packet coincide as in a free wave packet. As soon as the wave packet hits the wall it consists of incident and reflected parts. Its position expectation value cannot reach the wall because of the width of the packet. Thus it stays behind the classical expectation value. (q) Because of the dispersion, the wave packet becomes so wide that both ends are reflected at the left-hand and right-hand wall. It consists of right-moving and left-moving parts with equal probability. Thus its momentum expectation value vanishes. It also spreads over the whole width of the square well with – at average – equal probability density. Thus the position expectation value equals zero.

3.11.3 (d) The wave function possesses the period $T = T_1$ as calculated in Exercise 3.11.2 (n). Thus the wave packet has reassumed its initial shape. It represents a wave packet narrow compared to the width of the well. The classical particle follows a periodic motion between the walls of the well with a classical period $2d/v = 2dm/p = 3.858$. After 33 classical periods the time T_1 has elapsed and the classical position and the expectation value are at the same point. Because the absolute square of the wave packet reassumes its original well-localized shape after half the time T_1, the coincidence of the classical position and the expectation value occurs at $T_1/2$.

3.11.4 (b) The system of a point particle in an infinitely deep square well is conservative. Thus the expectation value of the energy is constant and thus all the time equal to its initial value. (c) The momentum expectation value of the particle is practically zero in periods where the wave packet fills the whole well and its position expectation value is at rest.

3.11.5 (h) The momentum of the particle, as well as its mass, have been increased by a factor of 10 leaving the velocity unchanged. However, because of the larger mass, the wave packet disperses much more slowly than in Exercise 3.11.4. Thus it oscillates more periods as a localized wave packet between the two walls of the square well.

Hints and Answers to the Exercises in Chapter 4

Note: You find descriptor files with solution descriptors in the directory `Solutions/1D_Scattering_(Chap_4)`.

4.2.1 (d) The kinetic energy of the third wave function is zero in the region of the repulsive square-well potential. Because the kinetic energy T is proportional to $\mathrm{d}^2\psi/\mathrm{d}x^2$, only the first derivative of the wave function can be different from zero in the region of the square well. (e) The trend of the transmission coefficient A_3 is most easily read off the plot of the absolute square (c). The transmission coefficient increases with increasing energy. For kinetic energies $E < 2$, the tunnel effect increases as the difference between the barrier height and the kinetic energy decreases. For energies above the barrier height, the reflection decreases.

4.2.2 (d) The tunnel probability increases exponentially as the barrier width decreases. (e) There is only an outgoing wave of the type $A'_N \exp(\mathrm{i}k_N x)$, $N = 3$, in the region N beyond the potential, see (4.18). Its absolute square is just the constant $|A_N|^2$. (f) In the region 1 the incident and reflected wave are superimposed, see (4.22). We use the representation $A'_1 = |A'_1|\exp(\mathrm{i}\alpha_1)$, $B'_1 = |B'_1|\exp(\mathrm{i}\beta_1)$. Then the absolute square of the superposition yields $|A'_1|^2 + |B'_1|^2 + 2|A'_1||B'_1|\cos(2k_1x + \alpha_1 - \beta_1)$. The third member is the interference term between incident and reflected waves. It exhibits twice the original wave number k_1, i.e., half the wavelength of the incident wave.

4.2.3 (d) The energy at which the wave pattern in the absolute square is smallest is $E = 1.18$. The wavelength of the incident wave is $\lambda = 2\pi/k = 2\pi\hbar/\sqrt{2mE} = 1.54$, $\hbar = 1$. Thus, half a wavelength fits roughly into the interval of the two walls forming the double barrier. Therefore, the reflected waves from the two barriers interfere destructively and the interference pattern with the incident wave almost vanishes.

4.2.4 (d) The width of the region of the repulsive double-well potential is $d = 2$. A nodeless wave function fitting into the well exactly has a wave number $k = \pi/d = \pi/2$. The corresponding kinetic energy is $E = (\hbar k)^2/2m$. With the values $\hbar = 1$, $m = 1$ we get $E = 1.2$, a value very close to the observed $E = 1.18$ of the resonance wave function without a node. (e) Any other wave function fitting into the region of the double-well potential has one or more nodes and thus a higher wave number and a higher kinetic energy. Thus there is no resonance at an energy lower than the value $E = 1.12$.

4.2.5 (a) Resonances occur at $E = 0.6$ and $E = 2.0$. (f) The first resonance at $E = 0.53$ has no node whereas the second resonance at $E = 2.28$ possesses one node. Because the wave function with the node has a second derivative much larger than the one of the lowest resonance, its kinetic energy – being proportional to the second derivative – is larger. Thus, the total energy of the second resonance (with one node) is larger than the one of the lowest resonance. As a general rule one may say that the resonance energy increases with the number of nodes in the resonance wave function between the two barriers.

4.2.6 (a) First resonance at $E_1 = 0.73$; large bump in the wave function in the left potential well. (b) Second resonance at $E_2 = 1.2$; large bump in the wave function in the right potential well. (c) Third resonance at $E_3 = 2.9$; wave function with two bumps, i.e., one node, in the left potential well. (d) Fourth resonance at $E_4 = 4.41$; wave function with two bumps, i.e., one node, in the right potential well. (e) Fifth resonance at $E_5 = 6.51$, wave function with three large bumps, i.e., two nodes, in the left potential well.

4.2.7 The maxima of the transmission coefficient occur at $E_1 = 0.73$, $E_2 = 1.2$, $E_3 = 2.95$, $E_4 = 4.5$, $E_5 = 6.52$.

4.2.8 (a) First resonance at $E_1 = 0.72$; no node. (c) Second resonance at $E_2 = 2.82$; one node. (g) Third resonance at $E_3 = 6.42$; two nodes.

4.2.9 The maxima of the transmission coefficient occur at $E_1 = 0.75$, $E_2 = 2.92$, $E_3 = 6.35$.

4.2.10 (b) First resonance at $E_1 = 1.1$. (e) Second resonance at $E_2 = 4.3$.

4.2.11 The maxima of the transmission coefficient occur at $E_1 = 1.15$, $E_2 = 4.4$.

4.2.12 (f) The trend of the differential resistance is understood as follows: (i) For a voltage $U \ll U_{\mathrm{res}}$ far below the resonance voltage $U_{\mathrm{res}} = V_1 - V_5 = 1$, the tunnel effect is very small; thus, the square of the coefficient A_5, $|A_5|^2$, is small. Therefore, the current through the quantum-well device is small. Also, the variation $\Delta|A_5|^2$ for the given increase $\Delta U = 0.5$ from (a) to (b) is small. Thus, the variation of the current ΔI associated with ΔU is small.

Therefore, the differential resistance $R = \Delta U/\Delta I = (1/\alpha)\Delta U/\Delta|A_5|^2$ is large. (ii) In the region close below the resonance voltage the absolute square $|A_5|^2$ grows quickly from (b) to (c) with the increase of the voltage $\Delta U = 0.5$ and thus the current ΔI; the differential resistance $R = \Delta U/\Delta I$ decreases quickly in the region close below the resonance region. (iii) In the region close above the resonance voltage (c,d) $U_{\rm res}$ the square $|A_5|^2$ of the coefficient A_5 decreases quickly from (c) to (d) for the given increase in voltage $\Delta U = 0.5$. Thus, the corresponding change in current ΔI is negative, so that the differential resistance $R = \Delta U/\Delta I$ becomes negative above the resonance voltage. At the resonance voltage $U_{\rm res}$ the differential resistance goes through zero.

4.2.13 (f) The potential walls ($V_2 = 60$) are much higher in this case than in Exercise 4.2.12 ($V_2 = 7$). Therefore, the resonance region is much more narrow than in Exercise 4.2.12. The variation of $|A_5|^2$ is much larger for the potential of Exercise 4.2.12. Therefore, the differential resistance varies much faster with increasing voltage.

4.2.14 (f) The second transmission resonance is much wider than the first resonance. Thus, the variation of the differential resistance is much slower.

4.2.15 (f) The discussion of the differential resistance follows the same lines of argument as in Exercise 4.2.12 (f).

4.2.17 (f) Because of the triple-wall structure of the potential, the transmission resonance is much more narrow than in the case of a double-wall potential as in Exercise 4.2.16 (f). Thus, the variation of the differential resistance $R = \Delta U/\Delta I = -\Delta V_3/\Delta I$ is much faster.

4.2.18 (a) The scattering wave functions outside the potential region with energies approaching the asymptotic minimum value $V_{\rm c}$, see Sect. 4.1.1, i.e., for momenta approaching zero, become constant or linear. Bound states are characterized by their number of nodes (in the potential region), increasing from zero to $N_{\rm bound} - 1$, the maximum number of nodes. In the adjacent continuum the (real or imaginary part of the) scattering states in this region begin with $N_{\rm bound}$ zeros (nodes). (b) Both real and imaginary part show two nodes, hence $N_{\rm bound} = 2$. (c) For this case one obtains $N_{\rm bound} = 47$.

4.2.19 (c) In the case of the triangular potential of part (a) one recognizes significant interference patterns on the left half-axis for almost the complete energy interval shown. In case (b), except for low energies, an interference pattern is almost invisible. Also for lower energies the interference is less significant for (b). This means that the bell-shaped potential from (b) shows very low reflection for a wide energy range. The potential indeed is an approximation by piecewise linear functions of a so-called reflectionless potential of the form $A\cosh^{-2}(ax+b)$ with especially chosen strength A and proportionality factor a. Such potentials are discussed, e.g., in: B. N. Zakhariev, A. A. Suzko, 'Direct and Inverse Problems' (Springer, Heidelberg, 1990).

4.2.20 The two potentials discussed here are similar approximations of the same spatial bell-shaped form as in Exercise 4.2.19 (b) but for two larger values of A. For these parameters reflection is also largely suppressed.

4.3.1 (d) The answer is obvious from (4.50) and the accompanying text of Sect. 4.1.

4.3.2 (d) The answer is obvious from (4.47) and the accompanying text of Sect. 4.1.

4.3.3 (d) The answer is obvious from (4.44) and the accompanying text of Sect. 4.1.

4.3.4 (d) Upon taking the absolute square of a harmonic wave (2.1), see Sect. 2.1, the time-dependent factor $\exp(-\mathrm{i}Et/\hbar)$ yields 1. Thus, the absolute square of a harmonic wave looks the same as the absolute square of a stationary scattering wave function at the same momentum.

4.3.5 (d) To the right of the potential barrier there is an outgoing wave propagating to the right. It possesses a time-independent amplitude A_3. (e) To the left of the potential barrier the time-dependent wave function is a superposition of the incident and reflected waves. The real part of the wave function thus exhibits a time-dependent amplitude.

4.3.6 (d) The wave packet is reflected completely for $E = 2$, an energy smaller than the height of the step potential. For $E = 6.5$ the reflection has not ceased to occur even though the kinetic energy is larger than the height of the step potential so that a classical particle would only be transmitted. For $E = 8$ the transmission probability has grown. Upon increasing the energy even further, the reflection disappears. (e) The transmitted part of the wave packet is faster than the classical particle, because its average energy corresponds to a value higher than the energy expectation value $E = 6.5$ of the incident wave packet. The potential step acts like a discriminator threshold.

4.3.8 (d) At the potential step at $x = 0$ the continuity conditions (3.41) are $A_1 + B_1 = A_2$, $k_1(A_1 - B_1) = k_2 A_2$. The solution is $A_2 = 2A_1k_1/(k_1 + k_2)$, $B_1 = A_1(k_1 - k_2)/(k_1 + k_2)$, with $k_1 = \sqrt{2mE}/\hbar$ and $k_2 = \sqrt{2m(E - V_2)}/\hbar = \sqrt{2m(E + |V_2|)}/\hbar$. The modulus $|B_1|$ of the reflection coefficient is a monotonically increasing function of $|V_2|$.

4.3.9 $E_1 = 16.3$, $E_2 = 17.25$, $E_3 = 18.8$, $E_4 = 20.95$.

4.3.10 (c) In an analogy to optics, the resonances occur because the reflection at a step increase of the potential corresponds to a reflection at a thinner medium whereas the subsequent reflection at a step decrease of the potential corresponds to a reflection on a denser medium. Thus, according to (4.54), resonances occur if the wavelength λ_2 of the wave function in region 2 is approximately equal to $\lambda_{2\ell} = 2d/\ell$, $\ell = 1, 2, 3, \ldots$. The corresponding energies are $E_\ell = V_2 + \ell^2\pi^2/2d^2$, see (4.56) with $\hbar = 1$, $m = 1$. The numerical values are $E_1 = 16.308$, $E_2 = 17.23$, $E_3 = 18.78$, $E_4 = 20.93$. (d) At resonance energy a large percentage of probability is stored in the region $0 \le x \le 4$. The resonance wave function decays exponentially in time after some time of the resonance formation. The exponential time dependence of the amplitude is equivalent to constant ratios of the amplitudes of the wave functions in the barrier region for equidistant instants in time.

4.3.12 (c) According to the formulae of Exercise 4.3.8 the reflection probability at the two descents of the potential at $x = 0$ and $x = 4$ decreases with increasing energy $E = \hbar^2 k_1^2/2m$. Thus, the exponential decay is faster for higher resonance energy. The uncertainty relation relates the lifetime τ and the width Γ of the resonance energy by $\tau \times \Gamma \approx h$. Thus, for the shorter-lived higher resonance the energy width Γ increases. This is obvious from the Argand diagram of Exercise 4.3.9.

4.3.13 (f) The low-lying transmission resonances occur as narrow maxima in the absolute square $|A_N|^2$. They correspond to fast sweeps of the pointer A_N in the complex Argand plot through the regions at which the pointer A_N has its maximal length $|A_N| = 1$.

4.3.14 (f) The relation of the prominent features of A_N to the T-matrix element T_T is most easily understood using (4.36).

4.3.15 (f) Compare (4.34).

4.3.16 (f) Compare (4.36).

4.3.17 $E = 0.08$.

4.3.18 There is one bound state at $E = -3.2$.

4.3.19 (d) The wave function inside the potential region possesses one node. The energy $E = 0.1$ corresponds to the lowest resonance energy (see Exercise 4.3.17). There is no resonance without a node because of the existence of one nodeless bound state in this potential, see Exercise 4.3.18.

4.3.20 (f) $E_1 = 0.9$, $E_2 = 3.4$, $E_3 = 7.8$, $E_4 = 14.1$.

4.3.26 (a) The least significant interference pattern or amount of reflected probability of the wave packet occurs for $V_2 \approx -2.8$. (b) The interference and thus the reflection becomes more pronounced for larger deviation of the average energy of the wave packet from the initial one, $E_0 = 0.25$, for which the potential value V_2 was chosen to minimize reflection. (The reflected parts can be resolved better by increasing the scale factor for the probability density.)

4.3.27 In all cases interference patterns are almost invisible and therefore almost no reflection of probability occurs. Within the potential region, because of the larger velocity there, a depression of the wave packet occurs. After the passage of the potential region the initially Gaussian wave packet rebuilds without a visible distortion, where it again shows the usual dispersion. See also Exercises 4.2.19 and 4.2.20.

4.3.28 The situation is similar to that of Exercise 4.3.27 but with larger potential strength.

4.4.1 (a) The resonance energies read off $|A_N|^2$ are $E_1 = 21$, $E_2 = 35.0$. (f) See Exercise 4.3.13 (f) and Sect. 12.2. (g) $E_1 = 20.93$, $E_2 = 35.74$.

4.4.2 (f) See Exercise 4.3.14 (f) and Sect. 12.2.

4.4.3 (f) See (4.34) and Sect. 12.2.

4.4.4 (f) See (4.36) and Sect. 12.2.

4.4.5 (a) The resonance energies read off $|A_N|^2$ are $E_1 = 17$, $E_2 = 21$, $E_3 = 27$, $E_4 = 36$. (f) See Exercise 4.3.13 (f) and Sect. 12.2. (g) $E_1 = 17.23$, $E_2 = 20.93$, $E_3 = 27.10$, $E_4 = 35.74$.

4.4.6 (f) See Exercise 4.3.14 (f) and Sect. 12.2.

4.4.7 (f) See (4.34) and Sect. 12.2.

4.4.8 (f) See (4.36) and Sect. 12.2.

4.4.9 The potential barrier of Exercise 4.4.5 is twice as wide as the one of Exercise 4.4.1. Thus, in Exercise 4.4.5 the low-lying resonances according to the relation (4.56) occurring at the energy values $E_\ell = V_2 + \ell^2\pi^2\hbar^2/(2md^2)$ show up at values $(E_\ell - V_2)$ that are a quarter of the values of Exercise 4.4.1. If we increase d again by a factor of 2, we expect the values $(E_\ell - V_2)$ at energies that are approximately another factor of 4 lower than the ones of Exercise 4.4.1. The transmission resonances show up as minima in the reflection coefficient B_1. The behavior of the coefficients T_T and T_R related to the transmission resonances is easily understood with the help of the relations (4.36).

4.4.10 (a) $E_1 = 16.2$, $E_2 = 17.2$, $E_3 = 18.8$, $E_4 = 20.9$, $E_5 = 23.7$, $E_6 = 27.1$, $E_7 = 31.1$, $E_8 = 35.9$. (f) See Exercise 4.3.13 (f) and Sect. 12.2. (g) $E_1 = 16.3$, $E_2 = 17.23$, $E_3 = 17.78$, $E_4 = 20.93$, $E_5 = 23.71$, $E_6 = 27.10$, $E_7 = 31.11$, $E_8 = 35.74$.

4.4.11 (f) See Exercise 4.3.14 and Sect. 12.2.

4.4.12 (f) See (4.34) and Sect. 12.2.

4.4.13 (f) See (4.36) and Sect. 12.2.

4.4.14 The potential consists of 4 wells. Therefore bound states form bands of 4 states and so do resonances. Particularly pronounced are those at low energies.

4.4.15 (a) A_N for low energies has a modulus significantly smaller than 1 but rapidly reaches the unitarity circle ($E \approx 0.15$), then departs noticeably from there, turns back, and from $E \approx 2$ on it stays approximately at the unitarity circle. Except for the very it turns clockwise in increasingly smaller steps asymptotically reaching 1. (b) Beginning at ≈ 1, A_N stays very near the unitarity circle, where it turns clockwise throughout in increasingly smaller steps and also reaches 1 asymptotically. (c) See (a), (b). While $|A_N|^2$ for case (a) has a minimum at ≈ 0.93 for $E \approx 0.6$, for (b) it stays near 1 with a maximum deviation of about 0.5%. The potential shows almost no reflection. See also Exercises 4.2.19 and 4.2.20.

4.5.1 (b) follows from (4.59).

4.5.2 $E = p^2/2m - mgx$, $p = 2m(E + mgx)^{1/2} = h/\lambda$. Therefore, with rising x (and for $g > 0$) p rises and λ falls.

4.7.1 (b) No (only in the limit $t \to \infty$). In the language of Bohm in addition to the constant force of the linear potential there is also the force caused by the quantum potential.

4.13.1 (d) For a refractive index n_2 of glass, the wavelength of light in the glass is given by $\lambda_2 = \lambda/n_2$, where λ is the wavelength in vacuum. Thus, the wavelength in region 2 is $\lambda/2$. (e) In region 2 there is only an outgoing transmitted wave $E_2 = A_2 \exp(ik_2x)$. Its absolute square is constant: $|E_2|^2 = |A_2|^2$. (f) The superposition of incident and reflected waves in region 1 causes the wiggly pattern. It indicates the contribution of the interference term of incident and reflected wave, see Exercise 4.2.2 (f).

4.13.2 (d) For the boundary between regions 1 and 2 at $x = 0$ the continuity equations (4.77) require $A'_1 + B'_1 = A'_2$ and $k_1(A'_1 - B'_1) = k_2A'_2$. With $k_2/k_1 = n_2$ we find $A'_2 = 2A'_1/(1+n_2)$ and $B'_1 = (1-n_2)A'_1/(1+n_2)$. Thus the transmission coefficient A'_2 is small for refractive index $n_2 \gg 1$ of a much denser medium in region 2.

4.13.3 (d) The expression $A'_2 = 2A'_1/(1 + n_2)$ obtained above in Exercise 4.13.2 (d) shows that for $n_2 \ll 1$ the amplitude A'_2 becomes larger than 1.

4.13.4 (a) The values of the physical electric field strength are the real parts of the complex field strength. The real part Re $E_{1+}(0)$ at the surface at $x = 0$ of the glass has the opposite sign to Re E_{1-} at $x = 0$. Thus, the phase shift upon reflection at a denser medium is π, since $\cos(\alpha+\pi) = -\cos\alpha$. (b) The real part Re E_{1+} at $x = 0$ has the same sign as the real part Re E_{1-} at $x = 0$. Thus, there is no phase shift upon reflection on a thinner medium.

4.13.5 (d) The absence of wiggles in the absolute square of the complex electric field strength in region 1 signals the vanishing or smallness of the interference term and thus the absence or smallness of the reflected wave at the wave numbers $k = \pi/2, \pi, 3\pi/2$. (e) At these wave numbers we observe transmission resonances upon transmission through a denser medium, see (4.87) of Sect. 4.12.

4.13.6 (d) The wavelengths of the resonances occurring now upon two-fold transmission into denser media follow from (4.83), Sect. 4.12. (e) The speed of light in region 3 is only $1/4$ of the speed in region 1. Because there is no reflection at a resonant wavelength, the energy-current density of the transmitted wave must be the same as the one of the incident wave. Therefore, the energy density, and thus the absolute square of the complex electric field strength, must change, see (4.78), (4.79) of Sect. 4.12.

4.14.1 (d) Only a transmitted outgoing harmonic wave propagates to the right in region 3. Its complex form is given by (4.67), see Sect. 4.12. The amplitudes of real and imaginary part are time independent. (f) In region 1 the superposition of incident and reflected waves leads to a time-dependent amplitude.

4.14.2 (g) Only the transmitted wave E_{3+} moving to the right propagates in region 3.

4.14.4 (g) The incident wave number $k = 3\pi/2$ leads to $k_2 = n_2 k = 3\pi$. The corresponding wavelength in region 2 is $\lambda_2 = 2\pi/k_2 = 2/3$. Thus, the thickness d of the region 2 fulfills $\lambda_2 = 2d/m$ for $m = 3$. Thus, at the incident wave number $k = 3\pi/2$ we find a transmission resonance, so that there is no reflected wave moving to the left in region 1. Therefore there is no constituent wave E_{1-} in region 1.

4.14.5 (c) The speed of light in the glass of refractive index $n = 4$ is $1/4$ of the vacuum speed of light. (d) See Exercise 4.13.6 (d). (e) The spatial extension of the light wave packet shrinks upon entering the glass sheet because the speed of light in glass is only $1/4$ of the vacuum speed of light. All wavelengths superimposed in the wave packet shrink according to $\lambda_2 = \lambda/n = \lambda/4$.

4.14.6 (c) The wave number of the incident wave packet, $k_0 = 7.854$, leads to a wave number $k_2 = 15.708$ in the layer with $n = 2$. This wave number fulfills the resonance condition $2k_2 d = (2m+1)\pi$ for the thickness $d_2 = 0.1$ of the layer with $n = 2$ and for $m = 0$, see (4.83) and accompanying text of Sect. 4.12. Thus, for light of wave number $k_0 = 7.854$ there is maximal destructive interference in region 1 between the waves reflected at the front and rear boundaries of the layer with $n = 2$. The refractive indices $n = 1$, $n_2 = 2$, and $n_3 = 4$ satisfy the condition (4.85) so that there is no reflection at the front surfaces of the three glass layers. The corresponding arguments hold true at the rear surfaces of the glass sheets.

4.14.7 (b) The wavelength in the region of the coating is given by $\lambda_1 = \lambda/n_1 = 449\,\mathrm{nm}$. The thickness of the coating is $d = \lambda_1/4 = 112\,\mathrm{nm}$.

4.14.8 (b) The modulus of the reflection coefficient B_1' of light is larger for larger refractive index n of the material, see Exercise 4.13.2 (d). We have $n = 1.5$, in Exercise 4.14.1 we had $n = 4$. The reflection is smaller for $n = 1.5$ than for $n = 4$.

4.14.9 (b) The simple coating for a lens is adjusted to an average wavelength within the spectrum of visible light. For larger and shorter wavelengths in the spectrum, some reflections at the surface of the lens remain. This is the reason for the bluish reflection of coated lenses.

4.15.1 (e) The resonant wave numbers read off $|A_N|^2$ are $k_1 = 7.85$, $k_2 = 23.6$. The resonant wave numbers calculated using (4.83) are $k_1 = 7.854$, $k_2 = 23.56$.

4.15.2 (e) The resonant wave number read off $|A_N|^2$ is $k_1 = 12.8$. The resonant wave number calculated using (4.83) is $k_1 = 12.83$.

Hints and Answers to the Exercises in Chapter 5

Note: You find descriptor files with solution descriptors in the directory `Solutions/Two_Particles_(Chap_5)`.

5.2.2 (e) The number of nodes in the center-of-mass coordinate $R = (x_1 + x_2)/2$ is equal to N, in the relative coordinate $r = x_2 - x_1$ it is equal to n. The nodes in R lead to node lines in the x_1, x_2 plane parallel to the diagonal in the first and third quadrant. The nodes in r show up as node lines parallel to the diagonal in the second and fourth quadrant of the x_1, x_2 plane.

5.2.3 (e) The doubling of the mass reduces the spatial extension of the wave function. The ground-state width $\sigma_0 = \sqrt{\hbar/m\omega}$ is reduced by a factor of $\sqrt{2}$ upon the doubling of the mass. The same factor $\sqrt{2}$ reduces the spatial extension in x and y also for the wave function of the higher states. The classical frequencies $\omega_R = \sqrt{k/m} = \omega_r$ for uncoupled oscillators are also reduced by a factor $\sqrt{2}$ upon doubling the mass. Thus, the energy eigenvalues are reduced by $\sqrt{2}$. Therefore, for a particle with twice the mass the eigenfunctions belonging to the same quantum numbers N, n are deeper down in the harmonic-oscillator potential and thus their spatial extension has shrunk.

5.2.4 (e) The coupling κ is positive. The coupling spring pulls the two particles toward each other. The oscillator potential in the relative variable r becomes steeper, see (5.12). Thus, the spatial extension of the stationary wave function in the direction of the diagonal $r = x_2 - x_1$ in the x_1, x_2 coordinate system shrinks. This effect also shows in the dashed curve in the x_1, x_2 plane, which marks the classically allowed region for the positions of the particles. In contrast to Exercises 5.2.2 and 5.2.3 it is no longer a circle but an ellipse with the shorter principal axis in r direction.

5.2.5 (e) The shrinking in the diagonal direction $r = x_2 - x_1$ becomes more prominent. (f) The probability of particle 1 being close to particle 2 is larger than the probability of it being far from particle 2. The particle coordinates exhibit a positive correlation.

5.2.6 (e) The coupling κ is negative, i.e., the coupling spring pushes the two particles apart. The potential in the relative variable becomes shallower, see (5.12). Thus, the spatial extension of the wave function in the diagonal direction $r = x_2 - x_1$ widens. This feature is also obvious from the dashed ellipse in the x_1, x_2 plane. It is more likely to find particle 2 at a position distant from that of particle 1 than close by. The particle coordinates exhibit a negative correlation.

5.2.7 (b) At the outer maxima of the probability density, the potential energy of the particles is larger than in the inner ones. Thus, the kinetic energy of the particles is lower in the outer maxima than in the inner ones. Hence, their speed is lower and therefore the time they need to pass the region of the outer maxima is larger. The probability density of the particles being in the outer maxima is larger than the probability density of being in the inner ones. (c) The particle speed and thus the particle momentum close to the origin in the x_1, x_2 plane is larger; thus, the wavelength of the wave function is smaller than in the outer regions of the plot. Hence, the widths of the inner maxima of the plot are smaller. (d) The system of two one-dimensional coupled oscillators has two uncoupled degrees of freedom: the center-of-mass motion and the relative motion. The energy remains constant in either degree of freedom. Its values are given by E_N and E_n, see (5.17). For the case $N = 0$, $n = 2$ the motion of the center of mass has the ground-state energy only. Its amplitude, or equivalently the spatial extension of the ground-state wave function in the center-of-mass coordinate R, is small. Thus, in this direction the classically allowed region is not completely exhausted by the probability density and the wave function.

5.2.8 (b) The lines of the rectangular grid confining the areas of the maxima in the plots of the probability density are the node lines in the center-of-mass coordinate R and in the relative coordinate r.

5.2.9 (e) $\omega_R = 1$, $\omega_r = 3$. The energy values of the states with the quantum numbers N, n are $E_{N,n} = \left(N + \frac{1}{2}\right) + 3\left(n + \frac{1}{2}\right) = N + 3n + 2$. (f) The graphs for $N \gg n$ represent stationary probability densities of the coupled oscillators where a large part of the total energy is in the degree of freedom R of the center-of-mass motion. This causes the wide extension of the wave function in the diagonal in the first and third quadrant of the x_1, x_2 plane.

5.3.1 (b) Because the two oscillators are uncoupled, the expectation value $\langle x_{20} \rangle$ being zero initially remains zero at all times. (c) The widths of the two uncoupled oscillators vary with time, since neither of the two initial widths σ_{10}, σ_{20} is equal to the ground-state width σ_0 divided by $\sqrt{2}$, see Exercise 3.7.4 and (3.35). (d) The initial correlation c_0 is equal to zero. Since the oscillators are uncoupled, no correlation is produced during the motion of the wave packet.

5.3.2 (b) The time dependence of the correlation reflects the periodic change of the widths in the two uncoupled oscillators, (3.35).

5.3.3 (a) $\sigma_0/\sqrt{2} = \sqrt{\hbar/(2m\omega)} = \sqrt{\hbar\sqrt{m}/2\sqrt{k}} = k^{-1/4}/\sqrt{2} = 0.5946$ for $\hbar = 1$, $m = 1$. (c) The ground-state width of the two uncoupled oscillators is σ_0. The initial widths have been chosen equal to $\sigma_0/\sqrt{2}$, the initial state is a coherent state of the two uncoupled oscillators, see Exercise 3.7.4 and (3.35). (d) For nonvanishing correlation the initial state is no longer a product of two Gaussian wave packets in the variables x_1, x_2. Thus, it is not a product of coherent states. The widths will no longer remain time independent.

5.3.4 (b) See Exercise 5.3.3 (d). (c) See Exercise 5.3.2 (b).

5.3.5 (b) Because of the positive initial correlation, the time dependence of the correlation of Exercise 5.3.4 starts with a positive value and oscillates. In this exercise the correlation is negative and thus the function $c(t)$ starts with the negative initial value. The time dependences of the two correlations differ only by a phase shift.

5.3.6 (b) The initial conditions for the position expectation values do not influence the correlation or the widths. Thus, they exhibit no time dependence.

5.3.7 (b) The oscillation of the expectation values $\langle x_1(t)\rangle$, $\langle x_2(t)\rangle$ is the same as for the position of classical particles. The two position expectation values exhibit a beat because of the coupling of the two oscillators. (c) Even though the initial correlation vanishes, a correlation emerges that is positive most of the time because of the attractive coupling.

5.3.9 (b) The motion of the expectation value of one particle is a mirror image of the motion of the other one. Thus, the center of mass of the two particles is at rest, the oscillation takes place in the relative coordinate. The coupled oscillator system is in a normal mode. Only one of the two uncoupled degrees of freedom – the center of mass and the relative motion – oscillates.

5.3.10 (b) The system of the two coupled oscillators is still in the normal mode of the relative motion. (c) Even though the initial correlation is negative, the time average of the correlation is positive because of the positive coupling.

5.3.12 (b) The motion of the two particles is the same in x_1 and x_2. The relative coordinate is zero all the time. The oscillation takes place in the center-of-mass coordinate R. The system of coupled oscillators is in the normal mode, which refers to the center-of-mass motion.

5.4.1 (c) The initial correlation $c_0 = 0$ vanishes. Thus, the initial Gaussian wave packet has axes parallel to the coordinate axes. Because the coupling of the two oscillators vanishes, no correlation is introduced during the motion.

5.4.2 (c) There is a positive initial correlation $c_0 = 0.8$. It shows up in the first graph of the multiple plot in which the orientation of the covariance ellipse is not parallel to the axes. (d) The covariance ellipse in the plot of the initial wave packet has its large principal axis oriented under a small angle with the diagonal direction $x_1 = x_2$. Thus, it is more likely than not that x_1 is close to x_2.

5.4.3 (c) The large principal axis is oriented under a small angle to the diagonal direction $x_1 = -x_2$.

5.4.4 (c) The initial expectation value $\langle x_2 \rangle$ at $t = t_0$ is zero. Without coupling between the two oscillators, $\langle x_2 \rangle$ would remain zero while $\langle x_1 \rangle$ moves. The attractive coupling pulls the particle 2 toward the position of particle 1, which is at positive x_1 values in the beginning. Thus, the expectation value $\langle x_2 \rangle$ takes on positive values for $t > t_0$.

5.4.5 (c) The coupling κ is larger, so the force exerted on particle 2 by particle 1 is larger. Thus, the acceleration of particle 2 out of its original position is bigger.

5.4.6 (c) The attractive coupling κ causes the appearance of a correlation in the initially uncorrelated wave packet.

5.4.7 (c) The coupling κ is now larger than the coupling constants in the former exercises. Thus, the frequency of the oscillation in the relative coordinate $r = x_2 - x_1$ is larger than before and there are more oscillations in r in the same time interval.

5.4.8 (c) The initial wave packet is wide compared to the ground state of the coupled oscillator. With the same arguments as in Exercise 3.7.2 we expect the width of the wave packet to decrease as it moves close to the center.

5.4.9 (c) The oscillation observed is the normal oscillation taking place in the relative coordinate $r = x_2 - x_1$.

5.4.10 (c) The oscillation observed is the normal oscillation taking place in the center-of-mass coordinate $R = (x_1 + x_2)/2$. (d) The normal oscillation in the relative motion in Exercise 5.4.9 is fast because of the large coupling $\kappa = 20$, which means that the oscillator of relative motion has a high eigenfrequency. The normal mode of the center-of-mass motion in this exercise has a spring constant of only $k = 2$. The oscillator of the center-of-mass motion has a much lower frequency.

5.4.11 (c) The repulsive coupling $\kappa < 0$ of the two oscillators pushes the particle 2 away from particle 1. Thus, particle 2 is accelerated into the negative x_2 direction instead of being pulled toward positive x_2 values, as in the case of an attractive coupling $\kappa > 0$.

5.4.12 (c) The effective spring constant in the oscillator of relative motion is $(k/2 + \kappa)$, see (5.12). For the values $k = 2$, $\kappa = -0.95$ the effective spring constant of relative motion is 0.1. Thus, the oscillator potential in the relative coordinate is very shallow. Therefore, the wave packet spreads into the direction of the relative coordinate. This way it gets anticorrelated.

5.4.13 (c) The wave function for bosons is symmetric, see (5.25). The symmetrization of the initial wave function of Exercise 5.4.1 with one hump with a nonsymmetric expectation value $\langle x_1 \rangle \neq \langle x_2 \rangle$ leads to an x_1, x_2-symmetric two-hump probability density. (d) The very narrow peak in plots where the two humps almost completely overlap is due to the symmetrization of the wave function.

5.4.14 (c) The initial Gaussian wave function is centered around a symmetric position $x_1 = x_2 = 3$ in the x_1, x_2 plane. Thus, symmetrization does not create a second hump.

5.4.15 (c) The direction of maximal width of the two humps in the initial probability distribution forms a small angle with the diagonal $x_1 = -x_2$ in the x_1, x_2 plane. Thus, the wave packet is anticorrelated.

5.4.17 (c) The oscillations in the relative coordinate $r = x_2 - x_1$ exhibit a much higher frequency than the oscillation in the center-of-mass motion; thus, the spring constant in the oscillator of the relative motion is larger, i.e., the coupling constant κ is large.

5.4.18 (c) The two-hump structure of the initial probability distribution says that it is just as likely that particle 1 is at $x_1 = 3$ and particle 2 at $x_2 = -3$ (hump to the left) as it is that particle 1 is at $x_1 = -3$ and particle 2 is at $x_2 = 3$ (hump to the right).

5.4.19 (c) The width in $r = x_2 - x_1$ is the width in the oscillator of relative motion. It has a strong attractive spring constant since the coupling $\kappa = 20$ is large. Thus, the frequency of this oscillator is high. Therefore, the width in the oscillator of relative motion changes very quickly. Even though we are looking at the normal mode of the center-of-mass oscillator, which keeps the expectation value of the relative coordinate time independent, the strong coupling shows in the oscillation of the width in the relative coordinate.

5.4.20 (c) See Exercise 5.4.11 (c).

5.4.21 (c) The two-particle wave function for fermions is antisymmetric. Therefore it has to vanish at $x_1 = x_2$. Hence, its absolute square, the probability distribution, vanishes.

5.4.22 (c) See Exercise 5.4.4.

5.4.25 (c) Because of vanishing correlation, equal widths, and equal initial positions, the initial wave function is obtained through the antisymmetrization of a product of two identical Gaussian wave packets, in x_1 and x_2, see (5.18). Thus the result of the antisymmetrization vanishes. The two fermions cannot be in the same state, as the Pauli principle requires.

5.4.26 (c) Also for nonvanishing correlation the initial wave function for distinguishable particles (5.18) remains symmetric in x_1 and x_2.

5.4.27 (c) The Pauli principle states that two fermions cannot occupy the same state. The different widths in the two coordinates allow for different states in the wave packet to be occupied by the two particles.

5.5.3 (d) The dips between the humps are slightly more pronounced for the fermions. (e) The marginal distributions are obtained as integrals in one variable x_1 (or x_2) over the joint probability distributions. Because the zero line $x_1 = x_2$ of the fermion probability distribution forms an angle of 45 degrees with either the x_1 axis or x_2 axis, the effect of the zero line gets washed out upon integration over x_1 or x_2.

Hints and Answers to the Exercises in Chapter 6

Note: You find descriptor files with solution descriptors in the directory `Solutions/3D_Free_Particle_(Chap_6)`.

6.3.1 (n) The minimal impact parameter b of the particle not entering the region $0 \le r \le 2\pi$ is $b = 2\pi$. The minimal classical angular momentum of the particle is $L = kb\hbar = 4\pi\hbar = 12.57\hbar$. (o) The partial wave with $\ell = 13$ starts becoming different from values close to zero only in the outer region adjacent to $r = 2\pi$. (p) The radial dependence of the ℓth partial wave is given by the spherical Bessel function $j_\ell(kr)$. Its behavior for $kr \ll 1$ is proportional to $(kr)^\ell$, see (11.40). Hence, the region in kr close to zero in which the partial wave is very small grows with ℓ.

6.3.2 (n) A partial sum up to N of the partial-wave decomposition (6.53) of the plane wave approximates the plane wave in a radial region close to $r = 0$. This region extends as far as the values of the partial wave of index $N + 1$ remain small compared to one. This is approximately the region $r < b$, where b is the classical impact parameter determined by $b = N/k$. A particle of momentum $p = \hbar k$ and impact parameter b has the classical angular momentum $L = bk\hbar = N\hbar$.

6.4.1 (d) The Gaussian wave packet of the form (6.24) widens independently in all three space coordinates. The time dependence of the width in every coordinate follows (2.10). The initial width is $\sigma_{x0} = 0.5$ so that with $\hbar = 1$, $M = 1$ we get $\sigma_x(t) = \sigma_{x0}\left(1 + t^2/4\sigma_{x0}^4\right)^{1/2} = 0.5\left(1 + 4t^2\right)^{1/2}$.

6.4.2 (e) The vector of the classical angular momentum is $\mathbf{L} = (0, 0, 4)$.

6.4.3 (d) The initial width in the x direction, σ_{x0}, is smaller than σ_{y0} in the y direction. The formula given in 6.4.1 (d) above shows that the coefficient of t^2 in $\sigma_x(t)$ is much larger than the corresponding coefficient in $\sigma_y(t)$. (e) The ripples occur along the direction of the momentum.

6.4.4 (d) The wave packet is at rest, thus there is no special direction besides the radial direction.

6.5.1 (b) The classical angular-momentum vector is $\mathbf{L} = (-0.5, -2, 5.5)$.

6.5.3 (b) The initial widths in all three coordinates are larger than in Exercise 6.5.2. Thus, the coefficients of t^2 in the formulae (2.10) for the three time-dependent widths $\sigma_x, \sigma_y, \sigma_z$ are smaller than for the situation in Exercise 6.5.2.

6.5.4 (b) Formula (2.10) can be rewritten to look like $\sigma_x = \hbar t \left(1 + 4\sigma_{x0}^4 m^2/\hbar^2 t^2\right)^{1/2} / (2\sigma_{x0} m)$, i.e., for large t the width σ_x grows approximately linearly in t.

6.6.1 (b) The classical angular momentum is $L = bp_0 = 1$, it points along the z direction. (c) The angular momentum for a particle with impact parameter $b' = b + \sigma_0$ is $L' = 2.5$; for $b'' = b - \sigma_0$ it is $L'' = -0.5$. (d) The maximum of the probabilities W_ℓ to find the angular momentum ℓ in the wave packet at $\ell = 1$ corresponds to the classical angular momentum of the wave packet. Because of the spatial extension of the Gaussian wave packet, there is a nonzero probability density for particles with larger and smaller impact parameters. As typical values for the width of the distribution of impact parameters we choose $b' = b + \sigma_0 = 1.6667$ and $b'' = b - \sigma_0 = -0.3333$. The corresponding angular momenta – as calculated above – are $L' = 2.5$ and $L'' = -0.5$. Actually, the distribution of m values in the plot shows that the values $\ell = 3$, $m = 3$ and $\ell = 1$, $m = -1$ have about half of the probability of the value $\ell = 1$, $m = 1$. This explains the features of the distribution of the $W_{\ell m}$.

6.6.2 (b) The average angular momentum of the distribution is the same as in Exercise 6.6.1. The halving of the impact parameter b is compensated by the doubling of the momentum p_0. The distribution of the angular momenta in the wave packet is the same as in Exercise 6.6.1 since the initial width σ_0 is halved in this exercise.

6.6.3 The Gaussian wave packet moves in the plane $z = 0$. It is an even function in the z coordinate. In polar coordinates it is thus an even function with respect to $\vartheta = \pi/2$. The spherical harmonics of $m = \ell - (2n+1)$, $n = 0, 1, \ldots, \ell - 1$, are odd with respect to $\vartheta = \pi/2$. Thus, they do not contribute to the partial-wave decomposition. The values $W_{\ell m}$ for $m = \ell - (2n+1)$ vanish.

Hints and Answers to the Exercises in Chapter 7

Note: You find descriptor files with solution descriptors in the directory `Solutions/3D_Bound_States_(Chap_7)`.

7.2.2 (e) The centrifugal barrier $\hbar^2\ell(\ell+1)/(2Mr^2)$ is a steeply increasing function as r goes to zero. It keeps the probability density low at small values of r. Because the barrier height at a given r grows with $\ell(\ell+1)$, the suppression becomes the more apparent the higher the angular momentum ℓ.

7.2.3 (e) Heisenberg's uncertainty principle $\Delta x \Delta p \geq \hbar/2$ can be adapted to the ground-state energy E_0 in a potential well of radius R to yield $R\sqrt{2ME_0} \gtrsim \hbar$. Thus the ground-state energy grows proportional to $1/R^2$, i.e., $E_0 \gtrsim \hbar^2/(2MR^2)$.

7.2.5 (e) The particle in a potential well of finite depth is not strictly confined to the inside of the well as is the case in the infinitely deep square well. Therefore, the curvature $\mathrm{d}^2\varphi/\mathrm{d}x^2$ of the wave functions $\varphi(x)$ of stationary states with the same number of nodes is smaller for the potential of finite depth than for one with infinite depth. Thus, the contributions of the radial energy and of the effective potential energy in a state with a given number of nodes are smaller for the potential of finite depth.

7.2.6 (e) For spherical square-well potentials of finite depth the Schrödinger equation is the same as for an infinitely deep square-well potential except for the constant value of the potential depth. This value can be absorbed into the energy eigenvalue of the stationary Schrödinger equation. Only the boundary condition at the value of the radius of the potential well for bound states in the infinitely deep well is different from the one for a potential well of finite depth. At the origin of the coordinate frame the boundary conditions are the same. Therefore, the wave functions look alike for small values of the radial coordinate r.

7.2.7 (e) Under the action of a harmonic force in three dimensions a particle of high angular momentum moves on an orbit far away from the center of the force.

7.2.8 (e) In a three-dimensional harmonic oscillator, the relation between energy and angular momentum is given by $E_{n_r\ell} = \left(2n_r + \ell + \frac{3}{2}\right)\hbar\omega$, see (7.41), (7.42). Thus, the lowest energy for fixed angular momentum ℓ is $E_{0\ell} = \left(\ell + \frac{3}{2}\right)\hbar\omega$. It is assumed for vanishing radial quantum number n_r. Except for the zero-point energy $3\hbar\omega/2$, the energy is solely rotational and potential energy $E = L\omega$, $L = \ell\hbar$.

7.2.10 (e) The effective potential is given by $V^{\mathrm{eff}}(r) = M\omega^2 r^2/2 + \ell(\ell+1)\hbar^2/(2Mr^2)$. Its minimum is at the value $r_0 = \left[\ell(\ell+1)\hbar^2/M^2\omega^2\right]^{1/4}$ of the radial variable. The potential at this value is $V^{\mathrm{eff}}(r_0) = \sqrt{\ell(\ell+1)}\hbar\omega$. Its curvature at r_0 is $\mathrm{d}^2V^{\mathrm{eff}}(r)/\mathrm{d}r^2\big|_{r=r_0} = 4M\omega^2$. Thus, the oscillator potential approximating the effective potential close to its minimum is $V_\ell^{\mathrm{app}}(r) = \sqrt{\ell(\ell+1)}\hbar\omega + \frac{M}{2}(2\omega)^2(r-r_0)^2$. (f) The wave functions in the approximating potential $V_\ell^{\mathrm{app}}(r)$ are harmonic-oscillator eigenfunctions centered about r_0. (g) The angular frequency of the approximating oscillator is 2ω, twice the frequency ω of the three-dimensional oscillator. It is independent of the angular momentum. (h) For fixed ℓ the spacing of the energy levels in the three-dimensional oscillator is $2\hbar\omega$. The eigenvalues are $E_{n_r\ell} = \left(2n_r + \ell + \frac{3}{2}\right)\hbar\omega$. Because the approximating potential has the frequency 2ω, its level spacing is also $2\hbar\omega$. The eigenvalues in the approximating oscillator are $E_{\rm app} = V(r_0) + 2(n' + 1/2)\hbar\omega = \left(2n' + \sqrt{\ell(\ell+1)} + 1\right)\hbar\omega$. The two expressions differ very little for large angular momentum $\ell \gg 1$.

7.2.11 (e) For vanishing angular momentum $\ell = 0$ the three lowest energy values are $E = -1/2, -1/8, -1/18$; for $\ell = 1$ the lowest three eigenvalues are $E = -1/8, -1/18, -1/32$.

7.2.12 (e) Eigenvalues for $\ell = 2$ are $E = -1/18, -1/32, -1/50$; eigenvalues for $\ell = 3$ are $E = -1/32, -1/50, -1/72$.

7.2.13 (e) The effective potential for the Coulomb interaction is given by $V^{\rm eff}(r) = -\alpha\hbar c/r + \ell(\ell+1)\hbar^2/(2Mr^2)$. The minimum occurs at the value $r_0 = \ell(\ell+1)\hbar c/(\alpha Mc^2)$. Here the effective potential has the value $V^{\rm eff}(r_0) = -\alpha^2 Mc^2/[2\ell(\ell+1)]$. The curvature at r_0 has the value $\mathrm{d}^2V^{\rm eff}(r)/\mathrm{d}r^2\big|_{r=r_0} = \alpha^4(Mc^2)^3/[\ell^3(\ell+1)^3\hbar^2c^2]$. The approximating oscillator potential centered about r_0 is $V^{\rm app}(r) = -\alpha^2Mc^2/[2\ell(\ell+1)] + (1/2)M\omega_\ell^2(r-r_0)^2$ with the square of the effective angular frequency $\omega_\ell^2 = \alpha^4(Mc^2)^2/[\ell^3(\ell+1)^3\hbar^2]$. (f) The energy eigenvalues of the approximating oscillator are $E_{\ell n'} = -\alpha^2Mc^2/[2\ell(\ell+1)] + (n'+1/2)\hbar\omega_\ell = -\alpha^2Mc^2[1 - 2(n'+1/2)/\sqrt{\ell(\ell+1)}]/[2\ell(\ell+1)]$. The exact formula $E = -\alpha^2Mc^2/2n^2$ can be expanded for large angular momentum ℓ if we use $n = n_{\rm r}+\ell+1$, the decomposition of the principal quantum number n into ℓ and the radial quantum number $n_{\rm r}$. For large ℓ we arrive at the approximation $E = -\alpha^2Mc^2[1-2(n'+1/2)/(\ell+1/2)]/2(\ell+1/2)^2$. Again for large ℓ the two expressions for E converge to the same value. (g) The exact Bohr radii are $a_n = n^2a_0$, where $a_0 = \hbar c/(\alpha Mc^2)$ is the innermost Bohr radius. For small $n_{\rm r} \ll \ell$ the above value for r_0 approaches the Bohr radius.

7.3.1 (e) In the neighborhood of the energy of the state, there are no eigenvalues of states in the first well. Therefore, there is no strong contribution of an eigenstate of the first well to the third eigenstate of the double well. Thus, the wave function is different from zero chiefly in the second well.

7.3.4 (c) Because of the strong repulsive potential in the innermost region $0 \le r < 0.5$, the potential is a hard-core potential. It keeps the values of the wave function in this region small as long as the energy of the state is small compared to the height of the hard core. Therefore, the influence of the repulsive centrifugal barrier on the wave functions and the energy eigenvalues is much smaller than in a potential without a hard core.

7.4.1 (b) The number of node lines at fixed polar angles is equal to $\ell - m$.

7.4.3 (b) The node half-circles are the nodes of the radial wave functions. With increasing number of node lines, the energy eigenvalue of the wave function increases.

7.4.4 The number of ϑ node lines is equal to $\ell - m$.

7.4.7 (b) Exercise 7.4.6 shows the probability density of the eigenstate with principal quantum number $n = 1$ and the angular-momentum quantum numbers $\ell = 15$, $m = 15$. This corresponds to a classical angular-momentum vector parallel to the quantization axis of angular momentum, which is the z axis in this case. Exercise 7.4.7 (a) exhibits the probability density for the angular-momentum quantum numbers $\ell = 15$, $m = 0$. A state with these quantum numbers is classically interpreted as representing an angular momentum in a direction perpendicular to the z axis. Because L_x and L_y do not commute with L_z, no particular direction in the x, y plane can be assigned to the classical vector of angular momentum in this case. In fact, the probability distribution is cylindrically symmetric about the z axis. The probability density is largest at the large values of z close to the wall of the spherically symmetric potential well. This can be interpreted as an assembly of classical orbits in the planes that contain the z axis. No particular one of these planes is distinguished, so that no special direction of angular momentum in the x, y plane can be assigned to the state with $m = 0$.

7.4.8 (b) The wall of the infinitely deep square well confines the wave function strictly to the range $r \leq a$. In the harmonic-oscillator potential, the wave function falls off with $\exp(-r^2/2\sigma_0^2)$. Thus, the decrease in the infinitely deep square-well potential is much faster.

7.4.12 (b) For eigenstates in the Coulomb potential, the quantum number ℓ of angular momentum satisfies the relation $\ell \leq n - 1$. Thus, no eigenstate exists with the quantum numbers $n = 1$, $\ell = 1$.

7.4.13 (b) The spherical harmonics $Y_{\ell m}(\vartheta, \varphi)$ possess $(\ell - m)$ nodes in the polar angle ϑ. The number of nodes in the radial variable r is $(n - 1 - \ell)$.

7.4.14 (b) The normalization of the radial wave function is given by an integral containing the measure $r^2\,\mathrm{d}r$. The wave function with the quantum numbers $\ell = n - 1$ has its large values in a region about the value $\ell(\ell + 1)a$ of the radial variable. Here r^2 is large; thus, the normalization of the wave function to one suppresses the height of the wave function. (c) For $m = \ell$ the wave function represents a particle with the vector of angular momentum in the direction of the axis of quantization, i.e., the z axis. This forces the particle to a far-out Bohr orbit.

7.5.1 (d) number of radial nodes: $n - 1$; number of polar nodes $\ell - m$.

7.5.2 (d) number of radial nodes: $n - 1$; number of polar nodes $\ell - m$.

7.5.3 (a) The lines are shifted away from the origin. (b) The lines are shifted still more to the outside. The potential is too shallow to accommodate states with $n = 3$, $\ell = 2$.

7.5.4 The radial extension increases as n_r grows.

7.7.1 (b) The classical angular-momentum vector has the components $L_x = -1$, $L_y = -7$, $L_z = 5$.

7.7.3 (b) The classical angular-momentum vector has the components $L_x = 0$, $L_y = 0$, $L_z = 0$. The motion of the wave packet is an oscillation through the center of the spherically symmetric potential.

7.7.5 (b) The ground-state width of the three-dimensional oscillator is $\sigma_0 = \sqrt{\hbar/M\omega}$. For $T = 1$ the angular frequency equals $\omega = 2\pi/T = 2\pi$. Thus, $\sigma_0 = (2\pi)^{-1/2} = 0.56$. The initial widths σ_{x0} and σ_{y0} are smaller than $\sigma_0/\sqrt{2} = 0.4$. In these directions the Gaussian wave packet is initially narrower than $\sigma_0/\sqrt{2}$. In the z direction the wave packet is initially wider than $\sigma_0/\sqrt{2}$. The probability ellipsoid of the initial state has prolate shape with the large principal axis in the z direction. After a quarter period the widths σ_x and σ_y have reached their maximum values, which are now larger than $\sigma_0/\sqrt{2}$, whereas σ_z is at its minimum, which is smaller than $\sigma_0/\sqrt{2}$, see Sect. 3.1, (3.35). After a quarter period $T/4$ the Gaussian wave packet possesses a probability ellipsoid of oblate shape with the smallest principal axis in z direction.

Hints and Answers to the Exercises in Chapter 8

Note: You find descriptor files with solution descriptors in the directory `Solutions/3D_Scattering_(Chap_8)`.

8.2.1 (d) The centrifugal barrier $V_\ell(r) = \hbar^2\ell(\ell+1)/(2Mr^2)$ grows quadratically with ℓ. For fixed energy of the incoming particle the wave function becomes more and more suppressed in the region of small r because of the r^ℓ behavior of the spherical Bessel functions $j_\ell(kr)$ for $kr \ll 1$. The range in which the wave function is smaller than a given value widens with increasing ℓ. This way the contribution of the rotational energy represented by the centrifugal barrier remains low enough to keep the total energy constant.

8.2.3 (d) The total energy contains three contributions, the radial, the rotational, and the potential energy. For growing total energy and fixed angular momentum the contribution of the rotational energy grows. This requires larger values of the wave function at low values of r where the centrifugal barrier is large.

8.2.5 (b) The energies of the first two resonances are $E_1 = 10.1$ and $E_2 = 10.5$.

8.2.9 For increasing ℓ the centrifugal barrier pushes the resonance wave function out of the region of small r, see Exercise 8.2.1. This increases the curvature of the wave function in the region close to the range of the potential. Thus, the kinetic-energy contribution grows and therefore so does the total energy.

8.3.4 For distances $r \gg r_N$ large compared to the range r_N of the potential, the radial scattering wave function has the form $R_\ell(kr) \sim \exp(i\delta_\ell)\sin(kr -$

$\ell\pi/2 + \delta_\ell)/kr$. For the scattering at a repulsive infinitely high potential of radius d, the phase shift δ_ℓ is simply given by $\delta_\ell = -kd$, because the wave function has to fulfill the condition $R_\ell(kd) = 0$ at the wall of the potential at $r = d$. For the scattering at a repulsive potential of finite height V_0, the wave function R_ℓ has the same form as above if $r \gg d$. However, the phase shift δ_ℓ is given by $\delta_\ell = -ka$ where the scattering length a replaces the radius of the repulsive potential. We have $a < d$ because the wave function penetrates somewhat into the repulsive square well of finite height. The free wave functions ($V = 0$) have the same form for the $R_\ell(kr)$ as above, however for $\delta_\ell = 0$. This phase shift determines the partial scattering amplitude f_ℓ, (8.50), and because of (8.41) and (8.45) both the differential cross section and the partial cross section as well.

8.4.1 (j) Summation of the partial scattering amplitudes f_ℓ to the scattering amplitude $f(k, \vartheta)$, (8.32), up to $L = 0$ includes the angular momentum $\ell = 0$ only. Thus only the zeroth Legendre polynomial $P_0(\cos\vartheta) = 1$ contributes. Hence, no ϑ dependence shows up for $L = 0$ in (a).

8.4.2 (c) For $E = 50$ the wavelength is $\lambda = 0.6282$, for $E = 5000$: $\lambda = 0.06282$. (d) The decrease of the differential cross section in the forward direction for energies increasing from the value $E = 0$ follows a $1/E$ behavior, see (8.41), (8.32).

8.4.4 (c) The two differential cross sections for the high energy $E = 5000$ look alike. This can be easily understood by using the lowest-order Born approximation, which is valid at high energies $E \gg |V_0|$. The scattering amplitude is given by a volume integral over a product of the free incoming and outgoing waves and the potential. The differential cross section, being proportional to the absolute square of the scattering amplitude, is in this high-energy approximation independent of the sign of the potential.

8.4.5 (i) For low energies of the incident plane wave, the zeroth partial wave yields the largest contribution to the differential cross section. The phase shift δ_0 for very low energies is determined by $\tan\delta_0 = -ka$, where a is the scattering length of the potential. For repulsive potentials we have $0 < a \leq d$ at low energies, see Exercise 8.3.4. For an attractive square-well potential the wave number κ within the range of the potential is $\kappa = \sqrt{2M|V_0|/\hbar^2 + k^2}$, where $|V_0|$ is the depth of the square-well potential. For $\ell = 0$, the radial wave function in region 1 has the form $R_{01} = (1/\kappa r)\sin\kappa r$. Outside the range of the potential, the radial wave function can be written as $R_{02} = (1/kr)\sin(kr + \delta_0)$, where δ_0 is the scattering phase for angular momentum zero. The continuity conditions yield $(1/k)\tan\delta_0 = (1/\kappa)\tan\kappa d$. For $k \to 0$ the scattering phase approaches zero, $\delta_0 \to 0$. Thus, for $k \to 0$ we get for the scattering length $a = -\delta_0/k \simeq -(1/k)\tan\delta_0$ the equation $a = -(1/\kappa)\tan\kappa d$. The numerical value of κ for $M = 1$ and $|V_0| = 3$ is $\kappa = 2.45$. For this value and $d = 2$

the scattering length for the attractive potential is obtained to be $a = 2.16$. For the repulsive potential we have $a \le 2$. The differential cross section for low energies is given by $\mathrm{d}\sigma/\mathrm{d}\Omega = (1/k^2)\sin^2\delta_0 = a^2$. Thus, for low energy the differential cross section for the repulsive potential of Exercise 8.4.3 is smaller than for the attractive potential of Exercise 8.4.1.

8.5.1 (f) The energy of the lowest resonance with vanishing angular momentum is $E_{01} = 3.9$.

8.5.2 (f) The two lowest resonances with angular momentum zero appear at the energies $E_{01} = 3.9$, $E_{02} = 8.2$. The values of the scattering phase at these resonances are $\delta_{01} = -\pi$, $\delta_{02} = -\pi/2$. (g) The partial scattering amplitudes at the two lowest resonances with vanishing angular momentum are calculated using (8.50). We get $f_{01} = 0$, $f_{02} = \mathrm{i}$.

8.5.3 (f) The energy of the lowest resonance with angular momentum $\ell = 1$ is $E_{11} = 5.5$.

8.5.4 (f) The energy of the lowest resonance with angular momentum $\ell = 2$ is $E_{21} = 7$. (g) The energy of the lowest resonance increases with angular momentum ℓ because of the positive energy contribution of the repulsive centrifugal barrier $V_\ell(r) = \hbar^2\ell(\ell+1)/(2Mr^2)$ and the increase of the radial energy with ℓ.

8.5.5 (f) The energies of the lowest resonances of angular momentum zero are $E_{01} = 1$, $E_{02} = 3.6$, $E_{03} = 7.7$. (g) In an infinitely deep square-well potential of radius 2 the energies of the lowest bound states of vanishing angular momentum are $E_1 = 1.23$, $E_2 = 4.93$, $E_3 = 11.10$.

8.5.9 (c) The small peaks in the plots of the partial cross sections σ_ℓ indicate the existence of resonances of angular momentum ℓ at the energies at which the peaks occur. (d) For vanishing angular momentum the energy eigenvalues of an infinitely deep square-well potential of width 2 are $E_{01} = 1.23$, $E_{02} = 4.93$, $E_{03} = 11.10$. The resonance peaks showing up in σ_0 are at $E_{01} = 1.1$, $E_{02} = 4.8$, $E_{03} = 8.6$. (e) The eigenfunctions in the infinitely deep square-well potential of radius d have to vanish at $r = d$. For the resonance wave functions in the barrier potential of this exercise, this is not so. Thus, the curvature of the eigenfunctions in the infinitely deep square well is larger than that of the corresponding resonance energies in the barrier potential.

8.5.10 (b) For the energy $E = 0.001$ of the incoming particle, the partial cross section σ_0 has practically the same value as the total cross section. This has to be so, because there are almost no contributions of higher angular momenta $\ell = 1, 2, \ldots$ to the incoming wave function of the low energy $E = 0.001$. (c) The classical total cross section for the elastic scattering of particles on a hard sphere of radius d is equal to its geometrical area $\sigma_{\mathrm{cl}} = \pi d^2$. For $d = 2.5$, we get $\sigma_{\mathrm{cl}} = 19.6$.

8.5.11 (a) $\sigma_{\mathrm{tot}} = 39.5$, (b) $\sigma_{\mathrm{tot}} = 46.8$, (c) $\sigma_{\mathrm{tot}} = 49.3$, (d) $\sigma_{\mathrm{tot}} = 49.8$, (e) $\sigma_{\mathrm{tot}} = 50.2$. (f) The total cross section for quantum-mechanical scattering on a hard sphere is $\sigma_{\mathrm{tot}} = 4\pi d^2$. For $d = 2$ the numerical value comes out to be $\sigma_{\mathrm{tot}} = 50.26$. It represents the upper limit for the series of numbers for σ_{tot} obtained in (a–e).

8.6.1 For the attractive potential the kinetic energy and therefore the momentum is increased for small r. Correspondingly, the local wavelength is shortened.

8.6.2 Because the $|V| \ll E$ near $r_0(E)$, we can use the following rough estimate: $r_0(E) \approx \ell/p = \ell/\sqrt{2ME}$. For $\ell = 10$, $M = 1$ we get $r_0(E) \approx 7(3)$ for $E = 1(5)$.

8.7.1 Clearly, many terms need be superimposed in the approximation (8.58). For the given energy $E = 5$ and the maximum radius $r_{\mathrm{end}} = 5$ a value of $\ell \approx kr_{\mathrm{end}} \approx 22$ is expected for a good approximation also at the outer radial bound, which can be verified by inspecting the plots.

8.7.2 The situation is comparable to that of Exercise 8.7.1.

Hints and Answers to the Exercises in Chapter 9

Note: You find descriptor files with solution descriptors in the directory `Solutions/Magnetic_Resonance_(Chap_9)`.

9.3.1 The width of the resonance increases as Ω_1/Ω_0 does.

Hints and Answers to the Exercises in Chapter 10

Note: You find descriptor files with solution descriptors in the directory `Solutions/Hybrid_States_(Chap_10)`.

10.3.1 The hybrid state with the hybridization parameter of $\lambda = \sqrt{3}$, i.e., the sp^3 state, has the largest extension in the direction $-z$.

Hints and Answers to the Exercises in Chapter 11

Note: You find descriptor files with solution descriptors in the directory `Solutions/Math_Functions_(Chap_11)`.

11.2.1 (d) The Hermite polynomial H_n possesses n zeros.

11.3.1 (d) The Legendre polynomial P_ℓ possesses ℓ zeros.

11.3.2 (c) The associated Legendre function P_ℓ^m possesses $(\ell - m)$ zeros.

11.17.1 The distribution widens and shifts towards larger k as λ increases.

11.18.1 The plots are easily understood by using the rules of complex algebra, in particular, $z = x + \mathrm{i}y = r\mathrm{e}^{\mathrm{i}\varphi}$ and Euler's formula $\mathrm{e}^{\mathrm{i}\alpha} = \cos\alpha + \mathrm{i}\sin\alpha$.

(a) $w = \mathrm{e}^z = \mathrm{e}^{x+\mathrm{i}y} = \mathrm{e}^x \mathrm{e}^{\mathrm{i}y} = \mathrm{e}^x(\cos y + \mathrm{i}\sin y)$. From this follows $\operatorname{Re} w = \mathrm{e}^x \cos y$, $\operatorname{Im} w = \mathrm{e}^x \sin y$, $|w| = \mathrm{e}^x$, $\arg w = y$. (b) Use $\log z = \log(r\mathrm{e}^{\mathrm{i}\varphi}) = \log r + \mathrm{i}\varphi$, $\sin z = \frac{1}{2\mathrm{i}}(\mathrm{e}^{\mathrm{i}z} - \mathrm{e}^{-\mathrm{i}z})$, $\cos z = \frac{1}{2}(\mathrm{e}^{\mathrm{i}z} + \mathrm{e}^{-\mathrm{i}z})$, $\sinh z = \frac{1}{2}(\mathrm{e}^z - \mathrm{e}^{-z})$, $\cosh z = \frac{1}{2}(\mathrm{e}^z + \mathrm{e}^{-z})$.

11.18.2 From $w = z^n = (r\mathrm{e}^{\mathrm{i}\varphi})^n = r^n \mathrm{e}^{\mathrm{i}n\varphi} = r^n(\cos n\varphi + \mathrm{i}\sin n\varphi)$ one gets $\operatorname{Re} w = r^n \cos n\varphi$, $\operatorname{Im} w = r^n \sin n\varphi$, $|w| = r^n$, $\arg w = n\varphi$.

A. A Systematic Guide to IQ

A.1 Overview

A.1.1 Starting IQ

We assume here that **IQ** is fully installed on your computer and that the **IQ** symbol featuring the letter $\hbar$ is displayed on the computer's desktop. (See Appendix B for installing **IQ** and the file `ReadMe.txt` on the CD-ROM for the case that the **IQ** symbol is not present on the desktop.) You start **IQ** simply by clicking (once or twice – depending on your operating system) on that symbol. The **IQ** *main frame* appears on the desktop. It contains the *main toolbar*, Fig. A.1.

Fig. A.1. The **IQ** main frame with the main toolbar containing buttons to start different actions. In its title bar the **IQ** main frame shows the name of the currently open descriptor file

A.1.2 Introductory Demonstration

Usually, right after the start of **IQ** a small frame, Fig. A.2, entitled *Interquanta Introduction* is displayed offering you to start a demonstration on how to use **IQ**. Before starting the demonstration you can choose between **Automatic** mode (demonstration is run off like a movie) and **Step-by-Step** mode (you have to press the button **Next** or the `Enter` key to go on to the next step). You can choose the demonstration to be **With Sound** (graphics displayed, explanations spoken) or **Without Sound** (graphics and explanations displayed on the screen).

The frequent user of **IQ** may not want the presentation of the frame in Fig. A.2 each time the program is started. It can be suppressed by *customizing* **IQ**, see Sect. A.1.14. The introductory demonstration can always be shown

S. Brandt et al., *Interactive Quantum Mechanics: Quantum Experiments on the Computer*,
DOI 10.1007/978-1-4419-7424-2_13, © Springer Science+Business Media, LLC 2011

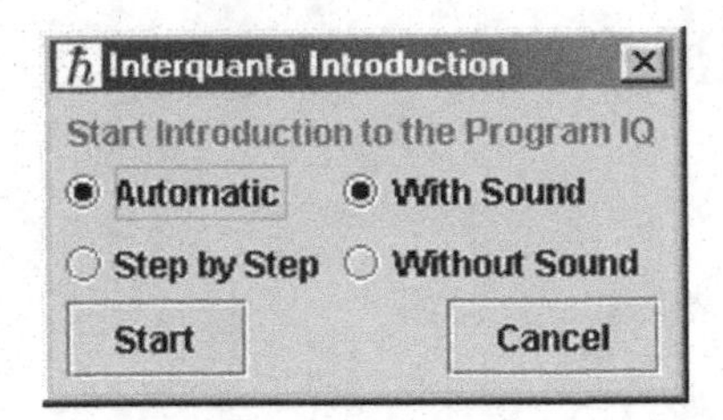

Fig. A.2. Frame offering the introductory demonstration

by pressing the **Run Demo** button in the **IQ** main frame and selecting the appropriate file, see Sect. A.1.13.

A.1.3 Selecting a Descriptor File

Each plot produced by **IQ** is completely defined through a set of parameters, called a *descriptor*. Such a set forms a *record* on a *descriptor file*, which thus contains a number of descriptors. A total of ten descriptor files forming the *descriptor library* are distributed as part of **IQ**. Each of the ten files contains descriptors corresponding to the physics topics treated in one of the ten chapters 2 through 11. You can create your own descriptor files with your own descriptors, see Sect. A.1.11.

Normally, **IQ** starts off with the *default descriptor file* open (see Sect. A.1.14 for setting the default). The name of the descriptor file that is currently open is shown in the title of the **IQ** main frame. Figure A.1 shows the name you will see as long as the default has not been changed. If, for some reason, **IQ** was unable to open the default descriptor file, no name is displayed and you must open another descriptor file.

To open a descriptor file press the button labeled **Descriptor File** on the main toolbar. A file chooser appears allowing you to select a descriptor file.

A.1.4 Selecting a Descriptor and Producing a Plot

To select a particular descriptor press the button **Descriptor** on the main toolbar. The *descriptor-selection panel* appears, which contains (in its upper part) a list of all the descriptors in the current descriptor file, Fig. A.3. (For the lower part – creation of a mother descriptor – see Sect. A.1.10.) Each line of the list corresponding to one descriptor contains three items, the *descriptor number*, the *descriptor title*, and the *descriptor time stamp*, which indicates the last time when the descriptor was changed.

You select a descriptor by clicking on the corresponding line. The selection panel vanishes and instead a *graphics frame* appears containing the *plot* which **IQ** produces using the parameters of the descriptor, Fig. A.4. The frame also contains at the top the descriptor title and below it a toolbar with

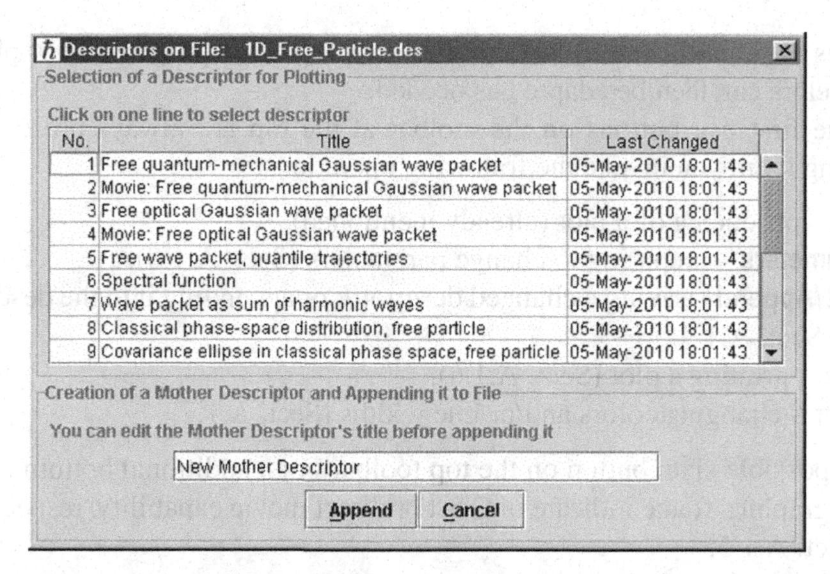

Fig. A.3. Descriptor-selection panel with, in its upper part, the list of descriptors on the current descriptor file. The lower part can be used to create a *mother descriptor*, see Sect. A.1.10

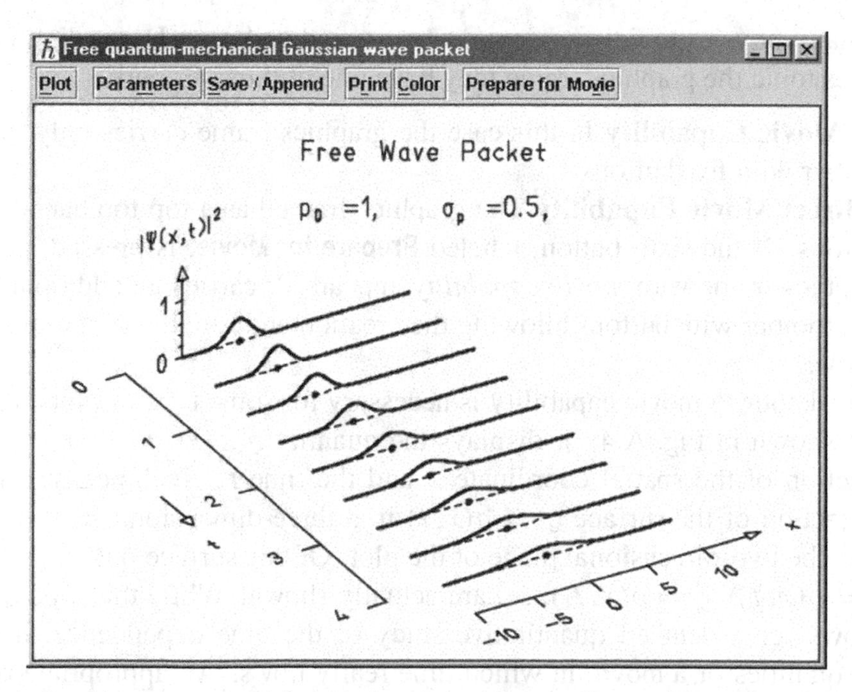

Fig. A.4. Graphics frame containing the descriptor title, a toolbar, and the plot corresponding to the descriptor

buttons. (The graphics frame does not show a plot if, while customizing, you unselected the possibility *plot graphics immediately*, see Sect. A.1.14. In that case you have to press the **Plot** button in the graphics frame.)

Within a descriptor you can change all parameters of the given physics topic, Sect. A.1.8. You cannot, however, change the physics topic itself. A physics topic is chosen by selecting an existing descriptor for that topic. Its parameters can then be adapted as needed.

The first five buttons on the toolbar at the top are always present. By pressing them you invoke the following operations:

- **Plot** – producing of a plot (already mentioned),
- **Parameters** – preparing to change parameters (Sect. A.1.8),
- **Save/Append** – saving a changed descriptor or appending it to the descriptor file (Sect. A.1.9),
- **Print** – printing a plot (Sect. A.1.6),
- **Color** – changing colors and/or line widths (Sect. A.1.7).

A possible sixth button on the top toolbar or an additional bottom toolbar of the graphics frame indicate indirect or direct movie capability, respectively, see Sect. A.1.5.

A.1.5 Creating and Running a Movie

For many physics topics movies can be created and run. Depending on the physics topic the graphics frame may have one of three properties.

- **No Movie Capability** In this case the graphics frame carries only the top toolbar with five buttons.
- **Indirect Movie Capability** The graphics frame has a top toolbar with six buttons. If the sixth button, labeled **Prepare for Movie**, is pressed, another graphics frame with *movie capability* appears. It carries an additional bottom toolbar with buttons allowing the creation and running of a movie, see below.
  This detour to movie capability is necessary for some type of plots like the one shown in Fig. A.4. It displays the quantity $\varrho(x,t) = |\Psi(x,t)|^2$ as a function of the spatial coordinate x and the time t. Technically it is the projection of the surface $\varrho = \varrho(x,t)$ in a three-dimensional x, t, ϱ space onto the two-dimensional plane of the plot. Of this surface only a few lines $\varrho = \varrho(x,t_0), \varrho = \varrho(x,t_1), \dots$ are actually shown. While this type of plot allows for a detailed quantitative study of the time dependence, it lacks the qualities of a movie in which time really flows. An appropriate movie is a series of plots of $\varrho = \varrho(x,t)$ in the x, ϱ plane for increasing values of t, displayed one after the other in quick succession. By pressing the button **Prepare for Movie** you create a descriptor and open a graphics frame corresponding to the physics parameters of the original descriptor but being of the plot type needed for movie capability. You can create and run movies with this new descriptor, change all parameters, and, if you wish, append it to the descriptor file of the original descriptor.

- **Direct Movie Capability** The graphics frame for many physics topics directly has movie capability. That is immediately apparent from the fact that it carries a top and a bottom toolbar as in Fig. A.5.

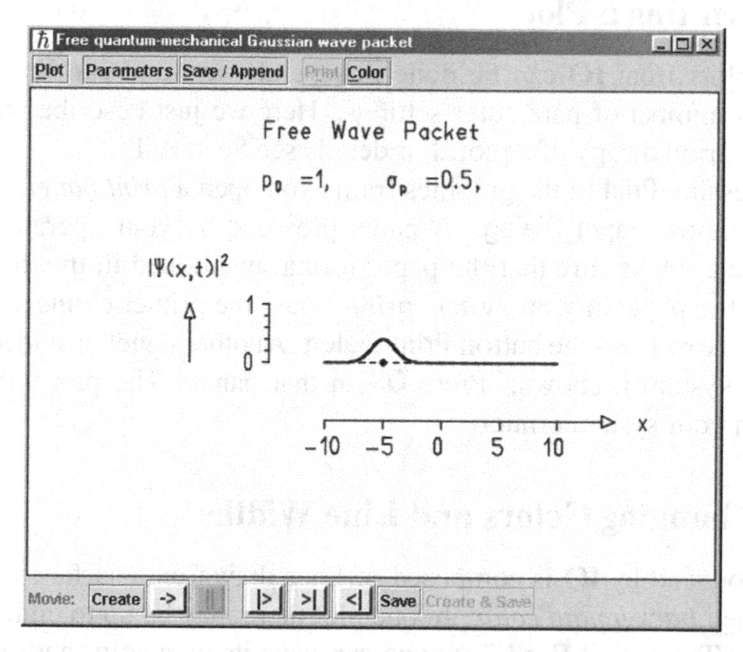

Fig. A.5. Graphics frame with movie capability. It carries an additional bottom toolbar with buttons for the creation and running of a movie

Perform the following steps to create and run that movie, in a graphics frame with direct movie capability:

- Press **Create**. All frames of the movie are produced and stored in memory. Once the movie is complete, the remaining buttons are enabled.
- Press ->| (run). The movie is shown and rerun until you press ||| (stop).
- You may go forward or backward by a single frame pressing >| or <|, respectively. Hitting |>| takes you back to the first frame.

You may change all parameters as described in Sect. A.1.8 and repeat creating the movie. For all descriptors with direct movie capability the corresponding parameter panel contains a subpanel **Movie** with physics information and technical data specific to the movie, see Sect. A.4.

A movie can be stored for later use independent of **IQ** by clicking on the **Save** button. This opens a standard file chooser in which you can assign the movie a file name and store it in the directory of your choice. The file is stored in the *animated GIF* format, a standardized format which allows viewing the movie in practically any browser.

There is also a button **Create and Save**. If you press that, a file chooser is opened. After you have entered a file name, the movie is computed and directly stored under the chosen name in the animated GIF format. If **Create**

and Save is used, each frame is directly stored away on disk and thus less main memory is required.

A.1.6 Printing a Plot

Printing plots from **IQ** can be done in several ways and can be controlled by quite a number of parameter settings. Here we just describe one way to produce a printed copy of a plot. For details see Sect. A.11.

By pressing **Print** in the graphics frame you open a *print panel*, Fig. A.24. Press the button **Paper Dialog**. A panel provided by your operating system is displayed. Make sure that the paper format mentioned in this panel is the format of the paper in your *system printer*, i.e., the printer connected to your computer. Now press the button **Print Dialog**. Another panel provided by your operating system is shown. Press **OK** in that panel. The plot will then be printed on your system printer.

A.1.7 Changing Colors and Line Widths

A plot produced by **IQ** is composed of lines drawn on a surface of uniform color, called *background color*. In one plot there may be up to eight different line types, Table A.1. Each line type can have its own color and line width. Colors and line widths for plots on the screen are determined by a *screen item record* in the *configuration file*, see Sect. A.1.14. They can be changed as follows.

Table A.1. The different line types and their numbers.

Number of Line Type	Name
1	3D surface (top side) 2D function graph
2	box, scales, numbers, arrows with texts
3	caption
4	additional text
5	depends on physics topic
6	depends on physics topic
7	depends on physics topic
8	3D surface (bottom side)

A *color-selection panel*, Fig. A.6, is displayed once you press **Color** in the graphics frame.

The essential field carries the header **IQ Items: Colors and Line Widths**. It contains, in nine small rectangles, the colors of background and line types

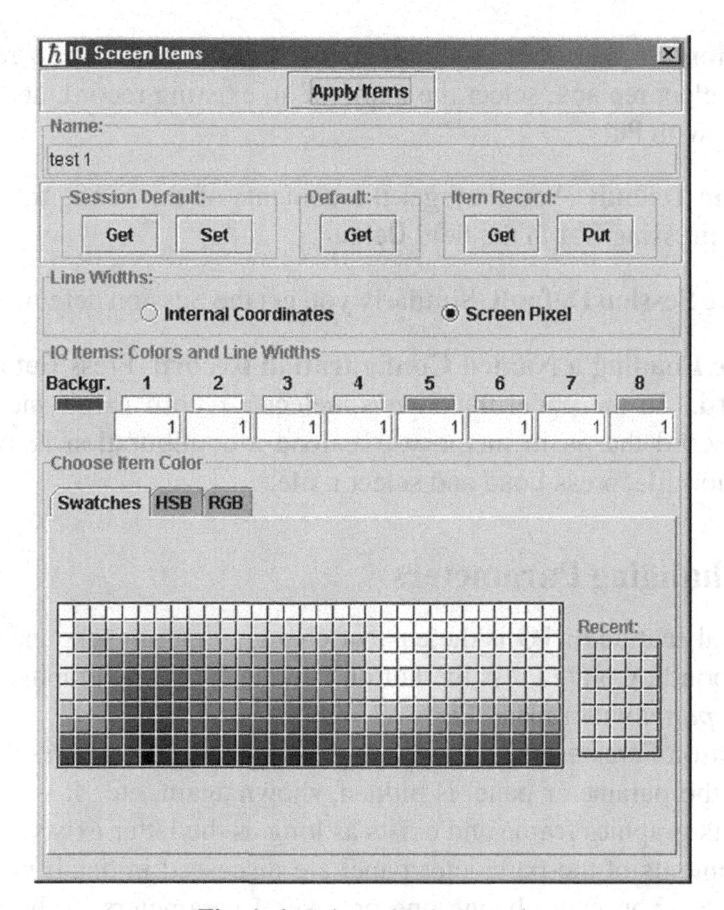

Fig. A.6. Color-selection panel

and, for the latter, also the widths. The **Units for Line Widths** are either **Internal Coordinates** (i.e., the line widths grow with the size of the graphics frame on the screen) or one **Screen Pixel**. To change a line width just change its numerical value (and its unit, if needed). To change a color you have first to select one of the nine colors by pressing on the corresponding rectangle. Next, you choose a new color using the *color chooser* at the bottom of the panel.

Applying Changes to a Single Plot Just press **Apply Items**.

Keeping Changes as Session Default If you want to keep the changes so that they are applied to plots produced later, you may press the **Set** button in the field **Session Default**. In order to set the session default from already existing records, use the *customize panel*, Sect. A.1.14.

Storing Changes in a Named Configuration Record Colors and line widths are grouped in *configuration records*, Sect. A.1.14. You may store your present setting by pressing **Put** in the field **Item Record**. Another panel opens in which you can define a **new** record, give it a name, press **Put** to store it in the memory (and, if you wish, press **Save** or **Save As** to also write it on a configuration file, see Sect. A.1.14). Rather than creating a new record, you may also select **replace**, select the name of an existing record, and overwrite that record with **Put**.

Getting the Default You may get the contents of the configuration named *default* by pressing **Get** in the field **Default**.

Getting the Session Default Similarly you get the Session default, see above.

Getting or Loading a Named Configuration Record Press **Get** in the field **Item Record**. In the panel that opens, select a record name and press **Get** to get a record that is in memory. To load a configuration record from a configuration file, press **Load** and select a file.

A.1.8 Changing Parameters

An essential feature of **IQ** is the ease of changing parameters and producing the corresponding plot. Press the button **Parameters** in the graphics frame. A *parameter panel* will appear, Fig. A.7.

The button **Parameters** acts as a toggle switch, i.e., upon pressing it again and again the parameter panel is hidden, shown again, etc. It is always connected to its graphics frame and exists as long as the latter exists.

The contents of the parameter panel are discussed in detail in Sects. A.5 through A.8. You may change one or several parameters in the panel. By pressing the **Plot** button in the graphics frame or in the parameter panel, you produce a plot corresponding to the changed set of parameters, i.e., the changed descriptor. You may, of course, change again, plot again, and also print the plots obtained in this way.

A.1.9 Saving a Changed Descriptor

Technically, all the changes are performed in a copy of the descriptor in the computer memory, not in the original descriptor, which is part of the current descriptor file residing on the hard disk. Therefore, if you simply close the graphics frame, the changes are lost. If you want to save the changed descriptor, press the button **Save / Append** in the graphics frame. A *descriptor-save panel* appears, Fig. A.8.

By pressing one of the buttons **Save** or **Append**, you save the descriptor in its original position on the file or you append it as an additional descriptor at the end of the descriptor file, respectively. The original position is highlighted in the list of descriptors on the file. You may change the descriptor's title before saving or appending. At the moment of saving or appending, the descriptor is given a new time stamp.

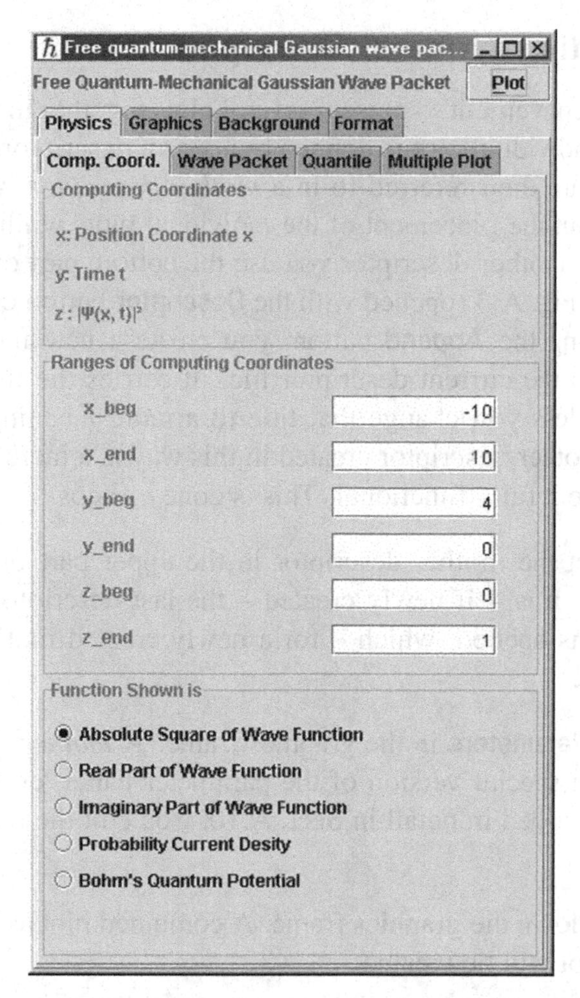

Fig. A.7. Parameter panel. It contains all parameters of a descriptor. Most of them can be changed. The parameters are arranged on several levels of subpanels, which are reached by pressing the appropriate 'tabs' near the top

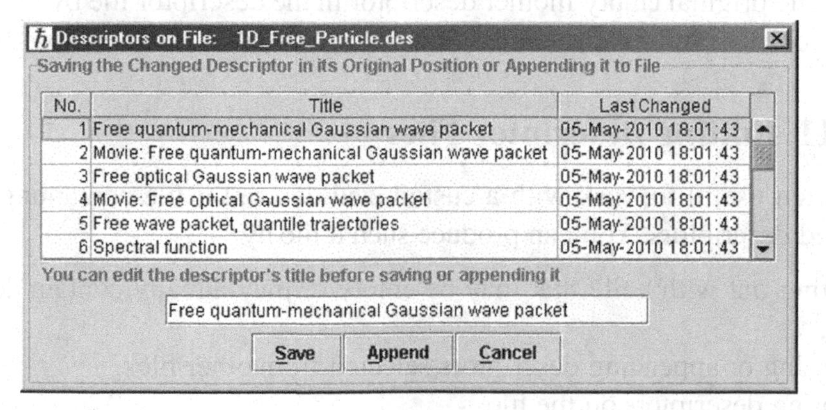

Fig. A.8. Descriptor-save panel. If the button **Save** is pressed, the descriptor in memory is saved in its original position in the descriptor file highlighted in the list shown at the top of the panel. Pressing the **Append** button results in appending the descriptor at the end of the descriptor file. Before saving or appending, the descriptor title may be changed in the line near the bottom

A.1.10 Creating a Mother Descriptor

It is sometimes convenient to present several plots together in a *combined plot* in which each individual plot is defined by its own descriptor. These individual descriptors are then referred to in a *mother descriptor*, which also contains details about the placement of the individual plots within the combined plot. To create a mother descriptor you use the bottom part of the descriptor-selection panel, Fig. A.3 (opened with the **Descriptor** button on the main toolbar). By pressing the **Append** button, you create a new mother descriptor and append it to the current descriptor file. It carries the title `New Mother Descriptor` unless you change that title to a more meaningful one before creation. The mother descriptor created in this way is a mere shell. You have to edit it to make it fully functional. This is done in steps.

Opening Select the mother descriptor in the upper part of the descriptor-selection panel. It is – if newly created – the last descriptor on the list. A graphics frame is opened, which – for a newly created mother descriptor – contains no plot.

Editing Press **Parameters** in the graphics frame. A *mother-descriptor panel* appears that is a special version of the parameter panel, cf. Sect. A.1.8. Its contents are described in detail in Sect. A.10. You edit the mother descriptor in this panel.

Testing Press **Plot** in the graphics frame. A combined plot as described in the mother descriptor will be created.

Saving Press **Save / Append** in the graphics frame. A *descriptor-save panel* appears in which you simply press the **Save** button at the bottom. This replaces the original empty mother descriptor in the descriptor file by the edited one. (For more information in the descriptor-save panel see Sect. A.1.9.)

A.1.11 Editing Descriptor Files

It is often useful to work with a customized descriptor file containing a few selected descriptors. You can produce such a file by

- starting out with a file that may be empty or may already contain descriptors,
- inserting or appending descriptors selected from other files,
- deleting descriptors on the file,
- rearranging the ordering of the descriptors on the file.

We recommend that you close all graphics frames before starting a file-editing session because only files not currently connected to a graphics frame can be edited. Then press the button **Edit Descriptor Files** on the main toolbar. An essentially empty *file-editing panel* is opened. Within that panel you may open two descriptor files, using the button **Open File A** and **Open File B** in the two halves of the panel. Lists for the two files are displayed, Fig. A.9. By entering a new name in the file chooser you may also open a hitherto nonexistent file, which does not yet contain descriptors. (Files for PC or Macintosh have different formats but **IQ** recognizes and accepts both formats on both types of computer. When you create a new file, you may select the format.)

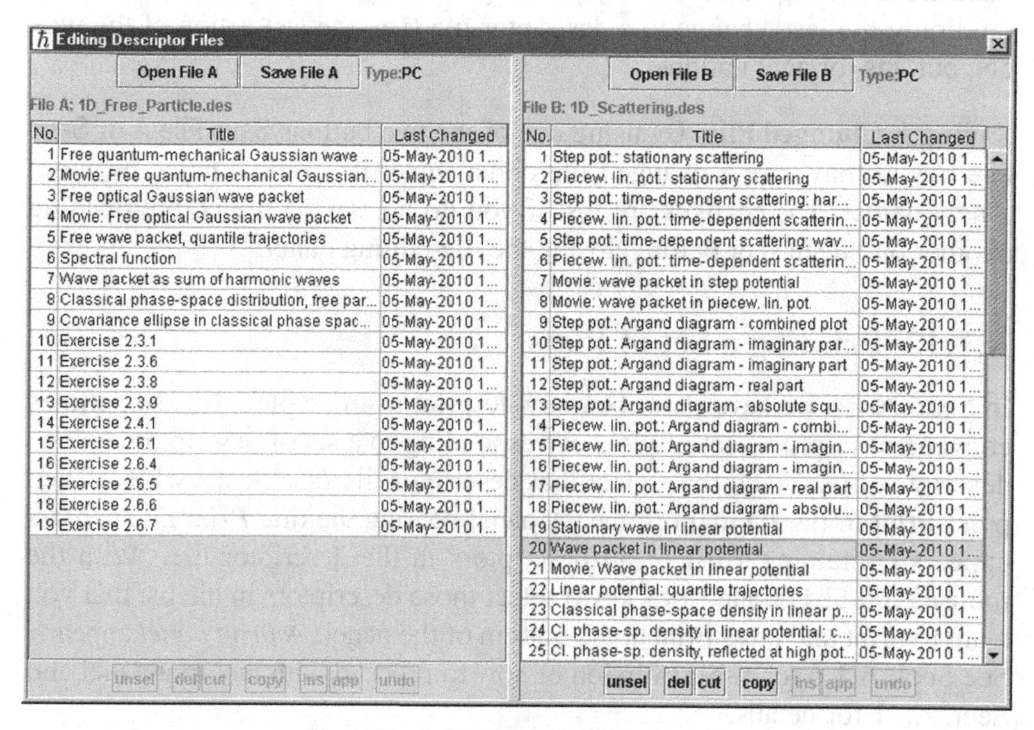

Fig. A.9. File-editing panel

You can now perform the following operations:

Selecting Using the mouse (and, if you want to select several descriptors in a file, also the `Shift` and `Ctrl` key), you select one or several descriptors.

Unselecting Pressing the **unsel** button you make the selection undone.

Deleting Pressing the **del** button deletes the selected descriptors.

Cutting Pressing the **cut** button copies the selected descriptors to a short-term memory called *clipboard* and deletes them from their file.

Copying Pressing the **copy** button copies the selected descriptors to the clipboard without deleting them. Note that the same clipboard is used for files A and B. It always contains the descriptors from the last cutting or copying operation irrespective of the file they belonged to originally.

Inserting With the **ins** button you insert the contents of the clipboard just before the first selected descriptor. If none is selected, insertion is done at the beginning of the file.

Appending With the **app** button you append the contents of the clipboard directly after the last selected descriptor. If none is selected, appending is done at the end of the file.

Making a Change Undone By pressing the **undo** button you make the last (and only the last) change to a descriptor file (i.e., the last action of the type **del**, **cut**, **ins**, or **app**) undone.

Saving a Changed File By using one of the two buttons **Save File A** or **Save File B** you may save a changed file either under its original name or under a new name. The changes are lost if you do not save a file before opening another one instead or before closing the file-editing panel.

A.1.12 Printing a Set of Plots

In Sect. A.1.6 we have described how to print a single plot. It is sometimes convenient to print a set of plots corresponding to a set of descriptors on one descriptor file. To do that select that descriptor file (see Sect. A.1.3). Then press **Print** in the **IQ** main frame. A panel bearing the title *Print Descriptors* appears containing a list of all descriptors in the descriptor file. With the mouse (and `Shift` and `Ctrl` keys) select those descriptors in the list that you want to print and press **Print** at the bottom of the panel. A *print panel* appears. See Sect. A.1.6 for a short version of how to proceed with the print panel and Sect. A.11 for details.

A.1.13 Running a Demonstration

To run a demonstration, press the button **Demo** on the main toolbar. A file chooser opens allowing you to choose a *demonstration file*, i.e., a file with the extension `demo`. For a first try we recommend to choose the file

```
1D_Free_Particle(NoSound).demo
```

Once a file is chosen, you are asked whether you want the demonstration to be automatic or step by step. By pressing the button corresponding to your choice, you start the demonstration. The demonstration is presented in the *demonstration frame*, Fig. A.10. Plots produced by **IQ** and explanatory text with a few formulae are shown one after another or side by side. Some demonstration files make use of prerecorded sound files so that the user listens to rather than reads the text and thus visually is fully focused on the plots.

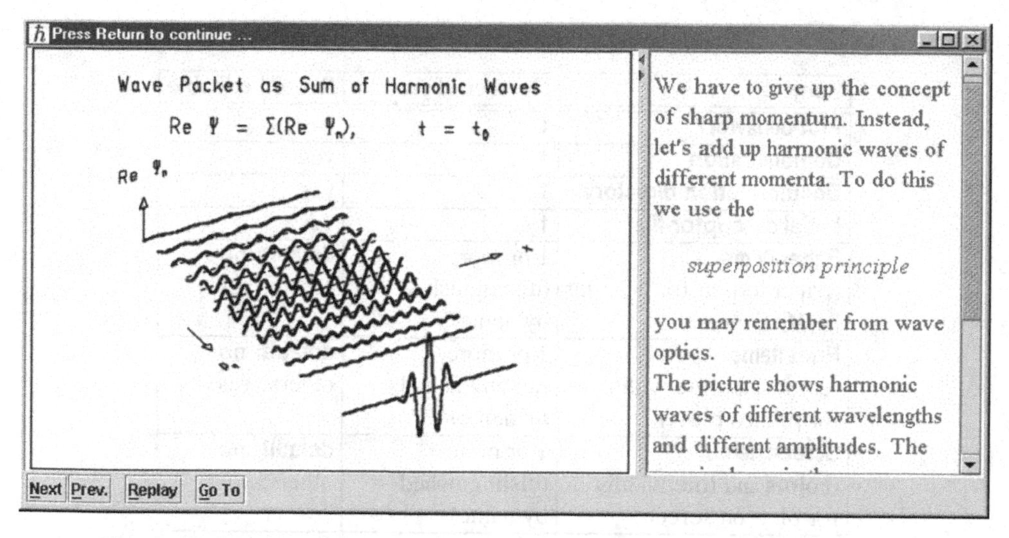

Fig. A.10. Demonstration frame containing an **IQ** plot and explanatory text

An *automatic demonstration* runs off from the beginning to the end following a predefined timing much like a movie or slide show. A *step-by-step demonstration* is divided into small steps. You start each step by clicking on the **Next** button or by simply pressing the `Enter` key. The button **Prev.** takes you back to the previous step. If you want to listen again from the beginning to the sound in the current step (if any), click on **Replay**. You may also proceed forward or backward to an arbitrary step by clicking on **Go To**. A pop-up window appears in which you can select the desired step and then present it by pressing the `Enter` key. (Double-clicking on the step reveals its substructure.) The pop-up window disappears with a mouse click anywhere outside it or if the `Escape` key is pressed.

If you wish to create your own demonstrations, consult Sect. A.12.

A.1.14 Customizing

The **IQ** program can be configured in several ways. When the program is started, the configuration is loaded into memory from the *configuration file* `iqini.cnf`. You may change and extend the configuration in memory and save it back to the file `iqini.cnf` or to some other configuration file.

The configuration information is summarized in Table A.2. It is structured in *configuration records* on five different subjects. For each of the first two subjects there is only one (unnamed) configuration record. For the other three subjects there may be several records with different names. One of them, bearing the name *default*, cannot be edited (and therefore stays unchanged). The others are editable.

By pressing the button **Customize**, you display the *customize panel*, Fig. A.11. Through this panel the configuration records are accessible and (at the

Table A.2. Configuration information

Subject	Number of records	Record editable
Plot behavior	1	yes
Demonstration	1	yes
Documentation directory	1	yes
Initial descriptor file	1	yes
Paper items (paper format for printing by **IQ** Export)	1 or more (distinguished by name)	**default**: no others: yes
Print items (colors and line widths for printed plots)	1 or more (distinguished by name)	**default**: no others: yes
Screen items (colors and line widths for plots on screen)	1 or more (distinguished by name)	**default**: no others: yes

bottom of the panel) tools are available to save an edited configuration file or to load another configuration file.

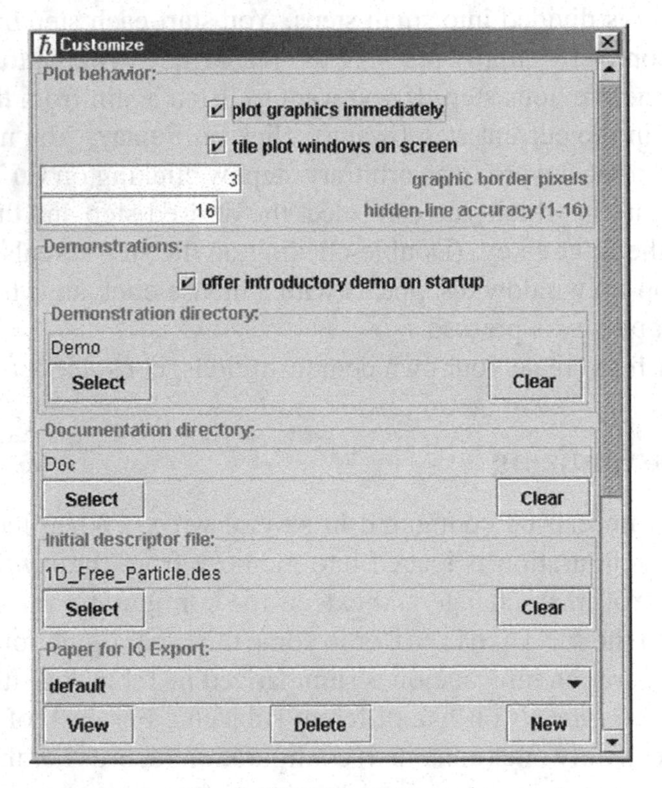

Fig. A.11. Customize panel

Editing an Unnamed Configuration Record

- **Plot behavior** – There are two *check boxes* and a numerical field:

- the check box **plot graphics immediately** is normally set *on*. (If it is not, after selection of a descriptor an empty graphics frame appears. A plot is created only after the plot button is pressed. This mode can be useful if you want to change descriptors without plotting.)
- Also the check box **tile windows on screen** should normally be *on*. (If it is not, all windows are created in the upper left corner of the screen if not otherwise determined by the operating system. There are, however, window managers under Linux that do not work well with Java, so that for them the check box has to be *off*.)
- The integer number **graphic border pixels** is usually left zero. (If it is not, then in the graphics frame pixels are added outside the border line of the plot. If you print the graphics using a system printer, the additional area forms part of the plot.)
- The integer number **hidden-line accuracy** is best left at `16`. It is a parameter of the hidden-line procedure used for plots of types *surface over Cartesian grid* (Sect. A.3.1) or *polar grid* (Sect. A.3.2) or of type *3D column plot* (Sect. A.3.8). In very rare cases a value smaller than `16` may yield better results.

- **Demonstration** – There is a check box and a field used to determine a directory:
  - The check box carries the title **offer introductory demo on start-up**. If it is *on*, the frame of Fig. A.1.2 will be shown when **IQ** is started.
  - The field **Demonstration directory** indicates the directory in which **IQ** looks for `*.demo` files, Sects. A.1.13, A.12. The default is the subdirectory `Demo` in the main directory of **IQ**. By pressing **Select** you get a *directory chooser* with which you can change the directory. With the **Clear** button you can re-establish the default.
- **Documentation directory** – There is a number of documentation files, in particular those needed to provide answers to **Help** requests. These files are stored in the documentation directory. The default is the subdirectory `Doc` in the main directory of **IQ**. To change the directory or to re-establish the default use the buttons **Select** or **Clear**, respectively.
- **Initial descriptor file** – The initial descriptor file is the file selected (appearing in the title bar of the **IQ** main frame) at program start-up. By pressing **Select** you can select an initial file. If **Clear** is pressed, no initial file is defined.

Editing a Named Configuration Record For the three types of named configuration records, there are three similar areas on the customize panel containing a combo box to select a configuration record and three buttons. With these you can do the following:

- *Editing an Existing Record:* Select the record in the combo box. Press **Edit**. Perform the desired changes in the panel that is opened. Close that panel.

(If the record is named *default* instead of an **Edit** button, there is a **View** button. You can inspect the panel but not change it.)

- *Creating a New Record:* Press **New**. A panel as described above but containing an additional field for the name of the new record is opened. Enter a name, perform the desired changes, and close the panel.
- *Deleting an Existing Record:* Select the record in the combo box. Press **Delete**.

Saving a Changed Configuration The methods described so far yield changes in the configuration in computer memory but are lost once **IQ** is closed. To save them to the file `iqini.cnf` press **Save**. To save them to another file press **Save as**, enter a file name (e.g., `myConfig.cnf`) in the panel that opens, and then press **Save** in that panel.

Loading Information from a Configuration File You may overwrite or extend the configuration in memory by loading information from a configuration file. To do this press **Load** and select the file. Records from the file with names different from those in memory are added to the configuration. Records with the same name and unnamed records may overwrite those in memory or be ignored – you are asked what you wish for each type of record. (If you do not want to be asked, you may select either **overwrite** or **keep** before pressing **Load**.)

A.1.15 Help and Context-Sensitive Help

Press the button **Help** in the **IQ** main frame. A window is opened in which a special version of the 'Acrobat® Reader' displays the text of this book. It is opened on the page with the beginning of Appendix A, *A Systematic Guide to* **IQ**. Using the tools of the reader, you can access every bit of information in the text.

The contents of the parameter panel depend very much on the physics topic chosen. Therefore, special help is provided for it. Select a subpanel of the parameter panel. (Selection is indicated graphically in the *tag field* by a thin frame around the name of the subpanel, e.g., **Physics** or **Comp. Coord.**.) Now, press the `F1` key on the keyboard. (On some keyboards press `Fn` plus `F1`.) A page relevant to the selected subpanel is displayed.

A.2 Coordinate Systems and Transformations

A.2.1 The Different Coordinate Systems

A.2.1.1 3D World Coordinates (W3 Coordinates) Figure A.13 shows a structure in three-dimensional space (3D space). Let us consider the whole structure to be built up of thin wires. We call the Cartesian coordinate system, in which this wire structure is described, the *3D world coordinate system* or *W3 coordinates* and denote a point in W3 coordinates by (X, Y, Z).

A.2.1.2 3D Computing Coordinates (C3 Coordinates) Looking at the scales in Fig. A.13, we observe that, although the x scale and the y scale have approximately equal lengths in W3 coordinates (i.e., if regarded as two pieces of wire suspended in space), their lengths are quite different if expressed by the numbers written next to the scales. The latter lengths are $\Delta x = 2$ and $\Delta y = 2\pi$.

The plot illustrates the function

$$z = f(x, y) \quad ,$$

where each point (x, y, z) is placed at the position (X, Y, Z) in W3 space. The coordinates x, y, z are called *3D computing coordinates* or simply *C3 coordinates*. They are connected to the W3 coordinates by a simple linear transformation given in Sect. A.2.2.

A.2.1.3 2D World Coordinates (W2 Coordinates) Our three-dimensional structure given in W3 coordinates must of course be projected onto the two-dimensional display screen or on the paper in a printer. This is done in two steps. We first project onto a plane placed in W3 space, and then (see Sect. A.2.1.4) from that plane onto the display screen or paper.

Let us consider an observer situated at some point in W3 space looking at the origin $(X = 0, \ Y = 0, \ Z = 0)$. The unit vector $\hat{\mathbf{n}}$ pointing from the origin to the observer is characterized by the *polar angle* ϑ (the angle between $\hat{\mathbf{n}}$ and the Z direction) and the *azimuthal angle* φ (the angle between the projection of $\hat{\mathbf{n}}$ onto the X, Y plane and the X axis). We now construct a plane somewhere in space perpendicular to $\hat{\mathbf{n}}$ and a two-dimensional ξ, η coordinate system in that plane. The ξ axis is chosen parallel to the X, Y plane.

We call the coordinates ξ, η the system of *2D world coordinates* or simply *W2 coordinates*. We can now perform a projection parallel to $\hat{\mathbf{n}}$ from W3 to W2 coordinates. The projection transformation is given in detail in Sect. A.2.2.

A.2.1.4 Device Coordinates (D Coordinates) A final transformation leads us to the sensitive plane of the plotting device, which can be the display screen or the paper. We call the system of u, v coordinates, used by the plotting device to address a point in the sensitive plane, the system of *device coordinates* or *D coordinates*. For the transformation from W2 to D coordinates see Sect. A.2.2.

A.2.2 Defining the Transformations

A.2.2.1 The Window–Viewport Concept A linear transformation from a variable x to another variable X can be uniquely defined by specifying just two points x_{beg}, x_{end} and the corresponding points X_{beg}, X_{end}, see Fig. A.12. A general point x is transformed to

$$X = X_{beg} + (x - x_{beg})\frac{X_{end} - X_{beg}}{x_{end} - x_{beg}} \quad .$$

If the range of the variable x is bounded, it is useful to choose the pairs (x_{beg}, x_{end}) and (X_{beg}, X_{end}) as the bounds of the variables. The interval

$$x_{beg} \leq x \leq x_{end}$$

is called the *window* in x, whereas the interval

$$X_{beg} \leq X \leq X_{end}$$

is called the *viewport* in X.

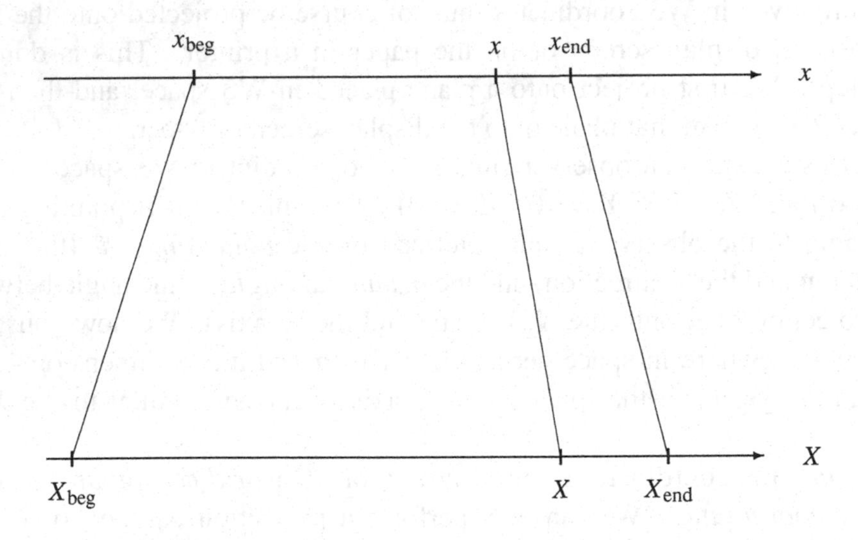

Fig. A.12. Linear transformation from x to X. The window in x is bounded by x_{beg}, x_{end}, the viewport in X by X_{beg}, X_{end}

A.2.2.2 The Chain of Transformations C3→W3→W2→D We have just seen that the transformation from C3 coordinates x to W3 coordinates X is

$$X = X_{beg} + (x - x_{beg})\frac{X_{end} - X_{beg}}{x_{end} - x_{beg}}$$

and therefore is completely defined once the window (x_{beg}, x_{end}) and the viewport (X_{beg}, X_{end}) are given. Of course, completely analogous formulae hold for the transformations $y \to Y$ and $z \to Z$.

The transformation from W3 coordinates (X, Y, Z) to W2 coordinates (ξ, η) is a parallel projection and is given by

$$\xi = -X \sin\varphi + Y \cos\varphi \quad ,$$

$$\eta = -X \cos\varphi \cos\vartheta - Y \sin\varphi \cos\vartheta + Z \sin\vartheta \quad .$$

It is completely determined by the polar angle ϑ and the azimuthal angle φ of the direction $\hat{\mathbf{n}}$ from the origin of W3 space to the observer.

Finally, we define a window

$$\xi_{beg} \le \xi \le \xi_{end} \quad , \quad \eta_{beg} \le \eta \le \eta_{end}$$

in W2 space and a viewport

$$u_{beg} \le u \le u_{end} \quad , \quad v_{beg} \le v \le v_{end}$$

in D space and project that part of W2 space that falls inside the window onto the viewport in D space (i.e., the sensitive plotting device surface).

A.2.2.3 Defining the Window and Viewport in C3 and W3 Coordinates
Figure A.13 is a plot of the type *surface over Cartesian grid in 3D*, cf. Sect. A.3.1. Such plots are illustrations of the function

$$z = f(x, y) \quad ,$$

in which x and y are varied over the ranges $x_{beg} \le x \le x_{end}$ and $y_{beg} \le y \le y_{end}$ of the windows x and y, respectively. The function $z = f(x, y)$ is drawn irrespective of the window (z_{beg}, z_{end}). Viewport (Z_{beg}, Z_{end}) and window (z_{beg}, z_{end}) are given to establish the relation between C3 and W3 coordinates. Their use is convenient to *magnify* or *reduce* the plot in the Z direction. The parameters $x_{beg}, x_{end}, \ldots$ and $X_{beg}, X_{end}, \ldots$ are given in the parameter panel. There they can be found under the heading **Ranges of Computing Coordinates** and **Ranges of World Coordinates** on the subpanels **Physics—Comp. Coord.** and **Graphics—Geometry**, respectively.

Exercise – Changing the Window in Computing Coordinates. Produce the plot of Fig. A.13. Open the parameter panel (by pressing the **Parameters** button). Shown is the subpanel **Physics—Comp. Coord.**. Change the value of **x_end** to `1.5`. The surface and the numbers at the scale in x are changed accordingly, although the apparent length of the scale (constructed in world coordinates) does not change.

Exercise – Changing the Viewport in World Coordinates. Now select the subpanel **Graphics—Geometry** and change **X_end** to 0. The surface and the apparent length of the scale in x are shrunk to half the original size.

Exercise – Scale and Offset in z. For the plot corresponding to Fig. A.13 the window $(z_{\mathrm{beg}}, z_{\mathrm{end}})$ in computing coordinates and the viewport $(Z_{\mathrm{beg}}, Z_{\mathrm{end}})$ in world coordinates are identical, namely $(-1, 1)$. Produce plots for

$$
\begin{aligned}
&\text{(a)}\ (z_{\mathrm{beg}}, z_{\mathrm{end}}) = (-2, 2);\ (Z_{\mathrm{beg}}, Z_{\mathrm{end}}) = (-1, 1);\\
&\text{(b)}\ (z_{\mathrm{beg}}, z_{\mathrm{end}}) = (-1, 1);\ (Z_{\mathrm{beg}}, Z_{\mathrm{end}}) = (-2, 2);\\
&\text{(c)}\ (z_{\mathrm{beg}}, z_{\mathrm{end}}) = (-1, 1);\ (Z_{\mathrm{beg}}, Z_{\mathrm{end}}) = (0, 2).
\end{aligned}
$$

The results are a reduction (a), a magnification (b), and a shifting (c) in z direction of the surface compared to the original plot.

A.2.2.4 Projecting from W3 to W2 The polar angle ϑ and the azimuthal angle φ, defining the position of the observer with respect to the origin of the W3 coordinate system, are found under the heading **Look-from Direction** on the subpanel **Graphics—Geometry**.

Exercise – Changing the projection. Produce the plot corresponding to Fig. A.13. Then plot again for different values of (ϑ, φ), e.g., $(30, -30)$, $(80, -30)$, $(60, -60)$.

A.2.2.5 Defining the Window in W2 Coordinates A window is defined through four quantities, the *width* W and the coordinates

$$X_{\text{look-at}},\ Y_{\text{look-at}},\ Z_{\text{look-at}}$$

of a *target point* or *look-at point* in W3 coordinates. The projection of that point is taken as the center of the rectangular window in W2. The window has the width W. The height of the window is chosen such that the window has the same width-to-height ratio (*aspect ratio*) as the viewport in D coordinates; see next section.

The four parameters of this section are displayed on the subpanel **Graphics —Geometry**. Changing them allows you to do *zooming* and *panning*. To *zoom in* you decrease W, to *zoom out* you increase W. Changing the look-at point you may move about your window in the W2 plane.

Exercise – Zooming and Panning. In the plot corresponding to Fig. A.13 change **width** to 25. (Result: zooming out.) Then change **Z_look_at** to 3. (Result: Panning. Because the look-at point moves up, the whole graphical structure moves down.)

A.2.2.6 Defining the Viewport in D Coordinates To determine the viewport

$$u_a \le u \le u_b \quad , \quad v_a \le v \le v_b$$

in D coordinates the subpanel **Format** of the parameter panel is used. It defines the *format* of the plot on the display screen or printer.

> **Exercise – Changing the Format.** In the plot corresponding to Fig. A.13 change from landscape to portrait format. Then produce a plot of equal width and height.

A.3 The Different Types of Plot

The plots created by **IQ** can be classified by their physics contents and also by the way they are shown graphically. The physics contents are mapped onto a mathematical function. This function is then presented graphically. Depending on the type of function, different forms of graphical representation – called *plot types* – are used. The plot type is chosen together with the physics topic. It cannot be changed directly.

Quantities influencing the graphical appearance are accessible in the three **Graphics** subpanels of the parameter panel:

- **Geometry** This subpanel in the first line names the plot type. The name is followed by
  - the **Ranges of World Coordinates**, cf. Sect. A.2.2.3,
  - the world coordinates of the **Look-at Point** and the **Width of the Plot** in world coordinates, cf. Sect. A.2.2.5,
  - the **Look-from Direction**, cf. Sect. A.2.2.4.
- **Accuracy** The parameters on this subpanel are discussed separately for each type in the following sections.
- **Hidden Lines** This subpanel exists only for 3D plots, for details see Sect. A.6.3.

A.3.1 Surface over Cartesian Grid in 3D

A function

$$z = f(x, y)$$

of two variables defines a surface embedded in three-dimensional x, y, z space (or 3D space). Figure A.13 shows the graphical representation of such a surface. In fact, only two sets of lines

$$z = f(x_i, y) \quad , \quad i = 1, \ldots, n_x \quad ,$$
$$z = f(x, y_i) \quad , \quad i = 1, \ldots, n_y \quad ,$$

are drawn, where the x_i or y_i are constants for a given line. The two sets of lines correspond to a *Cartesian grid* in the x, y plane. They are placed evenly in the x, y window in C3 coordinates, which corresponds to the X, Y viewport in W3 coordinates.

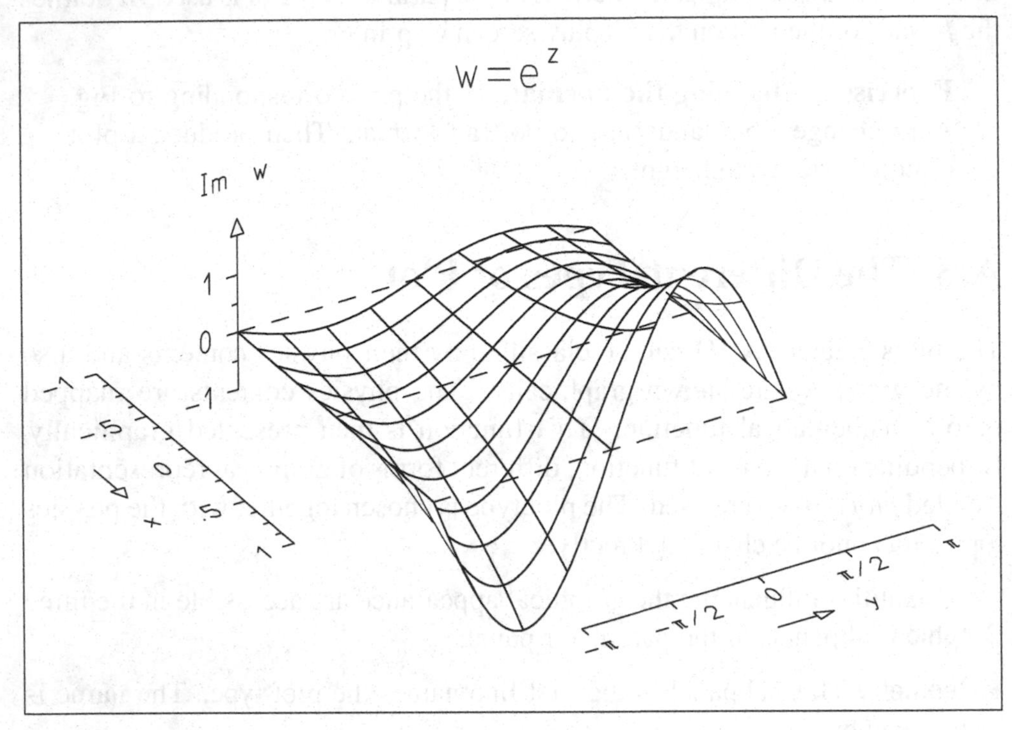

Fig. A.13. Plot produced with descriptor `Example for plot type "surface over Cartesian grid in 3D"` on file `Math_Functions.des`

The number of lines in each set can be changed in fields labeled **n_x** and **n_y** on the subpanel **Graphics—Accuracy**.

Exercise – Changing the Number of Lines. In the plot corresponding to Fig. A.13 try (a) $n_x = 11$, $n_y = 0$; (b) $n_x = 0$, $n_y = 9$; (c) $n_x = n_y = 41$.

Within one line only a certain number of points $z = f(x, y)$ is computed. They are connected by straight lines. You can alter the number of points by changing the value of the **Step Width** in the subpanel **Graphics—Accuracy**. The step width ΔX is the distance in the W3 coordinate X between two adjacent points. The corresponding quantity ΔY is set equal to ΔX. The default value for ΔX is $(X_{\text{end}} - X_{\text{beg}})/100$, i.e., one hundredth of the width of the viewport in W3. The values of ΔX and ΔY are still somewhat modified by **IQ** to match the grid defined by the X, Y viewport and the values of n_x and n_y.

Exercise – Changing the Step Width. In the plot corresponding to Fig. A.13 change **Step Width** to 1.

A.3.2 Surface over Polar Grid in 3D

So far we have used Cartesian coordinates only. We now introduce polar coordinates r, φ in the x, y plane

$$x = r\cos\varphi \quad , \quad y = r\sin\varphi$$

and construct two sets of lines in the x, y plane,

$$\text{rays } \varphi = \text{const} \quad ,$$
$$\text{arcs } r = \text{const} \quad .$$

We call these two sets of lines a *polar grid.* The function

$$z = f(x, y) = g(r, \varphi)$$

is then illustrated by lines

$$z = g(r_i, \varphi) \quad , \quad i = 1, \ldots, n_r \quad ,$$
$$z = g(r, \varphi_i) \quad , \quad i = 1, \ldots, n_\varphi \quad .$$

We then again have a surface in 3D space on which two sets of lines are drawn. The projection of these lines onto the x, y plane, however, forms a polar rather than a Cartesian grid. An example is given in Fig. A.14. Of course, the polar grid does not extend over the whole x, y plane.

For this plot type the **Ranges of Computing Coordinates** on the **Physics—Comp. Coord.** subpanel are given in the variables (r, φ) rather than in (x, y). The grid exists only within the ranges given.

Exercise – Changing of Ranges in r, φ. For the plot corresponding to Fig. A.14 change **r_beg** to 0.5 and plot. Then change **φ_end** to 360 and plot again.

The number n_r, n_φ of lines of the grid and the step widths of those lines are given on the subpanel **Graphics—Accuracy**.

Exercise: In the plot corresponding to Fig. A.14 change **n_φ** to 5 and plot. Then change **step width in φ** to 30 and plot again.

A.3.3 2D Function Graph

Once familiar with the way we present 3D graphics, it is convenient to think of a 2D plot as a picture of an object in 3D with all elements in the X, Y plane ($Z = 0$) and looked at from above, i.e., from the positive Z direction. Correspondingly, the subpanels **Physics—Comp. Coord.** and **Graphics—Geometry** contain only a subset of the parameters needed for 3D plots (see Sect. A.2.2,

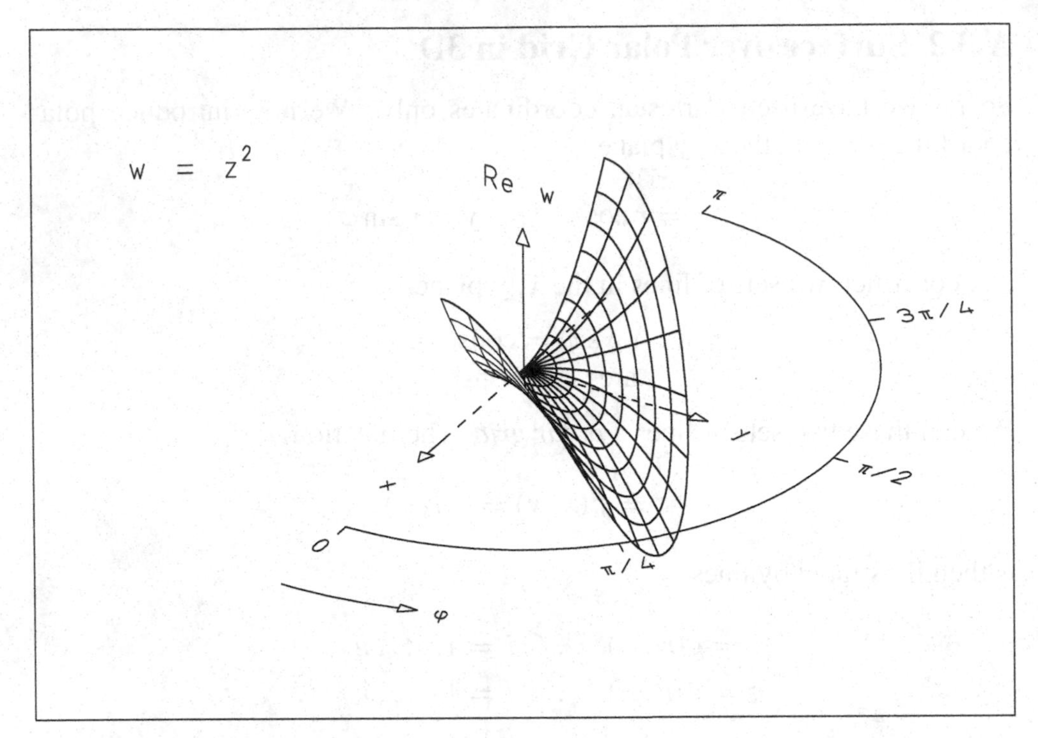

Fig. A.14. Plot corresponding to descriptor `Example for plot of the type "surface over polar grid in 3D"` on file `Math_Functions.des`

namely, the ranges of computing coordinates x and y, the ranges of the world coordinates X and Y, the X, Y coordinates of the look-at point, and the width of the plot).

There are various ways of plotting a function of a single variable in the x, y plane. They differ in the choice of the *independent variable*. In **IQ** the following four possibilities are implemented.

Cartesian Plot $y = y(x)$. This is the usual case with x being the independent and y the dependent variable.

Inverse Cartesian Plot $x = x(y)$. Here y is the independent and x the dependent variable.

Polar Diagram $r = r(\varphi)$. This diagram uses polar coordinates $r = (x^2 + y^2)^{1/2}$ and $\varphi = \arctan(y/x)$ with φ as independent and r as dependent variable.

Plot of Parameter Representation $x = x(p)$, $y = y(p)$. Here a *parameter* p is the independent variable and both coordinates x and y depend on it. The parameter is varied in small steps in the range $p_{\rm beg} \le p \le p_{\rm end}$ and a *trajectory* is drawn in the x, y plane as a succession of straight lines joining the points $(x(p), y(p))$.

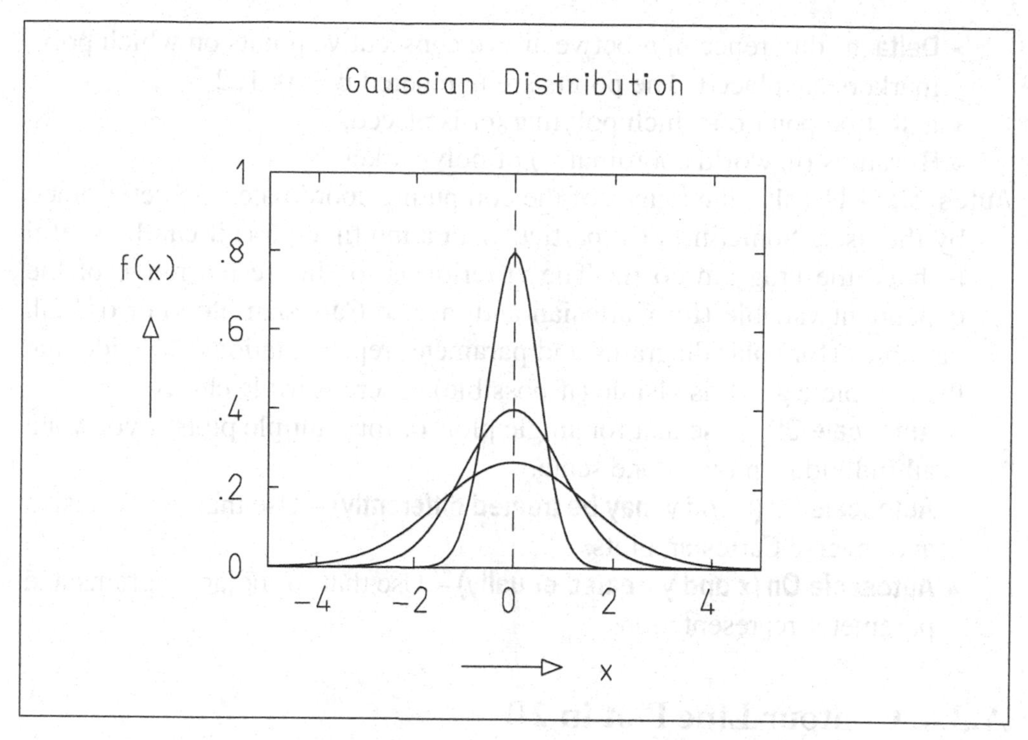

Fig. A.15. Plot produced with descriptor `Example for plot of type "2D function graph"` on file `Math_Functions.des`

An example for a Cartesian plot is Fig. A.15. An example for a polar diagram is Fig. 11.4. Figure 4.8 contains a plot of a parameter presentation (top left) and an inverse Cartesian plot (bottom left).

The *range of the independent variable*

- for the Cartesian and inverse Cartesian plots is, of course, the range of the corresponding computing coordinate and accessible on the subpanel **Physics—Comp. Coord.**,
- for the polar diagram is $0 \le \varphi \le 2\pi$ and cannot be changed,
- for the parameter representation is given at the bottom of the subpanel **Physics—Variables.**

On the subpanel **Graphics—Accuracy** you find some more items to influence the appearance of the plot.

Step Width of Independent Variable – The smaller this quantity **Delta** is chosen the more accurate the plot will be.

Polymarkers – In some cases it is useful to have marks placed on the graph that are equidistant in the independent variable. In particular, for the parameter representation this allows you to indicate the variation of the parameter p along the trajectory. In the jargon of computer graphics these marks are called *polymarkers*. You can change the following quantities:

- **Delta_n**: difference in n between two consecutive points on which polymarkers are placed. The points are numbered $n = 0, 1, 2, \ldots,$
- **n_0**: first point on which polymarker is placed,
- **R**: radius (in world coordinates) of polymarker.

Autoscale – Usually, the ranges of the computing coordinates are determined by the user. Sometimes (in particular in a multiple plot) it can be useful to have the program do it. The criterion is to choose the range of the dependent variable (for Cartesian and inverse Cartesian plots) or of both variables (for polar diagrams and parameter representations) so wide that the complete graph is visible (if possible). There is triple choice:

- **Autoscale Off** – Use that for single plots or for multiple plots if you want all individual plots in one scale.
- **Autoscale On (x and y may be treated differently)** – Use that for Cartesian and inverse Cartesian plots.
- **Autoscale On (x and y treated equally)** – Use that for polar diagrams and parameter representation.

A.3.4 Contour-Line Plot in 2D

Another type of 2D plot, Fig. A.16, is obtained by considering the function

$$z = f(x, y)$$

of the two independent variables x and y and requiring the function to have a constant value c. The locus of all points (x, y) fulfilling this requirement,

$$c = f(x, y) \quad ,$$

is called a *contour line* and can be thought of as projection onto the x, y plane of the line resulting from the intersection of the surface $z = f(x, y)$ and the plane $z = c$. You may plot only one such line or a set of n lines for

$$z = c \quad , \quad z = c + \Delta c \quad , \quad \ldots \quad , \quad z = c + (n-1)\Delta c \quad .$$

The values c and Δc are chosen at the bottom of the subpanel **Physics—Comp. Coord.**.

The contents of the subpanel **Graphics—Geometry** are as for 2D function graphs, see Sect. A.3.3.

The subpanel **Graphics—Accuracy** contains the **Number of Lines** n and the **Accuracy Parameter** Δx. The latter has the following meaning. The whole viewport in world coordinates is divided into squares of the size $\Delta X \times \Delta X$. Within each square the contour line is approximated by a straight line. The plot can become quite wrong if ΔX is too large. On the other hand, computing time increases fast as ΔX is decreased. In fact, it is proportional to $(\Delta X)^{-2}$.

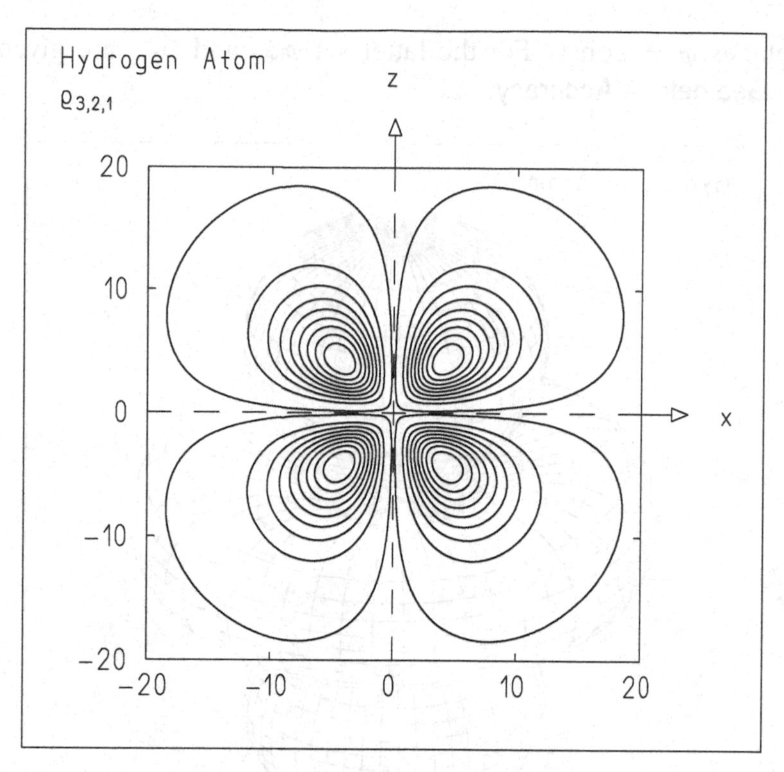

Fig. A.16. Plot produced with descriptor `Example for plot of type "contour-line plot in 2D"` on file `Math_Functions.des`

A.3.5 Contour-Surface Plot in 3D

An extension of the *contour-line* concept, Sect. A.3.4, is a *contour surface* of a function $f(x, y, z)$ of three independent variables x, y, z, Fig. A.17. It is defined by

$$f(x, y, z) = c = \text{const} \quad .$$

Technically, the contour surface is computed and presented as a set of contour lines, which themselves are obtained as intersections of the contour surface and some planes

$$x = \text{const} \quad , \quad y = \text{const} \quad , \quad z = \text{const} \quad , \quad \varphi = \text{const} \quad .$$

Here φ is the azimuth of spherical or cylindrical coordinates, i.e., the angle that a half-plane bounded by the z axis forms with the x, z plane. The subpanel **Geometry—Accuracy** contains the numbers n_x, n_y, n_z, n_φ of these planes. In practice only two (usually n_z and n_φ) should be nonzero. The planes $z = $ const are chosen such that the first one is $z = z_{\text{beg}}$, the last one $z = z_{\text{end}}$, and the others are placed equidistantly in between. Here z_{beg} and z_{end} are the limits of the ranges of computing coordinates, Sect. A.2.2.3. Analogous statements hold true for the planes $x = $ const, $y = $ const, and also for the planes φ = const. For the latter set φ_{beg} and φ_{end} are given on the subpanel **Geometry—Accuracy**.

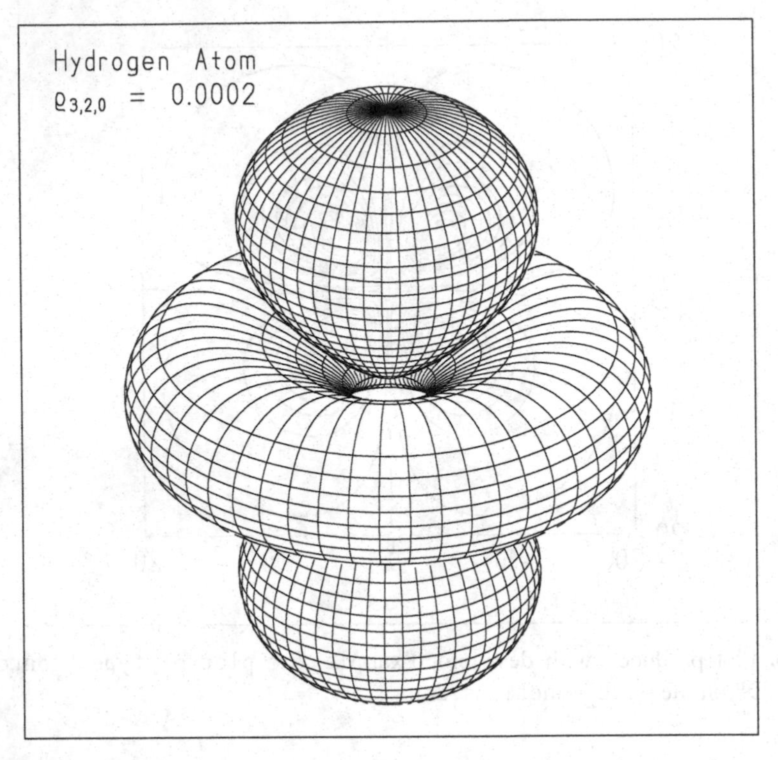

Fig. A.17. Plot produced with descriptor `Example for plot of type "contour-surface plot in 3D"` on file `Math_Functions.des`

Only that part of each plane that falls inside the rectangular box defined by the ranges of the computing coordinates x, y, z is considered and the contour line on that part of each plane is constructed. The accuracy of that construction is determined by the parameter ΔX just as in the case of contour lines in 2D.

Note that the box defined by the ranges of computing coordinates need not contain the contour surface completely. In such a case the surface appears to be 'cut open'.

It can be very time-consuming to produce a contour-surface plot. The computing time is proportional to the total number of lines and to $(\Delta X)^{-2}$. If the hidden-line technique, Sect. A.6.3, is used, it is increased by another large factor.

A.3.6 Polar Diagram in 3D

A non-negative function

$$r = r(\vartheta, \varphi)$$

of the polar angle ϑ and the azimuth φ of a spherical coordinate system can be visualized as a surface in 3D. A ray from the origin in the direction defined by ϑ and φ intersects the surface at the distance $r(\vartheta, \varphi)$. We call the picture of such a surface a *polar diagram in 3D*, Fig. A.18. In order not to distort the angles, world coordinates and computing coordinates are kept proportional to each other.

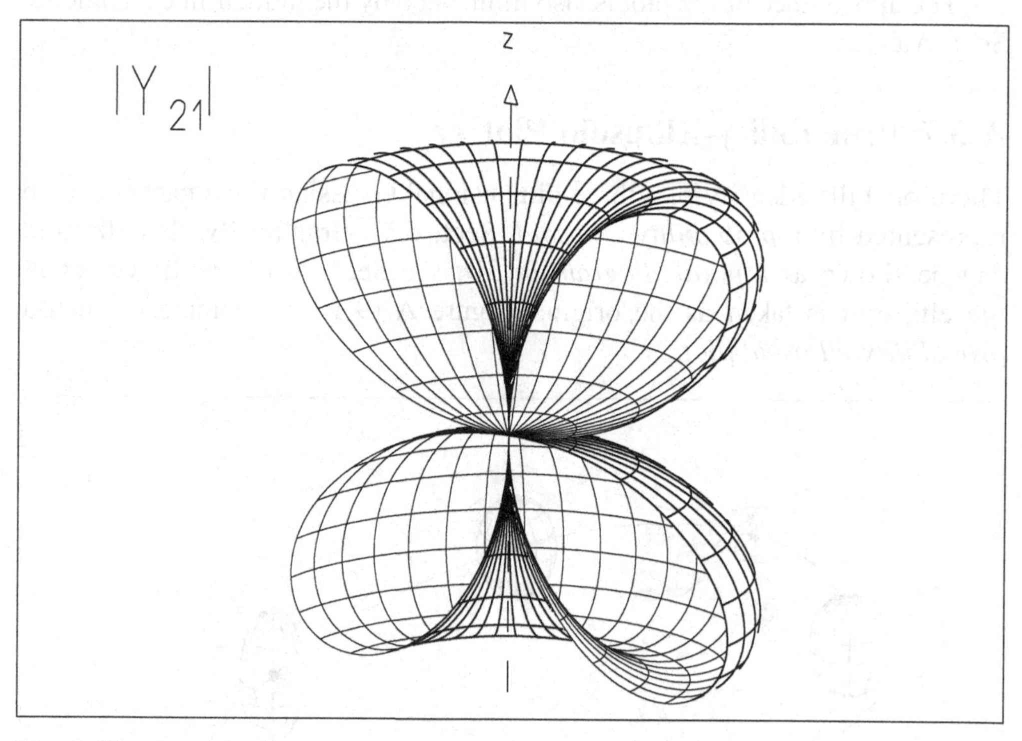

Fig. A.18. Plot produced with descriptor `Example for plot of type "polar diagram in 3D"` on file `Math_Functions.des`

The quantities

$$\vartheta_{\text{beg}},\ \vartheta_{\text{end}};\quad \varphi_{\text{beg}},\ \varphi_{\text{end}};\quad n_\vartheta,\ n_\varphi;\quad \Delta$$

on the subpanel **Graphics—Accuracy** determine in detail how the surface is shown.

A total of n_ϑ lines

$$r = r(\vartheta_i, \varphi) \quad , \quad i = 1, \ldots, n_\vartheta \quad ,$$

is constructed where ϑ_i is kept constant and only φ is varied in the range

$$\varphi_{\text{beg}} \le \varphi \le \varphi_{\text{end}} \quad .$$

The variation of φ from point to point on one line is given by $\boldsymbol{\Delta}$. The constant values ϑ_i are chosen as

$$\vartheta_1 = \vartheta_{\rm beg} \quad , \quad \vartheta_{n_\vartheta} = \vartheta_{\rm end} \quad ,$$

and equidistantly in between. For the n_φ lines

$$r = r(\vartheta, \varphi_i) \quad , \quad i = 1, \ldots, n_\varphi \quad ,$$

the same description holds but with ϑ and φ interchanged.

The appearance of the plot is also influenced by the hidden-line technique, Sect. A.6.3.

A.3.7 Probability-Ellipsoid Plot

The probability density of a three-dimensional Gaussian wave packet can be represented by a *probability ellipsoid*, Sect. 6.5. Graphically, this ellipsoid can be shown as a *polar diagram in 3D*, see Sect. A.3.6, if the center of the ellipsoid is taken as the origin. Figure A.19 is an example of such a *probability-ellipsoid plot*.

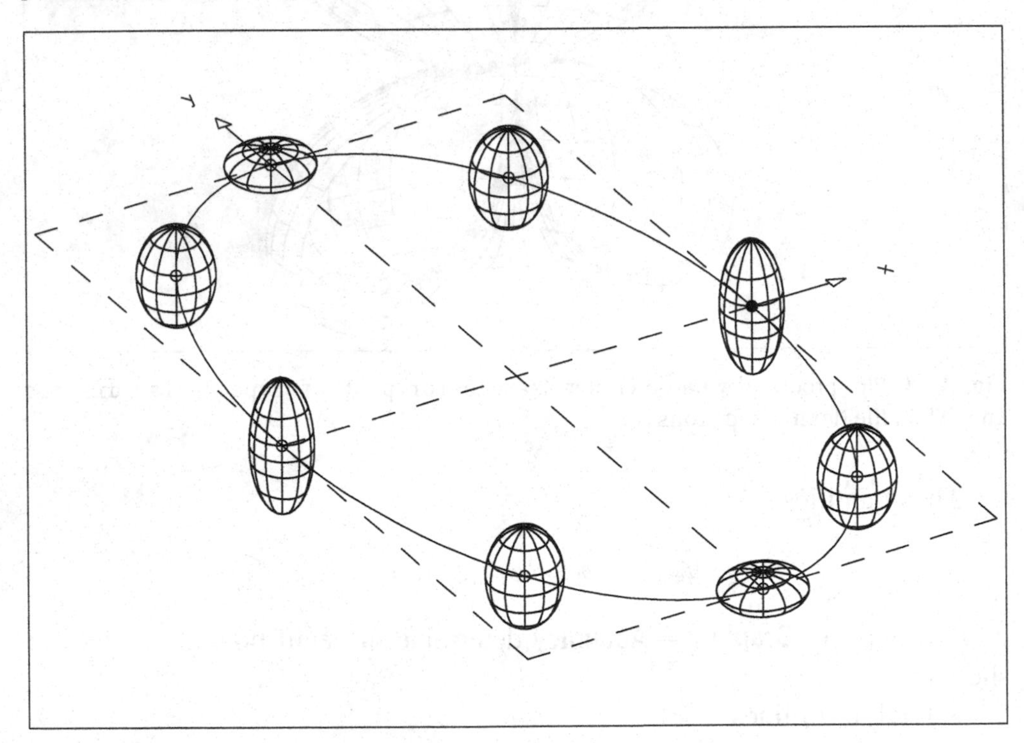

Fig. A.19. Plot produced with descriptor `Example for plot of type "probability-ellipsoid plot"` on file `Math_Functions.des`

In the subpanel **Graphics—Accuracy** the quantities

$$n_\vartheta, \quad n_\varphi, \quad \Delta$$

have the same meaning as for the *polar diagram in 3D*. The ranges of ϑ and φ fixed at

$$\vartheta_{\text{beg}} = 0^\circ, \quad \vartheta_{\text{end}} = 180^\circ; \quad \varphi_{\text{beg}} = 0^\circ, \quad \varphi_{\text{end}} = 360^\circ \quad .$$

The hidden-line procedure is always used.

A.3.8 3D Column Plot

A quantity that is defined only for certain discrete values of x and y is best represented by a set of columns placed on the x, y plane at those points where the quantity is defined. The height of each column indicates the size of the quantity, see Fig. A.20.

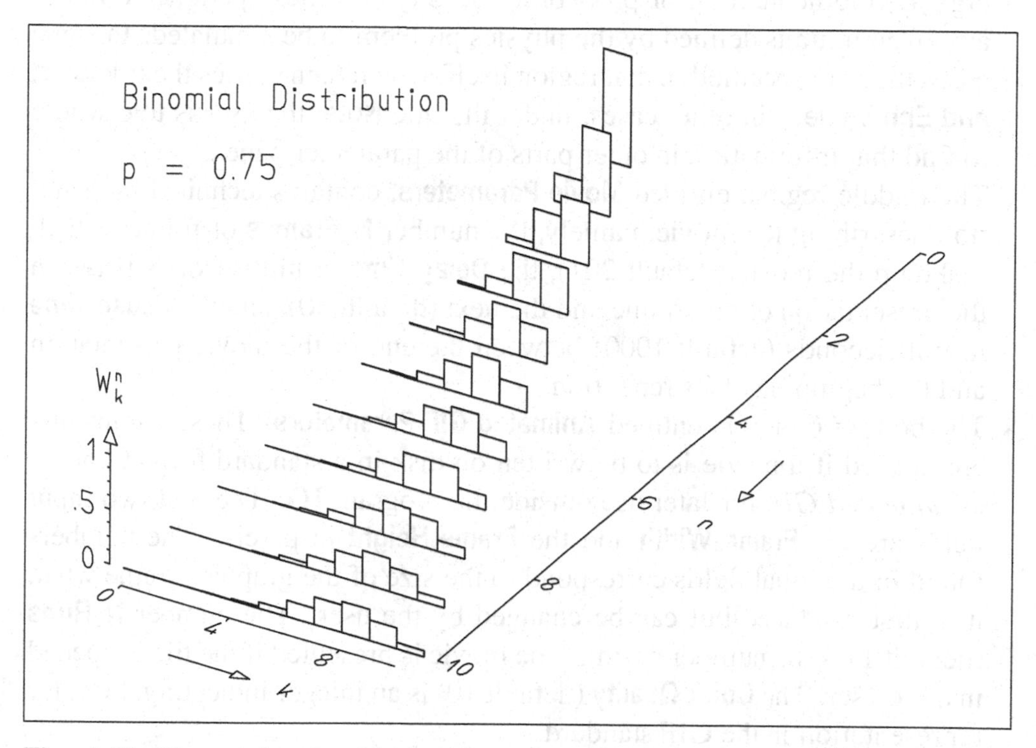

Fig. A.20. Plot produced with descriptor `Example for plot of type "3D column plot"` on file `Math_Functions.des`

For this *3D column plot* the ranges of computing coordinates in x and y are set automatically depending on the user's choice of physics parameters. The scale in z is adjusted as usual by the ranges in computing and world coordinates.

The subpanel **Graphics—Accuracy** allows you to adjust the lateral widths (in x and y) of the columns. The Figs. 11.15 and A.15 contain exactly the same information but the column widths are different.

A.4 Parameters – The Subpanel Movie

Many descriptors have a *movie capability*; the graphics frame created with such a descriptor carries a bottom toolbar with movie buttons, see Sect. A.1.5. The corresponding parameter panel contains a subpanel **Movie**, which is subdivided into three regions:

- The top region contains information on the start and end values of the parameter which is changed from frame to frame in the movie. This parameter mostly is the time t but may also be some other quantity, for instance, energy E, momentum p, or probability density ϱ. These parameter values are given in units defined by the physics problem to be simulated. In some cases they are specified in that region itself, which then carries the title **Start and End Values**. In other cases, under the title **Note**, the user is told where to find that information in other parts of the parameter panel.
- The middle region, entitled **Movie Parameters**, contains technical information describing the movie, namely, the number **N_Frames** of frames which make up the movie (default 201), the **Delay Time** in milliseconds between the presentation of one frame and the next (default 40), and the **Pause Time** in milliseconds (default 1000) between the end of the movie presentation and the beginning of its repetition.
- The bottom region is entitled **Animated GIF Parameters**. These parameters are needed if a movie is to be written on disk in a standard format known as *animated GIF* for later use outside the program **IQ**. The first two input fields are the **Frame Width** and the **Frame Height** in pixels. The numbers found in the input fields correspond to the size of the graphics frame when it is first produced but can be changed by the user. The number **N_Runs** (default 1) is the number of times the movie is presented if the file is opened in a browser. The **Color Quality** (default 10) is an integer indicating the color representation in the GIF standard.

A.5 Parameters – The Subpanel Physics

The subpanel **Physics** of the parameter panel itself contains one or more subpanels. Their contents depend on the physics topic chosen and is therefore explained in detail in the corresponding physics section of the book. General statements can be made about two of the subpanels. The name of the physics topic is always visible in the first line of the parameter panel.

A.5.1 The Subpanel Physics—Comp. Coord.

Under the heading **Computing Coordinates** the correspondence between computing coordinates (e.g., x, y, z) and the physics quantities represented by them is shown. An example is

x: Position Coordinate x,
y: Time t,
z: $|\Psi(x,t)|^2$.

Next, the **Ranges of Computing Coordinates** are given, see Sect. A.2.

A.5.2 The Subpanel Multiple Plot

For many (but not all) physics topics a *multiple plot* can be produced. This is a set of plots arranged in a regular array of rows and columns. The individual plots differ only in the values of one or two parameters that are incremented from plot to plot. Which parameters are incremented and by how much is indicated in the other subpanels of the **Physics** subpanel. On the subpanel **Multiple Plot** there are two groups of parameters.

Under the heading **Size of Multiple Plot** you find the number of rows and the number of columns of the multiple plot.

Under the heading **First Indices** you find two more quantities, **First Row Index** and **First Column Index**. These, in most cases are left at their default values zero. In some cases it may be useful to choose different values. Suppose two parameters p and q can be stepped up with initial values p_0, q_0 and increments Δp, Δq:

$$p_i = p_0 + i\Delta p \quad , \quad i = 0, 1, \ldots \quad ,$$
$$q_j = q_0 + j\Delta p \quad , \quad j = 0, 1, \ldots \quad .$$

Usually, the upper-left plot, i.e., the plot in the first row and the first column of a multiple plot, corresponds to the parameter pair p_0, q_0. If the first indices are chosen to be $i = I$, $j = J$, then it corresponds to p_I, q_J.

A.6 Parameters – The Subpanel Graphics

The subpanel **Graphics** of the parameter panel contains two (sometimes three) subpanels.

A.6.1 The Subpanel Graphics—Geometry

In its first line this subpanel carries the *name of the plot type*. It is followed by several groups of parameters:

- **Ranges of World Coordinates**, see Sect. A.2,
- **Look-at Point**, the world coordinates of a point which, after projection, will be exactly the center of the plot, see Sect. A.2.2.5,
- **Width of Plot** in world coordinates, see Sect. A.2.2.5,
- **Look-from Direction**, see Sect. A.2.2.4 (this group is present only for 3D plots).

A.6.2 The Subpanel Graphics—Accuracy

The contents of this subpanel depend very much on the type of plot. We list them here under the names of the plot types. For further details consult Sect. A.3.

- **Surface over Cartesian grid in 3D**, Sect. A.3.1:
  - **n_x** – number of lines $x =$ const,
  - **n_y** – number of lines $y =$ const,
  - **Step Width** – distances (in world coordinates X or Y) between two computed points on a line.
- **Surface over polar grid in 3D**, Sect. A.3.2:
  - **n_r** – number of lines $r =$ const,
  - **n_**φ – number of lines $\varphi =$ const,
  - **Step Width in R** – distance (in world coordinates) between two points computed on a line $\varphi =$ const,
  - **Step Width in** φ – distance (in degrees) between two points computed on a line $r =$ const.
- **2D function graph**, Sect. A.3.3:
  - **Delta** – step width of the independent variable. The unit is world coordinates for Cartesian and inverse Cartesian plots. It is radian for a polar diagram and it is the unit of the parameter for a parameter representation.

  For the features **Polymarkers** and **Autoscale** see Sect. A.3.3.
- **Contour-line plot in 2D**, Sect. A.3.4:
  - **ΔX** – resolution parameter (in world coordinates) for the algorithm finding the contour lines.
- **Contour-surface plot in 3D**, Sect. A.3.5:
  - **n_x** – number of contour lines on planes $x =$ const,
  - **n_y** – number of contour lines on planes $y =$ const,
  - **n_z** – number of contour lines on planes $z =$ const,
  - **n_**φ – number of contour lines on planes $\varphi =$ const,
  - φ**_beg** – azimuth (in degrees) for first plane $\varphi =$ const,
  - φ**_end** – azimuth (in degrees) for last plane $\varphi =$ const,
  - **ΔX** – resolution parameter (in world coordinates) for the algorithm finding the contour lines.
- **Polar diagram in 3D**, Sect. A.3.6:

- **n_ϑ** – number of lines ϑ = const,
- **n_φ** – number of lines φ = const,
- **Δ** – distance (in degrees) between points computed on a line ϑ = const or φ = const,
- **ϑ_beg** – lower end (in degrees) of range in ϑ,
- **ϑ_end** – upper end (in degrees) of range in ϑ,
- **φ_beg** – lower end (in degrees) of range in φ,
- **φ_end** – upper end (in degrees) of range in φ.

• **Probability-ellipsoid plot**, Sect. A.3.7:
- **n_ϑ** – number of lines ϑ = const,
- **n_φ** – number of lines φ = const,
- **Δ** – distance (in degrees) between points computed on a line ϑ = const or φ = const.

A.6.3 The Subpanel Graphics—Hidden Lines

This subpanel is present for some 3D plot types. A *hidden-line procedure* is a method not to show (i.e., to hide) lines on a surface in 3D space that are hidden to the observer by the surface itself. The procedure is quite different for the following two groups of plot types:

• **Surface over Cartesian or polar grid in 3D** – These surfaces have two *faces*. One is oriented toward the positive z direction (we call it the *top face*). The other one, the *bottom face*, is oriented toward the negative z direction. Normally, top and bottom face are given different colors. The user may choose one of the following possibilities for the hidden-line technique:
- **On**,
- **On – Top and Bottom in Same Color**,
- **On – Only Top Shown**,
- **Off**.

The latter two possibilities are sometimes useful, in particular for a *surface over Cartesian grid in 3D* on which only lines y = const are shown. This is done to present in one plot n_y graphs

$$f(x, y = y_i) \quad , \quad i = 1, 2, \ldots, n_y \quad ,$$

of a function of x with y as a parameter. For an example see Fig. 3.7.

• **Contour-surface plot in 3D and polar diagram in 3D** – Also the surfaces in these plots have two faces. In a polar diagram there is one face oriented toward the origin, the other face is oriented away from it. A contour surface, defined by $f(x, y, z) = c$, has a face toward the region of space where $f > c$ and one oriented toward $f < c$. The two faces are always shown in different colors. For these plot types the hidden-line procedure can be very time-consuming. You can therefore switch the procedure **On** or **Off**. In the **Off** state the program is much faster, no distinction can be made between the two faces, which, therefore, are shown in the same color. In the **On** state you may still influence the speed through the **Accuracy Parameter**. Both the quality of the procedure and the time it takes increase with the parameter. The default value 100 gives reasonable results. In particular, for contour-surface plots we recommend that you switch off the hidden-line technique (or at least reduce its accuracy) while you develop the plot and switch back on only after you found your final set of parameters.

A.7 Parameters – The Subpanel Background

Besides the graph of the function to be illustrated, our example plots contain a number of items that make them easier to understand, such as coordinate axes, scales, arrows, and text. We say that the *background* of our plots is made up of these items and discuss, in this section, how to control them. In Fig. A.21 the *function part* and the *background* of the plot of Fig. A.13 are shown separately.

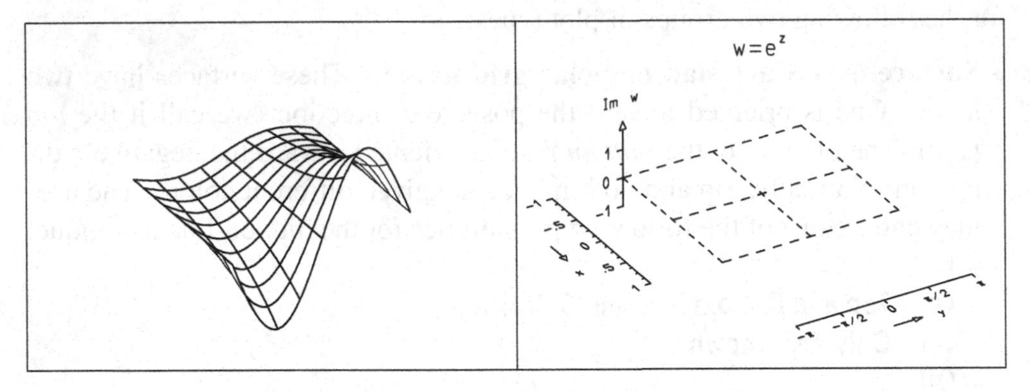

Fig. A.21. The function part (left) and the background (right) of Fig. A.13

All background items are controlled on the different **Background** subpanels on the parameter panel.

A.7.1 The Subpanel Background—Box

It is often useful to draw a *box* indicating the range in the X, Y plane for which a function is drawn. You may also wish to see the *coordinate axes*, i.e., the x axis and the y axis of the C3 coordinate system. Boxes and coordinate axes are controlled on the subpanel **Background—Box**.

For all plot types except *surface over polar grid in 3D* there are four sets of radio buttons (i.e., groups of possibilities of which one is selected). Two sets apply to the box and two to the coordinate axes. First, for the box, you may choose between

- **No Box**,
- **Box as Continuous Line**,
- **Box as Dashed Line**.

So far, the box is just a set of four lines. You can make these lines serve as scales of computing coordinates using the next set of buttons.

- **No Ticks** – just the box is drawn,
- **Ticks Pointing into Box** – ticks are added to the lines of the box turning them into scales,
- **Ticks and Numbers** – numbers are placed near some ticks on two sides of the box, so that the scales can be read off quantitatively.

For the coordinate axes there are two similar sets of buttons. Figure A.21 contains a box of continuous lines and coordinate axes as dashed lines. Neither box nor coordinate axes carry ticks or numbers. In the figure there are scales at some distance from the box, because for the figure shown they would otherwise interfere with the function graph. They are discussed in Sect. A.7.2.

Details of dash length, ticks, and numbers of box and coordinate axes are controlled on the subpanels of **Background—Scales**, see Sect. A.7.2.

For *2D function graphs* the functions are *clipped* (if necessary) at the upper and lower edges of the box, if you ask for the box to be drawn.

For plots of the type *surface over polar grid in 3D* the subpanel is somewhat different. The box given by the ranges of the computing coordinates r, φ is not rectangular, because the range for which the function is drawn is limited by rays and arcs. In the example of Fig. A.14 it is a semicircle.

We call it the *box in* r, φ. With the set of radio buttons in the field **Style of Box in r, φ** it can be

- **Not Shown**,
- shown with **Continuous Lines**,
- shown with **Dashed Lines**.

In addition there is a *box in* X, Y that is never shown but which is needed as a reference for the positioning of other background objects (coordinate axes, scales, arrows). Its **Extension in X** can be chosen to be

- $-R_{\text{end}} < X < R_{\text{end}}$,
- $-R_{\text{end}} < X < 0$,
- $0 < X < R_{\text{end}}$;

and similarly, its **Extension in Y**. There are two more sets of radio buttons concerning the coordinate axes as for the other types of plot.

A.7.2 The Subpanel Background—Scales

Figure A.21 contains a *scale in x* (Sect. A.7.2.1), a *scale in y* (Sect. A.7.2.2), and a *scale in z* (Sect. A.7.2.3). These scales are available in all 3D plots. For plots of the type *surface over polar grid in 3D* there can be in addition a *scale in φ*. For 2D plots, of course, only scales in x and in y are available. Every scale extends over the range of the corresponding coordinate.

A.7.2.1 The Scale in x The subpanel **Background—Scales—Scale in x** contains two sets of radio buttons and a group of numerical parameters. The use of the first set is obvious: It lets you choose between **No Scale**, **Scale as Continuous Line**, and **Scale as Dashed Line**. The second set defines the *notation of numbers* at the scale, **Decimal** (e.g., 4500) or **Exponential** (e.g., 4.5×10^3).

The numerical parameters are

- **d_X** – *distance* of scale from box (in W3 coordinates).
- **ld_X** – *length of dashes* (in W3 coordinates). Note that not only the dash length of the scales is determined by this parameter but also the dash length used for the box and the coordinate axes.
- **x_0** – a value of x (in C3 coordinates) at which you want a tick on the scale accompanied by a number.
- **delta_x** – a positive number δx (in C3 coordinates) defining the distance between two ticks with numbers. You will get ticks with numbers at $\ldots$, $x_0 - \delta x, x_0, x_0 + \delta x, \ldots$.
- **n_Interv_x** – the *number of intervals* between two numbered ticks, which are marked off by additional ticks. Default is 1, i.e., no additional ticks.
- **l_Ticks_X** – *length of ticks* (in W3 coordinates). If this parameter is set larger than the box size, it is restricted to the box size.

Note that these parameters also control the ticks and numbers on the x coordinate axis and on the edges of the box parallel to that axis.

A.7.2.2 The Scale in y The subpanel **Background—Scales—Scale in y** is completely analogous to the subpanel for the scale in x, Sect. A.7.2.1.

A.7.2.3 The Scale in z The subpanel **Background—Scales—Scale in z** contains a set of radio buttons to choose the graphical appearance of the scale and the notation of the numbers on the scale. In addition it carries two groups of parameters.

The first group, **Position in XY**, contains the X and Y coordinates of the scale, which itself is, of course, parallel to the Z axis.

The parameters in the second group, **Details**, are completely analogous to the corresponding ones for the x axis, see Sect. A.7.2.1.

A.7.2.4 The Scale in φ In plots of the type *surface over polar grid in 3D* you may want a scale for the polar angle φ. It is controlled through the subpanel **Background—Scales—Scale in φ.**

The scale is presented as a circle or circular arc around the origin in the plane $Z = 0$ extending over the range $\varphi_{\text{beg}} \leq \varphi \leq \varphi_{\text{end}}$. It will be placed outside the range in R, usually at $R = R_{\text{end}} + d_\varphi$, where d_φ is a distance in world coordinates. For $R_{\text{beg}} > 0$ it may occasionally be useful to place it at $R = R_{\text{beg}} - d_\varphi$.

The subpanel consists of three parts distinguished below by their headings.

- **Type of Scale, if Any** – Here you choose the graphical appearance of the scale and its positioning with respect to R, see above.
- **Select Ticks and Notation of Numbers** – Here you can choose whether the ticks on the scale point *outwards*, i.e., away from the range in R, or *inwards* and whether the numbers are in *decimal* or *exponential* notation.
- **Details** – The quantities under this heading are analogous to the ones for the scale in x, Sect. A.7.2.1. For simplicity the numerical values for **ld_φ**, **φ_0**, and **delta_φ** are given in degrees, although numbers at the scale will be given in radians.

A.7.3 The Subpanel Background—Arrows

As part of the background up to four *arrows* may be shown, two (called *arrow 1* and *arrow 2*) are straight arrows pointing in (or against) the X or Y direction and are placed in the X, Y plane. The *arrow in Z*, available in 3D plots, points in (or against) the Z direction. In plots of the type *surface over polar grid in 3D* an additional *arrow in φ* is available. It has the form of a circular arc in the X, Y plane.

Beyond the arrow tip a text is shown. This text depends on the *physics topic* chosen and cannot be set explicitly. The *text size* is controlled through the subpanel **Background—Texts**, Sect. A.7.4. The *arrow tip size* is adjusted to match the text size.

A.7.3.1 The Arrows 1 and 2 Each of the subpanels **Background—Arrows—Arrow 1** or **2** contains three items, **Orientation of Arrow, Box Side**, and **Details**. (Only if you choose *no arrow* in the first item the other two are missing.)

The arrow, in general, is oriented with respect to the *box*, i.e., the viewport in X, Y coordinates. A side of the box is chosen with the set of radio buttons **Chosen Box Side**. The possible sides are, see Fig. A.22,

- **Bottom (Low y)**,
- **Right (High x)**,
- **Top (High y)**,
- **Left (Low x)**.

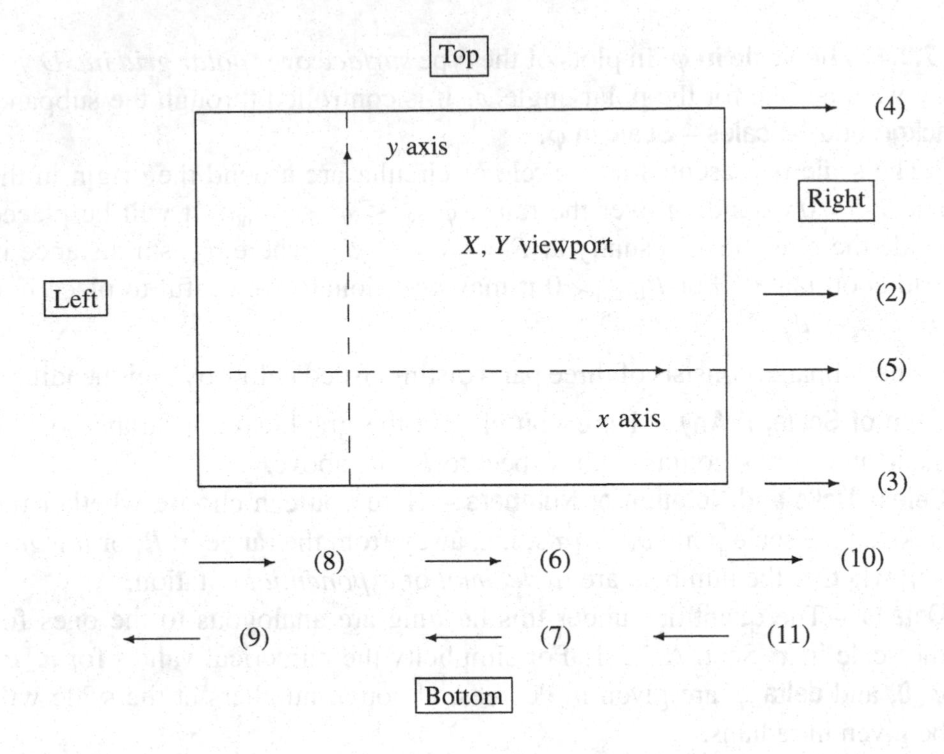

Fig. A.22. Possibilities to orient an arrow. For the numbers see text. The arrows 2 through 5 are placed with respect to the side *Right*, the arrows 6 through 11 with respect to the side *Bottom*

In the combo box **Orientation of Arrow** one of the following possibilities can be chosen:

1. no arrow,
2. perpendicular to box in middle of side,
3. perpendicular to box at lower end of side,
4. perpendicular to box at upper end of side,
5. along coordinate axis,
6. parallel to box in the middle of side,
7. antiparallel to box in the middle of side,
8. parallel to box at the lower end of side,
9. antiparallel to box at the lower end of side,
10. parallel to box at the upper end of side,
11. antiparallel to box at the upper end of side,
12. in x direction at a chosen point,
13. in $-x$ direction at a chosen point,
14. in y direction at a chosen point,
15. in $-y$ direction at a chosen point.

The possibilities 2 through 11 are illustrated in Fig. A.22. If you choose one of them under the heading **Details**, you find the *length* **l** of the arrow and its *distance from the box* **d**, both in world coordinates. An arrow of length zero is not shown but its text is.

In very rare cases none of the possibilities 2 through 11 may satisfy you. You can then place an arrow at a freely chosen point using one of the possibilities 12 through 15. In that case the parameter group **Details** contains the arrow length and the (world) coordinates (X, Y) of the arrow's base point.

A.7.3.2 The Arrow in the z Direction This arrow is controlled using the subpanel **Background—Arrows—Arrow in z**. Under **Type of Arrow** you have the choice between

- no arrow, no text,
- no arrow, but text at point (X, Y, Z),
- arrow in z direction,
- arrow in $-z$ direction.

Under **Details** you find the length **l_Z** of the arrow (if any) and the world coordinates **X_Z**, **Y_Z**, and **Z_Z** of the arrow's base (or – if only text is wanted – of the text).

A.7.3.3 The Arrow in φ In plots of the type *surface over polar grid in 3D*, it can be useful to have an arrow in the x, y plane pointing in the azimuthal direction. Graphically, the arrow is a section of a circle in the x, y plane centered at the origin of the W3 coordinate system. It is controlled through the subpanel **Background—Arrows—Arrow in φ**. Here, under **Type of Arrow** you may choose between

- no arrow,
- arrow in φ direction,
- arrow in $-\varphi$ direction.

Under **Details** you find

- the *length* of the arc forming the arrow (in degrees),
- the *radius* of the arc forming the arrow (in world coordinates),
- the *azimuth* (in degrees) of the arrow's base.

A.7.4 The Subpanel Background—Texts

Four different types of alphanumeric texts may show up in a plot,

- the texts at arrows,
- the numbers at scales,
- a caption,
- a line of additional text.

The *texts at the arrows* are determined by the physics topic chosen and cannot be changed. The *numbers at the scales* are controlled through the subpanels of **Background—Scales**, Sect. A.7.2.

Whereas the arrows and scales are part of the geometrical structure of the plot and define the position of their associated texts and numbers, the position of the *caption* and of the *additional text* has to be defined explicitly. This is done in the parameter groups **Caption** and **Additional Text** where the *horizontal* and *vertical position* of the (lower-left point of the rectangle circumscribing the) text is given. The coordinates are given in percent of the (horizontal or vertical) extension of the plot with respect to its lower-left corner. The fields **Caption** and **Additional Text** may contain a field with the text itself, which then can be edited. It may contain mathematical symbols and simple formulae, see Sect. A.9 for how to code them. If there is no such field, the text is set automatically with the physics topic. In that case you may occasionally want not to show the text at all. Just move it outside the frame by setting its vertical position to 200.

A.8 Parameters – The Subpanel Format

There you can choose between **A Format-Landscape, A Format-Portrait**, and **Free Format**. In the case of A format you can still change the **A Number**. For example, A5 (Landscape) corresponds to the format

$$0 \le u \le 21.0\,\mathrm{cm} \quad , \quad 0 \le v \le 14.8\,\mathrm{cm} \quad .$$

For the free format width and height are given explicitly in centimeters.

The format is taken literally only for plots to be printed. (But before actually starting the print process further changes can be done.) Plots on a display screen always make optimal use of the graphics frame.

In all cases the width-to-height ratio defined in the format is respected.

A.9 Coding Mathematical Symbols and Formulae

Text appearing in the plot may contain mathematical symbols and simple formulae. To achieve this the characters in a text string are interpreted by **IQ** before they are plotted. You have *three different alphabets* at your disposal: *Roman*, *Greek*, and *Math*, see Table A.3. You choose the alphabet with special characters called *selection flags*:

@ for Roman,
& for Greek,
% for Math.

Table A.3. The three alphabets available for plotting text

	Select Flags				Select Flags		
	Roman	Greek	Math		Roman	Greek	Math
Input	@	&	%	Input	@	&	%
A	A	A(ALPHA)	Ä	a	a	α(alpha)	ä
B	B	B(BETA)	B	b	b	β(beta)	b
C	C	X(CHI)	⌉	c	c	χ(chi)	c
D	D	Δ(DELTA)	Δ	d	d	δ(delta)	d
E	E	E(EPSILON)	E	e	e	ϵ(epsilon)	e
F	F	Φ(PHI)	F	f	f	φ(phi)	f
G	G	Γ(GAMMA)	$\neq$	g	g	γ(gamma)	g
H	H	H(ETA)	H	h	h	η(eta)	h
I	I	I(IOTA)	$\int$	i	i	ι(iota)	i
J	J	I(IOTA)	J	j	j	ι(iota)	j
K	K	K(KAPPA)	K	k	k	κ(kappa)	k
L	L	Λ(LAMBDA)	|	l	l	λ(lambda)	l
M	M	M(MU)	$\pm$	m	m	μ(mu)	m
N	N	N(NU)	N	n	n	ν(nu)	n
O	O	Ω(OMEGA)	Ö	o	o	ω(omega)	ö
P	P	Π(PI)	P	p	p	π(pi)	p
Q	Q	Θ(THETA)	Q	q	q	ϑ(theta)	q
R	R	R(RHO)	$\circ$	r	r	ρ(rho)	r
S	S	Σ(SIGMA)	ß	s	s	σ(sigma)	s
T	T	T(TAU)	¦	t	t	τ(tau)	t
U	U	O(OMICRON)	Ü	u	u	o(omicron)	ü
V	V		Ü	v	v		v
W	W	Ψ(PSI)	$\surd$	w	w	ψ(psi)	w
X	X	Ξ(XI)	X	x	x	ξ(xi)	x
Y	Y	Υ(UPSILON)	Å	y	y	υ(upsilon)	y
Z	Z	Z(ZETA)	Z	z	z	ζ(zeta)	z
~	~	~	~	−	−	−	−
!	!	!	!	=	=	=	$\equiv$
$	$	$	$	{	{	{	{
*	*	#	$\times$	}	}	}	}
(	(	$\uparrow$	$\leftarrow$	|	|	|	|
)	)	$\downarrow$	$\rightarrow$	[	[	&	[
+	+	+	+	]	]	@	]
‘	‘	‘	‘	\			
1	1	1	1	:	:	:	:
2	2	2	2	;	;	;	;
3	3	3	3	’	’	‘	’
4	4	4	4	<	<	$\subset$	$\leq$
5	5	5	5	>	>	$\supset$	$\geq$
6	6	6	6	?	?	§	$\sim$
7	7	7	7	,	,	,	,
8	8	8	8	.	.	.	.
9	9	9	9	/	/	\	%
0	0	0	0				

The selection flag closest to, and on the left of, a character determines the alphabet for that character. The default selection flag is @, i.e., characters not preceded by any selection flag are interpreted as Roman.

Besides selection flags there are *position flags*:

^ for superscript,
_ for subscript,
for normal line level,
" for backspace.

The default position flag is #, i.e., characters appear on the normal line level if not preceded by any position flag. You can use up to two consecutive steps of subscript or superscript leading you away from the normal line, e.g., A_{α_β}, A_{α^β}, A^{α^β}, A^{α_β}. The backspace flag (") causes the character following it to overwrite the preceding character rather than to be placed to the right of it. It allows the writing of expressions like W_n^k instead of $W_n{}^k$. For examples, see the plots in Sect. A.3.

A.10 A Combined Plot and Its Mother Descriptor

It is sometimes useful to combine plots corresponding to different descriptors. Such a *combined plot* is defined by a *mother descriptor*, which in turn *quotes* individual descriptors. For *creating a mother descriptor* see Sect. A.1.10. Here we discuss its contents and functionality.

Figure 4.8 is an example of a combined plot. You can display it by selecting descriptor `Argand diagram: combined plot` on the file `1D_Scattering.des` which is a mother descriptor. By pressing the button **Parameters** in the graphics frame you display its contents in a *mother-descriptor panel*, Fig. A.23. It consists of two subpanels which are discussed below.

A.10.1 The Subpanel Type and Format

This subpanel carries the following items:

- **Arrangement of Individual Plots** – a set of radio buttons. In practically all cases you will want to select the first button **Plots in Regular Matrix**. For other possibilities see Sect. A.10.3.
- **Size of Combined Plot** – a group of two numbers, namely, the **Number of Rows** and the **Number of Columns** in which the individual plots are arranged within the combined plot.
- **Appearance** – a check box determining whether the individual plots are separated by lines or not.

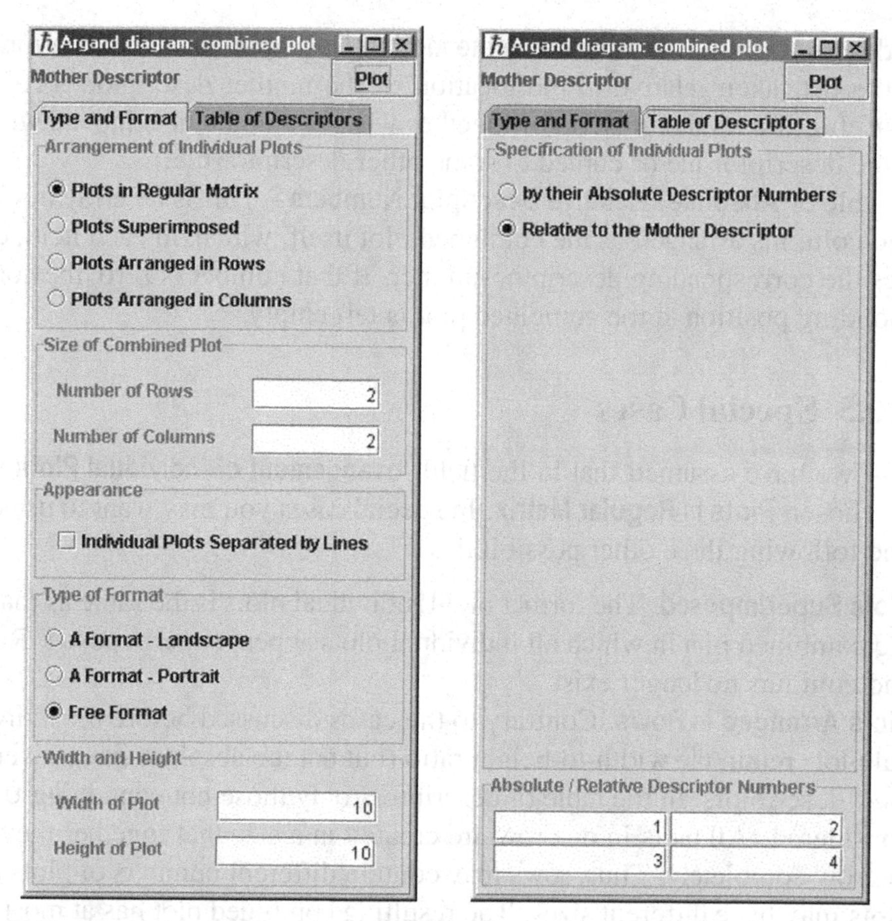

Fig. A.23. Mother-descriptor panel corresponding to descriptor `Argand diagram: combined plot` on file `1D_Scattering.des` and to Fig. 4.8. Both subpanels, **Type and Format** (left) and **Table of Descriptors** (right), are shown

- **Format** – The remaining items determine the *format*. You can choose between **A Format-Landscape, A Format-Portrait**, and **Free Format**. For A format you can still choose the **A Number**, e.g., A5. For free format the width and height are given explicitly in centimeters.
  Note: This format will be the format of each individual plot (the format defined in its descriptor will be ignored) except for some of the special cases discussed in Sect. A.10.3. The format of the combined plot itself is therefore, in general, considerably larger.

A.10.2 The Subpanel Table of Descriptors

Here you find two items:

- **Specification of Individual Plots** – You may choose between
  - **Absolute Descriptor Numbers**, i.e., the numbers appearing in the first column of the list of descriptors in the descriptor-selection panel, and

- descriptor numbers **Relative to the Mother Descriptor**, i.e., the location on the list taken relative to the location of the mother descriptor. (This is useful, if a mother descriptor together with its daughters is moved within the descriptor file or copied to some other descriptor file.)

- A table of **Absolute / Relative Descriptor Numbers** – This is an array of rows and columns arranged as the combined plot itself, which, in each field, carries the corresponding descriptor number. If that number is zero, the corresponding position in the combined plot is left empty.

A.10.3 Special Cases

So far we have assumed that in the field **Arrangement of Individual Plots** you have chosen **Plots in Regular Matrix**. In special cases you may want to use one of the following three other possibilities.

- **Plots Superimposed**. The format of all individual plots is the same as that of the combined plot in which all individual plots appear superimposed. Rows and columns no longer exist.
- **Plots Arranged in Rows**. Contrary to the cases discussed before the individual plots retain the width-to-height ratio (but not the absolute size), given in their descriptors. In the table of descriptors only those not containing 0 are considered. All plots in one row are created in a size that together they fill the row completely. Thus, rows may contain different numbers of plots and plots may have different sizes. The resulting combined plot has at most the size (width of plot $\times$ number of columns) $\times$ (height of plot $\times$ number of rows) but will usually be smaller.
- **Plots Arranged in Columns**. As described just above with the words row and column interchanged.

A.11 Details of Printing

The *print panel*, Fig. A.24, which appears when you want to print a single plot, Sect. A.1.6, or a set of plots, Sect. A.1.12, has two subpanels called **System Print** and **IQ Export**. The former is used to produce a plot on a *system printer*, i.e., a printer connected to your computer and known to your operating system. With the latter you can create a compact file in the PostScript language representing a plot.

A.11.1 Preview. Colors and Line Widths

At the bottom of the print panel you can select the **Units** in which lengths are given on the panel. You can also have a **Preview** of the printed plot in the

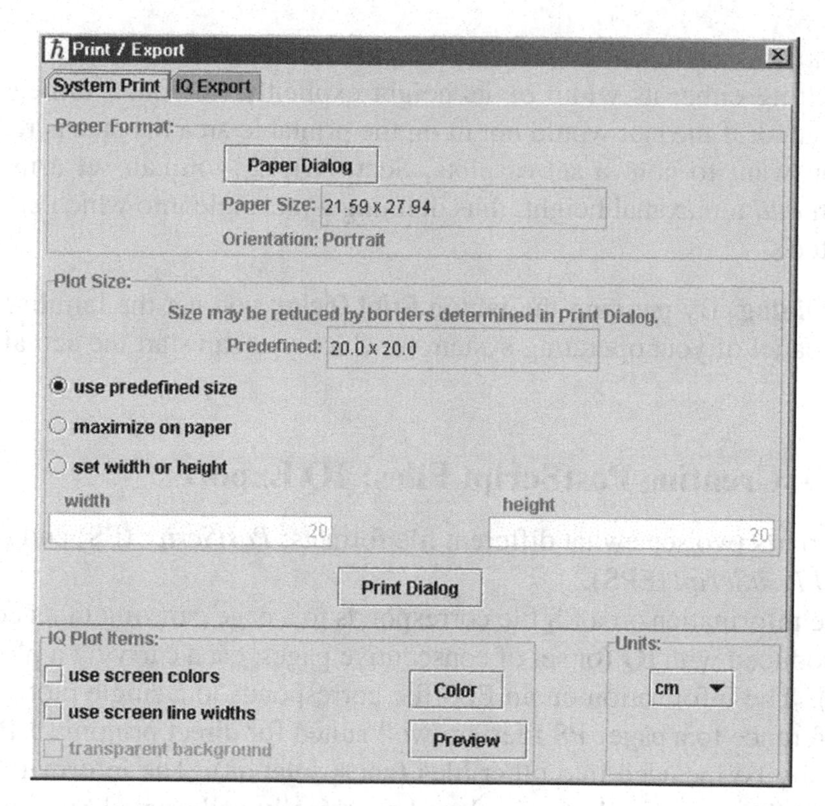

Fig. A.24. The print panel

graphics frame. For the preview print colors and line widths are used (rather than screen colors and line widths).

To change the print colors and line widths press **Color**. A *color-selection panel* appears. Its use is as described for the selection of screen colors, Sect. A.1.7. You may occasionally want to use screen colors and/or screen line widths also for printing. You can do that using the check boxes in the bottom-left corner of the print panel.

A.11.2 Using a System Printer

Paper Format Under **Paper Size** you find the dimensions of the paper. You also find the **Orientation** of the plot on the paper. The size must be that of the paper in your printer. If it is not, press **Paper Dialog**. You can then choose the correct size and also change the orientation.

Plot Size A plot is always printed in the width-to-height ratio as defined in its descriptor, see Sect. A.8. The size defined there is the *predefined size*. The size of the plot on paper can be defined in three ways:

- by using the predefined size (this can be done only if the predefined size fits on the area on the paper available for printing),

- by maximizing it, i.e., by printing the plot as large as possible,
- by setting either its width *or* its height explicitly (the other dimension is computed; if the plot would not fit on the printable area the size is reduced; if you want to print a set of plots, Sect. A.1.12, you can set a maximal width *and* a maximal height, thus defining a rectangle into which each plot is fitted).

Print Dialog By pressing the button **Print Dialog** you get the familiar print-dialog panel of your operating system in which you can start the actual printing.

A.11.3 Creating PostScript Files: IQ Export

IQ supports two somewhat different file formats: *PostScript* (PS) and *Encapsulated PostScript* (EPS).[1]

The information on a PS file corresponds to a page carrying in its center a plot produced with **IQ** (or set of consecutive pages, each carrying a plot in its center). The information on an EPS file corresponds to a single plot without any reference to a page. PS files are well suited for direct printing. EPS files can easily be included into other files (encapsulation). The information in a PS or EPS file can easily be displayed graphically and printed by a program such as *Ghostscript*. (Plots in EPS then appear in the lower-left corner of a page.)

On the top of the subpanel **IQ Export** you can choose between PS and EPS. The subpanel carries itself three more subpanels which we describe in reverse order.

Paper Size This subpanel (available for PS but not for EPS) contains the contents of the paper *configuration record*. That information can be changed, saved, and replaced as described for the color configuration record, Sect. A.1.7. It is best set once and for all in a customizing session, Sect. A.1.7.

Plot Size You can set the plot size as described for system printing, Sect. A.11.2. Moreover, you can indicate the *orientation* of the plot: If you **allow plot rotation**, the plot is rotated by 90° if the plot will then fit better onto the paper (i.e., it needs no or less demagnification). If, in addition, you allow **optimized rotation**, the plot is always oriented such that the long edge of the plot is parallel to the long edge of the paper.

Export If you want to create a set of plots on file in PS format, you can choose to print them on **Multiple files**, i.e., one file per plot, or on a **Single file**. After you made your choice, enter a **File name**, e.g., `myPlot.ps` or `myPlot.eps`. You may also change the directory into which the files are written.

[1] PostScript Language Reference Manual, 3rd ed., Adobe Systems Inc., 1999

Then press **Save** to create the file(s). In the case of multiple files these may, for example, have the names `myPlot3.ps`, `myPlot5.ps`, ... (or `.eps`). The numbers appearing in the file names are the descriptor numbers in the list from which you selected descriptors for printing. (You may also introduce the two characters `%d` anywhere in the file name. They will then be replaced by the descriptor number.)

If you want to create only a file for a single plot (i.e., you started out by pressing **Print** in the graphics frame), simply enter a file name, select another directory if you want, and press **Save**.

A.12 Preparing a Demonstration

In Sect. A.1.13 we have seen how to run a demonstration.

A demonstration consists of a set of *events* of the following types:

- the display of an **IQ** plot,
- the display of an HTML page,
- the display of an image or movie (e.g., in GIF, PNG, or animated GIF format),
- the playing of sound (spoken text).

The order in which events are started and the timing is controlled by a *demo file*. Before editing your demo file you have to prepare the following files and put them in one directory:

- a descriptor file on which the descriptors for the plots you want to show are collected, Sect. A.1.11,
- a set of HTML files with explanatory information you want to display,
- a set of sound files, e.g., in `WAV` format, if you want to include sound.

The demo file is best understood as a simple computer program with one command per line. The commands are executed consecutively. The demo file is a text file, which can be created with any editor. We suggest that you inspect one of the files `*.demo` in the `Demo` subdirectory.

There are 10 different commands:

- a command to open a descriptor file:
  `openDescriptorFile("`*`descriptorFileName`*`")`
- commands to display a plot corresponding to descriptor n on the descriptor file:
  `plot(`*n*`)`
  `plotLeft(`*n*`)`
  `plotRight(`*n*`)`
- commands to display an HTML file:

```
presentPage("htmlFileName")
presentPageLeft("htmlFileName")
presentPageRight("htmlFileName")
```

- commands to display an image file:
```
presentImage("imageFileName")
presentImageLeft("imageFileName")
presentImageRight("imageFileName")
```
- a command to start playing a sound file:
```
playSound("soundFileName")
```
- a command to wait a number m of milliseconds before executing the next command:
```
wait(m)
```
- a command to terminate the demonstration:
```
stop
```

The commands `plot`, `presentPage`, and `presentImage` use the full display surface of the demo frame, Fig. A.10, whereas the corresponding commands containing `Left` or `Right` use only the left or right part of that surface.

Lines beginning with `//` are ignored. They may serve as comments.

A sound file must not be started before the playing of the previous sound file has ended. The `wait` command has to be used to ensure that.

If a demonstration is run in the *step-by-step* mode, then the `wait` command simply halts the demonstration. The next command is executed after the button **Next** or the `Enter` key is pressed.

The HTML files and image files should be placed in the same directory as the demo file, because then only the file name has to be given as command argument. If a file is located in a different directory, the full path (either absolute or relative to the location of the demo file) has to be included, for instance

```
/.../htmfiles/demo.htm
pngfiles/demo.png
```

Note: If you work under Windows the backslash (\) has to be replaced by a double backslash (\\). An example is

```
C:\\...\\htmfiles\\demo.htm
pngfiles\\demo.png
```

B. How to Install IQ

B.1 Contents of the CD-ROM

The CD-ROM essentially contains three items:

- It contains the program package INTERQUANTA, version 4.0.
- The program needs the Java Runtime Environment including Swing classes, i.e., preferably version 1.3 or higher, which, most probably, is present on your machine. If it is not, it can, in most cases, be installed on your machine from the CD-ROM as part of the installation of INTERQUANTA. For details see Sect. B.2.2 below.
- To display help information INTERQUANTA uses a special version of the Adobe Acrobat® Reader which runs under Java. This particular reader is also installed on your machine from the CD-ROM as part of the installation of INTERQUANTA.

B.2 Computer Systems on which INTERQUANTA Can Be Used

B.2.1 Computers and Operating Systems

INTERQUANTA can be installed and run on the following types of computers:

- Personal Computers running under Windows, Linux, and OpenSolaris,
- Macintosh Computers running under Mac OS X,
- workstations (IBM RISC under AIX).

[INTERQUANTA was tested on 32-bit systems under Windows 98, NT 4.0, 2000, ME, XP, Vista, and 7, under Linux Kernel 2.6.13, and under Mac OS X 10.4.11 for PowerPC and x86 processors, under OpenSolaris 09.06 (SunOS 5.11 for i86pc), and AIX 4.3.3. On 64-bit systems it was tested under Windows 7, under Linux Kernel 2.6.28, and under Mac OS X 10.6.4 for x86_64 processors.]

S. Brandt et al., *Interactive Quantum Mechanics: Quantum Experiments on the Computer*,
DOI 10.1007/978-1-4419-7424-2_14,

B.2.2 Java Runtime Environment (JRE)

If the JRE, version 1.3 or higher, is not present on your computer, then it is brought there as part of the installation if you are working on a Personal Computer under Windows or Linux. (System requirements imposed by the JRE are listed in detail in `http://java.sun.com/javase/6/webnotes/install/-system-configurations.html`.) For other systems you have to download it from the web site of the manufacturer of your computer and install it before installing INTERQUANTA.

B.3 Installation with Options. The File `ReadMe.txt`

The installation procedure allows for various options. These are in particular

- the choice between single-user and multi-user installation (on operating systems supporting multiple users),
- the choice of directory names for program, documentation, and demo files. (By choosing the drive and directory names of the CD-ROM you can leave the spacious documentation and demo files there and save space on your hard disk.)

If you want to make use of these choices, consult the file `ReadMe.txt` or `ReadMe.html` placed in the top directory on the CD-ROM and proceed with the installation as described there.

B.4 Quick Installation for the Impatient User

In the rather common case that

- you want to install INTERQUANTA for a single user,
- you work on a Personal Computer under Windows, Linux, or OpenSolaris or on a Macintosh Computer under Mac OS X,
- you have 500 MB of free disk space,
- and you agree to the default directory names used by the installation program,

you can proceed as follows.

Step 1 Take notice of the copyright and license conditions in Sect. 2 of the file `ReadMe.txt` placed in the top directory on the CD-ROM. You must not proceed with the installation if you do not agree to these conditions.

Step 2 **For Windows:** With the Windows Explorer in the top directory of the CD-ROM you find the file `iqinst.bat`. Double-click on it. **For Linux, OpenSolaris, or Mac OS X:** There is a shell script `install.sh`, which you start from a terminal shell. For the Macintosh you can also double-click on `install(MacOSX).command` from the Finder.

Step 3 A panel will appear on your computer screen carrying four or five subpanels. Click on the subpanel **Install**. Then click on the button **Install** in that subpanel.

You can follow the progress of the installation in the text field of the panel. As last line the words "`Installation completed.`" appear. The top of the install panel now contains a **Close** button which you may press to make the whole panel disappear.

Step 4 You will find an icon labeled InterQuanta and featuring the symbol $\hbar$ ("hbar") on your desktop. By (double-)clicking on it you start the program.

Possible Updates

For possible additional release notes and/or updates see the web site

`http://www.springer.com/physics/quantum+physics/book/978-1-4419-7423-5`.

Index

S. Brandt et al., *Interactive Quantum Mechanics: Quantum Experiments on the Computer*,
DOI 10.1007/978-1-4419-7424-2,

Springer

Printed in the United States
By Bookmasters